String Theory,

Superstring Theory and Beyond

The two volumes that comprise *String Theory* provide an up-to-date, comprehensive, and pedagogic introduction to string theory.

Volume I, *An Introduction to the Bosonic String*, provides a thorough introduction to the bosonic string, based on the Polyakov path integral and conformal field theory. The first four chapters introduce the central ideas of string theory, the tools of conformal field theory and of the Polyakov path integral, and the covariant quantization of the string. The next three chapters treat string interactions: the general formalism, and detailed treatments of the tree-level and one loop amplitudes. Chapter eight covers toroidal compactification and many important aspects of string physics, such as T-duality and D-branes. Chapter nine treats higher-order amplitudes, including an analysis of the finiteness and unitarity, and various nonperturbative ideas. An appendix giving a short course on path integral methods is also included.

Volume II, *Superstring Theory and Beyond*, begins with an introduction to supersymmetric string theories and goes on to a broad presentation of the important advances of recent years. The first three chapters introduce the type I, type II, and heterotic superstring theories and their interactions. The next two chapters present important recent discoveries about strongly coupled strings, beginning with a detailed treatment of D-branes and their dynamics, and covering string duality, M-theory, and black hole entropy. A following chapter collects many classic results in conformal field theory. The final four chapters are concerned with four-dimensional string theories, and have two goals: to show how some of the simplest string models connect with previous ideas for unifying the Standard Model; and to collect many important and beautiful general results on world-sheet and spacetime symmetries. An appendix summarizes the necessary background on fermions and supersymmetry.

Both volumes contain an annotated reference section, emphasizing references that will be useful to the student, as well as a detailed glossary of important terms and concepts. Many exercises are included which are intended to reinforce the main points of the text and to bring in additional ideas.

An essential text and reference for graduate students and researchers in theoretical physics, particle physics, and relativity with an interest in modern superstring theory.

Joseph Polchinski received his Ph.D. from the University of California at Berkeley in 1980. After postdoctoral fellowships at the Stanford Linear Accelerator Center and Harvard, he joined the faculty at the University of Texas at Austin in 1984, moving to his present position of Professor of Physics at the University of California at Santa Barbara, and Permanent Member of the Institute for Theoretical Physics, in 1992.

Professor Polchinski is not only a clear and pedagogical expositor, but is also a leading string theorist. His discovery of the importance of D-branes in 1995 is one of the most important recent contributions in this field, and he has also made significant contributions to many areas of quantum field theory and to supersymmetric models of particle physics.

CAMBRIDGE MONOGRAPHS ON MATHEMATICAL PHYSICS

General Editors: P. V. Landshoff, D. R. Nelson, D. W. Sciama, S. Weinberg

CAMBRIDGE MONOGRAPHS ON
MATHEMATICAL PHYSICS

[†] Issued as a paperback

STRING THEORY
VOLUME II

Superstring Theory and Beyond

JOSEPH POLCHINSKI

Institute for Theoretical Physics
University of California at Santa Barbara

CAMBRIDGE
UNIVERSITY PRESS

PUBLISHED BY THE PRESS SYNDICATE OF THE UNIVERSITY OF CAMBRIDGE
The Pitt Building, Trumpington Street, Cambridge, United Kingdom

CAMBRIDGE UNIVERSITY PRESS
The Edinburgh Building, Cambridge CB2 2RU, UK
40 West 20th Street, New York, NY 10011–4211, USA
477 Williamstown Road, Port Melbourne, VIC 3207, Australia
Ruiz de Alarcón 13, 28014 Madrid, Spain
Dock House, The Waterfront, Cape Town 8001, South Africa

http://www.cambridge.org

First published 1998
Reprinted (with corrections) 1999, 2000, 2002, 2003

Printed in the United Kingdom at the University Press, Cambridge

Typeface Monotype Times 11/13pt   *System* LaTeX   [UPH]

A catalogue record for this book is available from the British Library

Library of Congress Cataloguing in Publication data
Polchinski, Joseph Gerard.
String theory / Joseph Polchinski.
p.   cm. – (Cambridge monographs on mathematical physics)
Includes bibliographical references.
Contents: v. 1. An introduction to the bosonic string –
v. 2. Superstring theory and beyond.
ISBN 0-521-63303-6 (v. I : hb). – ISBN 0-521-63304-4 (v. II : hb)
1. String models.   2. Superstring theories.   I. Title.
II. Series.
QC794.6.S85P65   1998
530.14–DC21   98-4545   CIP

ISBN 0 521 63304 4 hardback

To Dorothy, Steven, and Daniel

Contents

Outline of volume one

Foreword

From the beginning it was clear that, despite its successes, the Standard Model of elementary particles would have to be embedded in a broader theory that would incorporate gravitation as well as the strong and electroweak interactions. There is at present only one plausible candidate for such a theory: it is the theory of strings, which started in the 1960s as a not-very-successful model of hadrons, and only later emerged as a possible theory of all forces.

There is no one better equipped to introduce the reader to string theory than Joseph Polchinski. This is in part because he has played a significant role in the development of this theory. To mention just one recent example: he discovered the possibility of a new sort of extended object, the 'Dirichlet brane', which has been an essential ingredient in the exciting progress of the last few years in uncovering the relation between what had been thought to be different string theories.

Of equal importance, Polchinski has a rare talent for seeing what is of physical significance in a complicated mathematical formalism, and explaining it to others. In looking over the proofs of this book, I was reminded of the many times while Polchinski was a member of the Theory Group of the University of Texas at Austin, when I had the benefit of his patient, clear explanations of points that had puzzled me in string theory. I recommend this book to any physicist who wants to master this exciting subject.

Steven Weinberg
Series Editor
Cambridge Monographs on Mathematical Physics
1998

Preface

When I first decided to write a book on string theory, more than ten years ago, my memories of my student years were much more vivid than they are today. Still, I remember that one of the greatest pleasures was finding a text that made a difficult subject accessible, and I hoped to provide the same for string theory.

Thus, my first purpose was to give a coherent introduction to string theory, based on the Polyakov path integral and conformal field theory. No previous knowledge of string theory is assumed. I do assume that the reader is familiar with the central ideas of general relativity, such as metrics and curvature, and with the ideas of quantum field theory through non-Abelian gauge symmetry. Originally a full course of quantum field theory was assumed as a prerequisite, but it became clear that many students were eager to learn string theory as soon as possible, and that others had taken courses on quantum field theory that did not emphasize the tools needed for string theory. I have therefore tried to give a self-contained introduction to those tools.

A second purpose was to show how some of the simplest four-dimensional string theories connect with previous ideas for unifying the Standard Model, and to collect general results on the physics of four-dimensional string theories as derived from world-sheet and spacetime symmetries. New developments have led to a third goal, which is to introduce the recent discoveries concerning string duality, M-theory, D-branes, and black hole entropy.

In writing a text such as this, there is a conflict between the need to be complete and the desire to get to the most interesting recent results as quickly as possible. I have tried to serve both ends. On the side of completeness, for example, the various path integrals in chapter 6 are calculated by three different methods, and the critical dimension of the bosonic string is calculated in seven different ways in the text and exercises.

On the side of efficiency, some shorter paths through these two volumes are suggested below.

A particular issue is string perturbation theory. This machinery is necessarily a central subject of volume one, but it is somewhat secondary to the recent nonperturbative developments: the free string spectrum plus the spacetime symmetries are more crucial there. Fortunately, from string perturbation theory there is a natural route to the recent discoveries, by way of T-duality and D-branes.

One possible course consists of chapters 1–3, section 4.1, chapters 5–8 (omitting sections 5.4 and 6.7), chapter 10, sections 11.1, 11.2, 11.6, 12.1, and 12.2, and chapters 13 and 14. This sequence, which I believe can be covered in two quarters, takes one from an introduction to string theory through string duality, M-theory, and the simplest black hole entropy calculations. An additional shortcut is suggested at the end of section 5.1.

Readers interested in T-duality and related stringy phenomena can proceed directly from section 4.1 to chapter 8. The introduction to Chan–Paton factors at the beginning of section 6.5 is needed to follow the discussion of the open string, and the one-loop vacuum amplitude, obtained in chapter 7, is needed to follow the calculation of the D-brane tension.

Readers interested in supersymmetric strings can read much of chapters 10 and 11 after section 4.1. Again the introduction to Chan–Paton factors is needed to follow the open string discussion, and the one-loop vacuum amplitude is needed to follow the consistency conditions in sections 10.7, 10.8, and 11.2.

Readers interested in conformal field theory might read chapter 2, sections 6.1, 6.2, 6.7, 7.1, 7.2, 8.2, 8.3 (concentrating on the CFT aspects), 8.5, 10.1–10.4, 11.4, and 11.5, and chapter 15. Readers interested in four-dimensional string theories can follow most of chapters 16–19 after chapters 8, 10, and 11.

In a subject as active as string theory — by one estimate the literature approaches 10 000 papers — there will necessarily be important subjects that are treated only briefly, and others that are not treated at all. Some of these are represented by review articles in the lists of references at the end of each volume. The most important omission is probably a more complete treatment of compactification on curved manifolds. Because the geometric methods of this subject are somewhat orthogonal to the quantum field theory methods that are emphasized here, I have included only a summary of the most important results in chapters 17 and 19. Volume two of Green, Schwarz, and Witten (1987) includes a more extensive introduction, but this is a subject that has continued to grow in importance and clearly deserves an introductory book of its own.

This work grew out of a course taught at the University of Texas at Austin in 1987–88. The original plan was to spend a year turning the lecture notes into a book, but a desire to make the presentation clearer and more complete, and the distraction of research, got in the way. An early prospectus projected the completion date as June 1989 $\pm$ one month, off by 100 standard deviations. For eight years the expected date of completion remained approximately one year in the future, while one volume grew into two. Happily, finally, one of those deadlines didn't slip.

I have also used portions of this work in a course at the University of California at Santa Barbara, and at the 1994 Les Houches, 1995 Trieste, and 1996 TASI schools. Portions have been used for courses by Nathan Seiberg and Michael Douglas (Rutgers), Steven Weinberg (Texas), Andrew Strominger and Juan Maldacena (Harvard), Nathan Berkovits (Sâo Paola) and Martin Einhorn (Michigan). I would like to thank those colleagues and their students for very useful feedback. I would also like to thank Steven Weinberg for his advice and encouragement at the beginning of this project, Shyamoli Chaudhuri for a thorough reading of the entire manuscript, and to acknowledge the support of the Departments of Physics at UT Austin and UC Santa Barbara, the Institute for Theoretical Physics at UC Santa Barbara, and the National Science Foundation.

During the extended writing of this book, dozens of colleagues have helped to clarify my understanding of the subjects covered, and dozens of students have suggested corrections and other improvements. I began to try to list the members of each group and found that it was impossible. Rather than present a lengthy but incomplete list here, I will keep an updated list at the erratum website

http://www.itp.ucsb.edu/~joep/bigbook.html.

In addition, I would like to thank collectively all who have contributed to the development of string theory; volume two in particular seems to me to be largely a collection of beautiful results derived by many physicists. String theory (and the entire base of physics upon which it has been built) is one of mankind's great achievements, and it has been my privilege to try to capture its current state.

Finally, to complete a project of this magnitude has meant many sacrifices, and these have been shared by my family. I would like to thank Dorothy, Steven, and Daniel for their understanding, patience, and support.

Joseph Polchinski
Santa Barbara, California
1998

Notation

This book uses the +++ conventions of Misner, Thorne, & Wheeler (1973). In particular, the signature of the metric is $(-++\ldots+)$. The constants $\hbar$ and c are set to 1, but the Regge slope α' is kept explicit.

A bar $^-$ is used to denote the conjugates of world-sheet coordinates and moduli (such as z, τ and q), but a star * is used for longer expressions. A bar on a spacetime fermion field is the Dirac adjoint (this appears only in volume two), and a bar on a world-sheet operator is the Euclidean adjoint (defined in section 6.7). For the degrees of freedom on the string, the following terms are treated as synonymous:

$$\text{holomorphic} = \text{left-moving},$$
$$\text{antiholomorphic} = \text{right-moving},$$

as explained in section 2.1. Our convention is that the supersymmetric side of the heterotic string is right-moving. Antiholomorphic operators are designated by tildes $\tilde{\ }$; as explained in section 2.3, these are not the adjoints of holomorphic operators. Note also the following conventions:

$$d^2z \equiv 2dxdy , \quad \delta^2(z,\bar{z}) \equiv \frac{1}{2}\delta(x)\delta(y) ,$$

where $z = x + iy$ is any complex variable; these differ from most of the literature, where the coefficient is 1 in each definition.

Spacetime actions are written as $\boldsymbol{S}$ and world-sheet actions as S. This presents a problem for D-branes, which are T-dual to the former and S-dual to the latter; $\boldsymbol{S}$ has been used arbitrarily. The spacetime metric is $G_{\mu\nu}$, while the world-sheet metric is γ_{ab} (Minkowskian) or g_{ab} (Euclidean). In volume one, the spacetime Ricci tensor is $\boldsymbol{R}_{\mu\nu}$ and the world-sheet Ricci tensor is R_{ab}. In volume two the former appears often and the latter never, so we have changed to $R_{\mu\nu}$ for the spacetime Ricci tensor.

The following are used:

$\equiv$ defined as
$\cong$ equivalent to
$\approx$ approximately equal to
$\sim$ equal up to nonsingular terms (OPEs), or rough correspondence.

10

Type I and type II superstrings

Having spent volume one on a thorough development of the bosonic string, we now come to our real interest, the supersymmetric string theories. This requires a generalization of the earlier framework, enlarging the world-sheet constraint algebra. This idea arises naturally if we try to include spacetime fermions in the spectrum, and by guesswork we are led to *superconformal symmetry*. In this chapter we discuss the (1,1) superconformal algebra and the associated type I and II superstrings. Much of the structure is directly parallel to that of the bosonic string so we can proceed rather quickly, focusing on the new features.

10.1 The superconformal algebra

In bosonic string theory, the mass-shell condition

$$p_\mu p^\mu + m^2 = 0 \tag{10.1.1}$$

came from the physical state condition

$$L_0|\psi\rangle = 0 \,, \tag{10.1.2}$$

and also from $\tilde{L}_0|\psi\rangle = 0$ in the closed string. The mass-shell condition is the Klein–Gordon equation in momentum space. To get spacetime fermions, it seems that we need the Dirac equation

$$ip_\mu \Gamma^\mu + m = 0 \tag{10.1.3}$$

instead. This is one way to motivate the following generalization, and it will lead us to all the known consistent string theories.

Let us try to follow the pattern of the bosonic string, where L_0 and $\tilde{L}_0$ are the center-of-mass modes of the world-sheet energy-momentum tensor $(T_B, \tilde{T}_B)$. A subscript B for 'bosonic' has been added to distinguish these from the fermionic currents now to be introduced. It seems then that we

1

need new conserved quantities T_F and $\tilde{T}_F$, whose center-of-mass modes give the Dirac equation, and which play the same role as T_B and $\tilde{T}_B$ in the bosonic theory. Noting further that the spacetime momenta p^μ are the center-of-mass modes of the world-sheet current $(\partial X^\mu, \bar\partial X^\mu)$, it is natural to guess that the gamma matrices, with algebra

$$\{\Gamma^\mu, \Gamma^\nu\} = 2\eta^{\mu\nu} , \tag{10.1.4}$$

are the center-of-mass modes of an anticommuting world-sheet field ψ^μ.

With this in mind, we consider the world-sheet action

$$S = \frac{1}{4\pi} \int d^2z \left(\frac{2}{\alpha'} \partial X^\mu \bar\partial X_\mu + \psi^\mu \bar\partial \psi_\mu + \tilde\psi^\mu \partial \tilde\psi_\mu \right) . \tag{10.1.5}$$

For reference we recall from chapter 2 the XX operator product expansion (OPE)

$$X^\mu(z, \bar z) X^\nu(0, 0) \sim -\frac{\alpha'}{2} \eta^{\mu\nu} \ln |z|^2 . \tag{10.1.6}$$

The ψ conformal field theory (CFT) was described in section 2.5. The fields ψ^μ and $\tilde\psi^\mu$ are respectively holomorphic and antiholomorphic, and the operator products are

$$\psi^\mu(z)\psi^\nu(0) \sim \frac{\eta^{\mu\nu}}{z} , \quad \tilde\psi^\mu(\bar z)\tilde\psi^\nu(0) \sim \frac{\eta^{\mu\nu}}{\bar z} . \tag{10.1.7}$$

The *world-sheet supercurrents*

$$T_F(z) = i(2/\alpha')^{1/2}\psi^\mu(z)\partial X_\mu(z) , \quad \tilde{T}_F(\bar z) = i(2/\alpha')^{1/2}\tilde\psi^\mu(\bar z)\bar\partial X_\mu(\bar z) \tag{10.1.8}$$

are also respectively holomorphic and antiholomorphic, since they are just the products of (anti)holomorphic fields. The annoying factors of $(2/\alpha')^{1/2}$ could be eliminated by working in units where $\alpha' = 2$, and then be restored if needed by dimensional analysis. Also, throughout this volume the : : normal ordering of coincident operators will be implicit.

This gives the desired result: the modes ψ_0^μ and $\tilde\psi_0^\mu$ will satisfy the gamma matrix algebra, and the centers-of-mass of T_F and $\tilde{T}_F$ will have the form of Dirac operators. We will see that the resulting string theory has spacetime fermions as well as bosons, and that the tachyon is gone.

From the OPE and the Ward identity it follows (exercise 10.1) that the currents

$$j^\eta(z) = \eta(z)T_F(z) , \quad \tilde{j}^\eta(\bar z) = \bar\eta(\bar z)\tilde{T}_F(\bar z) \tag{10.1.9}$$

generate the *superconformal transformation*

$$\epsilon^{-1}(2/\alpha')^{1/2}\delta X^\mu(z, \bar z) = -\eta(z)\psi^\mu(z) - \eta(z)^*\tilde\psi^\mu(\bar z) , \tag{10.1.10a}$$

$$\epsilon^{-1}(\alpha'/2)^{1/2}\delta\psi^\mu(z) = \eta(z)\partial X^\mu(z) , \tag{10.1.10b}$$

$$\epsilon^{-1}(\alpha'/2)^{1/2}\delta\tilde\psi^\mu(\bar z) = \eta(z)^*\bar\partial X^\mu(\bar z) . \tag{10.1.10c}$$

This transformation mixes the commuting field X^μ with the anticommuting fields ψ^μ and $\tilde{\psi}^\mu$, so the parameter $\eta(z)$ must be anticommuting. As with conformal symmetry, the parameters are arbitrary holomorphic or antiholomorphic functions. That this is a symmetry of the action (10.1.5) follows at once because the current is (anti)holomorphic, and so conserved.

The commutator of two superconformal transformations is a conformal transformation,

$$\delta_{\eta_1}\delta_{\eta_2} - \delta_{\eta_2}\delta_{\eta_1} = \delta_v \,, \quad v(z) = -2\eta_1(z)\eta_2(z) \,, \tag{10.1.11}$$

as the reader can check by acting on the various fields. Similarly, the commutator of a conformal and superconformal transformation is a superconformal transformation. The conformal and superconformal transformations thus close to form the *superconformal algebra*. In terms of the currents, this means that the OPEs of T_F with itself and with

$$T_B = -\frac{1}{\alpha'}\partial X^\mu \partial X_\mu - \frac{1}{2}\psi^\mu \partial \psi_\mu \tag{10.1.12}$$

close. That is, only T_B and T_F appear in the singular terms:

$$T_B(z)T_B(0) \sim \frac{3D}{4z^4} + \frac{2}{z^2}T_B(0) + \frac{1}{z}\partial T_B(0) \,, \tag{10.1.13a}$$

$$T_B(z)T_F(0) \sim \frac{3}{2z^2}T_F(0) + \frac{1}{z}\partial T_F(0) \,, \tag{10.1.13b}$$

$$T_F(z)T_F(0) \sim \frac{D}{z^3} + \frac{2}{z}T_B(0) \,, \tag{10.1.13c}$$

and similarly for the antiholomorphic currents. The $T_B T_F$ OPE implies that T_F is a tensor of weight $(\frac{3}{2},0)$. Each scalar contributes 1 to the central charge and each fermion $\frac{1}{2}$, for a total

$$c = (1 + \tfrac{1}{2})D = \tfrac{3}{2}D \,. \tag{10.1.14}$$

This enlarged algebra with T_F and $\tilde{T}_F$ as well as T_B and $\tilde{T}_B$ will play the same role that the conformal algebra did in the bosonic string. That is, we will impose it on the states as a constraint algebra — it must annihilate physical states, either in the sense of old covariant quantization (OCQ) or of Becchi–Rouet–Stora–Tyutin (BRST) quantization. Because of the Minkowski signature of spacetime the timelike ψ^0 and $\tilde{\psi}^0$, like X^0, have opposite sign commutators and lead to negative norm states. The fermionic constraints T_F and $\tilde{T}_F$ will remove these states from the spectrum.

More generally, the $N = 1$ superconformal algebra in operator product form is

$$T_B(z)T_B(0) \sim \frac{c}{2z^4} + \frac{2}{z^2}T_B(0) + \frac{1}{z}\partial T_B(0) \,, \qquad (10.1.15a)$$

$$T_B(z)T_F(0) \sim \frac{3}{2z^2}T_F(0) + \frac{1}{z}\partial T_F(0) \,, \qquad (10.1.15b)$$

$$T_F(z)T_F(0) \sim \frac{2c}{3z^3} + \frac{2}{z}T_B(0) \,. \qquad (10.1.15c)$$

The Jacobi identity requires the same constant c in the $T_B T_B$ and $T_F T_F$ products (exercise 10.5). Here, $N = 1$ refers to the number of $(\frac{3}{2},0)$ currents. In the present case there is also an antiholomorphic copy of the same algebra, so we have an $(N, \tilde{N}) = (1,1)$ *superconformal field theory (SCFT)*. We will consider more general algebras in section 11.1.

Free SCFTs

The various free CFTs described in chapter 2 have superconformal generalizations. One free SCFT combines an anticommuting bc theory with a commuting $\beta\gamma$ system, with weights

$$h_b = \lambda \,, \quad h_c = 1 - \lambda \,, \qquad (10.1.16a)$$

$$h_\beta = \lambda - \tfrac{1}{2} \,, \quad h_\gamma = \tfrac{3}{2} - \lambda \,. \qquad (10.1.16b)$$

The action is

$$S_{BC} = \frac{1}{2\pi} \int d^2z \, (b\bar{\partial}c + \beta\bar{\partial}\gamma) \,, \qquad (10.1.17)$$

and

$$T_B = (\partial b)c - \lambda\partial(bc) + (\partial\beta)\gamma - \frac{1}{2}(2\lambda - 1)\partial(\beta\gamma) \,, \qquad (10.1.18a)$$

$$T_F = -\frac{1}{2}(\partial\beta)c + \frac{2\lambda - 1}{2}\partial(\beta c) - 2b\gamma \,. \qquad (10.1.18b)$$

The central charge is

$$[-3(2\lambda - 1)^2 + 1] + [3(2\lambda - 2)^2 - 1] = 9 - 12\lambda \,. \qquad (10.1.19)$$

Of course there is a corresponding antiholomorphic theory.

We can anticipate that the superconformal ghosts will be of this form with $\lambda = 2$, the anticommuting $(2,0)$ ghost b being associated with the commuting $(2,0)$ constraint T_B as in the bosonic theory, and the commuting $(\frac{3}{2},0)$ ghost β being associated with the anticommuting $(\frac{3}{2},0)$ constraint T_F. The ghost central charge is then $-26 + 11 = -15$, and the condition that the total central charge vanish gives the critical dimension

$$0 = \frac{3}{2}D - 15 \Rightarrow D = 10 \,. \qquad (10.1.20)$$

For $\lambda = 2$,

$$T_B = -(\partial b)c - 2b\partial c - \frac{1}{2}(\partial\beta)\gamma - \frac{3}{2}\beta\partial\gamma \,, \tag{10.1.21a}$$

$$T_F = (\partial\beta)c + \frac{3}{2}\beta\partial c - 2b\gamma \,. \tag{10.1.21b}$$

Another free SCFT is the superconformal version of the linear dilaton theory. This has again the action (10.1.5), while

$$T_B(z) = -\frac{1}{\alpha'}\partial X^\mu\partial X_\mu + V_\mu\partial^2 X^\mu - \frac{1}{2}\psi^\mu\partial\psi_\mu \,, \tag{10.1.22a}$$

$$T_F(z) = i(2/\alpha')^{1/2}\psi^\mu\partial X_\mu - i(2\alpha')^{1/2}V_\mu\partial\psi^\mu \,, \tag{10.1.22b}$$

each having an extra term as in the bosonic case. The reader can verify that these satisfy the $N = 1$ algebra with

$$c = \frac{3}{2}D + 6\alpha'V^\mu V_\mu \,. \tag{10.1.23}$$

10.2   Ramond and Neveu–Schwarz sectors

We now study the spectrum of the $X^\mu\psi^\mu$ SCFT on a circle. Much of this is as in chapter 2, but the new ingredient is a more general periodicity condition. It is clearest to start with the cylindrical coordinate $w = \sigma^1 + i\sigma^2$. The matter fermion action

$$\frac{1}{4\pi}\int d^2w \left(\psi^\mu\partial_{\bar{w}}\psi_\mu + \tilde{\psi}^\mu\partial_w\tilde{\psi}\right) \tag{10.2.1}$$

must be invariant under the periodic identification of the cylinder, $w \cong w + 2\pi$. This condition plus Lorentz invariance still allows two possible periodicity conditions for ψ^μ,

$$\text{Ramond (R):}\quad \psi^\mu(w + 2\pi) = +\psi^\mu(w)\,, \tag{10.2.2a}$$

$$\text{Neveu–Schwarz (NS):}\quad \psi^\mu(w + 2\pi) = -\psi^\mu(w)\,, \tag{10.2.2b}$$

where the sign must be the same for all μ. Similarly there are two possible periodicities for $\tilde{\psi}^\mu$. Summarizing, we will write

$$\psi^\mu(w + 2\pi) = \exp(2\pi i v)\,\psi^\mu(w)\,, \tag{10.2.3a}$$

$$\tilde{\psi}^\mu(\bar{w} + 2\pi) = \exp(-2\pi i\tilde{v})\,\tilde{\psi}^\mu(\bar{w})\,, \tag{10.2.3b}$$

where v and $\tilde{v}$ take the values 0 and $\frac{1}{2}$.

Since we are initially interested in theories with the maximum Poincaré invariance, X^μ must be periodic. (Antiperiodicity of X^μ is interesting, and we have already encountered it for the twisted strings on an orbifold, but it would break some of the translation invariance.) The supercurrent then has the same periodicity as the corresponding ψ,

$$T_F(w + 2\pi) = \exp(2\pi i v)\, T_F(w)\,, \tag{10.2.4a}$$

$$\tilde{T}_F(\bar{w} + 2\pi) = \exp(-2\pi i \tilde{v})\, \tilde{T}_F(\bar{w})\,. \tag{10.2.4b}$$

Thus there are four different ways to put the theory on a circle, each of which will lead to a different Hilbert space — essentially there are four different kinds of closed superstring. We will denote these by $(v, \tilde{v})$ or by NS–NS, NS–R, R–NS, and R–R. They are analogous to the twisted and untwisted sectors of the $\mathbf{Z}_2$ orbifold. Later in the chapter we will see that consistency requires that the full string spectrum contain certain combinations of states from each sector.

To study the spectrum in a given sector expand in Fourier modes,

$$\psi^\mu(w) = i^{-1/2} \sum_{r \in \mathbf{Z}+v} \psi_r^\mu \exp(irw)\,, \quad \tilde{\psi}^\mu(\bar{w}) = i^{1/2} \sum_{r \in \mathbf{Z}+\tilde{v}} \tilde{\psi}_r^\mu \exp(-ir\bar{w})\,,$$
$$\tag{10.2.5}$$

the phase factors being inserted to conform to convention later. On each side the sum runs over integers in the R sector and over (integers $+ \frac{1}{2}$) in the NS sector. Let us also write these as Laurent expansions. Besides replacing $\exp(-iw) \to z$ we must transform the fields,

$$\psi_{z^{1/2}}^\mu(z) = (\partial_z w)^{1/2} \psi_{w^{1/2}}^\mu(w) = i^{1/2} z^{-1/2} \psi_{w^{1/2}}^\mu(w)\,. \tag{10.2.6}$$

The clumsy subscripts are a reminder that these transform with half the weight of a vector. Henceforth the frame will be indicated implicitly by the argument of the field. The Laurent expansions are then

$$\psi^\mu(z) = \sum_{r \in \mathbf{Z}+v} \frac{\psi_r^\mu}{z^{r+1/2}}\,, \quad \tilde{\psi}^\mu(\bar{z}) = \sum_{r \in \mathbf{Z}+\tilde{v}} \frac{\tilde{\psi}_r^\mu}{\bar{z}^{r+1/2}}\,. \tag{10.2.7}$$

Notice that in the NS sector, the branch cut in $z^{-1/2}$ offsets the original antiperiodicity, while in the R sector it introduces a branch cut. Let us also recall the corresponding bosonic expansions

$$\partial X^\mu(z) = -i\left(\frac{\alpha'}{2}\right)^{1/2} \sum_{m=-\infty}^{\infty} \frac{\alpha_m^\mu}{z^{m+1}}\,, \quad \bar{\partial} X^\mu(\bar{z}) = -i\left(\frac{\alpha'}{2}\right)^{1/2} \sum_{m=-\infty}^{\infty} \frac{\tilde{\alpha}_m^\mu}{\bar{z}^{m+1}}\,,$$
$$\tag{10.2.8}$$

where $\alpha_0^\mu = \tilde{\alpha}_0^\mu = (\alpha'/2)^{1/2} p^\mu$ in the closed string and $\alpha_0^\mu = (2\alpha')^{1/2} p^\mu$ in the open string.

The OPE and the Laurent expansions (or canonical quantization) give the anticommutators

$$\{\psi_r^\mu, \psi_s^\nu\} = \{\tilde{\psi}_r^\mu, \tilde{\psi}_s^\nu\} = \eta^{\mu\nu} \delta_{r,-s}\,, \tag{10.2.9a}$$

$$[\alpha_m^\mu, \alpha_n^\nu] = [\tilde{\alpha}_m^\mu, \tilde{\alpha}_n^\nu] = m\eta^{\mu\nu} \delta_{m,-n}\,. \tag{10.2.9b}$$

For T_F and T_B the Laurent expansions are

$$T_F(z) = \sum_{r \in \mathbf{Z}+v} \frac{G_r}{z^{r+3/2}} , \quad \tilde{T}_F(\bar{z}) = \sum_{r \in \mathbf{Z}+\tilde{v}} \frac{\tilde{G}_r}{\bar{z}^{r+3/2}} , \quad (10.2.10a)$$

$$T_B(z) = \sum_{m=-\infty}^{\infty} \frac{L_m}{z^{m+2}} , \quad \tilde{T}_B(\bar{z}) = \sum_{m=-\infty}^{\infty} \frac{\tilde{L}_m}{\bar{z}^{m+2}} . \quad (10.2.10b)$$

The usual CFT contour calculation gives the mode algebra

$$[L_m, L_n] = (m-n)L_{m+n} + \frac{c}{12}(m^3 - m)\delta_{m,-n} , \quad (10.2.11a)$$

$$\{G_r, G_s\} = 2L_{r+s} + \frac{c}{12}(4r^2 - 1)\delta_{r,-s} , \quad (10.2.11b)$$

$$[L_m, G_r] = \frac{m-2r}{2}G_{m+r} . \quad (10.2.11c)$$

This is known as the *Ramond algebra* for r, s integer and the *Neveu–Schwarz algebra* for r, s half-integer. The antiholomorphic fields give a second copy of these algebras.

The superconformal generators in either sector are

$$L_m = \frac{1}{2} \sum_{n \in \mathbf{Z}} {}^\circ_\circ \alpha^\mu_{m-n} \alpha_{\mu n} {}^\circ_\circ + \frac{1}{4} \sum_{r \in \mathbf{Z}+v} (2r - m) {}^\circ_\circ \psi^\mu_{m-r} \psi_{\mu r} {}^\circ_\circ + a^m \delta_{m,0} ,$$
$$(10.2.12a)$$

$$G_r = \sum_{n \in \mathbf{Z}} \alpha^\mu_n \psi_{\mu r-n} . \quad (10.2.12b)$$

Again ${}^\circ_\circ \ {}^\circ_\circ$ denotes creation–annihilation normal ordering. The normal ordering constant can be obtained by any of the methods from chapter 2; we will use here the mnemonic from the end of section 2.9. Each periodic boson contributes $-\frac{1}{24}$. Each periodic fermion contributes $+\frac{1}{24}$ and each antiperiodic fermion $-\frac{1}{48}$. Including the shift $+\frac{1}{24}c = \frac{1}{16}D$ gives

$$\text{R:} \quad a^m = \frac{1}{16}D , \quad \text{NS:} \quad a^m = 0 . \quad (10.2.13)$$

For the open string, the condition that the surface term in the equation of motion vanish allows the possibilities

$$\psi^\mu(0, \sigma^2) = \exp(2\pi i v) \tilde{\psi}^\mu(0, \sigma^2) , \quad \psi^\mu(\pi, \sigma^2) = \exp(2\pi i v') \tilde{\psi}^\mu(\pi, \sigma^2) . \quad (10.2.14)$$

By the redefinition $\tilde{\psi}^\mu \to \exp(-2\pi i v')\tilde{\psi}^\mu$, we can set $v' = 0$. There are therefore two sectors, R and NS, as compared to the four of the closed string. To write the mode expansion it is convenient to combine ψ^μ and $\tilde{\psi}^\mu$ into a single field with the extended range $0 \leq \sigma^1 \leq 2\pi$. Define

$$\psi^\mu(\sigma^1, \sigma^2) = \tilde{\psi}^\mu(2\pi - \sigma^1, \sigma^2) \quad (10.2.15)$$

for $\pi \leq \sigma^1 \leq 2\pi$. The boundary condition $v' = 0$ is automatic, and the antiholomorphicity of $\tilde{\psi}^\mu$ implies the holomorphicity of the extended ψ^μ.

Finally, the boundary condition (10.2.14) at $\sigma^1 = 0$ becomes a periodicity condition on the extended ψ^μ, giving one set of R or NS oscillators and the corresponding algebra.

NS and R spectra

We now consider the spectrum generated by a single set of NS or R modes, corresponding to the open string or to one side of the closed string. The NS spectrum is simple. There is no $r = 0$ mode, so we define the ground state to be annihilated by all $r > 0$ modes,

$$\psi_r^\mu|0\rangle_{\text{NS}} = 0 , \quad r > 0 . \tag{10.2.16}$$

The modes with $r < 0$ then act as raising operators; since these are anticommuting, each mode can only be excited once.

The main point of interest is the R ground state, which is degenerate due to the ψ_0^μs. Define the ground states to be those that are annihilated by all $r > 0$ modes. The ψ_0^μ satisfy the Dirac gamma matrix algebra (10.1.4) with

$$\Gamma^\mu \cong 2^{1/2}\psi_0^\mu . \tag{10.2.17}$$

Since $\{\psi_r^\mu, \psi_0^\nu\} = 0$ for $r > 0$, the ψ_0^μ take ground states into ground states. The ground states thus form a representation of the gamma matrix algebra. This representation is worked out in section B.1; in $D = 10$ it has dimension 32. The reader who is not familiar with properties of spinors in various dimensions should read section B.1 at this point. We can take a basis of eigenstates of the Lorentz generators S_a, eq. (B.1.10):

$$|s_0, s_1, \ldots, s_4\rangle_{\text{R}} \equiv |\mathbf{s}\rangle_{\text{R}} , \quad s_a = \pm\tfrac{1}{2} . \tag{10.2.18}$$

The half-integral values show that these are indeed spacetime spinors. A more general basis for the spinors would be denoted $|\alpha\rangle_{\text{R}}$. In the R sector of the open string not only the ground state but all states have half-integer spacetime spins, because the raising operators are vectors and change the S_a by integers. In the NS sector, the ground state is annihilated by $S^{\mu\nu}$ and is a Lorentz singlet, and all other states then have integer spin.

The Dirac representation **32** is reducible to two Weyl representations **16** + **16**′, distinguished by their eigenvalue under Γ as in eq. (B.1.11). This has a natural extension to the full string spectrum. The distinguishing property of Γ is that it anticommutes with all Γ^μ. Since the Dirac matrices are now the center-of-mass modes of ψ^μ, we need an operator that anticommutes with the full ψ^μ. We will call this operator

$$\exp(\pi i F) , \tag{10.2.19}$$

where F, the *world-sheet fermion number,* is defined only mod 2. Since ψ^μ changes F by one it anticommutes with the exponential. It is convenient to write F in terms of spacetime Lorentz generators, which in either sector of the ψ CFT are

$$\Sigma^{\mu\lambda} = -\frac{i}{2} \sum_{r\in\mathbf{Z}+v} [\psi_r^\mu, \psi_{-r}^\lambda] . \qquad (10.2.20)$$

This is the natural extension of the zero-mode part (B.1.8). Define now

$$S_a = i^{\delta_{a,0}} \Sigma^{2a,2a+1} , \qquad (10.2.21)$$

the i being included to make S_0 Hermitean, and let

$$F = \sum_{a=0}^{4} S_a . \qquad (10.2.22)$$

This has the desired property. For example,

$$S_1(\psi_r^2 \pm i\psi_r^3) = (\psi_r^2 \pm i\psi_r^3)(S_1 \pm 1) , \qquad (10.2.23)$$

so these oscillators change F by ± 1. The definition (10.2.22) makes it obvious that F is conserved by the OPE of the vertex operators, as a consequence of Lorentz invariance.[1] When we include the ghost part of the vertex operator in section 10.4, we will see that it contributes to the total F, so that on the total matter plus ghost ground state one has

$$\exp(\pi i F)|0\rangle_{\text{NS}} = -|0\rangle_{\text{NS}} , \qquad (10.2.24a)$$

$$\exp(\pi i F)|\mathbf{s}\rangle_{\text{R}} = |\mathbf{s}'\rangle_{\text{R}} \Gamma_{\mathbf{s}'\mathbf{s}} . \qquad (10.2.24b)$$

The ghost ground state contributes a factor -1 in the NS sector and $-i$ in the R sector.

Closed string spectra

In the closed string, the NS–NS states have integer spin. Because the spins S_a are additive, the half-integers from the two sides of the R–R sector also combine to give integer spin. The NS–R and R–NS states, on the other hand, have half-integer spin.

Let us look in more detail at the R–R sector, where the ground states $|\mathbf{s}, \mathbf{s}'\rangle_{\text{R}}$ are degenerate on both the right and left. They transform as the product of two Dirac representations, which is worked out in section B.1:

$$\mathbf{32}_{\text{Dirac}} \times \mathbf{32}_{\text{Dirac}} = [0] + [1] + [2] + \ldots + [10]$$
$$= [0]^2 + [1]^2 + \ldots + [4]^2 + [5] , \qquad (10.2.25)$$

---

[1] Lorentz invariance of the OPE holds separately for the ψ and X CFTs (and the $\tilde{\psi}$ CFT in the closed string) because they are decoupled from one another. However, the world-sheet supercurrent is only invariant under the overall Lorentz transformation.

Table 10.1. *SO*(9, 1) *representations of massless R–R states.*

$(\exp(\pi i F), \exp(\pi i \tilde{F}))$			*SO*(9, 1) rep.
$(+1, +1)$:	$\mathbf{16} \times \mathbf{16}$	$=$	$[1] + [3] + [5]_+$
$(+1, -1)$:	$\mathbf{16} \times \mathbf{16'}$	$=$	$[0] + [2] + [4]$
$(-1, +1)$:	$\mathbf{16'} \times \mathbf{16}$	$=$	$[0] + [2] + [4]$
$(-1, -1)$:	$\mathbf{16'} \times \mathbf{16'}$	$=$	$[1] + [3] + [5]_-$

where $[n]$ denotes an antisymmetric rank n tensor. For the closed string there are separate world-sheet fermion numbers F and $\tilde{F}$, which on the ground states reduce to the *chirality* matrices Γ and $\tilde{\Gamma}$ acting on the two sides. The ground states thus decompose as in table 10.1.

10.3   Vertex operators and bosonization

Consider first the unit operator. Fields remain holomorphic at the origin, and in particular they are single-valued. From the Laurent expansion (10.2.7), the single-valuedness means that the unit operator must be in the NS sector; the conformal transformation that takes the incoming string to the point $z = 0$ cancels the branch cut from the antiperiodicity. The holomorphicity of ψ at the origin implies, via the contour argument, that the state corresponding to the unit operator satisfies

$$\psi_r^\mu |1\rangle = 0 \,, \quad r = \frac{1}{2}, \frac{3}{2}, \dots \,, \tag{10.3.1}$$

and therefore

$$|1\rangle = |0\rangle \,. \tag{10.3.2}$$

Since the $\psi\psi$ OPE is single-valued, all products of ψ and its derivatives must be in the NS sector. The contour argument gives the map

$$\psi_{-r}^\mu \to \frac{1}{(r - 1/2)!} \partial^{r-1/2} \psi^\mu(0) \,, \tag{10.3.3}$$

so that there is a one-to-one map between such products and NS states. The analog of the Noether relation (2.9.6) between the superconformal variation of an NS operator and the OPE is

$$\delta_\eta \mathscr{A}(z, \bar{z}) = -\epsilon \sum_{n=0}^{\infty} \frac{1}{n!} \left[\partial^n \eta(z) G_{n-1/2} + (\partial^n \eta(z))^* \tilde{G}_{n-1/2} \right] \cdot \mathscr{A}(z, \bar{z}) \,. \tag{10.3.4}$$

The R sector vertex operators must be more complicated because the Laurent expansion (10.2.7) has a branch cut. We have encountered this before, for the winding state vertex operators in section 8.2 and the orbifold twisted state vertex operators in section 8.5. Each of these introduces a branch cut (the first a log and the second a square root) into X^μ. For the winding state vertex operators there was a simple expression as the exponential of a free field. For the twisted state vertex operators there was no simple expression and their amplitudes are determined only with more effort. Happily, through a remarkable property of two-dimensional field theory, the R sector vertex operators can be related directly to the bosonic winding state vertex operators.

Let $H(z)$ be the holomorphic part of a scalar field,

$$H(z)H(0) \sim -\ln z . \tag{10.3.5}$$

For world-sheet scalars not associated directly with the embedding of the string in spacetime this is the normalization we will always use, corresponding to $\alpha' = 2$ for the embedding coordinates. As in the case of the winding state vertex operators we can be cavalier about the location of the branch cut as long as the final expressions are single-valued. We will give a precise oscillator definition below. Consider the basic operators $e^{\pm iH(z)}$. These have the OPE

$$e^{iH(z)}e^{-iH(0)} \sim \frac{1}{z} , \tag{10.3.6a}$$

$$e^{iH(z)}e^{iH(0)} = O(z) , \tag{10.3.6b}$$

$$e^{-iH(z)}e^{-iH(0)} = O(z) . \tag{10.3.6c}$$

The poles and zeros in the OPE together with smoothness at infinity determine the expectation values of these operators on the sphere, up to an overall normalization which can be set to a convenient value:

$$\left\langle \prod_i e^{i\epsilon_i H(z_i)} \right\rangle_{S_2} = \prod_{i<j} z_{ij}^{\epsilon_i\epsilon_j} , \quad \sum_i \epsilon_i = 0 . \tag{10.3.7}$$

The ϵ_i are ± 1 here, but this result holds more generally.

Now consider the CFT of two Majorana–Weyl fermions $\psi^{1,2}(z)$, and form the complex combinations

$$\psi = 2^{-1/2}(\psi^1 + i\psi^2) , \quad \overline{\psi} = 2^{-1/2}(\psi^1 - i\psi^2) . \tag{10.3.8}$$

These have the properties

$$\psi(z)\overline{\psi}(0) \sim \frac{1}{z} , \tag{10.3.9a}$$

$$\psi(z)\psi(0) = O(z) , \tag{10.3.9b}$$

$$\overline{\psi}(z)\overline{\psi}(0) = O(z) . \tag{10.3.9c}$$

Eqs. (10.3.6) and (10.3.9) are identical in form, and so the expectation values of $\psi(z)$ on the sphere are identical to those of $e^{iH(z)}$. We will write

$$\psi(z) \cong e^{iH(z)} , \quad \overline{\psi}(z) \cong e^{-iH(z)} \tag{10.3.10}$$

to indicate this. Of course, all of this extends to the antiholomorphic case,

$$\tilde{\psi}(\bar{z}) \cong e^{i\tilde{H}(\bar{z})} \ , \quad \bar{\tilde{\psi}}(z) \cong e^{-i\tilde{H}(\bar{z})} \ . \tag{10.3.11}$$

Since arbitrary local operators with integral k_R and k_L can be formed by repeated operator products of $e^{\pm iH(z)}$ and $e^{\pm i\tilde{H}(\bar{z})}$, and arbitrary local operators built out of the fermions and their derivatives can be formed by repeated operator products of $\psi(z)$, $\bar{\psi}(z)$, $\tilde{\psi}(\bar{z})$, and $\bar{\tilde{\psi}}(\bar{z})$, the equivalence of the theories can be extended to all local operators. Finally, in order for these theories to be the same as CFTs, the energy-momentum tensors must be equivalent. The easiest way to show this is via the operator products

$$e^{iH(z)}e^{-iH(-z)} = \frac{1}{2z} + i\partial H(0) + 2z T_B^H(0) + O(z^2) \ , \tag{10.3.12a}$$

$$\psi(z)\bar{\psi}(-z) = \frac{1}{2z} + \psi\bar{\psi}(0) + 2z T_B^\psi(0) + O(z^2) \ . \tag{10.3.12b}$$

With the result (10.3.10), this implies equivalence of the H momentum current with the ψ number current, and of the two energy-momentum tensors,

$$\psi\bar{\psi} \cong i\partial H \ , \quad T_B^\psi \cong T_B^H \ . \tag{10.3.13}$$

As a check, e^{iH} and ψ are both $(\frac{1}{2}, 0)$ tensors.

In the operator description of the theory, define

$$\psi(z) \cong \, {}^\circ_\circ e^{iH(z)} {}^\circ_\circ \ . \tag{10.3.14}$$

From the Campbell–Baker–Hausdorff (CBH) formula (6.7.23) we have for equal times $|z| = |z'|$

$$\begin{aligned} {}^\circ_\circ e^{iH(z)} {}^\circ_\circ \, {}^\circ_\circ e^{iH(z')} {}^\circ_\circ &= \exp\{-[H(z), H(z')]\} \, {}^\circ_\circ e^{iH(z')} {}^\circ_\circ \, {}^\circ_\circ e^{iH(z)} {}^\circ_\circ \\ &= - \, {}^\circ_\circ e^{iH(z')} {}^\circ_\circ \, {}^\circ_\circ e^{iH(z)} {}^\circ_\circ \ , \end{aligned} \tag{10.3.15}$$

where we have used the fact (8.2.21) that at equal times $[H(z), H(z')] = \pm i\pi$. Thus the bosonized operators do anticommute. This is possible for operators constructed purely out of bosons because they are nonlocal. In particular, note that the CBH formula gives the equal time commutator

$$\begin{aligned} H(z) \, {}^\circ_\circ e^{iH(z')} {}^\circ_\circ &= \, {}^\circ_\circ e^{iH(z')} {}^\circ_\circ \Big(H(z) + i[H(z), H(z')] \Big) \\ &= \, {}^\circ_\circ e^{iH(z')} {}^\circ_\circ \Big(H(z) - \pi \, \text{sign}(\sigma_1 - \sigma_1') \Big) \ , \end{aligned} \tag{10.3.16}$$

so that the fermion field operator produces a kink, a discontinuity, in the bosonic field.

This rather surprising equivalence is known as *bosonization*. Equivalence between field theories with very different actions and fields occurs frequently in two dimensions, especially in CFTs because holomorphicity puts strong constraints on the theory. (The great recent surprise is that it is also quite common in higher-dimensional field and string theories.) Many interesting CFTs can be constructed in several different ways. One form or another will often be more useful for specific purposes. Notice that there is no simple correspondence between one-boson and one-fermion states. The current, for example, is linear in the boson field but quadratic in the fermion field. A single boson is the same as one ψ fermion and one $\bar{\psi}$ fermion at the same point. On a Minkowski world-sheet, where holomorphic becomes left-moving, the fermions both move left at the speed of light and remain coincident, indistinguishable from a free boson. A single fermion, on the other hand, is created by an operator exponential in the boson field and so is a coherent state, which as we have seen is in the shape of a kink (10.3.16).

The complicated relationship between the bosonic and fermionic spectra shows up also in the partition function. Operator products of $e^{\pm iH(z)}$ generate all operators with integer k_L. The bosonic momentum and oscillator sums then give

$$\text{Tr}\,(q^{L_0}) = \left(\sum_{k_L \in \mathbf{Z}} q^{k_L^2/2} \right) \prod_{n=1}^{\infty} (1 - q^n)^{-1} \, . \tag{10.3.17}$$

In the NS sector of the fermionic theory, the oscillator sum gives

$$\text{Tr}\,(q^{L_0}) = \prod_{n=1}^{\infty} (1 + q^{n-1/2})^2 \, . \tag{10.3.18}$$

We know indirectly that these must be equal, since we can use the OPE to construct an analog in the fermionic theory for any local operator of the bosonic theory and vice versa. Expanding the products gives

$$1 + 2q^{1/2} + q + 2q^{3/2} + 4q^2 + 4q^{5/2} + \ldots \tag{10.3.19}$$

for each, and in fact the equality of (10.3.17) and (10.3.18) follows from the equality of the product and sum expressions for theta functions, section 7.2. Note that while bosonization was derived for the sphere, the sewing construction from chapter 9 guarantees that it holds on all Riemann surfaces, provided that we make equivalent projections on the spectra. In particular, we have seen that summing over integer k_L corresponds to summing over all local fermionic operators, the NS sector.

Bosonization extends readily to the R sector. In fact, once we combine two fermions into a complex pair we can consider the more general periodicity condition

$$\psi(w + 2\pi) = \exp(2\pi i v)\,\psi(w) \tag{10.3.20}$$

for any real v. In ten dimensions only $v = 0, \frac{1}{2}$ arose, but these more general periodicities are important in less symmetric situations. The Laurent expansion has the same form (10.2.7) as before,

$$\psi(z) = \sum_{r \in \mathbf{Z}+v} \frac{\psi_r}{z^{r+1/2}} , \quad \overline{\psi}(z) = \sum_{s \in \mathbf{Z}-v} \frac{\overline{\psi}_s}{z^{s+1/2}} , \tag{10.3.21}$$

with indices displaced from integers by $\pm v$. The algebra is

$$\{\psi_r, \overline{\psi}_s\} = \delta_{r,-s} . \tag{10.3.22}$$

Define a reference state $|0\rangle_v$ by

$$\psi_{n+v}|0\rangle_v = \overline{\psi}_{n+1-v}|0\rangle_v = 0 , \quad n = 0, 1, \dots . \tag{10.3.23}$$

The first nonzero terms in the Laurent expansions are then $r = -1 + v$ and $s = -v$, so for the corresponding local operator $\mathscr{A}_v$ the OPE is

$$\psi(z)\mathscr{A}_v(0) = O(z^{-v+1/2}) , \quad \overline{\psi}(z)\mathscr{A}_v(0) = O(z^{v-1/2}) . \tag{10.3.24}$$

The conditions (10.3.23) uniquely identify the state $|0\rangle_v$, and so the corresponding OPEs (10.3.24) determine the bosonic equivalent

$$\exp[i(-v + 1/2)H] \cong \mathscr{A}_v . \tag{10.3.25}$$

One can check the identification (10.3.25) by verifying that the weight is $h = \frac{1}{2}(v - \frac{1}{2})^2$. In the bosonic form this comes from the term $\frac{1}{2}p^2$ in L_0. In the fermionic form it follows from the usual commutator method (2.7.8) or the zero-point mnemonic.

The boundary condition (10.3.20) is the same for v and $v + 1$, but the reference state that we have defined is not. It is a ground state only for $0 \le v \le 1$. As we vary v, the state $|0\rangle_v$ changes continuously, and when we get back to the original theory at $v + 1$, by the definition (10.3.23) it has become the excited state

$$|0\rangle_{v+1} = \overline{\psi}_{-v}|0\rangle_v . \tag{10.3.26}$$

This is known as *spectral flow*. For the R case $v = 0$ there are the two degenerate ground states

$$|s\rangle \cong e^{isH}, \quad s = \pm\tfrac{1}{2} . \tag{10.3.27}$$

For the superstring in ten dimensions we need five bosons, H^a for $a = 0, \dots, 4$. Then[2]

$$2^{-1/2}(\pm\psi^0 + \psi^1) \cong e^{\pm iH^0} \tag{10.3.28a}$$

$$2^{-1/2}(\psi^{2a} \pm i\psi^{2a+1}) \cong e^{\pm iH^a} , \quad a = 1, \dots, 4 . \tag{10.3.28b}$$

---

[2] The precise operator definition has a subtlety when there are several species of fermion. The H^a for different a are independent and so the exponentials commute rather than anticommute. A cocycle is needed, as in eq. (8.2.22). A general expression will be given in the next section.

The vertex operator Θ_s for an R state $|s\rangle$ is

$$\Theta_s \cong \exp\left[i \sum_a s_a H^a\right] . \tag{10.3.29}$$

This operator, which produces a branch cut in ψ^μ, is sometimes called a *spin field*. For closed string states, this is combined with the appropriate antiholomorphic vertex operator, built from $\tilde{H}^a$.

The general bc CFT, renaming $\psi \to b$ and $\bar{\psi} \to c$, is obtained by modifying the energy-momentum tensor of the $\lambda = \frac{1}{2}$ theory to

$$T_B^{(\lambda)} = T_B^{(1/2)} - (\lambda - \tfrac{1}{2})\partial(bc) . \tag{10.3.30}$$

The equivalences (10.3.13) give the corresponding bosonic operator

$$T_B^{(\lambda)} \cong T_B^H - i(\lambda - \tfrac{1}{2})\partial^2 H . \tag{10.3.31}$$

This is the same as the linear dilaton CFT, with $V = -i(\lambda - \frac{1}{2})$. With this correspondence between V and λ, the linear dilaton and bc theories are equivalent,

$$b \cong e^{iH} , \quad c \cong e^{-iH} . \tag{10.3.32}$$

As a check, the central charges agree,

$$c = 1 - 3(2\lambda - 1)^2 = 1 + 12V^2 . \tag{10.3.33}$$

So do the dimensions of the fields (10.3.32), λ for b and $1-\lambda$ for c, agreeing with $k^2/2 + ikV$ for e^{ikH}. The nontensor behaviors of the currents bc and $i\partial H$ are also the same. Since the inner product for the reparameterization ghosts makes b and c Hermitean, the bosonic field H must be anti-Hermitean in this application. The bosonization of the ghosts is usually written in terms of a Hermitean field with the opposite sign OPE,

$$H \to i\rho ; \quad c \cong e^\rho , \quad b \cong e^{-\rho} . \tag{10.3.34}$$

10.4 The superconformal ghosts

To build the BRST current we will need, in addition to the anticommuting b and c ghosts of the bosonic string, commuting ghost fields β and γ of weight $(\frac{3}{2}, 0)$ and $(-\frac{1}{2}, 0)$, and the corresponding antiholomorphic fields. The action for this SCFT was given in eq. (10.1.17) and the currents T_B and T_F in eq. (10.1.21). The ghosts β and γ must have the same periodicity (10.2.4) as the generator T_F with which they are associated. This is necessary to make the BRST current periodic, so that it can be integrated to give the BRST charge. Thus,

$$\beta(z) = \sum_{r\in \mathbf{Z}+v} \frac{\beta_r}{z^{r+3/2}} \, , \quad \gamma(z) = \sum_{r\in \mathbf{Z}+v} \frac{\gamma_r}{z^{r-1/2}} \, , \qquad (10.4.1a)$$

$$b(z) = \sum_{m=-\infty}^{\infty} \frac{b_m}{z^{m+2}} \, , \quad c(z) = \sum_{m=-\infty}^{\infty} \frac{c_m}{z^{m-1}} \, , \qquad (10.4.1b)$$

and similarly for the antiholomorphic fields. The (anti)commutators are

$$[\gamma_r, \beta_s] = \delta_{r,-s} \, , \quad \{b_m, c_n\} = \delta_{n,-m} \, . \qquad (10.4.2)$$

Define the ground states $|0\rangle_{\rm NS,R}$ by

$$\beta_r|0\rangle_{\rm NS} = 0 \, , \, r \geq \tfrac{1}{2} \, , \quad \gamma_r|0\rangle_{\rm NS} = 0 \, , \, r \geq \tfrac{1}{2} \qquad (10.4.3a)$$

$$\beta_r|0\rangle_{\rm R} = 0 \, , \, r \geq 0 \, , \quad \gamma_r|0\rangle_{\rm R} = 0 \, , \, r \geq 1 \, , \qquad (10.4.3b)$$

$$b_m|0\rangle_{\rm NS,R} = 0 \, , \, m \geq 0 \, , \quad c_m|0\rangle_{\rm NS,R} = 0 \, , \, m \geq 1 \, . \qquad (10.4.3c)$$

We have grouped β_0 with the lowering operators and γ_0 with the raising ones, in parallel with the bosonic case. The spectrum is built as usual by acting on the ground states with the raising operators. The generators are

$$L_m^{\rm g} = \sum_{n\in \mathbf{Z}}(m+n)\substack{\circ\\\circ}b_{m-n}c_n\substack{\circ\\\circ} + \sum_{r\in \mathbf{Z}+v} \frac{1}{2}(m+2r)\substack{\circ\\\circ}\beta_{m-r}\gamma_r\substack{\circ\\\circ} + a^{\rm g}\delta_{m,0} \, , \qquad (10.4.4a)$$

$$G_r^{\rm g} = -\sum_{n\in \mathbf{Z}}\left[\frac{1}{2}(2r+n)\beta_{r-n}c_n + 2b_n\gamma_{r-n}\right] \, . \qquad (10.4.4b)$$

The normal ordering constant is determined by the usual methods to be

$$\text{R:} \quad a^{\rm g} = -\frac{5}{8} \, , \quad \text{NS:} \quad a^{\rm g} = -\frac{1}{2} \, . \qquad (10.4.5)$$

Vertex operators

We focus here on the $\beta\gamma$ CFT, as the bc parts of the vertex operators are already understood. Let us start by considering the state corresponding to the unit operator. From the Laurent expansions (10.4.1) it is in the NS sector and satisfies

$$\beta_r|1\rangle = 0 \, , \, r \geq -\frac{1}{2} \, , \quad \gamma_r|1\rangle = 0 \, , \, r \geq \frac{3}{2} \, . \qquad (10.4.6)$$

This is not the same as the ground state $|0\rangle_{\rm NS}$: the mode $\gamma_{1/2}$ annihilates $|0\rangle_{\rm NS}$ while its conjugate $\beta_{-1/2}$ annihilates $|1\rangle$. We found this also for the bc ghosts with c_1 and b_{-1}. Since anticommuting modes generate just two states, we had the simple relation $|0\rangle = c_1|1\rangle$ (focusing on the holomorphic side). For commuting oscillators things are not so simple: there is no state in the Fock space built on $|1\rangle$ by acting with $\gamma_{1/2}$ that has the properties of $|0\rangle_{\mathrm{NS}}$. The definition of the state $|0\rangle_{\mathrm{NS}}$ translates into

$$\gamma(z)\delta(\gamma(0)) = O(z) , \quad \beta(z)\delta(\gamma(0)) = O(z^{-1}) , \tag{10.4.7}$$

for the corresponding operator $\delta(\gamma)$. The notation $\delta(\gamma)$ reflects the fact that the field γ has a simple zero at the vertex operator. Recall that for the bc ghosts the NS ground state maps to the operator c, which is the anticommuting analog of a delta function. One can show that an insertion of $\delta(\gamma)$ in the path integral has the property (10.4.7).

To give an explicit description of this operator it is again convenient to bosonize. Of course β and γ are already bosonic, but bosonization here refers to a rewriting of the theory in a way that is similar to, but a bit more intricate than, the bosonization of the anticommuting bc theory.

Start with the current $\beta\gamma$. The operator product

$$\beta\gamma(z)\,\beta\gamma(0) \sim -\frac{1}{z^2} \tag{10.4.8}$$

is the same as that of $\partial\phi$, where $\phi(z)\phi(0) \sim -\ln z$ is a holomorphic scalar. Holomorphicity then implies that this equivalence extends to all correlation functions,

$$\beta\gamma(z) \cong \partial\phi(z) . \tag{10.4.9}$$

The OPE of the current with β and γ then suggests

$$\beta(z) \overset{?}{\cong} e^{-\phi(z)} , \quad \gamma(z) \overset{?}{\cong} e^{\phi(z)} . \tag{10.4.10}$$

For the bc system we would be finished: this approach leads to the same bosonization as before. For the $\beta\gamma$ system, however, the sign of the current–current OPE and therefore of the $\phi\phi$ OPE is changed. The would-be bosonization (10.4.10) gives the wrong OPEs: it would imply

$$\beta(z)\beta(0) \overset{?}{=} O(z^{-1}) , \quad \beta(z)\gamma(0) \overset{?}{=} O(z^1) , \quad \gamma(z)\gamma(0) \overset{?}{=} O(z^{-1}) , \tag{10.4.11}$$

whereas the correct OPE is

$$\beta(z)\beta(0) = O(z^0) , \quad \beta(z)\gamma(0) = O(z^{-1}) , \quad \gamma(z)\gamma(0) = O(z^0) . \tag{10.4.12}$$

To repair this, additional factors are added,

$$\beta(z) \cong e^{-\phi(z)}\partial\xi(z) , \quad \gamma \cong e^{\phi(z)}\eta(z) . \tag{10.4.13}$$

In order not to spoil the OPE with the current (10.4.9), the new fields $\eta(z)$ and $\xi(z)$ must be nonsingular with respect to ϕ, which means that the $\eta\xi$ theory is a new CFT, decoupled from the ϕ CFT. Further, the equivalence (10.4.13) will hold — all OPEs will be correct — if η and ξ satisfy

$$\eta(z)\xi(0) \sim \frac{1}{z} , \quad \eta(z)\eta(0) = O(z) , \quad \partial\xi(z)\partial\xi(0) = O(z) . \tag{10.4.14}$$

This identifies the $\eta\xi$ theory as a holomorphic CFT of the bc type: the OPE of like fields has a zero due to the anticommutativity.

It remains to study the energy-momentum tensor. We temporarily consider the general $\beta\gamma$ system, with β having weight λ'. The OPE

$$T(z)\beta\gamma(0) = \frac{1 - 2\lambda'}{z^3} + \dots \qquad (10.4.15)$$

determines the ϕ energy-momentum tensor,

$$T_B^\phi = -\frac{1}{2}\partial\phi\partial\phi + \frac{1}{2}(1 - 2\lambda')\partial^2\phi . \qquad (10.4.16)$$

The exponentials in the bosonization (10.4.13) thus have weights $\lambda'-1$ and $-\lambda'$ respectively, as compared with the weights λ' and $1 - \lambda'$ of β and γ. This fixes the weights of η and ξ as 1 and 0: this is a $\lambda = 1$ bc system, with

$$T_B^{\eta\xi} = -\eta\partial\xi \qquad (10.4.17)$$

and

$$T_B^{\beta\gamma} \cong T_B^\phi + T_B^{\eta\xi} . \qquad (10.4.18)$$

As a check, the central charges are $3(2\lambda' - 1)^2 + 1$ for T_B^ϕ and -2 for $T_B^{\eta\xi}$, adding to the $3(2\lambda' - 1)^2 - 1$ of the $\beta\gamma$ CFT. The need for extra degrees of freedom is not surprising. The $\beta\gamma$ theory has a greater density of states than the bc theory because the modes of a commuting field can be excited any number of times. One can check that the total partition functions agree, in the appropriate sectors.

If need be one can go further and represent the $\eta\xi$ theory in terms of a free boson, conventionally χ with $\chi(z)\chi(0) \sim \ln z$, as in the previous section. Thus

$$\eta \cong e^{-\chi} , \quad \xi \cong e^{\chi} , \qquad (10.4.19a)$$

$$\beta \cong e^{-\phi+\chi}\partial\chi , \quad \gamma \cong e^{\phi-\chi} . \qquad (10.4.19b)$$

The energy-momentum tensor is then

$$T_B = -\frac{1}{2}\partial\phi\partial\phi + \frac{1}{2}\partial\chi\partial\chi + \frac{1}{2}(1 - 2\lambda')\partial^2\phi + \frac{1}{2}\partial^2\chi . \qquad (10.4.20)$$

For the string, the relevant value is $\lambda' = \frac{3}{2}$. The properties (10.4.7) of $\delta(\gamma)$ determine the bosonization,

$$\delta(\gamma) \cong e^{-\phi} , \quad h = \frac{1}{2} . \qquad (10.4.21)$$

The fermionic parts of the tachyon and massless NS vertex operators are then

$$e^{-\phi} , \quad e^{-\phi}e^{\pm iH^a} \qquad (10.4.22)$$

respectively. For $\lambda' = \frac{3}{2}$, the exponential $e^{l\phi}$ has weight $-\frac{1}{2}l^2 - l$.

The operator Σ corresponding to $|0\rangle_R$ satisfies

$$\beta(z)\Sigma(0) = O(z^{-1/2}) , \quad \gamma(z)\Sigma(0) = O(z^{1/2}) . \tag{10.4.23}$$

This determines

$$\Sigma = e^{-\phi/2} , \quad h = \frac{3}{8} . \tag{10.4.24}$$

Adding the contribution -1 of the bc ghosts, the weight of $e^{-\phi/2}$ and of $e^{-\phi}$ agree with the values (10.4.5). The R ground state vertex operators are then

$$\mathscr{V}_s = e^{-\phi/2}\Theta_s , \tag{10.4.25}$$

with the spin field Θ_s having been defined in eq. (10.3.29).

We need to extend the definition of world-sheet fermion number F to be odd for β and γ. The ultimate reason is that it anticommutes with the supercurrent T_F and we will need it to commute with the BRST operator, which contains terms such as γT_F. The natural definition for F is then that it be the charge associated with the current (10.4.9), which is l for $e^{l\phi}$. Again, it is conserved by the OPE. This accounts for the ghost contributions in eq. (10.2.24). Note that this definition is based on spin rather than statistics, since the ghosts have the wrong spin-statistics relation; it would therefore be more appropriate to call F the *world-sheet spinor number*.

For completeness we give a general expression for the cocycle for exponentials of free fields, though we emphasize that for most purposes the details are not necessary. In general one has operators

$$\exp(ik_L \cdot H_L + ik_R \cdot H_R) , \tag{10.4.26}$$

with the holomorphic and antiholomorphic scalars not necessarily equal in number. The momenta k take values in some lattice Γ. The naive operator product has the phase of $z^{-k\circ k'}$, and for all pairs in Γ, $k \circ k'$ must be an integer. The notation is as in section 8.4, $k \circ k' = k_L \cdot k'_L - k_R \cdot k'_R$. When $k \circ k'$ is an odd integer the vertex operators anticommute rather than commute. A correctly defined vertex operator is

$$C_k(\alpha_0) {}^\circ_\circ \exp(ik_L \cdot H_L + ik_R \cdot H_R) {}^\circ_\circ \tag{10.4.27}$$

with the cocycle C_k defined as follows. Take a set of basis vectors k_α for Γ; that is, Γ consists of the integer linear combinations $n_\alpha k_\alpha$. Similarly write the vector of zero-mode operators in this basis, $\alpha_0 = \alpha_{0\alpha}k_\alpha$, Then for $k = n_\alpha k_\alpha$,

$$C_k(\alpha_0) = \exp\left(\pi i \sum_{\alpha > \beta} n_\alpha \alpha_{0\beta} k_\alpha \circ k_\beta \right) . \tag{10.4.28}$$

This generalizes the simple case (8.2.22). The reader can check that vertex operators with even $k \circ k$ now commute with all vertex operators, and those with odd $k \circ k$ anticommute among themselves. Note that a cocycle has no effect on the commutativity of a vertex operator with itself, so an exponential must be bosonic if $k \circ k$ is even and fermionic if $k \circ k$ is odd.

10.5   Physical states

In the bosonic string we started with a (diff×Weyl)-invariant theory. After fixing to conformal gauge we had to impose the vanishing of the conformal algebra as a constraint on the states. In the present case there is an analogous gauge-invariant form, and the superconformal algebra emerges as a constraint in the gauge-fixed theory. However, it is not necessary to proceed in this way, and it would require us to develop some machinery that in the end we do not need. Rather we can generalize directly in the gauge-fixed form, defining the superconformal symmetry to be a constraint and proceeding in parallel to the bosonic case to construct a consistent theory. We will first impose the constraint in the old covariant formalism, and then in the BRST formalism.

OCQ

In this formalism, developed for the bosonic string in section 4.1, one ignores the ghost excitations. We begin with the open string, imposing the physical state conditions

$$L_n^{\mathrm{m}}|\psi\rangle = 0 \,, \; n > 0 \,, \quad G_r^{\mathrm{m}}|\psi\rangle = 0 \,, \; r \geq 0 \,. \tag{10.5.1}$$

Only the matter part of any state is nontrivial — the ghosts are in their ground state — and the superscript 'm' denotes the matter part of each generator. There are also the equivalence relations

$$L_n^{\mathrm{m}}|\chi\rangle \cong 0 \,, \; n < 0 \,, \quad G_r^{\mathrm{m}}|\chi\rangle \cong 0 \,, \; r < 0 \,. \tag{10.5.2}$$

The mass-shell condition can always be written in terms of the total matter plus ghost Virasoro generator, which is the same as the world-sheet Hamiltonian H because the total central charge is zero:

$$L_0|\psi\rangle = H|\psi\rangle = 0 \,. \tag{10.5.3}$$

In ten flat dimensions this is

$$H = \begin{cases} \alpha' p^2 + N - \dfrac{1}{2} & \text{(NS)} \\ \alpha' p^2 + N & \text{(R)} \end{cases} . \tag{10.5.4}$$

The zero-point constants from the ghosts and longitudinal oscillators have canceled as usual, leaving the contribution of the transverse modes,

$$\text{NS: } 8\left(-\frac{1}{24}-\frac{1}{48}\right)=-\frac{1}{2}\,,\quad \text{R: } 8\left(-\frac{1}{24}+\frac{1}{24}\right)=0\,. \tag{10.5.5}$$

For the tachyonic and massless levels we need only the terms

$$G_0^{\text{m}}=(2\alpha')^{1/2}p_\mu\psi_0^\mu+\dots\,, \tag{10.5.6a}$$

$$G_{\pm 1/2}^{\text{m}}=(2\alpha')^{1/2}p_\mu\psi_{\pm 1/2}^\mu+\dots\,. \tag{10.5.6b}$$

The NS sector works out much as in the bosonic string. The lowest state is $|0;k\rangle_{\text{NS}}$, labeled by the matter state and momentum. The only nontrivial condition is from L_0, giving

$$m^2=-k^2=-\frac{1}{2\alpha'}\,. \tag{10.5.7}$$

This state is a tachyon. It has $\exp(\pi iF)=-1$, where F was given in eq. (10.2.24). The first excited state is

$$|e;k\rangle_{\text{NS}}=e\cdot\psi_{-1/2}|0;k\rangle_{\text{NS}}\,. \tag{10.5.8}$$

The nontrivial physical state conditions are

$$0=L_0|e;k\rangle_{\text{NS}}=\alpha'k^2|e;k\rangle_{\text{NS}}\,, \tag{10.5.9a}$$

$$0=G_{1/2}^{\text{m}}|e;k\rangle_{\text{NS}}=(2\alpha')^{1/2}k\cdot e|0;k\rangle_{\text{NS}}\,, \tag{10.5.9b}$$

while

$$G_{-1/2}^{\text{m}}|0;k\rangle_{\text{NS}}=(2\alpha')^{1/2}k\cdot\psi_{-1/2}|0;k\rangle_{\text{NS}} \tag{10.5.10}$$

is null. Thus

$$k^2=0\,,\quad e\cdot k=0\,,\quad e^\mu\cong e^\mu+k^\mu\,. \tag{10.5.11}$$

This state is massless, the half-unit of excitation canceling the zero-point energy, and has $\exp(\pi iF)=+1$. Like the first excited state of the bosonic string it is a massless vector, with $D-2$ spacelike polarizations. The constraints have removed the unphysical polarizations of ψ^μ, just as for X^μ in the bosonic case.

In the R sector the lowest states are

$$|u;k\rangle_{\text{R}}=|\mathbf{s};k\rangle_{\text{R}}u_{\mathbf{s}}. \tag{10.5.12}$$

Here $u_{\mathbf{s}}$ is the polarization, and the sum on $\mathbf{s}$ is implicit. The nontrivial physical state conditions are

$$0=L_0|u;k\rangle_{\text{R}}=\alpha'k^2|u;k\rangle_{\text{R}}\,, \tag{10.5.13a}$$

$$0=G_0^{\text{m}}|u;k\rangle_{\text{R}}=\alpha'^{1/2}|\mathbf{s}';k\rangle_{\text{R}}k\cdot\Gamma_{\mathbf{s}'\mathbf{s}}u_{\mathbf{s}}\,. \tag{10.5.13b}$$

Table 10.2. *Massless and tachyonic open string states.*

sector	$SO(8)$ spin	m^2
NS+	$\mathbf{8}_v$	0
NS−	$\mathbf{1}$	$-1/2\alpha'$
R+	$\mathbf{8}$	0
R−	$\mathbf{8}'$	0

The ground states are massless because the zero-point energy vanishes in the R sector. The G_0^{m} condition gives the massless Dirac equation

$$k \cdot \Gamma_{\mathrm{s's}} u_{\mathrm{s}} = 0 \,, \qquad (10.5.14)$$

which was our original goal in introducing the superconformal algebra. The G_0^{m} condition implies the L_0 condition, because $G_0^2 = L_0$ in the critical dimension and the ghost parts of G_0 annihilate the ghost vacuum.

In ten dimensions, massless particle states are classified by their behavior under the $SO(8)$ rotations that leave the momentum invariant. Take a frame with $k_0 = k_1$. In the NS sector, the massless physical states are the eight transverse polarizations forming the vector representation $\mathbf{8}_v$ of $SO(8)$. In the R sector, the massless Dirac operator becomes

$$k_0 \Gamma^0 + k_1 \Gamma^1 = -k_1 \Gamma^0 (\Gamma^0 \Gamma^1 - 1) = -2k_1 \Gamma^0 (S_0 - \tfrac{1}{2}) \,. \qquad (10.5.15)$$

The physical state condition is then

$$(S_0 - \tfrac{1}{2})|\mathbf{s}, 0; k\rangle_{\mathrm{R}} u_{\mathrm{s}} = 0 \,, \qquad (10.5.16)$$

so precisely the states with $s_0 = +\tfrac{1}{2}$ survive. As discussed in section B.1, we have under $SO(9,1) \to SO(1,1) \times SO(8)$ the decompositions

$$\mathbf{16} \to (+\tfrac{1}{2}, \mathbf{8}) + (-\tfrac{1}{2}, \mathbf{8}') \,, \qquad (10.5.17\mathrm{a})$$
$$\mathbf{16}' \to (+\tfrac{1}{2}, \mathbf{8}') + (-\tfrac{1}{2}, \mathbf{8}) \,. \qquad (10.5.17\mathrm{b})$$

Thus the Dirac equation leaves an $\mathbf{8}$ with $\exp(\pi iF) = +1$ and an $\mathbf{8}'$ with $\exp(\pi iF) = -1$.

The tachyonic and massless states are summarized in table 10.2. The open string spectrum has four sectors, according to the periodicity v and the world-sheet fermion number $\exp(\pi iF)$. We will use the notation NS$\pm$ and R$\pm$ to label these sectors. We will see in the next section that consistency requires us to keep only certain subsets of sectors, and that there are consistent string theories without the tachyon.

Table 10.3. *Products of SO(8) representations appearing at the massless level of the closed string. The R–NS sector has the same content as the NS–R sector.*

sector	$SO(8)$ spin		tensors		dimensions
(NS+,NS+)	$\mathbf{8}_v \times \mathbf{8}_v$	$=$	$[0] + [2] + (2)$	$=$	$\mathbf{1} + \mathbf{28} + \mathbf{35}$
(R+,R+)	$\mathbf{8} \times \mathbf{8}$	$=$	$[0] + [2] + [4]_+$	$=$	$\mathbf{1} + \mathbf{28} + \mathbf{35}_+$
(R+,R−)	$\mathbf{8} \times \mathbf{8}'$	$=$	$[1] + [3]$	$=$	$\mathbf{8}_v + \mathbf{56}_t$
(R−,R−)	$\mathbf{8}' \times \mathbf{8}'$	$=$	$[0] + [2] + [4]_-$	$=$	$\mathbf{1} + \mathbf{28} + \mathbf{35}_-$
(NS+,R+)	$\mathbf{8}_v \times \mathbf{8}$			$=$	$\mathbf{8}' + \mathbf{56}$
(NS+,R−)	$\mathbf{8}_v \times \mathbf{8}'$			$=$	$\mathbf{8} + \mathbf{56}'$

Closed string spectrum

The closed string is two copies of the open string, with the momentum rescaled $k \to \frac{1}{2}k$ in the generators. With $v, \tilde{v}$ taking the values 0 and $\frac{1}{2}$, the mass-shell condition can be summarized as

$$\frac{\alpha'}{4}m^2 = N - v = \tilde{N} - \tilde{v} . \qquad (10.5.18)$$

The tachyonic and massless closed string spectrum is obtained by combining one left-moving and one right-moving state, subject to the equality (10.5.18).

The (NS−,NS−) sector contains a closed string tachyon with $m^2 = -2/\alpha'$. At the massless level, combining the various massless left- and right-moving states from table 10.2 leads to the $SO(8)$ representations shown in table 10.3. Note that level matching prevents pairing of the NS− sector with any of the other three. As in the bosonic string, vector times vector decomposes into scalar, antisymmetric tensor, and traceless symmetric tensor denoted (2). The products of spinors are discussed in section B.1.

The 64 states in $\mathbf{8}_v \times \mathbf{8}$ and $\mathbf{8}_v \times \mathbf{8}'$ each separate into two irreducible representations. Denoting a state in $\mathbf{8}_v \times \mathbf{8}$ by $|i, \mathbf{s}\rangle$, we can form the eight linear combinations

$$|i, \mathbf{s}\rangle \Gamma^i_{\mathbf{ss}'} . \qquad (10.5.19)$$

These states transform among themselves under $SO(8)$, and they are in the $\mathbf{8}'$ representation because the chirality of the loose index $\mathbf{s}'$ is opposite to that of $\mathbf{s}$. The other 56 states form an irreducible representation $\mathbf{56}$. The product $\mathbf{8}_v \times \mathbf{8}'$ works in the same way. Note that there are several cases of distinct representations with identical dimensions: at dimension 8 a vector and two spinors, at dimension 56 an antisymmetric rank 3 tensor and two vector-spinors, at dimension 35 a traceless symmetric rank 2 tensor and self-dual and anti-self-dual rank 4 tensors.

BRST quantization

From the general structure discussed in chapter 4, in particular the expression (4.3.14) for the BRST operator for a general constraint algebra, the BRST operator can be constructed as a simple extension of the bosonic one:

$$Q_B = \frac{1}{2\pi i} \oint (dz\, j_B - d\bar{z}\, \tilde{j}_B) \,, \qquad (10.5.20)$$

where

$$
\begin{aligned}
j_B &= c T_B^m + \gamma T_F^m + \frac{1}{2}\left(c T_B^g + \gamma T_F^g\right) \\
&= c T_B^m + \gamma T_F^m + bc\partial c + \frac{3}{4}(\partial c)\beta\gamma + \frac{1}{4}c(\partial\beta)\gamma - \frac{3}{4}c\beta\partial\gamma - b\gamma^2 \,,
\end{aligned}
\qquad (10.5.21)
$$

and the same on the antiholomorphic side. As in the bosonic case, this is a tensor up to an unimportant total derivative term.

The BRST current has the essential property

$$j_B(z)b(0) \sim \ldots + \frac{1}{z}T_B(0) \,, \qquad j_B(z)\beta(0) \sim \ldots + \frac{1}{z}T_F(0) \,, \qquad (10.5.22)$$

so that the commutators of Q_B with the b, β ghosts give the corresponding constraints.[3] In modes,

$$\{Q_B, b_n\} = L_n \,, \qquad [Q_B, \beta_r] = G_r \,. \qquad (10.5.23)$$

From these one can verify nilpotence by the same steps as in the bosonic case (exercise 4.3) whenever the total central charge vanishes. Thus, we can replace some of the spacelike $X^\mu \psi^\mu$ SCFTs with any positive-norm SCFT such that the total matter central charge is $c^m = \tilde{c}^m = 15$. The BRST current must be periodic for the BRST charge to be well defined. The supercurrent of the SCFT must therefore have the same periodicity, R or NS, as the ψ^μ, β, and γ. The expansion of the BRST operator is

$$
\begin{aligned}
Q_B &= \sum_m c_{-m}L_m^m + \sum_r \gamma_{-r}G_r^m - \sum_{m,n}\frac{1}{2}(n-m)\,{}^\circ_\circ b_{-m-n}c_m c_n{}^\circ_\circ \\
&\quad + \sum_{m,r}\left[\frac{1}{2}(2r-m)\,{}^\circ_\circ\beta_{-m-r}c_m\gamma_r{}^\circ_\circ - \,{}^\circ_\circ b_{-m}\gamma_{m-r}\gamma_r{}^\circ_\circ\right] + a^g c_0 \,,
\end{aligned}
\qquad (10.5.24)
$$

where m and n run over integers and r over (integers $+ \nu$). The ghost normal ordering constant is as in eq. (10.4.5).

---

[3] The $bc\beta\gamma$ theory actually has a one-parameter family of superconformal symmetries, related by rescaling $\beta \to x\beta$ and $\gamma \to x^{-1}\gamma$. The general BRST construction (4.3.14) singles out the symmetry (10.1.21); this is most easily verified by noting that it correctly leads to the OPEs (10.5.22).

The observable spectrum is the space of BRST cohomology classes. As in the bosonic theory, we impose the additional conditions

$$b_0|\psi\rangle = L_0|\psi\rangle = 0 \ . \tag{10.5.25}$$

In addition, in the R sector we impose

$$\beta_0|\psi\rangle = G_0|\psi\rangle = 0 \ , \tag{10.5.26}$$

the logic being the same as for (10.5.25). The reader can again work out the first few levels by hand, the result being exactly the same as for OCQ. The no-ghost theorem is as in the bosonic case. The BRST cohomology has a positive definite inner product and is isomorphic to OCQ and to the transverse Hilbert space $\mathscr{H}^\perp$, which is defined to have no $\alpha^{0,1}$, $\psi^{0,1}$, b, c, β, or γ excitations. The proof is a direct imitation of the bosonic argument of chapter 4.

We have defined $\exp(\pi iF)$ to commute with Q_B. We can therefore consider subspaces with definite eigenvalues of $\exp(\pi iF)$ and the no-ghost theorem holds separately in each.

10.6 Superstring theories in ten dimensions

We now focus on the theory in ten flat dimensions. For the four sectors of the open string spectrum we will use in addition to the earlier notation NS±, R± the notation

$$(\alpha, F) \ , \tag{10.6.1}$$

where the combination

$$\alpha = 1 - 2\nu \tag{10.6.2}$$

is 1 in the R sector and 0 in the NS sector. Both α and F are defined only mod 2. The closed string has independent periodicities and fermion numbers on both sides, and so has 16 sectors labeled by

$$(\alpha, F, \tilde\alpha, \tilde F) \ . \tag{10.6.3}$$

Actually, six of these sectors are empty: in the NS− sector the level $L_0 - \alpha' p^2/4$ is half-integer, while in the sectors NS+, R+, and R− it is an integer. It is therefore impossible to satisfy the level-matching condition $L_0 = \tilde L_0$ if NS− is paired with one of the other three.

Not all of these states can be present together in a consistent string theory. Consider first the closed string spectrum. We have seen that the spinor fields have branch cuts in the presence of R sector vertex operators. Various pairs of vertex operators will then have branch cuts in their operator products — they are not *mutually local*. The operator F

counts the number of spinor fields in a vertex operator, so the net phase when one vertex operator circles another is

$$\exp \pi i \left(F_1 \alpha_2 - F_2 \alpha_1 - \tilde{F}_1 \tilde{\alpha}_2 + \tilde{F}_2 \tilde{\alpha}_1 \right) . \tag{10.6.4}$$

If this phase is not unity, the amplitude with both operators cannot be consistently defined.

A consistent closed string theory will then contain only some subset of the ten sectors. Thus there are potentially 2^{10} combinations of sectors, but only a few of these lead to consistent string theories. We impose three consistency conditions:

(a) From the above discussion, all pairs of vertex operators must be mutually local: if both $(\alpha_1, F_1, \tilde{\alpha}_1, \tilde{F}_1)$ and $(\alpha_2, F_2, \tilde{\alpha}_2, \tilde{F}_2)$ are in the spectrum then

$$F_1 \alpha_2 - F_2 \alpha_1 - \tilde{F}_1 \tilde{\alpha}_2 + \tilde{F}_2 \tilde{\alpha}_1 \in 2\mathbf{Z} . \tag{10.6.5}$$

(b) The OPE must close. The parameter α is conserved mod 2 under operator products (for example, $R \times R = NS$), as is F. Thus if $(\alpha_1, F_1, \tilde{\alpha}_1, \tilde{F}_1)$ and $(\alpha_2, F_2, \tilde{\alpha}_2, \tilde{F}_2)$ are in the spectrum then so is

$$(\alpha_1 + \alpha_2, F_1 + F_2, \tilde{\alpha}_1 + \tilde{\alpha}_2, \tilde{F}_1 + \tilde{F}_2) . \tag{10.6.6}$$

(c) For an arbitrary choice of sectors, the one-loop amplitude will not be modular-invariant. We will study modular invariance in the next section, but in order to reduce the number of possibilities it is useful to extract one simple necessary condition:

There must be at least one left-moving R sector ($\alpha = 1$) and at least one right-moving R sector ($\tilde{\alpha} = 1$).

We now solve these constraints. Assume first that there is at least one R–NS sector, $(\alpha, \tilde{\alpha}) = (1, 0)$. By the level-matching argument, it must either be (R+,NS+) or (R−,NS+). Further, by (a) only one of these can appear, because the product of the corresponding vertex operators is not single-valued. By (c), there must also be at least one NS–R or R–R sector, and because R–NS $\times$ R–R = NS–R, there must in any case be an NS–R sector. Again, this must be either (NS+,R+) or (NS+,R−), but not both. So we have four possibilities, (R+,NS+) or (R−,NS+) with (NS+,R+) or (NS+,R−). Applying closure and single-valuedness leads to precisely two additional sectors in each case, namely (NS+,NS+) and one R–R sector. The spectra which solve (a), (b), and (c) with at least one R–NS sector are

$$\begin{aligned}
\text{IIB:} \quad & (\text{NS+,NS+}) \;\; (\text{R+,NS+}) \;\; (\text{NS+,R+}) \;\; (\text{R+,R+}) \,, \\
\text{IIA:} \quad & (\text{NS+,NS+}) \;\; (\text{R+,NS+}) \;\; (\text{NS+,R−}) \;\; (\text{R+,R−}) \,, \\
\text{IIA}': \quad & (\text{NS+,NS+}) \;\; (\text{R−,NS+}) \;\; (\text{NS+,R+}) \;\; (\text{R−,R+}) \,, \\
\text{IIB}': \quad & (\text{NS+,NS+}) \;\; (\text{R−,NS+}) \;\; (\text{NS+,R−}) \;\; (\text{R−,R−}) \,.
\end{aligned}$$

Notice that none of these theories contains the tachyon, which lives in the sector (NS−,NS−).

These four solutions represent just two physically distinct theories. In the IIA and IIA′ theories the R–R states have the opposite chirality on the left and the right, and in the IIB and IIB′ theories they have the same chirality. A spacetime reflection on a single axis, say

$$X^2 \to -X^2 \,, \quad \psi^2 \to -\psi^2 \,, \quad \tilde{\psi}^2 \to -\tilde{\psi}^2 \,, \qquad (10.6.7)$$

leaves the action and the constraints unchanged but reverses the sign of $\exp(\pi i F)$ in the left-moving R sectors and the sign of $\exp(\pi i \tilde{F})$ in the right-moving R sectors. At the massless level this switches the Weyl representations, $\mathbf{16} \leftrightarrow \mathbf{16}'$. It therefore turns the IIA′ theory into IIA, and IIB′ into IIB.

Now suppose that there is no R–NS sector. By (c), there must be at least one R–R sector. In fact the combination of (NS+,NS+) with any single R–R sector solves (a), (b), and (c), but these turn out not to be modular-invariant. Proceeding further, one readily finds the only other solutions,

$$\text{0A:} \quad (\text{NS+,NS+}) \ (\text{NS−,NS−}) \ (\text{R+,R−}) \ (\text{R−,R+}) \,,$$

$$\text{0B:} \quad (\text{NS+,NS+}) \ (\text{NS−,NS−}) \ (\text{R+,R+}) \ (\text{R−,R−}) \,.$$

These are modular-invariant, but both have a tachyon and there are no spacetime fermions.

In conclusion, we have found two potentially interesting string theories, the *type IIA* and *IIB superstring theories*. Referring back to table 10.3, one finds the massless spectra

$$\text{IIA:} \quad [0] + [1] + [2] + [3] + (2) + \mathbf{8} + \mathbf{8}' + \mathbf{56} + \mathbf{56}' \,, \quad (10.6.8\text{a})$$

$$\text{IIB:} \quad [0]^2 + [2]^2 + [4]_+ + (2) + \mathbf{8}'^2 + \mathbf{56}^2 \,. \qquad (10.6.8\text{b})$$

The IIB theory is defined by keeping all sectors with

$$\exp(\pi i F) = \exp(\pi i \tilde{F}) = +1 \,, \qquad (10.6.9)$$

and the IIA theory by keeping all sectors with

$$\exp(\pi i F) = +1 \,, \quad \exp(\pi i \tilde{F}) = (-1)^{\tilde{\alpha}} \,. \qquad (10.6.10)$$

This projection of the full spectrum down to eigenspaces of $\exp(\pi i F)$ and $\exp(\pi i \tilde{F})$ is known as the *Gliozzi–Scherk–Olive (GSO) projection*. In the IIA theory the opposite GSO projections are taken in the NS–R and R–NS sectors, so the spectrum is nonchiral. That is, the spectrum is invariant under spacetime parity, which interchanges $\mathbf{8} \leftrightarrow \mathbf{8}'$ and $\mathbf{56} \leftrightarrow \mathbf{56}'$. On the world-sheet, this symmetry is the product of spacetime parity and world-sheet parity. In the IIB theory the same GSO projection is taken in each sector and the spectrum is chiral.

The type 0 theories are formed by a different method: for example, 0B is defined by keeping all sectors with

$$\alpha = \tilde{\alpha} , \quad \exp(\pi i F) = \exp(\pi i \tilde{F}) . \tag{10.6.11}$$

The projections that define the type II theories act separately on the left- and right-moving spinors, while the projections that define the type 0 theory tie the two together. The latter are sometimes called *diagonal GSO projections*.

The most striking features of the type II theories are the massless vector–spinor *gravitinos* in the NS–R and R–NS sectors. The terminology type II refers to the fact that these theories each have two gravitinos. In the IIA theory the gravitinos have opposite chiralities (Γ eigenvalues), and in the IIB theory they have the same chirality. The NS–R gravitino state is

$$\psi^{\mu}_{-1/2}|0; \mathbf{s}; k\rangle_{\text{NS–R}} u_{\mu\mathbf{s}} . \tag{10.6.12}$$

The physical state conditions are

$$k^2 = k^{\mu} u_{\mu\mathbf{s}} = k \cdot \Gamma_{\mathbf{s}\mathbf{s}'} u_{\mu\mathbf{s}'} = 0 , \tag{10.6.13}$$

as well as the equivalence relation

$$u_{\mu\mathbf{s}} \cong u_{\mu\mathbf{s}} + k_{\mu} \zeta_{\mathbf{s}} . \tag{10.6.14}$$

We have learned that such equivalence relations are the signature of a local spacetime symmetry. Here the symmetry parameter $\zeta_{\mathbf{s}}$ is a spacetime spinor so we have local *spacetime supersymmetry*. In flat spacetime there will be a conserved spacetime supercharge $Q^A_{\mathbf{s}}$, where A distinguishes the symmetries associated with the two gravitinos, and $\mathbf{s}$ is a spinor index of the same chirality as the corresponding gravitino. Thus the IIA theory has one supercharge transforming as the **16** of $SO(9,1)$ and one transforming as the **16**′, and the IIB theory has two transforming as the **16**.

The gravitino vertex operators are

$$\mathcal{V}_{\mathbf{s}} e^{-\tilde{\phi}} \tilde{\psi}^{\mu} e^{ik \cdot X} , \quad e^{-\phi} \psi^{\mu} \tilde{\mathcal{V}}_{\mathbf{s}} e^{ik \cdot X} . \tag{10.6.15}$$

The operators $\mathcal{V}_{\mathbf{s}}$ and $\tilde{\mathcal{V}}_{\mathbf{s}}$, defined in eq. (10.4.25), have weights $(1,0)$ and $(0,1)$ and so are world-sheet currents associated with the spacetime supersymmetries.

This is our first encounter with spacetime supersymmetry, and the reader should now study the appropriate sections of appendix B. Section B.2 gives an introduction to spacetime supersymmetry. Section B.4 discusses antisymmetric tensor fields, which we have in the massless IIA and IIB spectra. Section B.5 briefly discusses the IIA and IIB supergravity theories which describe the low energy physics of the IIA and IIB superstrings. In each of the type II theories, there is a unique massless representation, which has $2^8 = 256$ states. The massless superstring spectra are the massless representations of IIA and IIB $d = 10$ spacetime supersymmetry respectively. This is to be expected: if all requirements for a consistent string theory are met (and they are) then the existence of the gravitinos implies that the corresponding supersymmetries must be present.

The reader may feel that the construction in this section, which is the *Ramond–Neveu–Schwarz (RNS)* form of the superstring, is somewhat *ad hoc*. In particular one might expect that the spacetime supersymmetry should be manifest from the start. There is certainly truth to this, but the existing supersymmetric formulation (the *Green–Schwarz superstring*) seems to be even more unwieldy.

Note that the world-sheet and spacetime supersymmetries are distinct, and that the connection between them is indirect. The world-sheet supersymmetry parameter $\eta(z)$ is a spacetime scalar and world-sheet spinor, while the spacetime supersymmetry parameter ζ_s is a spacetime spinor and world-sheet scalar. The world-sheet supersymmetry is a constraint in the world-sheet theory, annihilating physical states. The spacetime supersymmetry is a global symmetry of the world-sheet theory, giving relations between masses and amplitudes, though it becomes a local symmetry in spacetime.

Let us note one more feature of the GSO projection. In bosonized form, all the R sector vertex operators have odd length-squared and all the NS sector vertex operators have even length-squared, in terms of the ∘ product defined in section 10.4. This can be seen at the lowest levels for the operators (10.4.22) and (10.4.25), the tachyon having been removed by the GSO projection. By the remark at the end of section 10.4, the spacetime spin is then correlated with the *world-sheet* statistics. In fact, this is the same as the *space-time* statistics. The world-sheet statistics governs the behavior of the world-sheet amplitude under simultaneous exchange of world-sheet position, spacetime momentum, and other quantum numbers. After integrating over position, this determines the symmetry of the spacetime S-matrix. The result is the expected spacetime spin-statistics connection. Note that operators with the wrong spin-statistics connection, such as ψ^μ and $e^{-\phi}$, appear at intermediate stages but the projections that produce a consistent theory also give the spin-statistics connection. This is certainly a rather technical way for the spin-statistics theorem to arise, but it is worth noting that all string theories seem to obey the usual spin-statistics relation.

Unoriented and open superstrings

The IIB superstring, with the same chiralities on both sides, has a world-sheet parity symmetry Ω. We can gauge this symmetry to obtain an

unoriented closed string theory.[4] In the NS–NS sector, this eliminates the
[2], leaving [0] + (2), just as it does in the unoriented bosonic theory.
The fermionic NS–R and R–NS sectors of the IIB theory have the same
spectra, so the Ω projection picks out the linear combination (NS–R) +
(R–NS), with massless states $\mathbf{8'} + \mathbf{56}$. In particular, one gravitino survives
the projection. Finally, the existence of the gravitino means that there
must be equal numbers of massless bosons and fermions, so a consistent
definition of the world-sheet parity operator must select the [2] from the
R–R sector to give 64 of each. One can understand this as follows. The
R–R vertex operators

$$\mathcal{V}_s \tilde{\mathcal{V}}_{s'} \tag{10.6.16}$$

transform as $\mathbf{8} \times \mathbf{8} = [0] + [2] + [4]_+$. The [0] and $[4]_+$ are symmetric
under interchange of s and s' and the [2] antisymmetric (one can see this
by counting states, 36 versus 28, or in more detail by considering the S_a
eigenvalues of the representations). World-sheet parity adds or subtracts
a tilde to give

$$\tilde{\mathcal{V}}_s \mathcal{V}_{s'} = -\mathcal{V}_{s'} \tilde{\mathcal{V}}_s \,, \tag{10.6.17}$$

where the final sign comes from the fermionic nature of the R vertex
operators. Thus, projecting onto $\Omega = +1$ picks out the antisymmetric [2].
The result is the *type I closed unoriented theory*, with spectrum

$$[0] + [2] + (2) + \mathbf{8'} + \mathbf{56} \; = \; \mathbf{1} + \mathbf{28} + \mathbf{35} + \mathbf{8'} + \mathbf{56} \,. \tag{10.6.18}$$

However, this theory by itself is inconsistent, as we will explain further
below.

Now consider open string theories. Closure of the OPE in open + open
$\rightarrow$ closed scattering implies that any open string that couples consistently
to type I or type II closed superstrings must have a GSO projection in
the open string sector. The two possibilities and their massless spectra are

$$\mathrm{I:} \quad \mathrm{NS+, \; R+} = \mathbf{8}_v + \mathbf{8} \,,$$
$$\mathrm{\tilde{I}:} \quad \mathrm{NS+, \; R-} = \mathbf{8}_v + \mathbf{8'} \,.$$

Adding Chan–Paton factors, the gauge group will again be $U(n)$ in the
oriented case and $SO(n)$ or $Sp(k)$ in the unoriented case. The $\mathbf{8}$ or $\mathbf{8'}$ spinors
are known as *gauginos* because they are related to the gauge bosons by
supersymmetry. They must be in the adjoint representation of the gauge
group, like the gauge bosons, because supersymmetry commutes with the
gauge symmetry.

---

[4] The analogous operation in the IIA theory would be to gauge the symmetry which is the product
of world-sheet and spacetime parity, but this breaks some of the Poincaré invariance. We will
encounter this in section 13.2.

We can already anticipate that not all of these theories will be consistent. The open string multiplets, with 16 states, are representations of $d = 10$, $N = 1$ supersymmetry but not of $N = 2$ supersymmetry. Thus the open superstring cannot couple to the oriented closed superstring theories, which have two gravitinos.[5] It can only couple to the unoriented closed string theory (10.6.18) and so the open string theory must also be unoriented for consistent interactions. With the chirality (10.6.18), the massless open string states must be $\mathbf{8}_v + \mathbf{8}$. This is required by spacetime supersymmetry, or by conservation of $\exp(\pi i F)$ on the world-sheet. The result is the unoriented *type I open plus closed superstring theory*, with massless content

$$[0] + [2] + (2) + \mathbf{8}' + \mathbf{56} + (\mathbf{8}_v + \mathbf{8})_{SO(n) \text{ or } Sp(k)} . \qquad (10.6.19)$$

There is a further inconsistency in all but the $SO(32)$ theory. We will see in section 10.8 that for all other groups, as well as the purely closed unoriented theory, there is a one-loop divergence and superconformal anomaly. We will also see, in chapter 12, that the spacetime gauge and coordinate symmetries have an anomaly at one loop for all but the $SO(32)$ theory.

Thus we have found precisely three tachyon-free and nonanomalous string theories in this chapter: type IIA, type IIB, and type I $SO(32)$.

10.7 Modular invariance

Superstring interactions are the subject of the next chapter, but there is one important amplitude that involves no interactions, only the string spectrum. This is the one-loop vacuum amplitude, studied for the bosonic string in chapter 7. We study the vacuum amplitude for the closed superstring in this section and for the open string in the next.

We make the guess, correctly it will turn out, that the torus amplitude is again given by the Coleman–Weinberg formula (7.3.24) with the region of integration replaced by the fundamental region for the moduli space of the torus:

$$Z_{T_2} = V_{10} \int_F \frac{d^2\tau}{4\tau_2} \int \frac{d^{10}k}{(2\pi)^{10}} \sum_{i \in \mathscr{H}^\perp} (-1)^{\mathbf{F}_i} q^{\alpha'(k^2 + m_i^2)/4} \bar{q}^{\alpha'(k^2 + \tilde{m}_i^2)/4} , \qquad (10.7.1)$$

with $q = \exp(2\pi i\tau)$. We have included the minus sign for spacetime

---

[5] At the world-sheet level the problem is that the total derivative null gravitino vertex operators give rise to nonzero world-sheet boundary terms. Only one linear combination of the two null gravitinos decouples, so we must make the world-sheet parity projection in order to eliminate the other.

fermions from the Coleman–Weinberg formula, distinguishing the space-time fermion number **F** from the world-sheet fermion number F. The masses are given in terms of the left- and right-moving parts of the transverse Hamiltonian by

$$m^2 = 4H^\perp/\alpha' , \quad \tilde{m}^2 = 4\tilde{H}^\perp/\alpha' . \tag{10.7.2}$$

The trace includes a sum over the different $(\alpha, F; \tilde{\alpha}, \tilde{F})$ sectors of the superstring Hilbert space. In each sector it breaks up into a product of independent sums over the transverse X, ψ, and $\tilde{\psi}$ oscillators, and the transverse Hamiltonian similarly breaks up into a sum. Each transverse X contributes as in the bosonic string, the total contribution of the oscillator sum and momentum integration being as in eq. (7.2.9),

$$Z_X(\tau) = (4\pi^2\alpha'\tau_2)^{-1/2}(q\bar{q})^{-1/24} \prod_{n=1}^{\infty} \left(\sum_{N_n, \tilde{N}_n=1}^{\infty} q^{nN_n}\bar{q}^{n\tilde{N}_n} \right)$$

$$= (4\pi^2\alpha'\tau_2)^{-1/2}|\eta(q)|^{-2} , \tag{10.7.3}$$

where $\eta(\tau) = q^{1/24} \prod_{n=1}^{\infty}(1 - q^n)$. In addition there is a factor $i(4\pi^2\alpha'\tau_2)^{-1}$ from the $k^{0,1}$ integrations.

For the ψs, the mode sum in each sector depends on the spatial periodicity α and includes a projection operator $\frac{1}{2}[1 \pm \exp(\pi iF)]$. Although for the present we are interested only in R and NS periodicities, let us work out the partition functions for the more general periodicity (10.3.20),

$$\psi(w + 2\pi) = \exp[\pi i(1 - \alpha)] \, \psi(w) \tag{10.7.4}$$

where again $\alpha = 1 - 2v$. By the definition (10.3.23) of the ground state, the raising operators are

$$\psi_{-m+(1-\alpha)/2} , \quad \bar{\psi}_{-m+(1+\alpha)/2} , \quad m = 1, 2, \ldots . \tag{10.7.5}$$

The ground state weight was found to be $\alpha^2/8$. Then

$$\mathrm{Tr}_\alpha\left(q^H\right) = q^{(3\alpha^2-1)/24} \prod_{m=1}^{\infty} \left[1 + q^{m-(1-\alpha)/2}\right]\left[1 + q^{m-(1+\alpha)/2}\right] . \tag{10.7.6}$$

To define the general boundary conditions we have joined the fermions into complex pairs. Thus we can define a fermion number Q which is $+1$ for ψ and -1 for $\bar{\psi}$. To be precise, define Q to be the H-momentum in the bosonization (10.3.10) so that it is conserved by the OPE. The bosonization (10.3.25) then gives the charge of the ground state as $\alpha/2$.

Thus we can define the more general trace

$$Z^{\alpha}{}_{\beta}(\tau) = \mathrm{Tr}_{\alpha}\left[q^{H}\exp(\pi i\beta Q)\right] \tag{10.7.7a}$$

$$= q^{(3\alpha^2-1)/24}\exp(\pi i\alpha\beta/2)$$

$$\times \prod_{m=1}^{\infty}\left[1+\exp(\pi i\beta)q^{m-(1-\alpha)/2}\right]\left[1+\exp(-\pi i\beta)q^{m-(1+\alpha)/2}\right] \tag{10.7.7b}$$

$$= \frac{1}{\eta(\tau)}\vartheta\left[\begin{matrix}\alpha/2\\\beta/2\end{matrix}\right](0,\tau)\,. \tag{10.7.7c}$$

The notation in the final line was introduced in section 7.2, but our discussion of these functions in the present volume will be self-contained.

The charge Q modulo 2 is the fermion number F that appears in the GSO projection. Thus the traces that are relevant for the ten-dimensional superstring are

$$Z^{0}{}_{0}(\tau) = \mathrm{Tr}_{NS}\left[q^{H}\right]\,, \tag{10.7.8a}$$

$$Z^{0}{}_{1}(\tau) = \mathrm{Tr}_{NS}\left[\exp(\pi iF)\,q^{H}\right]\,, \tag{10.7.8b}$$

$$Z^{1}{}_{0}(\tau) = \mathrm{Tr}_{R}\left[q^{H}\right]\,, \tag{10.7.8c}$$

$$Z^{1}{}_{1}(\tau) = \mathrm{Tr}_{R}\left[\exp(\pi iF)\,q^{H}\right]\,. \tag{10.7.8d}$$

We should emphasize that these traces are for a *pair* of dimensions.

Tracing over all eight fermions, the GSO projection keeps states with $\exp(\pi iF) = +1$. This is $Z_{\psi}^{+}(\tau)$, where

$$Z_{\psi}^{\pm}(\tau) = \frac{1}{2}\left[Z^{0}{}_{0}(\tau)^4 - Z^{0}{}_{1}(\tau)^4 - Z^{1}{}_{0}(\tau)^4 \mp Z^{1}{}_{1}(\tau)^4\right]\,. \tag{10.7.9}$$

The half is from the projection operator, the minus sign in the second term is from the ghost contribution to $\exp(\pi iF)$, and the minus signs in the third and fourth (R sector) terms are from spacetime spin-statistics. For $\tilde{\psi}$ in the IIB theory one obtains the conjugate $Z_{\psi}^{+}(\tau)^*$. In the IIA theory, $\tilde{F} = -1$ in the R sector so the result is $Z_{\psi}^{-}(\tau)^*$. In all,

$$Z_{T_2} = iV_{10}\int_{F}\frac{d^2\tau}{16\pi^2\alpha'\tau_2^2}Z_X^8 Z_{\psi}^{+}(\tau)Z_{\psi}^{\pm}(\tau)^*\,. \tag{10.7.10}$$

We know from the discussion of bosonic amplitudes that modular invariance is necessary for the consistency of string theory. In the superstring this works out in an interesting way. The combination $d^2\tau/\tau_2^2$ is modular-invariant, as is Z_X. To understand the modular transformations of the fermionic traces, note that $Z^{\alpha}{}_{\beta}$ is given by a path integral on the torus over fermionic fields ψ with periodicities

$$\psi(w + 2\pi) = -\exp(-\pi i\alpha)\,\psi(w)\,, \tag{10.7.11a}$$

$$\psi(w + 2\pi\tau) = -\exp(-\pi i\beta)\,\psi(w)\,. \tag{10.7.11b}$$

This gives

$$\psi[w + 2\pi(\tau + 1)] = \exp[-\pi i(\alpha + \beta)]\,\psi(w)\,. \tag{10.7.12}$$

Naively then, $Z^{\alpha}{}_{\beta}(\tau) = Z^{\alpha}{}_{\alpha+\beta-1}(\tau+1)$, since both sides are given by the same path integral. Also, defining $w' = w/\tau$ and $\psi'(w') = \psi(w)$,

$$\psi'(w' + 2\pi) = -\exp(-\pi i\beta)\,\psi'(w') \tag{10.7.13a}$$

$$\psi'(w' - 2\pi/\tau) = -\exp(\pi i\alpha)\,\psi'(w')\,, \tag{10.7.13b}$$

so that naively $Z^{\alpha}{}_{\beta}(\tau) = Z^{\beta}{}_{-\alpha}(-1/\tau)$. It is easy to see that by these two transformations one can always reach a path integral with $\alpha = 1$, accounting for rule (c) from the previous section.

The reason these modular transformations are naive is that there is no diff-invariant way to define the phase of the path integral for purely left-moving fermions. For left- plus right-moving fermions with matching boundary conditions, the path integral can be defined by Pauli–Villars or other regulators. This is the same as the absolute square of the left-moving path integral, but leaves a potential phase ambiguity in that path integral separately.[6] The naive result is correct for $\tau \to -1/\tau$, but under $\tau \to \tau + 1$ there is an additional phase,

$$
\begin{aligned}
Z^{\alpha}{}_{\beta}(\tau) &= Z^{\beta}{}_{-\alpha}(-1/\tau) \\
&= \exp[-\pi i(3\alpha^2 - 1)/12]\,Z^{\alpha}{}_{\alpha+\beta-1}(\tau + 1)\,.
\end{aligned}
\tag{10.7.14}
$$

The $\tau \to \tau + 1$ transformation follows from the explicit form (10.7.7b), the phase coming from the zero-point energy with the given boundary conditions. The absence of a phase in $\tau \to -1/\tau$ can be seen at once for $\tau = i$. Note that $Z^{1}{}_{1}$ actually vanishes due to cancellation between the two R sector ground states, but we have assigned a formal transformation law for a reason to be explained below.

The phase represents a *global gravitational anomaly*, an inability to define the phase of the path integral such that it is invariant under large coordinate transformations. Of course, a single left-moving fermion has $c \neq \tilde{c}$ and so has an anomaly even under infinitesimal coordinate transformations, but the global anomaly remains even when a left- and right-moving fermion are combined. For example, the product $Z^{1}{}_{0}(\tau)^{*}Z^{0}{}_{0}(\tau)$ has no infinitesimal anomaly and should come back to itself under $\tau \to \tau + 2$,

---

[6] The phase factor is a holomorphic function of τ, because the $Z^{\alpha}{}_{\beta}$ are. Since it has magnitude 1, this implies that it is actually independent of τ.

but in fact picks up a phase $\exp(-\pi i/2)$. This phase arises from the *level mismatch*, the difference of zero-point energies in the NS and R sectors.

The reader can verify that with the transformations (10.7.14), the combinations $Z_\psi^\pm$ are invariant under $\tau \to -1/\tau$ and are multiplied by $\exp(2\pi i/3)$ under $\tau \to \tau + 1$. Combined with the conjugates from the right-movers, the result is modular-invariant and the torus amplitude consistent. It is necessary for the construction of this invariant that there be a multiple of *eight* transverse fermions. Recall from section 7.2 that invariance under $\tau \to \tau + 1$ requires that $L_0 - \tilde{L}_0$ be an integer for all states. For a single real fermion in the R–NS sector the difference in ground state energies is $\frac{1}{16}$. For eight fermions this becomes $\frac{1}{2}$, so that states with an odd number of NS excitations (as required by the GSO projection) are level-matched. Note also that modular invariance forces the minus signs in the combination (10.7.9), in particular the relative sign of $(Z^0_0)^4$ and $(Z^1_0)^4$ which corresponds to Fermi statistics for the R sector states.

In the type 0 superstrings the fermionic trace is

$$\frac{1}{2}\left[|Z^0_0(\tau)|^N + |Z^0_1(\tau)|^N + |Z^1_0(\tau)|^N \mp |Z^1_1(\tau)|^N\right] \tag{10.7.15}$$

with $N = 8$. This is known as the *diagonal modular invariant*, and it is invariant for any N because the phases cancel in the absolute values.

The type II theories have spacetime supersymmetry. This implies equal numbers of bosons and fermions at each mass level, and so Z_{T_2} should *vanish* in these theories by cancellation between bosons and fermions. Indeed it does, as a consequence of $Z^1_1 = 0$ and the 'abstruse identity' of Jacobi,

$$Z^0_0(\tau)^4 - Z^0_1(\tau)^4 - Z^1_0(\tau)^4 = 0 . \tag{10.7.16}$$

The same cancellation occurs in the open and unoriented theories.

Although we have focused on the path integral without vertex operators, amplitudes with vertex operators must also be modular-invariant. In the present case the essential issue is the path integral measure, and one can show by explicit calculation (or by indirect arguments) that the modular properties are the same with or without vertex operators. However, with a general vertex operator insertion the $\alpha = \beta = 1$ path integral will no longer vanish, nor will the sum of the other three. The general amplitude will then be modular-invariant provided that the vacuum is modular-invariant *without* using the vanishing of Z^1_1 or the abstruse identity (10.7.16) — as we have required.

More on c = 1 CFT

The equality of the bosonic and fermionic partition functions (10.3.17) and (10.3.18) was one consequence of bosonization. These partition functions are not modular-invariant and so do not define a sensible string background. The fermionic spectrum consists of all NS–NS states. The bosonic spectrum consists of all states with integer k_R and k_L; this is not the spectrum of toroidal compactification at any radius. The simplest modular-invariant fermionic partition function is the diagonal invariant, taking common periodicities for the left- and right-movers. In terms of the states, this amounts to projecting

$$\alpha = \tilde{\alpha} , \quad \exp(\pi i F) = \exp(\pi i \tilde{F}) . \tag{10.7.17}$$

The NS–NS sector consists of the local operators we have been considering, and the chirality projection $\exp(\pi i F) = \exp(\pi i \tilde{F})$ means that on the bosonic side $k_R = k_L$ mod 2. The bosonic equivalents for the R–R sector states have half-integral k_R and k_L with again $k_R = k_L$ mod 2. In all,

$$(k_R, k_L) = (n_1, n_2) \text{ or } (n_1 + \tfrac{1}{2}, n_2 + \tfrac{1}{2}) \tag{10.7.18}$$

for integers n_1 and n_2 such that $n_1 - n_2 \in 2\mathbf{Z}$. This is the spectrum of a boson on a circle of radius 2, or 1 by T-duality, which we see is equivalent to a complex fermion with the diagonal modular-invariant projection. (The dimensionless radius r for the H scalar corresponds to the radius $R = r(\alpha'/2)^{1/2}$ for X^μ, so $r = 2^{1/2}$ is self-dual.)

To obtain an equivalent fermionic theory at arbitrary radius, add

$$\partial H \bar{\partial} H \cong -\overline{\psi} \psi \tilde{\overline{\psi}} \tilde{\psi} \tag{10.7.19}$$

to the world-sheet Lagrangian density. The H theory is still free, but the equivalent fermionic theory is now an interacting field theory known as the *Thirring model*. The Thirring model has a nontrivial perturbation series but is solvable precisely because of its equivalence to a free boson. Actually, for any *rational* r, the bosonic theory is also equivalent to a *free* fermion theory with a more complicated twist (exercise 10.15).

Another interesting CFT consists of the set of vertex operators with

$$k_R = m/3^{1/2} , \quad k_L = n/3^{1/2} , \quad m - n \in 3\mathbf{Z} . \tag{10.7.20}$$

(This discussion should actually be read after section 11.1.) It is easy to check that this has the same properties as the set of vertex operators with integer $k_{R,L}$. That is, it is a single-valued operator algebra, but does not correspond to the spectrum of the string for any value of r, and does not have a modular-invariant partition function. Its special property is the existence of the operators

$$\exp\left[\pm i 3^{1/2} H(z)\right] , \quad \exp\left[\pm i 3^{1/2} \tilde{H}(\bar{z})\right] . \tag{10.7.21}$$

These have weights $(\tfrac{3}{2}, 0)$ and $(0, \tfrac{3}{2})$: they are world-sheet supercurrents!

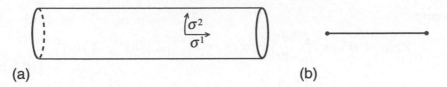

(a)   (b)

Fig. 10.1.   (a) Cylinder in the limit of small t. (b) Analogous field theory graph.

This CFT has $(2,2)$ world-sheet supersymmetry. The standard representation, in which the supercurrent is quadratic in free fields, has two free X and two free ψ fields for central charge 3. This is rather more economical, with one free scalar and central charge 1. The reader can readily check that with appropriate normalization the supercurrents generate the $N = 2$ OPE (11.1.4).

This theory becomes modular-invariant if one twists by the symmetry

$$(H, \tilde{H}) \to (H, \tilde{H}) + \frac{2\pi}{2 \times 3^{1/2}}(1, -1) \,. \tag{10.7.22}$$

This projects the spectrum onto states with $m - n \in 6\mathbf{Z}$ and adds in a twisted sector with $m, n \in \mathbf{Z} + \frac{1}{2}$. The resulting spectrum is the string theory at $r = 2 \times 3^{1/2}$. This twist is a diagonal GSO projection, in that the supercurrent is odd under the symmetry.

10.8   Divergences of type I theory

The cylinder, Möbius strip, and Klein bottle have no direct analog of the modular group, but the condition that the tadpole divergences cancel among these three graphs plays a similar role in restricting the possible consistent theories. The cancellation is very similar to what we have already seen in the bosonic theory in chapter 7. The main new issue is the inclusion of the various sectors in the fermionic path integral, and in particular the separate contributions of closed string NS–NS and R–R tadpoles.

The cylinder

Consider first the cylinder, shown in figure 10.1(a). One can immediately write down the amplitude by combining the bosonic result (7.4.1), converted to ten dimensions, with the fermionic trace (10.7.9) from one side of the type II string. We write it as a sum of two terms,

$$Z_{C_2} = Z_{C_2,0} + Z_{C_2,1} \,, \tag{10.8.1}$$

where

$$Z_{C_2,0} = iV_{10}\, n^2 \int_0^\infty \frac{dt}{8t} (8\pi^2\alpha' t)^{-5} \eta(it)^{-8} \left[Z^0{}_0(it)^4 - Z^1{}_0(it)^4 \right],$$

(10.8.2a)

$$Z_{C_2,1} = iV_{10}\, n^2 \int_0^\infty \frac{dt}{8t} (8\pi^2\alpha' t)^{-5} \eta(it)^{-8} \left[-Z^0{}_1(it)^4 - Z^1{}_1(it)^4 \right].$$

(10.8.2b)

Note that the GSO and Ω projection operators each contribute a factor of $\frac{1}{2}$. We have separated the terms according to whether $\exp(\pi i F)$ appears in the trace. In $Z_{C_2,0}$ it does not, and so the ψ^μ are antiperiodic in the σ^2 direction. In $Z_{C_2,1}$ it does appear and the ψ^μ are periodic. We can also regard the cylinder as a closed string appearing from and returning to the vacuum as in figure 10.1(b); we have used this idea in chapters 7 and 8. The periodicities of the ψ^μ mean that in terms of the closed string exchange, the part $Z_{C_2,0}$ comes from NS–NS strings and the part $Z_{C_2,1}$ from R–R strings.

We know from the previous section that the total fermionic partition function vanishes by supersymmetry, so that $Z_{C_2,1} = -Z_{C_2,0}$; we concentrate then on $Z_{C_2,0}$. Using the modular transformations

$$\eta(it) = t^{-1/2}\eta(i/t), \quad Z^\alpha{}_\beta(it) = Z^\beta{}_{-\alpha}(i/t)$$

(10.8.3)

and defining $s = \pi/t$, this becomes

$$Z_{C_2,0} = i\frac{V_{10}n^2}{8\pi(8\pi^2\alpha')^5} \int_0^\infty ds\, \eta(is/\pi)^{-8} \left[Z^0{}_0(is/\pi)^4 - Z^0{}_1(is/\pi)^4 \right]$$

$$= i\frac{V_{10}n^2}{8\pi(8\pi^2\alpha')^5} \int_0^\infty ds\, [16 + O(\exp(-2s))].$$

(10.8.4)

The divergence as $s \to \infty$ is due to a massless closed string tadpole, which as noted must be an NS–NS state. Thus we identify this as a dilaton plus graviton interaction $(-G)^{1/2}e^{-\Phi}$ coming from the disk, as in the bosonic string.

However, there is a paradox here: the $d = 10$, $N = 1$ supersymmetry algebra does not allow such a term. Even more puzzling, $Z_{C_2,1}$ has an equal and opposite divergence which must be from a tadpole of an R–R state, but the only massless R–R state is the rank 2 tensor which cannot have a Lorentz-invariant tadpole.

One can guess the resolution of this as follows. The type IIB string has rank n potentials for all even n, with n and $8 - n$ equivalent by Poincaré duality. The Ω projection removes $n = 0$ and its equivalent $n = 8$, as well as $n = 4$: all the multiples of four. This leaves $n = 2$, its equivalent $n = 6$ — and $n = 10$. A 10-form potential C_{10} can exist in ten dimensions but its 11-form field strength dC_{10} is identically zero. The integral of the

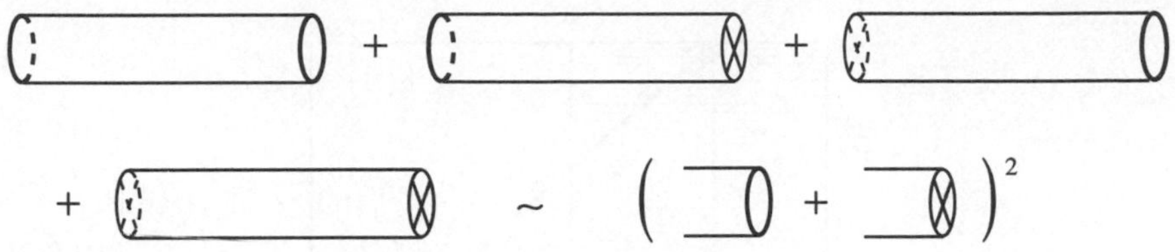

Fig. 10.2. Schematic illustration of cancellation of tadpoles.

potential over spacetime

$$\mu_{10} \int C_{10} \tag{10.8.5}$$

is invariant under $\delta C_{10} = d\chi_9$ and so can appear in the action. Since there is no kinetic term the propagator for this field is $1/0$, and the effect of the tadpole is a divergence

$$\frac{\mu_{10}^2}{0}. \tag{10.8.6}$$

This must be the origin of the divergence in $Z_{C_2,1}$, as indeed a more detailed analysis does show. The equation of motion from varying C_{10} is just $\mu_{10} = 0$, so unlike the divergences encountered previously this one cannot be removed by a correction to the background fields. It represents an actual inconsistency.

The Klein bottle

We know from the study of the bosonic string divergences that there is still the possibility of canceling this tadpole as shown in figure 10.2. The cylinder, Möbius strip, and Klein bottle each have divergences from the massless closed string states, the total being proportional to square of the sum of the disk and RP_2 tadpoles. The relative size of the two tadpoles depends on the Chan–Paton factors, and cancels for a particular gauge group.[7]

The relation of the Möbius strip and Klein bottle as depicted in figure 10.2 to the twisted-strip and twisted-cylinder pictures was developed in section 7.4, and is shown in figure 10.3. In order to sum as in figure 10.2, one must rescale the surfaces so that the circumference in the σ^2 direction

---

[7] In the vacuum amplitude the sum of the NS–NS and R–R divergences is zero for each topology separately because the trace vanishes by supersymmetry. This is not sufficient, because they will no longer cancel when vertex operators are added near one end of each surface. The NS–NS and R–R tadpoles must vanish separately when summed over topologies.

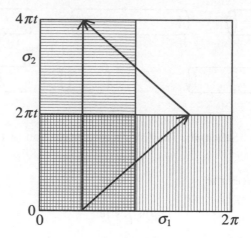

Fig. 10.3. Two fundamental regions for the Klein bottle. The right- and left-hand edges are periodically identified, as are the upper and lower edges. In addition the diagonal arrow shows an orientation-reversing identification. The vertically hatched region is a fundamental region for the twisted-cylinder picture, as is the horizontally hatched region for the decription with two crosscaps. As shown by the arrows, the periodicity of fields in the σ_2-direction of the latter description can be obtained by applying the orientation-reversing periodicity twice. The same picture applies to the Möbius strip, with the right- and left-hand edges boundaries, and with the range of σ_1 changed to π.

and length in the σ^1 directions are uniform; we have taken these to be 2π and s respectively. From figures 10.1 and 10.3 it follows that s is related to the usual modulus t for these surfaces by $s = \pi/t$, $\pi/4t$, and $\pi/2t$ for the cylinder, Möbius strip, and Klein bottle respectively.

Each amplitude is obtained as a sum of traces, from summing over the various periodicity conditions and from expanding out the projection operators. We need to determine which terms contribute to the NS–NS exchange and which to the R–R exchange by examining the boundary conditions on the fermions in the world-sheet path integral. On the Klein bottle the GSO projection operator is

$$\frac{1 + \exp(\pi i F)}{2} \cdot \frac{1 + \exp(\pi i \tilde{F})}{2} . \tag{10.8.7}$$

With $R = \Omega \exp(\pi i \beta F + \pi i \tilde{\beta} \tilde{F})$ in the trace, the path integral boundary conditions are

$$\psi(w + 2\pi it) = -R\psi(w)R^{-1} = -\exp(\pi i \beta)\,\tilde{\psi}(\bar{w}) , \tag{10.8.8a}$$
$$\tilde{\psi}(\bar{w} + 2\pi it) = -R\tilde{\psi}(\bar{w})R^{-1} = -\exp(\pi i \tilde{\beta})\,\psi(w) , \tag{10.8.8b}$$

with the usual extra sign for fermionic fields. As indicated by the arrows in figure 10.3, these imply that

$$\psi(w + 4\pi it) = \exp[\pi i(\beta + \tilde{\beta})]\,\psi(w)\,. \tag{10.8.9}$$

The NS–NS exchange, from the sectors antiperiodic under $\sigma^2 \to \sigma^2 + 4\pi t$, then comes from traces weighted by $\Omega\exp(\pi iF)$ or $\Omega\exp(\pi i\tilde{F})$; further, these two traces are equal. Both NS–NS and R–R states contribute to the traces, making the separate contributions[8]

$$\text{NS–NS:} \quad q^{-1/3}\prod_{m=1}^{\infty}(1 + q^{2m-1})^8 = Z^0_0(2it)^4\,, \tag{10.8.10a}$$

$$\text{R–R:} \quad -16q^{2/3}\prod_{m=1}^{\infty}(1 + q^{2m})^8 = -Z^1_0(2it)^4\,, \tag{10.8.10b}$$

where $q = \exp(-2\pi t)$.

The full Klein bottle contribution to the NS–NS exchange is then

$$\begin{aligned}
Z_{K_2,0} &= iV_{10}\int_0^\infty \frac{dt}{8t}\,(4\pi^2\alpha't)^{-5}\eta(2it)^{-8}\Big[Z^0_0(2it)^4 - Z^1_0(2it)^4\Big] \\
&= i\frac{2^{10}V_{10}}{8\pi(8\pi^2\alpha')^5}\int_0^\infty ds\,\eta(is/\pi)^{-8}\Big[Z^0_0(is/\pi)^4 - Z^0_1(is/\pi)^4\Big] \\
&= i\frac{2^{10}V_{10}}{8\pi(8\pi^2\alpha')^5}\int_0^\infty ds\,[16 + O(\exp(-2s))]\,, \tag{10.8.11}
\end{aligned}$$

and $Z_{K_2,1} = -Z_{K_2,0}$. The bosonic part is (7.4.15) converted to $D = 10$.

The Möbius strip

In the open string Ω acts as

$$\Omega\psi^\mu(w)\Omega^{-1} = \tilde{\psi}^\mu(\pi - \bar{w}) = \psi^\mu(w - \pi)\,, \tag{10.8.12}$$

using the doubling trick (10.2.15). In terms of the modes this is

$$\Omega\psi^\mu_r\Omega^{-1} = \exp(-\pi ir)\,\psi^\mu_r\,. \tag{10.8.13}$$

The phase is imaginary in the NS sector and squares to -1. Thus

$$\Omega^2 = \exp(\pi iF)\,. \tag{10.8.14}$$

Since $\exp(\pi iF) = 1$ by the GSO projection, this is physically the same as squaring to the identity, but the combined Ω and GSO projections require

---

[8] In evaluating these, note that only states with identical ψ and $\tilde{\psi}$ excitations contribute to traces containing Ω. The signs from $\exp(\pi iF)$ or $\exp(\pi i\tilde{F})$ just cancel the signs from anticommuting ψs past $\tilde{\psi}$s, so that all terms in each trace have the same sign. The overall sign in the NS–NS trace (positive) can be determined from the graviton, and the overall sign in the R–R trace (negative) from the argument (10.6.17).

a single projection operator

$$\frac{1 + \Omega + \Omega^2 + \Omega^3}{4} \, . \tag{10.8.15}$$

With $R = \Omega \exp(\pi i \beta F)$ in the trace, the fields have the periodicities

$$\psi^\mu(w + 4\pi it) = -\exp(\pi i \beta)\psi^\mu(w + 2\pi it - \pi) = \psi^\mu(w - 2\pi) \, . \tag{10.8.16}$$

It follows that in the R sector of the trace the fields are periodic in the σ^2-direction, corresponding to the R–R exchange, while the NS sector of the trace gives the NS–NS exchange.

It is slightly easier to focus on the R–R exchange, where the traces with Ω and $\Omega \exp(\pi i F)$ sum to

$$-16q^{1/3} \prod_{m=1}^{\infty} [1 + (-1)^m q^m]^8 - (1-1)^4 q^{1/3} \prod_{m=1}^{\infty} [1 - (-1)^m q^m]^8$$
$$= Z^0{}_1(2it)^4 Z^1{}_0(2it)^4 \, . \tag{10.8.17}$$

The full Möbius amplitude, rewriting the bosonic part slightly, is

$$\begin{aligned}
Z_{M_2,1} &= \pm inV_{10} \int_0^\infty \frac{dt}{8t} (8\pi^2 \alpha' t)^{-5} \frac{Z^0{}_1(2it)^4 Z^1{}_0(2it)^4}{\eta(2it)^8 Z^0{}_0(2it)^4} \\
&= \pm 2in \frac{2^5 V_{10}}{8\pi(8\pi^2\alpha')^5} \int_0^\infty ds \, \frac{Z^0{}_1(2is/\pi)^4 Z^1{}_0(2is/\pi)^4}{\eta(2is/\pi)^8 Z^0{}_0(2is/\pi)^4} \\
&= \pm 2in \frac{2^5 V_{10}}{8\pi(8\pi^2\alpha')^5} \int_0^\infty ds \, [16 + O(\exp(-2s))] \, , \tag{10.8.18}
\end{aligned}$$

where the upper sign is for $SO(n)$. We have used (7.4.22) in $D = 10$.

The total divergence from R–R exchange is

$$Z_1 = -i(n \mp 32)^2 \frac{V_{10}}{8\pi(8\pi^2\alpha')^5} \int_0^\infty ds \, [16 + O(\exp(-2s))] \, . \tag{10.8.19}$$

The R–R tadpole vanishes only for the gauge group $SO(32)$. For each world-sheet topology the NS–NS divergence is the negative of the R–R divergence, so the dilaton–graviton tadpole also vanishes for $SO(32)$. This calculation does not determine the sign of the tadpole, but it should be $n \mp 32$. That is, changing from a symplectic to orthogonal projection changes the sign of RP_2, not of the disk. This is necessary for unitarity: the number of cross-caps is conserved mod 2 when a surface is cut open, so the sign is not determined by unitarity; this is not the case for the number of boundaries.

Exercises

10.1 (a) Find the OPE of T_F with X^μ and ψ^μ.
(b) Show that the residues of the OPEs of the currents (10.1.9) are proportional to the superconformal variations (10.1.10).

10.2 (a) Verify the commutator (10.1.11), up to terms proportional to the equations of motion.
(b) Verify that the commutator of a conformal and a superconformal transformation is a superconformal transformation.

10.3 (a) Verify the OPE (10.1.13).
(b) Extend this to the linear dilaton SCFT (10.1.22).

10.4 Obtain the R and NS algebras (10.2.11) from the OPE.

10.5 From the Jacobi identity for the R–NS algebra, show that the coefficients of the central charge terms in $T_B T_B$ and $T_F T_F$ are related.

10.6 Express $\exp(\pi i F)$ explicitly in terms of mode operators in the R and NS sectors of the ψ^μ CFT.

10.7 Verify that the expectation value (10.3.7) has the appropriate behavior as $z_i \to \infty$, and show that together with the OPE this determines the result up to normalization.

10.8 Verify the weight of the fermionic ground state $\mathscr{A}_v$ for general real v:
(a) from the commutator (2.7.8);
(b) from the mnemonic of section 2.9.
The most direct, but most time-consuming, method would be to find the relation between conformal and creation–annihilation normal ordering as in eq. (2.7.11).

10.9 By any of the above methods, determine the ghost normal ordering constants (10.4.5).

10.10 Enumerate the states corresponding to each term in the expansion (10.3.19), in both fermionic and bosonic form.

10.11 Find the fermionic operator F_n equivalent to $e^{\pm inH(z)}$. Here are two possible methods: build F_n iteratively in n by taking repeated operator products with $e^{\pm iH(z)}$; or deduce $\psi_m^\pm \cdot F_n$ directly from the OPE. Check your answer by comparing dimensions and fermion numbers.

10.12 By looking at the eigenvalues of S_a, verify the spinor decompositions (10.5.17).

10.13 (a) Verify the operator products (10.5.22).
(b) Using the Jacobi identity as in exercise 4.3, verify nilpotence of Q_B.

10.14 Work out the massless level of the open superstring in BRST quantization.

10.15 Consider a single complex fermion, with the spectrum summed over all sectors such that $v = \tilde{v}$ is a multiple of $1/2p$ for integer p. Impose the projection that the numbers of left- and right-moving excitations differ by a multiple of $2p$. Show that the spectrum is the same as that of a periodic scalar at radius $r = 1/p$. Show that this can be understood as a $\mathbf{Z}_p$ twist of the $r = 1$ theory. A further $\mathbf{Z}_{2q}$ twist of the T-dual $r = 2p$ theory produces an arbitrary rational value.

11

The heterotic string

11.1 World-sheet supersymmetries

In the last chapter we were led by guesswork to the idea of enlarging the world-sheet constraint algebra, adding the supercurrents $T_F(z)$ and $\tilde{T}_F(\bar{z})$. Now let us see how much further we can generalize this idea. We are looking for sets of holomorphic and antiholomorphic currents whose Laurent coefficients form a closed algebra.

Let us start by emphasizing the distinction between global symmetries and constraints. Global symmetries on the world-sheet are just like global symmetries in spacetime, implying relations between masses and between amplitudes. However, we are also singling out part of the symmetry to impose as a *constraint,* meaning that physical states must be annihilated by it, either in the OCQ or BRST sense. In the bosonic string, the spacetime Poincaré invariance was a global symmetry of the world-sheet theory, while the conformal symmetry was a constraint. Our present interest is in constraint algebras. In fact we will find only a very small set of possibilities, but some of the additional algebras we encounter will appear later as global symmetries.

To begin we should note that the set of candidate world-sheet symmetry algebras is very large. In the bosonic string, for example, any product of factors $\partial^n X^\mu$ is a holomorphic current. In most cases the OPE of such currents will generate an infinite number of new currents, which is probably too big an algebra to be useful. However, even restricting to algebras with finite numbers of currents leaves an infinite number of possibilities.

Let us focus first on the holomorphic currents. We have seen in section 2.9 that in a unitary CFT an operator is holomorphic if and only if it is of weight $(h, 0)$ with $h \geq 0$. Although the complete world-sheet theory with ghosts and timelike oscillators does not have a positive norm, the spatial part does and so is a unitary representation of the symmetry.

45

Because $\tilde{h} = 0$, the spin of the current is also equal to h. Also, by taking real and imaginary parts we can assume the currents to be Hermitian. Now let us consider some possibilities:

Spin $h \geq 2$. Algebras with spin > 2 currents are often referred to collectively as *W algebras*. Many are known, including several infinite families, but there is no complete classification. We will encounter one example in chapter 15, as a global symmetry of a CFT. There have been attempts to use some of these as constraint algebras. One complication is that the commutator of generators is in general a nonlinear function of the generators, making the construction of the BRST operator nontrivial. The few examples that have been constructed appear, upon gauge fixing, to be special cases of bosonic strings. Further, the geometric interpretation, analogous to the Riemann surface construction used to formulate bosonic string perturbation theory, is not clear. So we will restrict our attention to constraint algebras with $h \leq 2$. Also, CFTs can have multiple $(2,0)$ currents as global symmetries. The bosonic string has at least 27, namely the ghost energy-momentum tensor and the energy-momentum tensor for each X^{μ} field. However, only the sum of these has a geometric interpretation, in terms of conformal invariance, and so we will assume that there is precisely one $(2,0)$ constraint current which is the overall energy-momentum tensor.

Spin h not a multiple of $\frac{1}{2}$. For a current j of spin h,

$$j(z)j(0) \sim z^{-2h} \tag{11.1.1}$$

with a coefficient that can be shown by a positivity argument not to vanish. This is multi-valued if $2h$ is not an integer. Although there are again many known CFTs with such currents, the nonlocality of these currents leads to substantial complications if one tries to impose them as constraints. Attempts to construct such *fractional strings* have led only to partial results and it is not clear if such theories exist. So we will restrict our attention to h a multiple of $\frac{1}{2}$.

With these assumptions the possible algebras are very limited, with spins 0, $\frac{1}{2}$, 1, $\frac{3}{2}$, and 2. Solution of the Jacobi identities allows only the algebras shown in table 11.1. The first two entries are of course the conformal and $N = 1$ superconformal algebras that we have already studied. The three $N = 4$ algebras are related. The second algebra is a special case of the first where the $U(1)$ current becomes the gradient of a scalar. The third is a subalgebra of the second.

The ghost central charge is determined by the number of currents of each spin. The central charge for the ghosts associated with a current of spin h is

$$c_h = (-1)^{2h+1}[3(2h-1)^2 - 1] , \tag{11.1.2a}$$
$$c_2 = -26 , \quad c_{3/2} = +11 , \quad c_1 = -2 , \quad c_{1/2} = -1 , \quad c_0 = -2 . \tag{11.1.2b}$$

Table 11.1. *World-sheet superconformal algebras. The number of currents of each spin and the total ghost central charge are listed, as are the global symmetry generated by the spin-1 currents and the transformation of the supercharges under these.*

$n_{3/2} \equiv N$	n_1	$n_{1/2}$	n_0	c^{g}	symmetry	T_F rep.
0	0	0	0	-26		
1	0	0	0	-15		
2	1	0	0	-6	$U(1)$	± 1
3	3	1	0	0	$SU(2)$	**3**
4	7	4	0	0	$SU(2) \times SU(2) \times U(1)$	**(2,2,0)**
4	6	4	1	0	$SU(2) \times SU(2)$	**(2,2)**
4	3	0	0	12	$SU(2)$	**2**

The sign $(-1)^{2h+1}$ takes into account the statistics of the ghosts, anticommuting for integer spin and commuting for half-integer spin. Since the matter central charge c^{m} is $-c^{\mathrm{g}}$, there is only one new algebra, $N = 2$, that can have a positive critical dimension.

Actually, for $N = 0$ and $N = 1$ there can also be additional spin-1 and spin-$\frac{1}{2}$ constraints, provided the supercurrent is neutral under the corresponding symmetry. However, these larger algebras are not essentially different. The negative central charges of the ghosts allow additional matter, but the additional constraints precisely remove the added states so that these reduce to the old $N = 0$ and $N = 1$ theories. Nevertheless this construction is sometimes useful, as we will see in section 15.5.

For $N = 2$ it is convenient to join the two real supercurrents into one complex supercurrent

$$T_F^{\pm} = 2^{-1/2}(T_{F1} \pm iT_{F2}) \, . \tag{11.1.3}$$

The $N = 2$ algebra in operator product form is then

$$T_B(z)T_F^{\pm}(0) \sim \frac{3}{2z^2}T_F^{\pm}(0) + \frac{1}{z}\partial T_F^{\pm}(0) \, , \tag{11.1.4a}$$

$$T_B(z)j(0) \sim \frac{1}{z^2}j(0) + \frac{1}{z}\partial j(0) \, , \tag{11.1.4b}$$

$$T_F^{+}(z)T_F^{-}(0) \sim \frac{2c}{3z^3} + \frac{2}{z^2}j(0) + \frac{2}{z}T_B(0) + \frac{1}{z}\partial j(0) \, , \tag{11.1.4c}$$

$$T_F^{+}(z)T_F^{+}(0) \sim T_F^{-}(z)T_F^{-}(0) \sim 0 \, , \tag{11.1.4d}$$

$$j(z)T_F^{\pm}(0) \sim \pm\frac{1}{z}T_F^{\pm}(0) \, , \tag{11.1.4e}$$

$$j(z)j(0) \sim \frac{c}{3z^2} \, . \tag{11.1.4f}$$

In particular this implies that $T_F^{\pm}$ and j are primary fields and that $T_F^{\pm}$ has charge ± 1 under the $U(1)$ generated by j. The constant c in $T_F^{+}T_F^{-}$

and jj must be the central charge. This follows from the Jacobi identity for the modes, but we will not write out the mode expansion in full until chapter 19, where we will have more need of it.

The smallest linear representation of the $N = 2$ algebra has two real scalars and two real fermions, which we join into a complex scalar Z and complex fermion ψ. The action is

$$S = \frac{1}{2\pi} \int d^2z \left(\partial \overline{Z} \bar{\partial} Z + \overline{\psi} \bar{\partial} \psi + \tilde{\psi} \partial \tilde{\psi} \right) . \qquad (11.1.5)$$

The currents are

$$T_B = -\partial \overline{Z} \partial Z - \frac{1}{2}(\overline{\psi} \partial \psi + \psi \partial \overline{\psi}) , \quad j = -\overline{\psi}\psi , \qquad (11.1.6a)$$

$$T_F^+ = 2^{1/2} i \psi \partial \overline{Z} , \quad T_F^- = 2^{1/2} i \overline{\psi} \partial Z . \qquad (11.1.6b)$$

There is also a set of antiholomorphic currents, so this $Z\psi\tilde{\psi}$ CFT has $(2,2)$ superconformal symmetry.

The central charge of the $Z\psi\tilde{\psi}$ CFT is 3, so two copies will cancel the ghost central charge. Since there are two real scalars in each CFT the critical dimension is 4. However, these dimensions come in complex pairs, so that the spacetime signature can be purely Euclidean, or $(2,2)$, but not the Minkowski $(3,1)$. Further, while the theory has four-dimensional translational invariance it does not have four-dimensional Lorentz invariance — the dimensions are paired together in a definite way in the supercharges. Instead the symmetry is $U(2)$ or $U(1,1)$, complex rotations on the two Zs. Finally, the spectrum is quite small. The constraints fix two full sets of $Z\psi\tilde{\psi}$ oscillators (the analog of the light-cone gauge), leaving none. Thus there is just the center-of-mass motion of a single state. This has some mathematical interest, but whether it has physical applications is more conjectural.

Thus we have reduced what began as a rather large set of possible algebras down to the original $N = 0$ and $N = 1$. There is, however, another generalization, which is to have different algebras on the left- and right-moving sides of the closed string. The holomorphic and antiholomorphic algebras commute and there is no reason that they should be the same. In the open string, the boundary conditions relate the holomorphic and antiholomorphic currents so there is no analogous construction.

This allows the one new possibility, the $(N, \tilde{N}) = (0, 1)$ *heterotic string*; $(N, \tilde{N}) = (1, 0)$ would be the same on redefining $z \to \bar{z}$. We study this new algebra in detail in the remainder of the chapter. In addition the $(0, 2)$ and $(1, 2)$ heterotic string theories are mathematically interesting and may have a less direct physical relevance.

It should be emphasized that the analysis in this section had many explicit and implicit assumptions, and one should be cautious in assuming that all string theories have been found. Indeed, there are some string theories that do not fall into this classification. One is the Green–Schwarz form of the superstring. This has no simple covariant gauge-fixing, but in the light-cone gauge it is in fact equivalent to the RNS superstring, via bosonization. We will not have space to develop this in detail, but will see a hint of it in chapter 12. Another exception is *topological string theory*, where in a covariant gauge the constraints do not satisfy spin-statistics as we have assumed. This string theory has no physical degrees of freedom, but is of mathematical interest in that its observables are topological.

In fact, we will find the same set of physical string theories from an entirely different and nonperturbative point of view in chapter 14, suggesting that all have been found. To be precise, there are other theories with stringlike excitations, but the theories found in this and the previous chapter seem to be the only ones which have a limit where they become weakly coupled, so that a string perturbation theory exists.

11.2   The $SO(32)$ and $E_8 \times E_8$ heterotic strings

The $(0, 1)$ heterotic string combines the constraints and ghosts from the left-moving side of the bosonic string with those from the right-moving side of the type II string. We could try to go further and combine the whole left-moving side of the bosonic string, with 26 flat dimensions, with the ten-dimensional right-moving side of the type II string. In fact this can be done, but since its physical meaning is not so clear we will for now keep the same number of dimensions on both sides. The maximum is then ten, from the superconformal side. We begin with the fields

$$X^\mu(z, \bar{z}) , \quad \tilde{\psi}^\mu(\bar{z}) , \quad \mu = 0, \ldots, 9 , \tag{11.2.1}$$

with total central charge $(c, \tilde{c}) = (10, 15)$. The ghost central charges add up to $(c^g, \tilde{c}^g) = (-26, -15)$, so the remaining matter has $(c, \tilde{c}) = (16, 0)$. The simplest possibility is to take 32 left-moving spin-$\frac{1}{2}$ fields

$$\lambda^A(z) , \quad A = 1, \ldots, 32 . \tag{11.2.2}$$

The total matter action is

$$S = \frac{1}{4\pi} \int d^2z \left(\frac{2}{\alpha'} \partial X^\mu \bar{\partial} X_\mu + \lambda^A \bar{\partial} \lambda^A + \tilde{\psi}^\mu \partial \tilde{\psi}_\mu \right) . \tag{11.2.3}$$

The operator products are

$$X^\mu(z, \bar{z}) X^\nu(0, 0) \sim -\eta^{\mu\nu} \frac{\alpha'}{2} \ln |z|^2 , \tag{11.2.4a}$$

$$\lambda^A(z) \lambda^B(0) \sim \delta^{AB} \frac{1}{z} , \tag{11.2.4b}$$

$$\tilde{\psi}^\mu(\bar{z}) \tilde{\psi}^\nu(0) \sim \eta^{\mu\nu} \frac{1}{\bar{z}} . \tag{11.2.4c}$$

The matter energy-momentum tensor and supercurrent are

$$T_B = -\frac{1}{\alpha'}\partial X^\mu \partial X_\mu - \frac{1}{2}\lambda^A \partial \lambda^A \,, \tag{11.2.5a}$$

$$\tilde{T}_B = -\frac{1}{\alpha'}\bar{\partial} X^\mu \bar{\partial} X_\mu - \frac{1}{2}\tilde{\psi}^\mu \bar{\partial}\tilde{\psi}_\mu \,, \tag{11.2.5b}$$

$$\tilde{T}_F = i(2/\alpha')^{1/2}\tilde{\psi}^\mu \bar{\partial} X_\mu \,. \tag{11.2.5c}$$

The world-sheet theory has symmetry $SO(9,1) \times SO(32)$. The $SO(32)$, acting on the λ^A, is an internal symmetry. In particular, none of the λ^A can have a timelike signature because there is no fermionic constraint on the left-moving side to remove states of negative norm. So while the action for the λ^A is the same as for the $\tilde{\psi}^\mu$ of the RNS superstring, the resulting theory is very different because of the constraints.

The right-moving ghosts are the same as in the RNS superstring, and the left-movers the same as in the bosonic string. It is straightforward to construct the nilpotent BRST charge and show the no-ghost theorem, with any BRST-invariant periodicity conditions. As usual this still holds if we replace any of the spatial $(X^\mu, \tilde{\psi}^\mu)$ and the λ^A with a unitary $(0,1)$ SCFT of the equivalent central charge.

To finish the description of the theory, we need to give the boundary conditions on the fields and specify which sectors are in the spectrum. This is more complicated than in the type II strings, because now neither Poincaré nor BRST invariance require common boundary conditions on all the λ^A. Periodicity of T_B only requires that the λ^A be periodic up to an arbitrary $O(32)$ rotation,

$$\lambda^A(w + 2\pi) = O^{AB}\lambda^B(w) \,. \tag{11.2.6}$$

We will not carry out a systematic search for consistent theories as we did for the RNS string, but will describe all the known theories. Nine ten-dimensional theories based on the action (11.2.3) are known, though six have tachyons and so are consistent only in the same sense as the bosonic string. Of the three tachyon-free theories, two have spacetime supersymmetry and these are our main interest. In this section we construct the two supersymmetric theories and in the next the seven nonsupersymmetric theories.

In the IIA and IIB superstrings the GSO projection acted separately on the left- and right-moving sides. This will be also true in any supersymmetric heterotic theory. The world-sheet current associated with spacetime symmetry is $\mathcal{V}_s$ as in eq. (10.4.25), with $\mathbf{s}$ in the $\mathbf{16}$. In order for the corresponding charge to be well defined, the OPE of this current with any vertex operator must be single-valued. For the right-moving spinor part of the vertex operator, the spin eigenvalue s' must then satisfy

$$\mathbf{s} \cdot \mathbf{s}' + \frac{l}{2} \in \mathbf{Z} \tag{11.2.7}$$

for all $\mathbf{s} \in \mathbf{16}$, where l is -1 in the NS sector and $-\frac{1}{2}$ in the R sector. Taking $\mathbf{s} = (\frac{1}{2}, \frac{1}{2}, \frac{1}{2}, \frac{1}{2}, \frac{1}{2})$, this condition is precisely the right-moving GSO projection

$$\exp(\pi i \tilde{F}) = 1 \; ; \tag{11.2.8}$$

any other $\mathbf{s} \in \mathbf{16}$ gives the same condition.

Now let us try a GSO projection on the left-moving spinors also. That is, we take periodicities

$$\lambda^A(w + 2\pi) = \pm \lambda^A(w) \tag{11.2.9}$$

with the same sign on all 32 components, and impose

$$\exp(\pi i F) = 1 \tag{11.2.10}$$

for the left-moving fermion number. It is easily verified by means of bosonization that the OPE is local and closed, just as in the IIA and IIB strings. Combine the 32 real fermions into 16 complex fermions,

$$\lambda^{K\pm} = 2^{-1/2}(\lambda^{2K-1} \pm i \lambda^{2K}) , \quad K = 1, \dots, 16 . \tag{11.2.11}$$

These can then be bosonized in terms of 16 left-moving scalars $H^K(z)$. By analogy to the definition of F in the type II string define

$$F = \sum_{K=1}^{16} q_K , \tag{11.2.12}$$

where $\lambda^{K\pm}$ has $q_K = \pm 1$. Then F is additive so the OPE is closed, and the projection (11.2.10) guarantees that there are no branch cuts with the R sector vertex operators. Note that in the bosonized description we have 26 left-moving and 10 right-moving bosons, so the theory (11.2.3) really is a fusion (heterosis) of the bosonic and type II strings. We will emphasize the fermionic description in the present section, returning to the bosonic description later.

Modular invariance is straightforward. The partition function for the λ is

$$Z_{16}(\tau) = \frac{1}{2}\left[Z^0_{\ 0}(\tau)^{16} + Z^0_{\ 1}(\tau)^{16} + Z^1_{\ 0}(\tau)^{16} + Z^1_{\ 1}(\tau)^{16} \right] . \tag{11.2.13}$$

The modular transformations just permute the four terms, with no phase under $\tau \to -1/\tau$ and a phase of $\exp(2\pi i/3)$ under $\tau \to \tau + 1$. The latter cancels the opposite phase from the partition function $Z^+_\psi(\tau)^*$ of $\tilde{\psi}$. The form (11.2.13) parallels that of $Z^+_\psi(\tau)$ in the type II string but with all $+$

signs. This is necessary from several points of view. With 32 rather than 8 fermions, the signs in the modular transformations are raised to the fourth power and so the first three terms must enter with a common sign. As usual the $Z^1{}_1$ term transforms only into itself and its sign depends on the chirality in the R sector. Three other theories, defined by flipping the chirality in one or both R sectors, are physically equivalent. Also, the relative minus sign in the first and second terms of $Z_\psi^+(\tau)$ came from the F of the superconformal ghosts, which we do not have on the left-moving side of the heterotic string. The relative minus sign in the first and third terms came from spacetime statistics, but the λ are spacetime scalars and so are their R sector states. So modular invariance, conservation of F by the OPE, and spacetime spin-statistics are all consistent with the partition function (11.2.13).

We now find the lightest states. The right-moving side is the same as in the type II string, with no tachyon and $\mathbf{8}_v + \mathbf{8}$ at the massless level. On the left-moving side, the normal ordering constant in the left-moving transverse Hamiltonian $H^\perp = \alpha' m^2/4$ is

$$\text{NS}: \ -\frac{8}{24} - \frac{32}{48} = -1 \ , \quad \text{R}: \ -\frac{8}{24} + \frac{32}{24} = +1 \ . \tag{11.2.14}$$

The left-moving NS ground state is therefore a tachyon. The first excited states

$$\lambda^A_{-1/2}|0\rangle_{\text{NS}} \tag{11.2.15}$$

have $H^\perp = -\frac{1}{2}$ but are removed by projection (11.2.10): the NS ground state now has $\exp(\pi i F) = +1$ because there is no contribution from ghosts. A state of $H^\perp = 0$ can be obtained in two ways:

$$\alpha^i_{-1}|0\rangle_{\text{NS}} \ , \quad \lambda^A_{-1/2}\lambda^B_{-1/2}|0\rangle_{\text{NS}} \ . \tag{11.2.16}$$

The λ^A transform under an $SO(32)$ internal symmetry. Under the full symmetry $SO(8)_{\text{spin}} \times SO(32)$, the NS ground state is invariant, $(\mathbf{1}, \mathbf{1})$. The second state in (11.2.16) is antisymmetric under $A \leftrightarrow B$, so the massless states (11.2.16) transform as $(\mathbf{8}_v, \mathbf{1}) + (\mathbf{1}, [\mathbf{2}])$. The antisymmetric tensor representation is the adjoint of $SO(32)$, with dimension $32 \times 31/2 = \mathbf{496}$.

Table 11.2 summarizes the tachyonic and massless states on each side. The left-movers are given with their $SO(8) \times SO(32)$ quantum numbers and the right-movers with their $SO(8)$ quantum numbers. Closed string states combine left- and right-moving states at the same mass. The left-moving side, like the bosonic string, has a would-be tachyon, but there is no right-mover to pair it with so the theory is tachyon-free. At the massless level, the product

$$(\mathbf{8}_v, \mathbf{1}) \times (\mathbf{8}_v + \mathbf{8}) = (\mathbf{1}, \mathbf{1}) + (\mathbf{28}, \mathbf{1}) + (\mathbf{35}, \mathbf{1}) + (\mathbf{56}, \mathbf{1}) + (\mathbf{8}', \mathbf{1}) \tag{11.2.17}$$

Table 11.2. *Low-lying heterotic string states.*

m^2	NS	R	$\tilde{N}S$	$\tilde{R}$
$-4/\alpha'$	$(\mathbf{1}, \mathbf{1})$	-	-	-
0	$(\mathbf{8}_v, \mathbf{1}) + (\mathbf{1}, \mathbf{496})$	-	$\mathbf{8}_v$	$\mathbf{8}$

is the type I supergravity multiplet. The product

$$(\mathbf{1}, \mathbf{496}) \times (\mathbf{8}_v + \mathbf{8}) = (\mathbf{8}_v, \mathbf{496}) + (\mathbf{8}, \mathbf{496}) \tag{11.2.18}$$

is an $N = 1$ gauge multiplet in the adjoint of $SO(32)$. The latter is therefore a gauge symmetry in spacetime.

This is precisely the same massless content as the type I open plus closed $SO(32)$ theory. However, these two theories have different massive spectra. In the open string, the gauge quantum numbers are carried by an $SO(32)$ vector at each endpoint, so even at the massive levels there will never be more than a rank 2 tensor representation of the gauge group. In the heterotic string, the gauge quantum numbers are carried by fields that propagate on the whole world sheet. At massive levels any number of these can be excited, allowing arbitrarily large representations of the gauge group. Remarkably, however, we will see in chapter 14 that the type I and heterotic $SO(32)$ theories are one and the same.

The second heterotic string theory is obtained by dividing the λ^A into two sets of 16 with independent boundary conditions,

$$\lambda^A(w + 2\pi) = \begin{cases} \eta \lambda^A(w), & A = 1, \ldots, 16, \\ \eta' \lambda^A(w), & A = 17, \ldots, 32, \end{cases} \tag{11.2.19}$$

with η and η' each ± 1. Correspondingly, there are the operators

$$\exp(\pi i F_1), \quad \exp(\pi i F_1'), \tag{11.2.20}$$

which anticommute with λ^A for $A = 1, \ldots, 16$ and $A = 17, \ldots, 32$ respectively. Take separate GSO projections on the right-movers and the two sets of left-movers. That is, sum over the $2^3 = 8$ possible combinations of periodicities with the projections

$$\exp(\pi i F_1) = \exp(\pi i F_1') = \exp(\pi i \tilde{F}) = 1. \tag{11.2.21}$$

Again closure and locality of the OPE and modular invariance are easily verified. In particular the partition function

$$Z_8(\tau)^2 = \frac{1}{4} \left[Z^0_0(\tau)^8 + Z^0_1(\tau)^8 + Z^1_0(\tau)^8 + Z^1_1(\tau)^8 \right]^2 \tag{11.2.22}$$

transforms in the same way as $Z_\psi^\pm$ and Z_{16}. It is important here that the fermions are in groups of 16, so that the minus signs from $Z_\psi^\pm$ (which was for eight fermions) are squared.

As before, the lightest states on the right-hand side are the massless $\mathbf{8}_v + \mathbf{8}$. On the left-hand side, the sector NS–NS′ again has a normal ordering constant of -1, so the ground state is tachyonic but finds no matching state on the right. The first excited states, at $m^2 = 0$, are

$$\alpha^i_{-1}|0\rangle_{\text{NS,NS}'} \,,$$
$$\lambda^A_{-1/2}\lambda^B_{-1/2}|0\rangle_{\text{NS,NS}'} \,, \quad 1 \leq A, B \leq 16 \text{ or } 17 \leq A, B \leq 32 \,. \quad (11.2.23)$$

There is a difference here from the $SO(32)$ case: because there are separate GSO projections on each set of 16, A and B must come from the same set. Since the $SO(32)$ symmetry is partly broken by the boundary conditions, we classify states by the surviving $SO(16) \times SO(16)$. The states (11.2.23) include the antisymmetric tensor adjoint representation for each $SO(16)$, with dimension $16 \times 15/2 = \mathbf{120}$.

In the left-moving R–NS′ sector the normal ordering constant is

$$-\frac{8}{24} + \frac{16}{24} - \frac{16}{48} = 0 \,, \quad (11.2.24)$$

so the ground states are massless. Making eight raising and eight lowering operators out of the 16 λ^A zero modes produces a 256-dimensional spinor representation of the first $SO(16)$. The GSO projection separates it into two irreducible representations, $\mathbf{128} + \mathbf{128}'$, the former being in the spectrum. The NS–R′ sector produces a $\mathbf{128}$ of the other $SO(16)$, and the R–R′ sector again has no massless states.

In all, the $SO(8)$ spin $\times SO(16) \times SO(16)$ content of the massless level of the left-hand side is

$$(\mathbf{8}_v, \mathbf{1}, \mathbf{1}) + (\mathbf{1}, \mathbf{120}, \mathbf{1}) + (\mathbf{1}, \mathbf{1}, \mathbf{120}) + (\mathbf{1}, \mathbf{128}, \mathbf{1}) + (\mathbf{1}, \mathbf{1}, \mathbf{128}) \,. \quad (11.2.25)$$

Combining these with the right-moving $\mathbf{8}_v$ gives, for each $SO(16)$, massless vector bosons which transform as $\mathbf{120} + \mathbf{128}$. Consistency of the spacetime theory requires that massless vectors transform under the *adjoint* representation of the gauge group. There is indeed a group, the exceptional group E_8, that has an $SO(16)$ subgroup under which the E_8 adjoint $\mathbf{248}$ transforms as $\mathbf{120} + \mathbf{128}$. Evidently this second heterotic string theory has gauge group $E_8 \times E_8$. The world-sheet theory has a full $E_8 \times E_8$ symmetry, even though only an $SO(16) \times SO(16)$ symmetry is manifest in the present description. The additional currents are given by bosonization as

$$\exp\left[i \sum_{K=1}^{16} q_K H^K(z)\right] \,. \quad (11.2.26)$$

This is a spin field, just as in the fermion vertex operator (10.4.25). For the first E_8 the charges are

$$q_K = \begin{cases} \pm\frac{1}{2}, & K = 1, \ldots 8 \\ 0, & K = 9, \ldots 16 \end{cases}, \qquad \sum_{K=1}^{16} q_K \in 2\mathbf{Z}, \qquad (11.2.27)$$

and vice versa for the second. These are indeed $(1,0)$ operators. The massless spectrum is the $d = 10$, $N = 1$ supergravity multiplet plus an $E_8 \times E_8$ gauge multiplet. The $SO(8)_{\text{spin}} \times E_8 \times E_8$ quantum numbers of the massless fields are

$$(\mathbf{1}, \mathbf{1}, \mathbf{1}) + (\mathbf{28}, \mathbf{1}, \mathbf{1}) + (\mathbf{35}, \mathbf{1}, \mathbf{1}) + (\mathbf{56}, \mathbf{1}, \mathbf{1}) + (\mathbf{8'}, \mathbf{1}, \mathbf{1})$$
$$+ (\mathbf{8}_v, \mathbf{248}, \mathbf{1}) + (\mathbf{8}, \mathbf{248}, \mathbf{1}) + (\mathbf{8}_v, \mathbf{1}, \mathbf{248}) + (\mathbf{8}, \mathbf{1}, \mathbf{248}) . \qquad (11.2.28)$$

Consistency requires the fermions to be in groups of 16. We could make a modular-invariant theory using groups of eight, the left-moving partition function being $(Z_{\tilde{\psi}}^\pm)^4$. However, we have seen that modular invariance requires minus signs in $Z_{\tilde{\psi}}^\pm$. These signs would give negative weight to left-moving R sector states and would correspond to the projection $\exp(\pi i F) = -1$ in the NS sector. The first is inconsistent with spin-statistics because these states are spacetime scalars, and the second is inconsistent with closure of the OPE thus making the interactions inconsistent. The $SO(32)$ and $E_8 \times E_8$ theories are the only supersymmetric heterotic strings in ten dimensions.

11.3   Other ten-dimensional heterotic strings

The other heterotic string theories can all be constructed from a single theory by the twisting construction introduced in section 8.5. The 'least twisted' theory, in the sense of having the smallest number of path integral sectors, corresponds to the diagonal modular invariant

$$\frac{1}{2}\Big[Z_0^0(\tau)^{16} Z_0^0(\tau)^{*4} - Z_1^0(\tau)^{16} Z_1^0(\tau)^{*4}$$
$$- Z_0^1(\tau)^{16} Z_0^1(\tau)^{*4} - Z_1^1(\tau)^{16} Z_1^1(\tau)^{*4} \Big] . \qquad (11.3.1)$$

This invariant corresponds to taking all fermions, λ^A and $\tilde{\psi}^\mu$, to be simultaneously periodic or antiperiodic on each cycle of the torus. In terms of the spectrum the world-sheet fermions are either all R or all NS, with the diagonal projection

$$\exp[\pi i(F + \tilde{F})] = 1 . \qquad (11.3.2)$$

This theory is consistent except for a tachyon, the state

$$\lambda^A_{-1/2}|0\rangle_{\text{NS,NS}}, \quad m^2 = -\frac{2}{\alpha'}, \quad \exp(\pi i F) = \exp(\pi i \tilde{F}) = -1, \qquad (11.3.3)$$

which transforms as a vector under $SO(32)$. At the massless level are the states

$$\alpha^i_{-1}\tilde\psi^j_{-1/2}|0\rangle_{\text{NS,NS}} , \quad \lambda^A_{-1/2}\lambda^B_{-1/2}\tilde\psi^j_{-1/2}|0\rangle_{\text{NS,NS}} , \qquad (11.3.4)$$

which are the graviton, dilaton, antisymmetric tensor and $SO(32)$ gauge bosons. There are fermions in the spectrum, but the lightest are at $m^2 = 4/\alpha'$.

Now let us twist by various symmetries. Consider first the $\mathbf{Z}_2$ generated by $\exp(\pi i\tilde F)$. Combined with the diagonal projection (11.3.2) this gives the total projection

$$\frac{1+\exp[\pi i(F+\tilde F)]}{2}\cdot\frac{1+\exp(\pi i\tilde F)}{2} = \frac{1+\exp(\pi iF)}{2}\cdot\frac{1+\exp(\pi i\tilde F)}{2} . \quad (11.3.5)$$

This is the same as the projections (11.2.8) plus (11.2.10) defining the supersymmetric $SO(32)$ heterotic string. Also, the spatial twist by $\exp(\pi i\tilde F)$ adds in the sectors in which the λ^A and $\tilde\psi^\mu$ have opposite periodicities. The twisted theory is thus the $SO(32)$ heterotic string. Twisting by $\exp(\pi iF)$ has the same effect.

Now consider twisting the diagonal theory by $\exp(\pi iF_1)$, which flips the sign of the first 16 λ^A and which was used to construct the $E_8\times E_8$ heterotic string. The resulting theory is nonsupersymmetric — as in eq. (11.2.8), a theory will be supersymmetric if and only if the projections include $\exp(\pi i\tilde F) = 1$. It has gauge group $E_8\times SO(16)$ and a tachyon in the $(\mathbf{1},\mathbf{16})$. We leave it to the reader to verify this. A further twist by $\exp(\pi i\tilde F)$ produces the supersymmetric $E_8\times E_8$ heterotic string.

One can carry this further by dividing the λ^A into groups of 8, 4, 2, and 1 as follows. Write the $SO(32)$ index A in binary form, $A = 1+d_1d_2d_3d_4d_5$, where each of the digits d_i is zero or one. Define the operators $\exp(\pi iF_i)$ for $i = 1,\ldots,5$ to anticommute with those λ^A having $d_i = 0$ and commute with those having $d_i = 1$. There are essentially five possible twist groups, with 2, 4, 8, 16, or 32 elements, generated respectively by choosing one, two, three, four or five of the $\exp(\pi iF_i)$ and forming all products. The first of these produces the $E_8\times SO(16)$ theory just described; the further twists produce the gauge groups $SO(24)\times SO(8)$, $E_7\times E_7\times SO(4)$, $SU(16)\times SO(2)$, and E_8. None of these theories is supersymmetric, and all have tachyons. A further twist by $\exp(\pi i\tilde F)$ produces a supersymmetric theory which in each case is either the $SO(32)$ theory or the $E_8\times E_8$ theory. These gauge symmetries are less manifest in this construction, with more of the currents coming from R sectors.

Let us review the logic of the twisting construction. The vertex operator corresponding to a sector twisted by a group element h produces branch cuts in the fields, but the projection onto h-invariant states means that these branch cuts do not appear in the products of vertex operators.

Since h is a symmetry the projection is preserved by interactions. On the torus, the sum over spatial and timelike twists is modular-invariant, and this generalizes to any genus. However, we have learned in section 10.7 that naive modular invariance of the sum over path integral boundary conditions is not enough, because in general there are anomalous phases in the modular transformations. Only for a right–left symmetric path integral do the phases automatically cancel.

At one loop the anomalous phases appear only in the transformation $\tau \to \tau + 1$, where they amount to a failure of the level-matching condition $L_0 - \tilde{L}_0 \in \mathbf{Z}$. It is further a theorem that for an Abelian twist group (like the products of $\mathbf{Z}_2$s considered here), the one-loop amplitude and in fact all amplitudes are modular-invariant precisely if in every twisted sector, before imposing the projection, there is an infinite number of level-matched states. The projection can then be defined to satisfy level matching. In the heterotic string, taking a sector in which k of the λ^A satisfy R boundary conditions and $32 - k$ satisfy NS boundary conditions, the zero-point energy is

$$-\frac{8}{24} + \frac{k}{24} - \frac{(32 - k)}{48} = -1 + \frac{k}{16} \, . \tag{11.3.6}$$

The oscillators raise the energy by a multiple of $\frac{1}{2}$, so the energies on the left-moving side are $\frac{1}{16}k$ mod $\frac{1}{2}$. On the right-moving side we are still taking the fermions to have common boundary conditions for Lorentz invariance, so the energies are multiples of $\frac{1}{2}$. Thus the level-matching condition is satisfied precisely if k is a multiple of eight. Closure of the OPE and spacetime spin-statistics actually require k to be a multiple of 16, as we have seen. The twists $\exp(\pi i F_i)$ were defined so that any product of them anticommutes with exactly 16 of the λ^A, satisfying this condition.

When the level-matching condition is satisfied, there can in fact be more than one modular-invariant and consistent theory. Consider a twisted theory with partition function

$$Z = \frac{1}{\text{order}(H)} \sum_{h_1, h_2 \in H} Z_{h_1, h_2} \, , \tag{11.3.7}$$

where h_1 and h_2 are the spatial and timelike periodicities on the torus. Then the theory with partition function

$$Z = \frac{1}{\text{order}(H)} \sum_{h_1, h_2 \in H} \epsilon(h_1, h_2) Z_{h_1, h_2} \tag{11.3.8}$$

is also consistent (modular-invariant with closed and local OPE) provided that the phases $\epsilon(h_1, h_2)$ satisfy

$$\epsilon(h_1, h_2) = \epsilon(h_2, h_1)^{-1} , \qquad (11.3.9a)$$

$$\epsilon(h_1, h_2)\epsilon(h_1, h_3) = \epsilon(h_1, h_2 h_3) , \qquad (11.3.9b)$$

$$\epsilon(h, h) = 1 . \qquad (11.3.9c)$$

In terms of $\hat{h}_2$ defined in the original twisted theory, the new twisted theory is no longer projected onto H-invariant states, but onto states satisfying

$$\hat{h}_2|\psi\rangle_{h_1} = \epsilon(h_1, h_2)^{-1}|\psi\rangle_{h_1} \qquad (11.3.10)$$

in the sector twisted by h_1. In other words, states are now eigenvectors of $\hat{h}$, with a sector-dependent phase; equivalently we have made a sector-dependent redefinition

$$\hat{h} \rightarrow \epsilon(h_1, h)\hat{h} . \qquad (11.3.11)$$

The phase factor $\epsilon(h_1, h_2)$ is known as *discrete torsion*.

There is one interesting possibility for discrete torsion in the theories above, in the group generated by $\exp(\pi i F_1)$ and $\exp(\pi i \tilde{F})$ that produces the $E_8 \times E_8$ string from the diagonal theory. For

$$(h_1, h_2) = \left(\exp[\pi i(k_1 F_1 + l_1 \tilde{F})], \exp[\pi i(k_2 F_1 + l_2 \tilde{F})]\right) \qquad (11.3.12)$$

the phase

$$\epsilon(h_1, h_2) = (-1)^{k_1 l_2 + k_2 l_1} \qquad (11.3.13)$$

satisfies the conditions (11.3.9). It modifies the projection from

$$\exp(\pi i F_1) = \exp(\pi i F_1') = \exp(\pi i \tilde{F}) = 1 , \qquad (11.3.14)$$

which produces to the supersymmetric $E_8 \times E_8$ string, to

$$\exp[\pi i(F_1 + \alpha_1' + \tilde{\alpha})] = \exp[\pi i(F_1' + \alpha_1 + \tilde{\alpha})] = \exp[\pi i(\tilde{F} + \alpha_1 + \alpha_1')] = 1 . \qquad (11.3.15)$$

The notation parallels that in eq. (10.7.11): under $w \rightarrow w + 2\pi$, the $\tilde{\psi}^\mu$, the first 16 λ^A, and the second 16 λ^A pick up phases $-\exp(-\pi i \tilde{\alpha})$, $-\exp(\pi i \alpha_1)$, and $-\exp(\pi i \alpha_1')$ respectively. The αs are conserved by the OPE so the projections are consistent. In other words, the spectrum consists of the sectors

$$(\text{NS}+, \text{NS}+, \text{NS}+) ,$$
$$(\text{NS}-, \text{NS}-, \text{R}+) , \ (\text{NS}-, \text{R}+, \text{NS}-) , \ (\text{NS}+, \text{R}-, \text{R}-) ,$$
$$(\text{R}+, \text{NS}-, \text{NS}-) , \ (\text{R}-, \text{R}-, \text{NS}+) , \ (\text{R}-, \text{NS}+, \text{R}-) ,$$
$$(\text{R}+, \text{R}+, \text{R}+)$$

where the notation refers respectively to the $\tilde{\psi}^\mu$, the first 16 λ^A, and the second 16 λ^A.

Gravitinos, in the sectors (R±, NS+, NS+), are absent from the spectrum. So also are tachyons, which are in the (NS−, NS−, NS+) and (NS−, NS+, NS−) sectors. The twists leave an $SO(16) \times SO(16)$ gauge symmetry. Classifying states by their $SO(8)_{\text{spin}} \times SO(16) \times SO(16)$ quantum numbers, one finds the massless spectrum

$$(\text{NS+}, \text{NS+}, \text{NS+}) : \quad (\mathbf{1}, \mathbf{1}, \mathbf{1}) + (\mathbf{28}, \mathbf{1}, \mathbf{1}) + (\mathbf{35}, \mathbf{1}, \mathbf{1})$$
$$+ (\mathbf{8}_v, \mathbf{120}, \mathbf{1}) + (\mathbf{8}_v, \mathbf{1}, \mathbf{120}) \,,$$
$$(\text{R+}, \text{NS}-, \text{NS}-) : \quad (\mathbf{8}, \mathbf{16}, \mathbf{16}) \,,$$
$$(\text{R}-, \text{R}-, \text{NS+}) : \quad (\mathbf{8}', \mathbf{128}', \mathbf{1}) \,,$$
$$(\text{R}-, \text{NS+}, \text{R}-) : \quad (\mathbf{8}', \mathbf{1}, \mathbf{128}') \,.$$

This shows that a tachyon-free theory without supersymmetry is possible.

11.4 A little Lie algebra

In the open string the gauge charges are carried by the Chan–Paton degrees of freedom at the endpoints. In the closed string the charges are carried by fields that move along the string. We saw this earlier for the Kaluza–Klein gauge symmetry and the enhanced gauge symmetries that appear when the bosonic string is compactified, and now we see it again in the heterotic string. In the following sections we will discuss these closed string gauge symmetries in a somewhat more systematic way, but first we need to introduce a few ideas from Lie algebra. Space forbids a complete treatment; we focus on some basic ideas and some specific results that will be needed later.

Basic definitions

A Lie algebra is a vector space with an antisymmetric product $[T, T']$. In terms of a basis T^a the product is

$$[T^a, T^b] = if^{ab}{}_c T^c \tag{11.4.1}$$

with $f^{ab}{}_c$ the *structure constants*. The product is required to satisfy the Jacobi identity

$$[T^a, [T^b, T^c]] + [T^b, [T^c, T^a]] + [T^c, [T^a, T^b]] = 0 \,. \tag{11.4.2}$$

The associated Lie group is generated by the exponentials

$$\exp(i\theta_a T^a) \,, \tag{11.4.3}$$

with the θ_a real. For a compact group, the associated compact Lie algebra has a positive inner product

$$(T^a, T^b) = d^{ab} \,, \tag{11.4.4}$$

which is invariant,

$$([T, T'], T'') + (T', [T, T'']) = 0 . \tag{11.4.5}$$

This invariance is equivalent to the statement that f^{abc} is completely antisymmetric, where d^{ab} is used to raise the index.

We are interested in simple Lie algebras, those having no nontrivial invariant subalgebras (ideals). A general compact algebra is a sum of simple algebras and $U(1)$s. For a simple algebra the inner product is unique up to normalization, and there is a basis of generators in which it is simply δ^{ab}. For any representation r of the Lie algebra (any set of matrices $t^a_{r,ij}$ satisfying (11.4.1) with the given $f^{ab}{}_c$), the trace is invariant and so for a simple Lie algebra is proportional to d^{ab},

$$\mathrm{Tr}(t^a_r t^b_r) = T_r d^{ab} \tag{11.4.6}$$

from some constant T_r. Also, $t^a_r t^b_r d_{ab}$ commutes with all the t^c_r and so is proportional to the identity,

$$t^a_r t^b_r d_{ab} = Q_r \tag{11.4.7}$$

with Q_r the *Casimir invariant* of the representation r.

The classical Lie algebras are

- $SU(n)$: Traceless Hermitean $n \times n$ matrices. The corresponding group consists of unitary matrices with unit determinant.[1] This algebra is also denoted A_{n-1}.

- $SO(n)$: Antisymmetric Hermitean $n \times n$ matrices. The corresponding group $SO(n, \mathbf{R})$ consists of real orthogonal matrices with unit determinant. For $n = 2k$ this algebra is also denoted D_k. For $n = 2k + 1$ it is denoted B_k.

- $Sp(k)$: Hermitean $2k \times 2k$ matrices with the additional condition

$$MTM^{-1} = -T^T . \tag{11.4.8}$$

Here the superscript T denotes the transpose, and

$$M = i \begin{bmatrix} 0 & I_k \\ -I_k & 0 \end{bmatrix} \tag{11.4.9}$$

with I_k the $k \times k$ identity matrix. The corresponding group consists of unitary matrices U with the additional property

$$MUM^{-1} = (U^T)^{-1} . \tag{11.4.10}$$

---

[1] To be precise the Lie algebra determines only the local structure of the group. Many groups, differing only by discrete identifications, will have a common Lie algebra.

Confusingly, the notation $Sp(2k)$ is also used for this group. It is also denoted C_k.

From each of the compact groups one obtains various noncompact groups by multiplying some generators by i. For example, the traceless *imaginary* matrices generate the group $SL(n, \mathbf{R})$ of real matrices of unit determinant. The group $SO(m, n, \mathbf{R})$ preserving a Lorentzian inner product is similarly obtained from $SO(m+n)$. Another noncompact group is generated by imaginary rather than Hermitean matrices satisfying the symplectic condition (11.4.8) and consists of real matrices satisfying (11.4.10). This noncompact group is also denoted $Sp(k)$ or $Sp(2k)$; occasionally $USp(2k)$ is used to distinguish the compact unitary case.

Such noncompact groups do not appear in Yang–Mills theory (the result would not be unitary) but they have other applications. Some of the $SL(n, \mathbf{R})$ and $SO(m, n, \mathbf{R})$ appear as low energy symmetries in compactified string theory, as discussed in section B.5 and chapter 14. The real form of $Sp(k)$ is an invariance of the Poisson bracket in classical mechanics.

Roots and weights

A useful description of any Lie algebra h begins with a maximal set of commuting generators H^i, $i = 1, \ldots, \text{rank}(g)$. The remaining generators E^α can be taken to have definite charge under the H^i,

$$[H^i, E^\alpha] = \alpha^i E^\alpha . \tag{11.4.11}$$

The rank(g)-dimensional vectors α^i are known as *roots*. It is a theorem that there is only one generator for a given root so the notation E^α is unambiguous. The Jacobi identity determines the rest of the algebra to be

$$[E^\alpha, E^\beta] = \begin{cases} \epsilon(\alpha, \beta) E^{\alpha+\beta} & \text{if } \alpha + \beta \text{ is a root} , \\ 2\alpha \cdot H / \alpha^2 & \text{if } \alpha + \beta = 0 , \\ 0 & \text{otherwise} . \end{cases} \tag{11.4.12}$$

The products $\alpha \cdot H$ and α^2 are defined by contraction with d_{ij}, the inverse of the inner product (11.4.4) restricted to the commuting subalgebra. Taking the trace with H^i, the second equation determines the normalization $(E^\alpha, E^{-\alpha}) = 2/\alpha^2$. The constants $\epsilon(\alpha, \beta)$ take only the values ± 1.

The matrices t^i_r that represent H^i can all be taken to be diagonal. Their simultaneous eigenvalues w^i, combined into vectors

$$(w^1, \ldots, w^{\text{rank}(g)}) , \tag{11.4.13}$$

are the weights, equal in number to the dimension of the representation. The roots are the same as the weights of the adjoint representation.

Examples:

- For $SO(2k) = D_k$, consider the k 2×2 blocks down the diagonal and let H^i be

$$\begin{bmatrix} 0 & i \\ -i & 0 \end{bmatrix} \tag{11.4.14}$$

  in the ith block and zero elsewhere. This is a maximal commuting set. The $2k$-vectors $(1, \mp i, 0, \ldots, 0)$ have eigenvalues

$$(\pm 1, 0^{k-1}) \tag{11.4.15}$$

  under the k H^i; these are weights of the vector representation. The other weights are the same with the ± 1 in any other position.

  The adjoint representation is the antisymmetric tensor, which is contained in the product of two vector representations. The weights are additive so the roots are obtained by adding any distinct (because of the antisymmetry) pair of vector weights. This gives

$$(+1, +1, 0^{k-2}) , \ (+1, -1, 0^{k-2}) , \ (-1, -1, 0^{k-2}) , \tag{11.4.16}$$

  and all permutations of these. The k zero roots obtained by adding any weight and its negative are just the H^i.

  The diagonal generators (11.4.14) are the same as are used in section B.1 to construct the spinor representations. In the spinor representation the weights w^i are given by all k-vectors with components $\pm \frac{1}{2}$, with the $\mathbf{2^{k-1}}$ having an even number of $-\frac{1}{2}$s and the $\mathbf{2^{k-1'}}$ an odd number.

- For $SO(2k+1) = B_k$, one can take the same set of diagonal generators with a final row and a final column of zeros. The weights in the vector representation are the same as above plus (0^k) from the added row. The additional generators have roots

$$(\pm 1, 0^{k-1}) \tag{11.4.17}$$

  and all permutations.

- The adjoint of $Sp(k) = C_k$ is the *symmetric* tensor, so one can obtain the roots as for $SO(2k)$ except that the vector weights need not be distinct. The resulting roots are those of $SO(2k)$ together with

$$(\pm 2, 0^{k-1}) \tag{11.4.18}$$

  and permutations. It is usually conventional to normalize the generators such that the longest root has length-squared two, so we must divide all the roots by $2^{1/2}$.

- For $SU(n) = A_{n-1}$ it is useful first to consider $U(n)$, even though this algebra is not simple. The n commuting generators H^i can be taken to have a 1 in the ii position and zeros elsewhere. The charged generator with a 1 in the ij position then has eigenvalue $+1$ under H^i and -1 under H^j: the roots are all permutations of

$$(+1, -1, 0^{n-2}) . \tag{11.4.19}$$

Note that all roots lie in the hyperplane $\sum_i \alpha^i = 0$; this is because all eigenvalues of the $U(1)$ generator are zero. The roots of $SU(n)$ are just the roots of $U(n)$ regarded as lying in this hyperplane.

- We have stated that the E_8 generators decompose into the adjoint plus one spinor of $SO(16)$. The commuting generators of $SO(16)$ can also be taken as commuting generators of E_8, so the roots of E_8 are the roots of $SO(16)$ plus the weights of the spinor, namely all permutations of the roots (11.4.16) plus

$$(+\tfrac{1}{2}, +\tfrac{1}{2}, +\tfrac{1}{2}, +\tfrac{1}{2}, +\tfrac{1}{2}, +\tfrac{1}{2}, +\tfrac{1}{2}, +\tfrac{1}{2}) \tag{11.4.20}$$

and the roots obtained from this by an even number of sign flips. Equivalently this set is described by

$$\alpha^i \in \mathbf{Z} \text{ for all } i , \text{ or } \alpha^i \in \mathbf{Z} + \tfrac{1}{2} \text{ for all } i , \tag{11.4.21a}$$

and

$$\sum_{i=1}^{8} \alpha^i \in 2\mathbf{Z} , \qquad \sum_{i=1}^{8}(\alpha^i)^2 = 2 . \tag{11.4.21b}$$

For A_n, D_k, and E_8 (and E_6 and E_7, which we have not yet described), all roots are of the same length. These are referred to as *simply-laced* algebras. For B_k and C_k (and F_4 and G_2) there are roots of two different lengths so one refers to long and short roots.

A quantity that will be useful later is the *dual Coxeter number* $h(g)$ of the Lie algebra g, defined by

$$-\sum_{c,d} f^{ac}{}_d f^{bd}{}_c = h(g)\psi^2 d^{ab} . \tag{11.4.22}$$

Here ψ is any long root. For reference, we give the values for all simple Lie algebras in table 11.3. The definition (11.4.22) makes $h(g)$ independent of the arbitrary normalization of the inner product d^{ab} because the inverse appears in $\psi^2 = \psi^i \psi^j d_{ij}$.

Useful facts for grand unification

The exceptional group E_8 is connected to the groups appearing in grand unification through a series of subgroups. This will play a role in the com-

Table 11.3. *Dimensions and Coxeter numbers for simple Lie algebras.*

	$SU(n)$	$SO(n), n \geq 4$	$Sp(k)$	E_6	E_7	E_8	F_4	G_2
$\dim(g)$	$n^2 - 1$	$n(n-1)/2$	$2k^2 + k$	78	133	248	52	14
$h(g)$	n	$n - 2$	$k + 1$	12	18	30	9	4

pactification of the heterotic string, and so we record without derivation the necessary results.

The first subgroup is

$$E_8 \rightarrow SU(3) \times E_6 . \tag{11.4.23}$$

We have not described E_6 explicitly, but the reader can reproduce this and the decomposition (11.4.24) from the known properties of spinor representations, as well as the further decomposition of the E_6 representations in table 11.4 (exercise 11.5). In simple compactifications of the $E_8 \times E_8$ string, the fermions of the Standard Model can all be thought of as arising from the **248**-dimensional adjoint representation of one of the E_8s. It is therefore interesting to trace the fate of this representation under the successive symmetry breakings. Under $E_8 \rightarrow SU(3) \times E_6$,

$$\mathbf{248} \rightarrow (\mathbf{8},\mathbf{1}) + (\mathbf{1},\mathbf{78}) + (\mathbf{3},\mathbf{27}) + (\bar{\mathbf{3}},\overline{\mathbf{27}}) . \tag{11.4.24}$$

That is, the adjoint of E_8 contains the adjoints of the subgroups, with half the remaining 162 generators transforming as a triplet of $SU(3)$ and a complex **27**-dimensional representation of E_6 and half as the conjugate of this. Further subgroups are shown in table 11.4. The first three subgroups correspond to successive breaking of E_6 down to the Standard Model group through smaller grand unified groups; the fourth is an alternate breaking pattern.

It is a familiar fact from grand unification that precisely one $SU(3) \times SU(2) \times U(1)$ generation of quarks and leptons is contained in the **10** plus $\bar{\mathbf{5}}$ of $SU(5)$. Tracing back further, we see that a generation fits into the single representation **16** of $SO(10)$, together with an additional state $\mathbf{1}_{-5}$. This extra state is neutral under $SU(5)$, and so under $SU(3) \times SU(2) \times U(1)$, and can be regarded as a right-handed neutrino. Going back to E_6, the **27** contains the 15 states of a single generation plus 12 additional states. Relative to $SU(5)$ unification, $SO(10)$ and E_6 are more unified in the sense that a generation is contained within a single representation, but less economical in that the representation contains additional unseen states as well. In fact, the latter may not be such a

Table 11.4. *Subgroups and representations of grand unified groups.*

$E_6 \rightarrow SO(10) \times U(1)$

$\mathbf{78} \rightarrow \mathbf{45}_0 + \mathbf{16}_{-3} + \overline{\mathbf{16}}_3 + \mathbf{1}_0$

$\mathbf{27} \rightarrow \mathbf{1}_4 + \mathbf{10}_{-2} + \mathbf{16}_1$

$SO(10) \rightarrow SU(5) \times U(1)$

$\mathbf{45} \rightarrow \mathbf{24}_0 + \mathbf{10}_4 + \overline{\mathbf{10}}_{-4} + \mathbf{1}_0$

$\mathbf{16} \rightarrow \mathbf{10}_{-1} + \bar{\mathbf{5}}_3 + \mathbf{1}_{-5}$

$\mathbf{10} \rightarrow \mathbf{5}_2 + \bar{\mathbf{5}}_{-2}$

$SU(5) \rightarrow SU(3) \times SU(2) \times U(1)$

$\mathbf{10} \rightarrow (\mathbf{3}, \mathbf{2})_1 + (\bar{\mathbf{3}}, \mathbf{1})_{-4} + (\mathbf{1}, \mathbf{1})_6$

$\bar{\mathbf{5}} \rightarrow (\bar{\mathbf{3}}, \mathbf{1})_2 + (\mathbf{1}, \mathbf{2})_{-3}$

$E_6 \rightarrow SU(3) \times SU(3) \times SU(3)$

$\mathbf{78} \rightarrow (\mathbf{8}, \mathbf{1}, \mathbf{1}) + (\mathbf{1}, \mathbf{8}, \mathbf{1}) + (\mathbf{1}, \mathbf{1}, \mathbf{8}) + (\mathbf{3}, \mathbf{3}, \mathbf{3}) + (\bar{\mathbf{3}}, \bar{\mathbf{3}}, \bar{\mathbf{3}})$

$\mathbf{27} \rightarrow (\mathbf{3}, \bar{\mathbf{3}}, \mathbf{1}) + (\mathbf{1}, \mathbf{3}, \bar{\mathbf{3}}) + (\bar{\mathbf{3}}, \mathbf{1}, \mathbf{3})$

difficulty. To see why, consider the decomposition of the **27** of E_6 under $SU(3) \times SU(2) \times U(1) \subset SU(5) \subset SO(10) \subset E_6$:

$$
\begin{aligned}
\mathbf{27} \rightarrow\ & (\mathbf{3}, \mathbf{2})_1 + (\bar{\mathbf{3}}, \mathbf{1})_{-4} + (\mathbf{1}, \mathbf{1})_6 + (\bar{\mathbf{3}}, \mathbf{1})_2 + (\mathbf{1}, \mathbf{2})_{-3} \\
& + [\mathbf{1}_0] \\
& + [(\bar{\mathbf{3}}, \mathbf{1})_2 + (\mathbf{3}, \mathbf{1})_{-2}] + [(\mathbf{1}, \mathbf{2})_{-3} + (\mathbf{1}, \mathbf{2})_3] + [\mathbf{1}_0] .
\end{aligned}
\tag{11.4.25}
$$

The first line lists one generation, the second the extra state appearing in the **16** of $SO(10)$, and the third the additional states in the **27** of E_6. The subset within each pair of square brackets is a real representation of $SU(3) \times SU(2) \times U(1)$. The significance of this is that for a real representation r, the CPT conjugate also is in the representation r, and so the combined gauge plus $SO(2)$ helicity representation for the particles plus their antiparticles is $(r, +\frac{1}{2}) + (r, -\frac{1}{2})$. This is the same as for a *massive* spin-$\frac{1}{2}$ particle in representation r, so it is consistent with the gauge and spacetime symmetries for these particles to be massive. In the most general invariant action, all particles in [ ] brackets will have large (of order the unification scale) masses. It is notable that for any of the $\mathbf{10} + \bar{\mathbf{5}}$ of $SU(5)$, the **16** of $SO(10)$, or the **27** of E_6, the natural $SU(3) \times SU(2) \times U(1)$ spectrum is precisely a standard generation of quarks and leptons.

11.5   Current algebras

The gauge boson vertex operators in the heterotic string are of the form $j(z)\bar{\psi}^\mu(\bar{z})e^{ik\cdot X}$, where $j(z)$ is either a fermion bilinear $\lambda^A\lambda^B$ or a spin field (11.2.26). Similarly the gauge boson vertex operators for the toroidally compactified bosonic string were of the form $j(z)\bar{\partial}X^\mu(\bar{z})e^{ik\cdot X}$ with j being ∂X^m for the Kaluza–Klein gauge bosons or an exponential for the enhanced gauge symmetry (or the same with right and left reversed). All these currents are holomorphic $(1,0)$ operators. In this section we consider general properties of such currents.

Let us consider in a general CFT the set of $(1,0)$ currents $j^a(z)$. Conformal invariance requires their OPE to be of the form

$$j^a(z)j^b(0) \sim \frac{k^{ab}}{z^2} + \frac{ic^{ab}{}_c}{z}j^c(0) \tag{11.5.1}$$

with k^{ab} and $c^{ab}{}_c$ constants. Dimensionally, the z^{-2} term must be a c-number and the z^{-1} term must be proportional to a current. The Laurent coefficients

$$j^a(z) = \sum_{m=-\infty}^{\infty} \frac{j_m^a}{z^{m+1}} \tag{11.5.2}$$

thus satisfy a closed algebra

$$[j_m^a, j_n^b] = mk^{ab}\delta_{m,-n} + ic^{ab}{}_c j_{m+n}^c . \tag{11.5.3}$$

In particular,

$$[j_0^a, j_0^b] = ic^{ab}{}_c j_0^c . \tag{11.5.4}$$

That is, the $m=0$ modes form a Lie algebra g, and

$$c^{ab}{}_c = f^{ab}{}_c . \tag{11.5.5}$$

We focus first on the case of simple g. The $j_1^a\, j_0^b\, j_{-1}^c$ Jacobi identity requires that

$$f^{bc}{}_d k^{ad} + f^{ba}{}_d k^{dc} = 0 . \tag{11.5.6}$$

This is the same relation as that defining the Lie algebra inner product d^{ab}, and since we are assuming g to be simple it must be that

$$k^{ab} = \hat{k}d^{ab} \tag{11.5.7}$$

for some constant $\hat{k}$. The algebra (11.5.3) is variously known as a *current algebra*, an *affine Lie algebra*, or an *(affine) Kac–Moody algebra*. The currents are $(1,0)$ tensors, so

$$[L_m, j_n^a] = -nj_{m+n}^a . \tag{11.5.8}$$

Physically, the j_n^a generate position-dependent g-transformations. This is possible in quantum field theory because there is a local current. The *central extension* or *Schwinger term* $\hat{k}$ must always be positive in a unitary theory. To show this, note that

$$\hat{k}d^{aa} = \langle 1| \, [\, j_1^a, j_{-1}^a \,] \, |1\rangle = \| \, j_{-1}^a|1\rangle \|^2 \qquad (11.5.9)$$

(no sum on a). For a compact Lie algebra d^{aa} is positive and so $\hat{k}$ must be nonnegative. It can vanish only if $j_{-1}^a|1\rangle = 0$, but the vertex operator for $j_{-1}^a|1\rangle$ is precisely the current j^a: any matrix element of j^a can be obtained by gluing $j_{-1}^a|1\rangle$ into the world-sheet. Thus $\hat{k} = 0$ only if the current vanishes identically.

The coefficient $\hat{k}$ is quantized. To show this, consider any root α. Defining

$$J^3 = \frac{\alpha \cdot H}{\alpha^2} \, , \qquad J^{\pm} = E^{\pm\alpha} \, , \qquad (11.5.10)$$

one finds from the general form (11.4.12) that these satisfy the $SU(2)$ algebra

$$[J^3, J^{\pm}] = J^{\pm} \, , \qquad [J^+, J^-] = 2J^3 \, . \qquad (11.5.11)$$

The reader can verify that the two sets

$$\frac{\alpha \cdot H_0}{\alpha^2} \, , \qquad E_0^{\alpha} \, , \qquad E_0^{-\alpha} \, , \qquad (11.5.12a)$$

$$\frac{\alpha \cdot H_0 + \hat{k}}{\alpha^2} \, , \qquad E_1^{\alpha} \, , \qquad E_{-1}^{-\alpha} \qquad (11.5.12b)$$

also satisfy the $SU(2)$ algebra. The first is just the usual center-of-mass Lie algebra, while the second is known as *pseudospin*. From familiar properties of $SU(2)$, $2J_3$ must be an integer, and so $2\hat{k}/\alpha^2$ must be an integer. This condition is most stringent if α is taken to be one of the long roots of the algebra (denoted ψ). The *level*

$$k = \frac{2\hat{k}}{\psi^2} \qquad (11.5.13)$$

is then a nonnegative integer, and positive for a nontrivial current.

It is common to normalize the Lie algebra inner product to give the long roots length-squared two, so that $\hat{k} = k$ is the coefficient of the leading term in the OPE. We will usually do this in examples, as we have done in giving the roots of various Lie algebras in the previous section. Incidentally, it follows that with this normalization the generators (11.4.14) are normalized, so the $SO(n)$ inner product is half of the vector representation trace. Similarly the inner product for $SU(n)$ such that the long roots have length-squared two is equal to the trace in the fundamental representation. In general expressions we will keep the inner product arbitrary, inserting explicit factors of ψ^2 so that results are independent of the normalization.

We will, however, take henceforth a basis for the generators such that $d^{ab} = \delta^{ab}$.

The level represents the relative magnitude of the z^{-2} and z^{-1} terms in the OPE. For $U(1)$ the structure constant is zero and only the z^{-2} term appears. Hence there is no analog of the level. It is convenient to normalize all the $U(1)$ currents to

$$ j^a(z)j^b(0) \sim \frac{\delta^{ab}}{z^2} \ . \tag{11.5.14} $$

From this OPE and holomorphicity it follows that each $U(1)$ current algebra is isomorphic to a free boson CFT,

$$ j^a = i\partial H^a \ . \tag{11.5.15} $$

We will often use this equivalence.

The current algebra in the heterotic string consisted of n real fermions $\lambda^A(z)$. The currents

$$ i\lambda^A \lambda^B \tag{11.5.16} $$

form an $SO(n)$ algebra. The maximal set of commuting currents is $i\lambda^{2K-1}\lambda^{2K}$ for $K = 1,\ldots,[n/2]$. These correspond to the generators (11.4.14), which are normalized such that roots (11.4.16) have length-squared two. The level is then the coefficient of the leading term in the OPE; this is $1/z^2$, so the level is $k = 1$. The case $n = 3$ is an exception: there are no long roots, only the short roots ± 1, so we must rescale the diagonal current to $2^{1/2}i\lambda^1\lambda^2$ and the level is $k = 2$.

For any real representation r of any Lie algebra, one can construct from $\dim(r)$ real fermions the currents

$$ \frac{i}{2}\lambda^A \lambda^B t^a_{r,AB} \ . \tag{11.5.17} $$

These satisfy a current algebra with level $k = T_r/\psi^2$, with T_r defined in eq. (11.4.6). The case in the previous paragraph is the n-dimensional vector representation of $SO(n)$, for which $T_R = \psi^2$. As another example, nk fermions transforming as k copies of the vector representation give level k.

As a final example consider the $SU(2)$ symmetry at the self-dual point of toroidal compactification. The current is $\exp[2^{1/2}iH(z)]$. The current $i\partial H$ is then normalized so that the weight (from the OPE) is $2^{1/2}$, with length-squared two. The OPE of $i\partial H$ with itself starts as $1/z^2$, so the level is again $k = 1$.

In some cases one may have sectors in which some currents are not periodic, $j^a(w + 2\pi) = R^{ab} j^b(w)$, where R^{ab} is any automorphism of the algebra. In these, the modes of the currents are fractional and satisfy a *twisted* affine Lie algebra.

The Sugawara construction

In current algebras with conformal symmetry, there is a remarkable connection between the energy-momentum tensor and the currents, which leads to a great deal of interesting structure. Define the operator

$$:jj(z_1) := \lim_{z_2 \to z_1} \left(j^a(z_1)j^a(z_2) - \frac{\hat{k}\,\dim(g)}{z_{12}^2} \right) , \tag{11.5.18}$$

with the sum on a implicit. We first wish to show that up to normalization the OPE of $:jj:$ with j^a is the same as that of T_B with j^a. This takes a bit of effort; the same calculation is organized in a different way in exercise 11.7.

The OPE of the product $:jj:$ is not the same as the product of the OPEs, because the two currents in $:jj:$ are closer to each other than they are to the third current; we must make a less direct argument using holomorphicity. Consider the following product:

$$j^a(z_1)j^a(z_2)j^c(z_3) = \frac{\hat{k}}{z_{31}^2} j^c(z_2) + \frac{if^{cad}}{z_{31}} j^d(z_1)j^a(z_2) + \frac{\hat{k}}{z_{32}^2} j^c(z_1)$$

$$+ \frac{if^{cad}}{z_{32}} j^a(z_1)j^d(z_2) + \text{terms holomorphic in } z_3 . \tag{11.5.19}$$

We have used the current–current OPE to determine the singularities as z_3 approaches z_1 or z_2, with a holomorphic remainder. In this relation take $z_2 \to z_1$ and make a Laurent expansion in z_{21}, being careful to expand both the operator products and the explicit z_2 dependence. Keep the term of order z_{21}^0 (there is some cancellation from the antisymmetry of f^{cad}) to obtain

$$:jj(z_1): j^c(z_3) \sim \frac{2\hat{k}}{z_{13}^2} j^c(z_1) + \frac{f^{cad}f^{ead}}{z_{13}^2} j^e(z_1)$$

$$= \frac{2\hat{k} + h(g)\psi^2}{z_{13}^2} j^c(z_1)$$

$$= (k + h(g))\psi^2 \left[\frac{1}{z_{13}^2} j^c(z_3) + \frac{1}{z_{13}} \partial j^c(z_3) \right] . \tag{11.5.20}$$

Here $h(g)$ is again the dual Coxeter number. Define

$$T_B^{\rm s}(z) = \frac{1}{(k + h(g))\psi^2} :jj(z): . \tag{11.5.21}$$

The OPE of $T_B^{\rm s}$ with the current is the same as that of the energy-momentum tensor $T_B(z)$,

$$T_B^{\rm s}(z)j^c(0) \sim T_B(z)j^c(0) . \tag{11.5.22}$$

Now repeat the above with $j^c(z_3)$ replaced by $T_B^s(z_3)$,

$$j^a(z_1)j^a(z_2)T_B^s(z_3) = \frac{1}{z_{31}^2}j^a(z_1)j^a(z_2) + \frac{1}{z_{31}}\partial j^a(z_1)j^a(z_2)$$

$$+ \frac{1}{z_{32}^2}j^a(z_1)j^a(z_2) + \frac{1}{z_{32}}j^a(z_1)\partial j^a(z_2) + \text{terms holomorphic in } z_3 \,.$$
$$\text{(11.5.23)}$$

Again expand in z_{21} and keep the term of order z_{21}^0 to obtain

$$T_B^s(z_1)T_B^s(z_3) \sim \frac{c^{g,k}}{2z_{13}^4} + \frac{2}{z_{13}^2}T_B^s(z_3) + \frac{1}{z_{13}}\partial T_B^s(z_3) \qquad \text{(11.5.24)}$$

with

$$c^{g,k} = \frac{k \dim(g)}{k + h(g)} \,. \qquad \text{(11.5.25)}$$

This is of the standard form for an energy-momentum tensor, with central charge $c^{g,k}$. The Laurent coefficients

$$L_0^s = \frac{1}{(k + h(g))\psi^2}\left(j_0^a j_0^a + 2\sum_{n=1}^{\infty} j_{-n}^a j_n^a\right) , \qquad \text{(11.5.26a)}$$

$$L_m^s = \frac{1}{(k + h(g))\psi^2}\sum_{n=-\infty}^{\infty} j_n^a j_{m-n}^a \,, \quad m \neq 0 \,, \qquad \text{(11.5.26b)}$$

satisfy a Virasoro algebra with this central charge. The vanishing of the normal ordering constant in L_0^s can be deduced by noting that holomorphicity requires L_0^s and also j_n^a for $n \geq 0$ to annihilate the state $|1\rangle$.

We have used the jj OPE to determine the $: jj :: jj :$ OPE. We could not do this directly, because the jj OPE is valid only for two operators close compared to all others, and in this case there are two additional currents in the vicinity. Naive application of the OPE would give the wrong normalization for T^s and $c^{g,k}$. The argument above uses the OPE only where it is valid, and then takes advantage of holomorphicity. The operator T_B^s constructed from the product of two currents is known as a *Sugawara energy-momentum tensor*.

Finding the Sugawara tensor for a $U(1)$ current algebra is easy. With the normalization (11.5.14) it is simply

$$T_B^s = \frac{1}{2} :jj: \,, \qquad \text{(11.5.27)}$$

as one sees by writing the current in terms of a free boson, $j = i\partial H$.

The tensor T_B^s may or may not be equal to the total T_B of the CFT. Define

$$T_B' = T_B - T_B^s \,. \qquad \text{(11.5.28)}$$

Since the T_B and T_B^s OPEs with j^a have the same singular terms, the product

$$T_B'(z_1)j^a(z_2) \sim 0 \qquad (11.5.29)$$

is nonsingular. Since T_B^s itself is constructed from the currents, this implies $T_B^s T_B' \sim 0$. Then

$$T_B'(z)T_B'(0) = T_B(z)T_B(0) - T_B^s(z)T_B^s(0) - T_B'(z)T_B^s(0) - T_B^s(z)T_B'(0)$$

$$\sim \frac{c'}{2z^4} + \frac{2}{z^2}T_B'(0) + \frac{1}{z}\partial T_B'(0) , \qquad (11.5.30)$$

the standard TT OPE with central charge

$$c' = c - c^{g,k} . \qquad (11.5.31)$$

The internal theory thus separates into two decoupled CFTs. One has an energy-momentum tensor T_B^s constructed entirely from the current, and the other an energy-momentum tensor T_B' that commutes with the current. We will use the term *current algebra* to refer to the first factor alone, since the two CFTs are completely independent. For a unitary CFT c' must be nonnegative and so

$$c^{g,k} \leq c , \qquad (11.5.32)$$

and T_B' is trivial precisely if

$$c^{g,k} = c , \qquad (11.5.33)$$

in which case $T_B = T_B^s$.

We now consider examples. The dual Coxeter number can be written as a sum over the roots. For any simply-laced algebra, $h(g) + 1 = \dim(g)/\mathrm{rank}(g)$, and so

$$c^{g,k} = \frac{k \dim(g) \, \mathrm{rank}(g)}{\dim(g) + (k-1)\mathrm{rank}(g)} . \qquad (11.5.34)$$

For any simply-laced algebra at $k = 1$, the central charge is therefore

$$c^{g,1} = \mathrm{rank}(g) . \qquad (11.5.35)$$

For the $E_8 \times E_8$ and $SO(32)$ heterotic strings, this is the same as the central charge of the free fermion representation, and for the free boson representation of the next section: these are Sugawara theories. The operator $:jj:$ looks as though it should be quartic in the fermions, but by using the OPE and the antisymmetry of the fermions one finds that T_B^s reduces to the usual $-\frac{1}{2}\lambda^A\partial\lambda^A$.

Another example is $SU(2) = SO(3)$, for which

$$c^{g,k} = \frac{3k}{2+k} = 1, \frac{3}{2}, \frac{9}{5}, 2, \frac{15}{7}, \ldots \to 3 . \qquad (11.5.36)$$

We have seen the first CFT in this series (the self-dual point of toroidal compactification) and the second (free fermions). Most levels do not have a free-field representation. For any current algebra the central charge lies in the range

$$\text{rank}(g) \le c^{g,k} \le \dim(g) \ . \tag{11.5.37}$$

The first equality holds only for a simply-laced algebra at level one, and the second only for an Abelian algebra or in the limit $k \to \infty$.

Primary fields

By acting repeatedly with the lowering operators j_n^a with $n > 0$, one reaches a *highest weight* or *primary* state of the current algebra, a state annihilated by all the j_n^a for $n > 0$. It is therefore also annihilated by the L_n^s for $n > 0$, eq. (11.5.26), so is a highest weight state of the Virasoro algebra. The center-of-mass generators j_0^a take primary states into primary states, so the latter form a representation of the algebra g,

$$j_0^a |r, i\rangle = |r, j\rangle t_{r,ji}^a \ , \tag{11.5.38}$$

with r (not summed) labeling the representation. It then follows that

$$L_0^s |r, i\rangle = \frac{1}{(k + h(g))\psi^2} |r, k\rangle t_{r,kj}^a t_{r,ji}^a$$

$$= \frac{Q_r}{(k + h(g))\psi^2} |r, i\rangle \ , \tag{11.5.39}$$

with Q_r the Casimir (11.4.7). The weights of the primary fields are thus determined in terms of the algebra, level, and representation,

$$h_r = \frac{Q_r}{(k + h(g))\psi^2} = \frac{Q_r}{2\hat{k} + Q_g} \ , \tag{11.5.40}$$

where Q_g is the Casimir for the adjoint representation. For $SU(2)$ at level k, the weight of the spin-j primary is

$$h_j = \frac{j(j+1)}{k+2} \ . \tag{11.5.41}$$

It is also true that at any given level, only a finite number of representations are possible for the primary states. For any root α of g and any weight λ of r, the $SU(2)$ algebra (11.5.12b) implies that

$$\langle r, \lambda | [E_1^\alpha, E_{-1}^{-\alpha}] |r, \lambda\rangle = 2 \langle r, \lambda|(\alpha \cdot H_0 + \hat{k})|r, \lambda\rangle /\alpha^2$$

$$= 2(\alpha \cdot \lambda + \hat{k})/\alpha^2 \ . \tag{11.5.42}$$

The left-hand side is $\|E_{-1}^{-\alpha}|r, \lambda\rangle\|^2 \ge 0$, and so $\hat{k} \ge -\alpha \cdot \lambda$. Combining this with the same for $-\alpha$ gives

$$\hat{k} \ge |\alpha \cdot \lambda| \tag{11.5.43}$$

for all weights λ of r. Taking α to be a long root ψ, the level must satisfy

$$k \geq \frac{2|\psi \cdot \lambda|}{\psi^2} = 2|J^3| \,, \tag{11.5.44}$$

where J^3 refers to the $SU(2)$ algebra (11.5.12a) constructed from the charges j_0^a and the root ψ. For $g = SU(2)$ the statement is that the spin j of any primary state can be at most $\frac{1}{2}k$. For example at $k = 1$, only the representations **1** and **2** are possible. For $g = SU(3)$ at $k = 1$, only the **1**, **3**, and $\bar{\mathbf{3}}$ can appear. For $g = SU(n)$ at level k, only representations whose Young tableau has k or fewer columns can appear.

The expectation values of primary fields are completely determined by symmetry. We defer the details to chapter 15.

Finally, let us briefly discuss the gauge symmetries of the type I theory in this same abstract language. The matter part of the gauge boson vertex operator is

$$\dot{X}^\mu \lambda^a e^{ik \cdot X} \tag{11.5.45}$$

on the boundary, where the λ^a are weight 0 fields. In a unitary CFT such λ^a must be constant by the equations of motion. The OPE is then

$$\lambda^a(y_1)\lambda^b(y_2) = \left[\theta(y_1 - y_2)d^{ab}{}_c + \theta(y_2 - y_1)d^{ba}{}_c \right] \lambda^c(y_2) \,, \tag{11.5.46}$$

so the λ^a form a multiplicative algebra with structure constants $d^{ab}{}_c$. The antisymmetric part of $d^{ab}{}_c$ is the structure constant of the gauge Lie algebra. This is an abstract description of the Chan–Paton factor. The requirement that the λ^a algebra be associative has been shown to forbid the gauge group $E_8 \times E_8$.

11.6 The bosonic construction and toroidal compactification

We have seen in the construction of winding state vertex operators in section 8.2 that we may consider independent left- and right-moving scalars. Let us try to construct a heterotic string with 26 left-movers and 10 right-movers, which together with the ψ^μ give the correct central charge. The main issue is the spectrum of $k_{L,R}$; as in section 8.4 we use dimensionless momenta

$$l_{L,R} = (\alpha'/2)^{1/2} k_{L,R} \tag{11.6.1}$$

in much of the discussion. Recall that an ordinary noncompact dimension corresponds to a left- plus a right-mover with $l_L^\mu = l_R^\mu = l^\mu$ taking continuous values; let there be $d \leq 10$ noncompact dimensions. The remaining momenta,

$$(l_L^m, l_R^n) \,, \quad d \leq m \leq 25 \,, \, d \leq n \leq 9 \,, \tag{11.6.2}$$

take values in some lattice Γ. From the discussion of Narain compactification in section 8.4, we know that the requirements for a consistent CFT are locality of the OPE plus modular invariance. After taking the GSO projection on the right-movers, the conditions on Γ are precisely as in the bosonic case. Defining the product

$$l \circ l' = l_L \cdot l'_L - l_R \cdot l'_R \,, \qquad (11.6.3)$$

the lattice must be an *even self-dual Lorentzian lattice of signature* $(26 - d, 10 - d)$,

$$l \circ l \in 2\mathbf{Z} \quad \text{for all } l \in \Gamma \,, \qquad (11.6.4a)$$

$$\Gamma = \Gamma^* \,. \qquad (11.6.4b)$$

As in the bosonic case, where the signature was $(26 - d, 26 - d)$, all such lattices have been classified. Consider first the maximum possible number of noncompact dimensions, $d = 10$. In this case, the $\circ$ product has only positive signs, so the l_L^m form an even self-dual *Euclidean* lattice of dimension 16. Even self-dual Euclidean lattices exist only when the dimension is a multiple of 8, and for dimension 16 there are exactly two such lattices, Γ_{16} and $\Gamma_8 \times \Gamma_8$. The lattice Γ_{16} is the set of all points of the form

$$(n_1, \ldots, n_{16}) \quad \text{or} \quad (n_1 + \tfrac{1}{2}, \ldots, n_{16} + \tfrac{1}{2}) \,, \qquad (11.6.5a)$$

$$\sum_i n_i \in 2\mathbf{Z} \qquad (11.6.5b)$$

for any integers n_i. The lattice Γ_8 is similarly defined to be all points

$$(n_1, \ldots, n_8) \quad \text{or} \quad (n_1 + \tfrac{1}{2}, \ldots, n_8 + \tfrac{1}{2}) \,, \qquad (11.6.6a)$$

$$\sum_i n_i \in 2\mathbf{Z} \,. \qquad (11.6.6b)$$

The left-moving zero-point energy is -1 as in the bosonic string, so the massless states would have left-moving vertex operators ∂X^μ, ∂X^m, or $e^{ik_L \cdot X(z)}$ with $l_L^2 = 2$. Tensored with the usual right-moving $\mathbf{8}_v + \mathbf{8}$, the first gives the usual graviton, dilaton, and antisymmetric tensor. The 16 ∂X^m currents form a maximal commuting set corresponding to the m-momenta. The momenta l_L^m are the charges under these and so are the roots of the gauge group. For Γ_{16}, the points of length-squared two are just the $SO(32)$ roots (11.4.16). For Γ_8 the points of length-squared two are the E_8 roots (11.4.21). Thus the two possible lattices give the same two gauge groups, $SO(32)$ and $E_8 \times E_8$, found earlier. The commuting currents have singularity $1/z^2$, so $k = 1$ again.

It is easy to see that the earlier fermionic construction and the present bosonic one are equivalent under bosonization. The integral points on the lattices (11.6.5) and (11.6.6) map to the NS sectors of the current algebra and the half-integral points to the R sectors. The constraint that the total

k_L^m be even is the GSO projection on the left-movers in each theory. We have seen in the previous section that the dynamics of a current algebra is completely determined by its symmetry, so we can give a representation-independent description of the left-movers as an $SO(32)$ or $E_8 \times E_8$ level one current algebra.[2]

Let us note some general results about Lie algebras and lattices. The set of all integer linear combinations of the roots of a Lie algebra g is known as the *root lattice* Γ_g of g. Now take any representation r and let λ be any weight of r. The set of points $\lambda + v$ for all $v \in \Gamma_g$ is denoted Γ_r. It can be shown by considering various $SU(2)$ subgroups that for a simply-laced Lie algebra with roots of length-squared two,

$$\Gamma_r \subset \Gamma_g^* . \tag{11.6.7}$$

The union of all Γ_r is the weight lattice Γ_w, and[3]

$$\Gamma_w = \Gamma_g^* . \tag{11.6.8}$$

For example, the weight lattice of $SO(2n)$ has four sublattices:

$$(0) : \quad 0 + \text{any root} ; \tag{11.6.9a}$$

$$(v) : \quad (1, 0, 0, \ldots, 0) + \text{any root} ; \tag{11.6.9b}$$

$$(s) : \quad (\tfrac{1}{2}, \tfrac{1}{2}, \tfrac{1}{2}, \ldots, \tfrac{1}{2}) + \text{any root} ; \tag{11.6.9c}$$

$$(c) : \quad (-\tfrac{1}{2}, \tfrac{1}{2}, \tfrac{1}{2}, \ldots, \tfrac{1}{2}) + \text{any root} . \tag{11.6.9d}$$

These are respectively the root lattice, the lattice containing the weights of the vector representation, and the lattices containing the weights of the two 2^{n-1}-dimensional spinor representations. The lattice Γ_8 is the root lattice of E_8 and is also the weight lattice because it is self-dual. The root lattice of $SO(32)$ gives the integer points in Γ_{16}. The full Γ_{16} is the root lattice plus one spinor lattice of $SO(32)$.

The level one current algebra for any simply-laced Lie algebra g can similarly be represented by rank(g) left-moving bosons, the momentum lattice being the root lattice of g with the roots scaled to length-squared two. The constants $\epsilon(\alpha, \beta)$ appearing in the Lie algebra (11.4.12) can then be determined from the vertex operator OPE; this is one situation where the explicit form of the cocycle is needed. A modular-invariant CFT can be obtained by taking also rank(g) right-moving bosons, with the momentum lattice being

$$\Gamma = \sum_r \Gamma_r \times \tilde{\Gamma}_r . \tag{11.6.10}$$

---

[2] To be precise it is still necessary to specify the spectrum, which amounts to specifying which primary fields appear. Modular invariance generally restricts the possibilities greatly.

[3] For the nonsimply-laced algebras $Sp(k)$ and $SO(2k + 1)$, these same relations hold between the weight lattice of one and the dual of the root lattice of the *other*.

That is, the spectrum runs over all sublattices of the weight lattice, with the left- and right-moving momenta taking values in the same sublattice.

Toroidal compactification

In parallel to the bosonic case, all even self-dual lattices of signature $(26-d, 10-d)$ can be obtained from any single lattice by $O(26-d, 10-d, \mathbf{R})$ transformations. Again start with any given solution Γ_0; for example, this could be either of the ten-dimensional theories with all compact dimensions orthogonal and at the $SU(2) \times SU(2)$ radius. Then any lattice

$$\Gamma = \Lambda \Gamma_0 , \quad \Lambda \in O(26 - d, 10 - d, \mathbf{R}) \qquad (11.6.11)$$

defines a consistent heterotic string theory. As in the bosonic case there is an equivalence

$$\Lambda_1 \Lambda \Lambda_2 \Gamma_0 \cong \Lambda \Gamma_0 , \qquad (11.6.12)$$

where

$$\Lambda_1 \in O(26 - d, \mathbf{R}) \times O(10 - d, \mathbf{R}) , \quad \Lambda_2 \in O(26 - d, 10 - d, \mathbf{Z}) . \quad (11.6.13)$$

The moduli space is then

$$\frac{O(26 - d, 10 - d, \mathbf{R})}{O(26 - d, \mathbf{R}) \times O(10 - d, \mathbf{R}) \times O(26 - d, 10 - d, \mathbf{Z})} . \qquad (11.6.14)$$

The discrete T-duality group $O(26 - d, 10 - d, \mathbf{Z})$ of invariances of Γ_0 is understood to act on the right.

Now consider the unbroken gauge symmetry. There are $26 - d$ gauge bosons with vertex operators $\partial X^m \tilde{\psi}^\mu$ and $10 - d$ with vertex operators $\partial X^\mu \tilde{\psi}^m$. These are the original 16 commuting symmetries of the ten-dimensional theory plus $10 - d$ Kaluza–Klein gauge bosons and $10 - d$ more from compactification of the antisymmetric tensor. In addition there are gauge bosons $e^{i k_L \cdot X_L} \tilde{\psi}^\mu$ for every point on the lattice Γ such that

$$l_L^2 = 2 , \quad l_R = 0 . \qquad (11.6.15)$$

There are no gauge bosons from points with $l_R \neq 0$ because the mass of such a state will be at least $\frac{1}{2} l_R^2$. For generic boosts Λ, giving generic points in the moduli space, there are no points in Γ with $l_R = 0$ and so no additional gauge bosons; the gauge group is $U(1)^{36-2d}$. At special points the gauge symmetry is enhanced. Obviously one can get $SO(32) \times U(1)^{20-2d}$ or $E_8 \times E_8 \times U(1)^{20-2d}$ from compactifying the original ten-dimensional theory on a torus without Wilson lines, just as in field theory. However, as in the bosonic string, there are stringy enhanced gauge symmetries at special points in moduli space. For example, the lattice $\Gamma_{26-d,10-d}$, defined by analogy to the lattices Γ_8 and Γ_{16}, gives rise to $SO(52 - 2d) \times U(1)^{10-d}$. As in the bosonic case, the low energy physics near the point of enhanced symmetry is the Higgs mechanism. All groups obtained in this way have rank $36 - 2d$. This is the maximum in perturbation theory, but we will see in chapter 19 that nonperturbative effects can lead to larger gauge symmetries.

The number of moduli, from the dimensions of the SO groups, is

$$\frac{1}{2}\left[(36-2d)(35-2d)-(26-d)(25-d)-(10-d)(9-d)\right] = (26-d)(10-d) .$$

$$\text{(11.6.16)}$$

As in the bosonic string these can be interpreted in terms of backgrounds for the fields of the ten-dimensional gauge theory. The compact components of the metric and antisymmetric tensor give a total of $(10 - d)^2$ moduli just as before. In addition there can be Wilson lines, constant backgrounds for the gauge fields A_m. As discussed in chapter 8, due to the potential $\text{Tr}([A_m, A_n]^2)$ the fields in different directions commute along flat directions and so can be chosen to lie in a $U(1)^{16}$ subgroup. Thus there are $16(10 - d)$ parameters in A_m for $(26 - d)(10 - d)$ in all.

In chapter 8 we studied quantization with antisymmetric tensor and open string Wilson line backgrounds. Here we leave the details to the exercises and quote the result. If we compactify $x^m \cong x^m + 2\pi R$ with constant backgrounds G_{mn}, B_{mn}, and A_m^I, then canonical quantization gives

$$k_{Lm} = \frac{n_m}{R} + \frac{w^n R}{\alpha'}(G_{mn} + B_{mn}) - q^I A_m^I - \frac{w^n R}{2}A_n^I A_m^I , \qquad \text{(11.6.17a)}$$

$$k_L^I = (q^I + w^m R A_m^I)(2/\alpha')^{1/2} , \qquad \text{(11.6.17b)}$$

$$k_{Rm} = \frac{n_m}{R} + \frac{w^n R}{\alpha'}(-G_{mn} + B_{mn}) - q^I A_m^I - \frac{w^n R}{2}A_n^I A_m^I , \qquad \text{(11.6.17c)}$$

where n_m and w^m are integers and q^I is on the Γ_{16} or $\Gamma_8 \times \Gamma_8$ lattice depending on which string has been compactified. The details are left to exercise 11.10. Let us note that with the gauge fields set to zero this reduces to the bosonic result (8.4.7). The terms in k_{Lm} and k_{Rm} that are linear in A^I come from the effect of the Wilson line on the periodicity, as in eq. (8.6.7). The term in k_L^I that is linear in A^I comes about as follows. For a string that winds around the compact dimension, the Wilson line implies that the current algebra fermions are no longer periodic. The corresponding vertex operator (10.3.25) shows that the momentum is shifted. Finally, the terms quadratic in A^I can be most easily checked by verifying that $\alpha' k \circ k/2$ is even.

To compare this spectrum with the Narain description one must go to coordinates in which $G_{m'n'} = \delta_{m'n'}$ so that $k_{m'} = e_{m'}{}^n k_n$, the tetrad being defined by $\delta_{p'q'} = e_p{}^m e_q{}^n G_{mn}$. The discrete T-duality group is generated by T-dualities on the separate axes, large spacetime coordinate transformations, and quantized shifts of the antisymmetric tensor background and Wilson lines.

There is an interesting point here. Because the coset space (11.6.14) is the general solution to the consistency conditions, we must obtain this same set of theories whether we compactify the $SO(32)$ theory or the $E_8 \times E_8$ theory. From another point of view, note that the coset space is noncompact because of the Lorentzian signature — one can go to the limit of infinite Narain boost. Such a limit corresponds physically to taking one or more of the compact dimensions to infinite radius. Then one such limit gives the ten-dimensional $SO(32)$ theory, while another gives the ten-dimensional $E_8 \times E_8$ theory. Clearly one should think of all the different toroidally compactified heterotic strings as different states in a single theory. The two ten-dimensional theories are then distinct limits of this single theory.

Let us make the connection between these theories more explicit. Compactify the $SO(32)$ theory on a circle of radius R, with $G_{99} = 1$ and Wilson line

$$RA_9^I = \mathrm{diag}\left(\tfrac{1}{2}^8, 0^8\right) . \tag{11.6.18}$$

Adjoint states with one index from $1 \leq A \leq 16$ and one from $17 \leq A \leq 32$ are antiperiodic due to the Wilson line, so the gauge symmetry is reduced to $SO(16) \times SO(16)$. Now compactify the $E_8 \times E_8$ theory on a circle of radius R' with $G_{99} = 1$ and Wilson line

$$R'A_9^I = \mathrm{diag}\left(1, 0^7, 1, 0^7\right) . \tag{11.6.19}$$

The integer-charged states from the $SO(16)$ root lattice in each E_8 remain periodic while the half-integer charged states from the $SO(16)$ spinor lattices become antiperiodic. Again the gauge symmetry is $SO(16) \times SO(16)$. To see the relation between these two theories, focus on the states that are neutral under $SO(16) \times SO(16)$, those with $k_L^I = 0$. In both theories these are present only for $w = 2m$ even, because of the shift in k_L^I. The respective neutral spectra are

$$k_{L,R} = \frac{\tilde{n}}{R} \pm \frac{2mR}{\alpha'} , \quad k'_{L,R} = \frac{\tilde{n}'}{R'} \pm \frac{2m'R'}{\alpha'} , \tag{11.6.20}$$

with the subscript 9 suppressed. The primes denote the $E_8 \times E_8$ theory, and $\tilde{n} = n + 2m$, $\tilde{n}' = n' + 2m'$. We have used the explicit form of the Wilson line in each case, as well as the fact that $k_L^I = 0$. Under $(\tilde{n}, m) \leftrightarrow (m', \tilde{n}')$ and $(k_L, k_R) \leftrightarrow (k'_L, -k'_R)$, the spectra are identical if $RR' = \alpha'/2$. This symmetry extends to the full spectrum.

Finally let us ask how realistic a theory one obtains by compactification down to four dimensions. At generic points of moduli space the massless spectrum is given by dimensional reduction, simply classifying states by their four-dimensional symmetries. Analyzing the spectrum in terms of the four-dimensional $SO(2)$ helicity, the $SO(8)$ spins decompose as

$$8_v \to +1,\ 0^6,\ -1\,, \tag{11.6.21a}$$

$$8 \to +\tfrac{1}{2}^4,\ -\tfrac{1}{2}^4\,, \tag{11.6.21b}$$

and so

$$8_v \times 8_v \to +2,\ +1^{12},\ 0^{38},\ -1^{12},\ -2\,, \tag{11.6.22a}$$

$$8 \times 8_v \to \tfrac{3}{2}^4,\ \tfrac{1}{2}^{28},\ -\tfrac{1}{2}^{28},\ -\tfrac{3}{2}^4\,. \tag{11.6.22b}$$

From the supergravity multiplet there is a graviton, with helicities ± 2. There are four gravitinos, each with helicities $\pm\tfrac{3}{2}$. Toroidal compactification does not break any supersymmetry. Since in four dimensions the supercharge has four components, the 16 supersymmetries reduce to $d = 4$, $N = 4$ supersymmetry. The supergravity multiplet also includes 12 Kaluza–Klein and antisymmetric tensor gauge bosons, some fermions, and 36 moduli for the compactification. The final two spin-zero states are the dilaton and the axion. In four dimensions a two-tensor $B_{\mu\nu}$ is equivalent to a scalar (section B.4). This is the axion, whose physics we will discuss further in chapter 18.

In ten dimensions the only fields carrying gauge charge are the gauge field and gaugino. These reduce as discussed in section B.6 to an $N = 4$ vector multiplet — a gauge field, four Weyl spinors, and six scalars, all in the adjoint. For enhanced gauge symmetries, which are not present in ten dimensions, one still obtains the same $N = 4$ vector multiplet because of the supersymmetry. Compactification with $N = 4$ supersymmetry cannot give rise to the Standard Model because the fermions are necessarily in the adjoint of the gauge group. One gravitino is good, as we will explain in more detail in section 16.2, but four are too much of a good thing. We will see in chapter 16 that a fairly simple orbifold twist reduces the supersymmetry to $N = 1$ and gives a realistic spectrum.

Supersymmetry and BPS states

A little thought shows that the supersymmetry algebra of the toroidally compactified theory must be of the form[4]

$$\{Q_\alpha, Q_\beta^\dagger\} = 2P_\mu(\Gamma^\mu\Gamma^0)_{\alpha\beta} + 2P_{Rm}(\Gamma^m\Gamma^0)_{\alpha\beta}\,. \tag{11.6.23}$$

This differs from the simple dimensional reduction of the ten-dimensional algebra in that we have replaced P_m with P_{Rm}, the total right-moving

---

[4] For clarity a projection operator is omitted — all spinor indices in this equation must be in the **16**.

momentum k_{Rm} of all strings in a given state. These are equal only for a state of total winding number zero. To obtain the algebra (11.6.23) directly from a string calculation requires some additional machinery that we will not develop until the next chapter. However, it is clear that the algebra must take this form because the spacetime supersymmetry involves only the right-moving side of the heterotic string.

Let us look for Bogomolnyi–Prasad–Sommerfield (BPS) states, states that are annihilated by some of the Q_α. Take the expectation value of the algebra (11.6.23) in any state $|\psi\rangle$ of a single string of mass M in its rest frame. The left-hand side is a nonnegative matrix. The right-hand side is

$$2(M + k_{Rm}\Gamma^m\Gamma^0)_{\alpha\beta} \,. \tag{11.6.24}$$

The zero eigenvectors of this matrix are the supersymmetries that annihilate $|\psi\rangle$. Since $(k_{Rm}\Gamma^m\Gamma^0)^2 = k_R^2$, the eigenvalues of the matrix (11.6.24) are

$$2(M \pm |k_R|) \,, \tag{11.6.25}$$

with half having each sign. A BPS state therefore has $M^2 = k_R^2$. Recalling the heterotic string mass-shell conditions on the right-moving side,

$$M^2 = \begin{cases} k_R^2 + 4(\tilde{N} - \tfrac{1}{2})/\alpha' & \text{(NS)} \,, \\ k_R^2 + 4\tilde{N}/\alpha' & \text{(R)} \,, \end{cases} \tag{11.6.26}$$

the BPS states are those for which the right-movers are in an R ground state or in an NS state with one $\psi_{-1/2}$ excited. The latter are the lowest NS states to survive the GSO projection, so it makes sense to change terminology at this point and call them ground states as well. The BPS states are then precisely those states for which the right-moving side is in its $\mathbf{8}_v + \mathbf{8}$ ground state, but with arbitrarily large k_R. These can be paired with many possible states on the left-moving side. The left-moving mass-shell condition is

$$M^2 = k_L^2 + 4(N - 1)/\alpha' \tag{11.6.27}$$

or

$$N = 1 + \alpha'(k_R^2 - k_L^2)/4 = 1 - n_m w^m - q^I q^I/2 \,. \tag{11.6.28}$$

Any left-moving oscillator state is possible, as long as the compact momenta and winding satisfy the condition (11.6.28). For any given left-moving state, the 16 right-moving states $\mathbf{8}_v + \mathbf{8}$ form an ultrashort multiplet of the supersymmetry algebra, as compared to the 256 states in a normal massive multiplet.

It is interesting to look at the ten-dimensional origin of the modified supersymmetry algebra (11.6.23). Rewrite the algebra as

$$\{Q_\alpha, Q_\beta^\dagger\} = 2P_M(\Gamma^M\Gamma^0)_{\alpha\beta} - 2\frac{\Delta X_m}{2\pi\alpha'}(\Gamma^m\Gamma^0)_{\alpha\beta} \; , \tag{11.6.29}$$

where ΔX^m is the total winding of the string. Consider the limit that the compactification radii become macroscopic, so that a winding string is macroscopic as well. The central charge term in the supersymmetry algebra must be proportional to a conserved charge, so we are looking for a charge proportional to the length ΔX of a string. Indeed, the string couples to the antisymmetric two-tensor field as

$$\frac{1}{2\pi\alpha'}\int_M B = \frac{1}{2}\int d^{10}x\, j^{MN}(x)B_{MN}(x) \tag{11.6.30a}$$

$$j^{MN}(x) = \frac{1}{2\pi\alpha'}\int_M d^2\sigma\,(\partial_1 X^M \partial_2 X^N - \partial_1 X^N \partial_2 X^M)\delta^{10}(x - X(\sigma)) \; . \tag{11.6.30b}$$

This is the natural generalization of the gauge coupling of a point particle, as discussed in section B.4. Integrating the current at fixed time gives the charge

$$Q^M = \int d^9x\, j^{M0} = \frac{1}{2\pi\alpha'}\int dX^M \; , \tag{11.6.31}$$

the integral running along the world-line of the string. The full supersymmetry algebra is then

$$\{Q_\alpha, Q_\beta^\dagger\} = 2(P_M - Q_M)(\Gamma^M\Gamma^0)_{\alpha\beta} \; . \tag{11.6.32}$$

In ten noncompact dimensions the charge (11.6.31) vanishes for any finite closed string but can be carried by an infinite string, for example an infinite straight string which would arise as the $R \to \infty$ limit of a winding string. It is often useful to contemplate such macroscopic strings, which of course have infinite total mass and charge but finite values per unit length. Under compactification the combination $P_m - Q_m$ is the right-moving gauge charge. The left-moving charges do not appear in the supersymmetry algebra.

It is natural to wonder whether the algebra (11.6.32) is now complete, and in fact it is not. Consider compactification to four dimensions at a generic point in the moduli space where the gauge symmetry is broken to $U(1)^{28}$. Grand unified theories in which the $U(1)$ of the Standard Model is embedded in a simple group always have magnetic monopoles arising from the quantization of topologically nontrivial classical solutions. String theory is not an ordinary grand unified theory but it also has magnetic monopoles. Compactification of the heterotic string leads to three kinds of gauge symmetry: the ten-dimensional symmetries, the Kaluza–Klein symmetries, and the antisymmetric tensor symmetries. For each there is a corresponding monopole solution: the 't Hooft–Polyakov monopole, the Kaluza–Klein monopole, and the H-monopole respectively. Of course since the various charges are interchanged by the $O(22, 6, \mathbf{Z})$ T-duality, the monopoles must be as well. Monopole charges appear in the supersymmetry algebra; in the present case it is again the right-moving charges that appear. In the low energy supergravity theory there is a symmetry that interchanges the electric and magnetic charges, so they must appear in the supersymmetry algebra in a symmetric way. We will discuss similar central charge terms extensively in chapters 13 and 14.

Exercises

11.1 Show that the operators (10.7.21) with appropriate normalization generate the full $N = 2$ superconformal algebra (11.1.4).

11.2 Show that if a $(\frac{3}{2}, 0)$ constraint j_F is not tensor, then $L_1 \cdot j_F$ is a nonvanishing $(\frac{1}{2}, 0)$ constraint, and a linear combination of $L_{-1} \cdot L_1 \cdot j_F$ and j_F is a tensor $(\frac{3}{2}, 0)$ constraint.

11.3 Show that if we take the GSO projection on the λ^A in groups of eight, modular invariance is inconsistent with spacetime spin-statistics. Show that the OPE does not close.

11.4 (a) Find the massless and tachyonic states in the theory obtained by twisting the diagonal theory on the group generated by $\exp(\pi i F_1)$.
(b) Do the same for the group generated by $\exp(\pi i F_1)$ and $\exp(\pi i F_2)$.

11.5 (a) The decompositions of the spinor representation under $SO(16) \to SO(6) \times SO(10)$ and under $SO(6) \to SU(3) \times U(1)$ are obtained in section B.1. Use this to show that the adjoint of E_8 decomposes into $SU(3)$ representations with the degeneracies (11.4.24). The 78 generators neutral under $SU(3)$ must form a closed algebra: this is E_6.
(b) Use the same decompositions to show that the E_6 representations decompose as shown in table 11.4 under $E_6 \to SO(10) \times U(1)$.
(c) In a similar way obtain the decompositions shown in table 11.4 for $SO(10) \to SU(5) \times U(1)$.

11.6 Repeat parts (a) and (b) of the previous exercise for $SO(16) \to SO(4) \times SO(12)$ and $SO(4) \to SU(2) \times U(1)$ to obtain the analogous properties of E_7.

11.7 Show that $:jj(0): = j^a_{-1} \cdot j^a_{-1} \cdot 1$. Act with the Laurent expansion (11.5.2) for $j^c(z)$ and verify the OPE (11.5.20) in the Sugawara construction. Similarly verify the OPE (11.5.24).

11.8 For the free-fermion currents (11.5.16) for $SO(n)$, verify that the Sugawara construction gives the usual bilinear T_B.

11.9 Show that the lattice

$$\Gamma = \sum_r \Gamma_r \times \tilde{\Gamma}_r$$

is even and self-dual, where Γ_r is a weight sublattice of $SO(22)$, $\tilde{\Gamma}_r$ the same weight sublattice for $SO(6)$, and the sum runs over the four sublattices of $SO(2n)$. Show that this gives a four-dimensional compactification of the heterotic string with $SO(44) \times U(1)^6$ gauge symmetry.

11.10 (a) Verify the spectrum (11.6.17) for one compact dimension with a Wilson line background only.
(b) For the full spectrum (11.6.17), verify that $\alpha' k \circ k/2$ is even for any state and that $\alpha' k \circ k'/2$ is integral for any pair of states. The $\circ$ product is

$$k \circ k = k_L^I k_L'^I + G^{mn}(k_{Lm} k_{Ln}' - k_{Rm} k_{Rn}') \; .$$

(c) (Optional) Verify the full result (11.6.17) by canonical quantization. Recall that the antisymmetric tensor background has already been treated in chapter 8. Reference: Narain, Sarmadi, & Witten (1987).

11.11 In the $E_8 \times E_8$ string, the currents $i\partial H^I$ plus the vertex operators for the points of length two form a set of $(1,0)$ currents satisfying the $E_8 \times E_8$ algebra. From the $1/z$ term in the OPE, find the commutation relations of E_8. Be sure to include the cocycle in the vertex operator.

11.12 Find the Hagedorn temperatures of the type I, II, and heterotic string theories. Use the result (7.2.30) for the asymptotics of the partition function to express the Hagedorn temperature in general form.

12

Superstring interactions

In this chapter we will examine superstring interactions from two complementary points of view. First we study the interactions of the massless degrees of freedom, which are highly constrained by supersymmetry. The first section discusses the tree-level interactions, while the second discusses an important one-loop effect: the anomalies in local spacetime symmetries. We then develop superstring perturbation theory. We introduce superfields and super-Riemann surfaces to give superconformal symmetry a geometric interpretation, and calculate a variety of tree-level and one-loop amplitudes.

12.1 Low energy supergravity

The ten-dimensional supersymmetric string theories all have 32 or 16 supersymmetry generators. This high degree of supersymmetry completely determines the low energy action.

Type IIA superstring

We begin by discussing the field theory that has the largest possible spacetime supersymmetry and Poincaré invariance, namely eleven-dimensional supergravity. As explained in the appendix, the upper limit on the dimension arises because nontrivial consistent field theories cannot have massless particles with spins greater than two.

This theory would seem to have no direct connection to superstring theory, which requires ten dimensions. Our immediate interest in it is that, as discussed in section B.5, its supersymmetry algebra is the same as that of the IIA theory. The action of the latter can therefore be obtained by dimensional reduction, toroidal compactification keeping only fields that are independent of the compact directions. For now this is just a trick to take advantage of the high degree of supersymmetry, but in chapter 14 we will see that there is much more going on.

The eleven-dimensional supergravity theory has two bosonic fields, the metric G_{MN} and a 3-form potential $A_{MNP} \equiv A_3$ with field strength F_4. Higher-dimensional supergravities contain many different p-form fields; to distinguish these from one another we will denote the rank by an italicized subscript, as opposed to numerical tensor indices which are written in roman font. In terms of the $SO(9)$ spin of massless states, the metric gives a traceless symmetric tensor with 44 states, and the 3-form gives a rank 3 antisymmetric tensor with 84 states. The total number of bosonic states is then 128, equal to the dimension of the $SO(9)$ vector-spinor gravitino.

The bosonic part of the action is given by

$$2\kappa_{11}^2 S_{11} = \int d^{11}x (-G)^{1/2} \left(R - \frac{1}{2}|F_4|^2 \right) - \frac{1}{6} \int A_3 \wedge F_4 \wedge F_4 . \quad (12.1.1)$$

The form action, written out fully, is proportional to

$$\int d^d x (-G)^{1/2}|F_p|^2 = \int d^d x \frac{(-G)^{1/2}}{p!} G^{M_1 N_1} \dots G^{M_p N_p} F_{M_1 \dots M_p} F_{N_1 \dots N_p} .$$
$$(12.1.2)$$

The $p!$ cancels the sum over permutations of the indices, so that each independent component appears with coefficient 1. Forms are written as tensors with lower indices in order that their gauge transformations do not involve the metric.

We will take such results from the literature without derivation. Our interest is only in certain general features of the various actions, and we will not write out the full fermionic terms or supersymmetry transformations. For the supergravities arising from string theories, one can verify the action by comparison with the low energy limits of string amplitudes; a few such calculations are given later in the chapter and in the exercises. Also, many important features, such as the coupling of the dilaton, will be understood from general reasoning.

Now dimensionally reduce as in section 8.1. The general metric that is invariant under translations in the 10-direction is

$$ds^2 = G_{MN}^{11}(x^\mu)dx^M dx^N$$
$$= G_{\mu\nu}^{10}(x^\mu)dx^\mu dx^\nu + \exp(2\sigma(x^\mu))[dx^{10} + A_\nu(x^\mu)dx^\nu]^2 . \quad (12.1.3)$$

Here M, N run from 0 to 10 and μ, ν from 0 to 9. We have added a superscript 11 to the metric appearing in the earlier supergravity action and introduced a new ten-dimensional metric $G_{\mu\nu}^{10} \neq G_{\mu\nu}^{11}$. The ten-dimensional metric will appear henceforth, so the superscript 10 will be omitted.

The eleven-dimensional metric (12.1.3) reduces to a ten-dimensional metric, a gauge field A_1, and a scalar σ. The potential A_3 reduces to two potentials A_3 and A_2, the latter coming from components where one index is along the compact 10-direction. The three terms (12.1.1) become

$$S_1 = \frac{1}{2\kappa_{10}^2} \int d^{10}x \, (-G)^{1/2} \left(e^\sigma R - \frac{1}{2} e^{3\sigma} |F_2|^2 \right) , \qquad (12.1.4a)$$

$$S_2 = -\frac{1}{4\kappa_{10}^2} \int d^{10}x \, (-G)^{1/2} \left(e^{-\sigma} |F_3|^2 + e^\sigma |\widetilde{F}_4|^2 \right) , \qquad (12.1.4b)$$

$$S_3 = -\frac{1}{4\kappa_{10}^2} \int A_2 \wedge F_4 \wedge F_4 = -\frac{1}{4\kappa_{10}^2} \int A_3 \wedge F_3 \wedge F_4 . \qquad (12.1.4c)$$

We have compactified the theory on a circle of coordinate period $2\pi R$ and defined $\kappa_{10}^2 = \kappa_{11}^2 / 2\pi R$. The normalization of the kinetic terms is canonical for $2\kappa_{10}^2 = 1$.

In the action (12.1.4) we have defined

$$\widetilde{F}_4 = dA_3 - A_1 \wedge F_3 , \qquad (12.1.5)$$

the second term arising from the components $G^{\mu\,10}$ in the 4-form action (12.1.2). We will use $F_{p+1} = dA_p$ to denote the simple exterior derivative of a potential, while field strengths with added terms are distinguished by a tilde as in eq. (12.1.5). Note that the action contains several terms where p-form potentials appear, rather than their exterior derivatives, but which are still gauge invariant. These are known as *Chern–Simons* terms, and we see that they are of two types. One involves the wedge product of one potential with any number of field strengths, and it is gauge invariant as a consequence of the Bianchi identities for the field strengths. The other appears in the kinetic term for the modified field strength (12.1.5). The second term in $\widetilde{F}_4$ has a gauge variation

$$- d\lambda_0 \wedge F_3 = -d(\lambda_0 \wedge F_3). \qquad (12.1.6)$$

It is canceled by a transformation

$$\delta' A_3 = \lambda_0 \wedge F_3 , \qquad (12.1.7)$$

which is in addition to the usual $\delta A_3 = d\lambda_2$. In the present case, the Kaluza–Klein gauge transformation λ_0 originates from reparameterization of x^{10}, and the transformation (12.1.7) is simply part of the eleven-dimensional tensor transformation. Since the combination $\widetilde{F}_4$ is invariant under both λ_0 and λ_2 transformations we should regard it as the physical field strength, but with a nonstandard Bianchi identity

$$d\widetilde{F}_4 = -F_2 \wedge F_3 . \qquad (12.1.8)$$

Poincaré duality of the form theory, developed in section B.4 for forms without Chern–Simons terms, interchanges these two kinds of Chern–Simons term.

The fields of the reduced theory are the same as the bosonic fields of the IIA string, as they must be. In particular the scalar σ must be the dilaton Φ, up to some field redefinition. The terms in the action have a variety of σ-dependences. Recall that the string coupling constant is determined by the value of the dilaton. As discussed in section 3.7, this means that after appropriate field redefinitions the tree-level spacetime action is multiplied by an overall factor $e^{-2\Phi}$, and otherwise depends on Φ only through its derivatives. 'Appropriate redefinitions' means that the fields are the same as those appearing in the string world-sheet sigma model action.

Since we have arrived at the action (12.1.4) without reference to string theory, we have no idea as yet how these fields are related to those in the world-sheet action. We will proceed by guesswork, and then explain the result in world-sheet terms. First redefine

$$G_{\mu\nu} = e^{-\sigma} G_{\mu\nu}(\text{new}), \quad \sigma = \frac{2\Phi}{3} . \qquad (12.1.9)$$

The original metric will no longer appear, so to avoid cluttering the equations we do not put a prime on the new metric. Then

$$S_{\text{IIA}} = S_{\text{NS}} + S_{\text{R}} + S_{\text{CS}} , \qquad (12.1.10a)$$

$$S_{\text{NS}} = \frac{1}{2\kappa_{10}^2} \int d^{10}x \, (-G)^{1/2} e^{-2\Phi} \left(R + 4\partial_\mu \Phi \partial^\mu \Phi - \frac{1}{2}|H_3|^2 \right) , \qquad (12.1.10b)$$

$$S_{\text{R}} = -\frac{1}{4\kappa_{10}^2} \int d^{10}x \, (-G)^{1/2} \left(|F_2|^2 + |\tilde{F}_4|^2 \right) , \qquad (12.1.10c)$$

$$S_{\text{CS}} = -\frac{1}{4\kappa_{10}^2} \int B_2 \wedge F_4 \wedge F_4 . \qquad (12.1.10d)$$

Note that $R \to e^\sigma R + \ldots$, that $(-G)^{1/2} \to e^{-5\sigma}(-G)^{1/2}$, and that the form action (12.1.2) scales as $e^{(p-5)\sigma}$.

We have regrouped terms according to whether the fields are in the NS–NS or R–R sector of the string theory; the Chern–Simons action contains both. It will be useful to distinguish R–R from NS–NS forms, so for the R–R fields we henceforth use C_p and F_{p+1} for the potential and field strength, and for the NS–NS fields B_2 and H_3. Also, we will use A_1 and F_2 for the open string and heterotic gauge fields, and B_2 and H_3 for the heterotic antisymmetric tensor.

The NS action now involves the dilaton in standard form. Eq. (12.1.9) is the unique redefinition that does this. The R action does not have the expected factor of $e^{-2\Phi}$, but can be brought to this form by the further redefinition

$$C_1 = e^{-\Phi} C_1' , \qquad (12.1.11)$$

after which

$$\int d^{10}x\,(-G)^{1/2}|F_2|^2 = \int d^{10}x\,(-G)^{1/2}e^{-2\Phi}|F_2'|^2 \,, \qquad (12.1.12a)$$

$$F_2' \equiv dC_1' - d\Phi \wedge C_1' \,, \qquad (12.1.12b)$$

and similarly for F_3 and C_4. The action (12.1.12) makes explicit the dilaton dependence of the loop expansion, but at the cost of complicating the Bianchi identity and gauge transformation,

$$dF_2' = d\Phi \wedge F_2' \,, \quad \delta C_1' = d\lambda_0' - \lambda_0'\,d\Phi \,. \qquad (12.1.13)$$

For this reason the form (12.1.10) is usually used. For example, in a time-dependent dilaton field, it is the charge to which the unprimed fields couple that will be conserved.

Let us now make contact with string theory and see why the background R–R fields appearing in the world-sheet action have the more complicated properties (12.1.13). We work at the linearized level, in terms of the vertex operators

$$\mathcal{V}_\alpha \tilde{\mathcal{V}}_\beta (C\Gamma^{\mu_1\cdots\mu_p})_{\alpha\beta}e_{\mu_1\ldots\mu_p}(X) \,. \qquad (12.1.14)$$

Here $\mathcal{V}_\alpha$ is the R ground state vertex operator (10.4.25) and $\Gamma^{\mu_1\cdots\mu_p} = \Gamma^{[\mu_1}\ldots\Gamma^{\mu_p]}$. The nontrivial physical state conditions are from $G_0 \sim p_\mu\psi_0^\mu$ and $\tilde{G}_0 \sim p_\mu\tilde{\psi}_0^\mu$, and amount to two Dirac equations, one acting on the left spinor index and one on the right:

$$\Gamma^\nu\Gamma^{\mu_1\cdots\mu_p}\partial_\nu e_{\mu_1\ldots\mu_p}(X) = \Gamma^{\mu_1\cdots\mu_p}\Gamma^\nu\partial_\nu e_{\mu_1\ldots\mu_p}(X) = 0 \,. \qquad (12.1.15)$$

By antisymmetrizing all $p+1$ gamma matrices and keeping anticommutators one obtains

$$\Gamma^\nu\Gamma^{\mu_1\cdots\mu_p} = \Gamma^{\nu\mu_1\cdots\mu_p} + p\eta^{\nu[\mu_1}\Gamma^{\mu_2\cdots\mu_p]} \,, \qquad (12.1.16a)$$

$$\Gamma^{\mu_1\cdots\mu_p}\Gamma^\nu = (-1)^p\Gamma^{\nu\mu_1\cdots\mu_p} + (-1)^{p+1}p\eta^{\nu[\mu_1}\Gamma^{\mu_2\cdots\mu_p]} \,. \qquad (12.1.16b)$$

The Dirac equations (12.1.15) are then equivalent to

$$de_p = d*e_p = 0 \,. \qquad (12.1.17)$$

These are first order equations, unlike the second order equations encountered previously for bosonic fields. In fact, they have the same form as the field equation and Bianchi identity for a *p*-form field strength. Thus we identify the function $e_{\mu_1\ldots\mu_p}(X)$ appearing in the vertex operator as the R–R *field strength* rather than potential. To confirm this, observe that in the IIA theory the spinors in the R–R vertex operator (12.1.14) have opposite chirality and so their product in table 10.1 contains forms of even rank, the same as the IIA R–R field strengths.

This has one consequence that will be important later on. Amplitudes for R–R forms will always contain a power of the momentum and so vanish at zero momentum. The zero-momentum coupling of a gauge field measures the charge, so this means that *strings are neutral under all R–R gauge fields*.

The derivation of the field equations (12.1.17) was for a flat background. Now let us consider the effect of a dilaton gradient. It is convenient that the linear dilaton background gives rise to the free CFT (10.1.22),

$$T_F = i(2/\alpha')^{1/2}\psi^\mu \partial X_\mu - 2i(\alpha'/2)^{1/2}\Phi_{,\mu}\partial\psi^\mu \,, \qquad (12.1.18a)$$

$$G_0 \sim (\alpha'/2)^{1/2}\psi_0^\mu(p_\mu + i\Phi_{,\mu}) \,, \qquad (12.1.18b)$$

and similarly for $\tilde{T}_F$ and $\tilde{G}_0$. The field equations are modified to

$$(d - d\Phi\wedge)e_p = (d - d\Phi\wedge) * e_p = 0 \,. \qquad (12.1.19)$$

Thus the Bianchi identity and field equation for the string background fields are modified in the fashion deduced from the action. There is no such modification for the NS–NS tensor. It couples to the world-sheet through its potential,

$$\frac{1}{2\pi\alpha'}\int_M B_2 \,. \qquad (12.1.20)$$

This is invariant under $\delta B_2 = d\lambda_1$ independent of the dilaton, and so $H_3 = dB_2$ is invariant and $dH_3 = 0$.

Massive IIA supergravity

There is a generalization of the IIA supergravity theory which has no simple connection with eleven-dimensional supergravity but which plays a role in string theory. The IIA theory has a 2-form and a 4-form field strength, and by Poincaré duality a 6-form and an 8-form as well,

$$\tilde{F}_6 = *\tilde{F}_4 \,, \quad \tilde{F}_8 = *F_2 \,; \qquad (12.1.21)$$

again, a tilde denotes a field strength with a nonstandard Bianchi identity. The pattern suggests we also consider a 10-form $F_{10} = dC_9$. The free field equation would be

$$d*F_{10} = 0 \,, \qquad (12.1.22)$$

and since $*F_{10}$ is a scalar this means that

$$* F_{10} = \text{constant} \,. \qquad (12.1.23)$$

Thus there are no propagating degrees of freedom. Nevertheless, such a field would have a physical effect, since it would carry energy density. This is closely analogous to an electric field F_2 in two space-time dimensions, where there are no propagating photons but there is an energy density and a linear potential that confines charges.

Such a field can indeed be included in IIA supergravity. The action is

$$S'_{\text{IIA}} = \tilde{S}_{\text{IIA}} - \frac{1}{4\kappa_{10}^2} \int d^{10}x\,(-G)^{1/2}M^2 + \frac{1}{2\kappa_{10}^2} \int MF_{10}\,. \qquad (12.1.24)$$

Here $\tilde{S}_{\text{IIA}}$ is the earlier IIA action (12.1.10) with the substitutions

$$F_2 \to F_2 + MB_2\,, \quad F_4 \to F_4 + \frac{1}{2}MB_2 \wedge B_2\,, \quad \tilde{F}_4 \to \tilde{F}_4 + \frac{1}{2}MB_2 \wedge B_2\,.$$
$$(12.1.25)$$

The scalar M is an auxiliary field, meaning that it appears in the action without derivatives (and in this case only quadratically). Thus it can be integrated out, at the cost of introducing a rather nonlinear dependence on B_2.

We will see in the next chapter that this massive supergravity does arise in the IIA string. To put the 9-form potential in perspective, observe that the maximum-rank potential that gives rise to a propagating field in ten dimensions is an 8-form, whose 9-form field strength is dual to a 1-form. The latter is just the gradient of the R–R scalar field C_0. A 10-form potential also fits in ten dimensions but does not give rise to propagating states. We saw in section 10.8 that this does exist in the type I string, so we should not be surprised that the 9-form will appear in string theory as well.

Type IIB superstring

For low energy IIB supergravity there is a problem due to the self-dual field strength $F_5 = *F_5$. As discussed in section B.4 there is no covariant action for such a field, but the following comes close:

$$S_{\text{IIB}} = S_{\text{NS}} + S_{\text{R}} + S_{\text{CS}}\,, \qquad (12.1.26a)$$

$$S_{\text{NS}} = \frac{1}{2\kappa_{10}^2} \int d^{10}x\,(-G)^{1/2}e^{-2\Phi}\left(R + 4\partial_\mu\Phi\partial^\mu\Phi - \frac{1}{2}|H_3|^2\right)\,, \qquad (12.1.26b)$$

$$S_{\text{R}} = -\frac{1}{4\kappa_{10}^2} \int d^{10}x\,(-G)^{1/2}\left(|F_1|^2 + |\tilde{F}_3|^2 + \frac{1}{2}|\tilde{F}_5|^2\right)\,, \qquad (12.1.26c)$$

$$S_{\text{CS}} = -\frac{1}{4\kappa_{10}^2} \int C_4 \wedge H_3 \wedge F_3\,, \qquad (12.1.26d)$$

where

$$\tilde{F}_3 = F_3 - C_0 \wedge H_3\,, \qquad (12.1.27a)$$

$$\tilde{F}_5 = F_5 - \frac{1}{2}C_2 \wedge H_3 + \frac{1}{2}B_2 \wedge F_3\,. \qquad (12.1.27b)$$

The NS–NS action is the same as in IIA supergravity, while the R–R and Chern–Simons actions are closely parallel in form. The equation of motion and Bianchi identity for $\tilde{F}_5$ are

$$d * \tilde{F}_5 = d\tilde{F}_5 = H_3 \wedge F_3 \, . \tag{12.1.28}$$

Recall that the spectrum of the IIB string includes the degrees of freedom of a self-dual 5-form field strength. The field equations from the action (12.1.26) are consistent with

$$* \tilde{F}_5 = \tilde{F}_5 \tag{12.1.29}$$

but they do not imply it. This must be imposed as an added constraint on the *solutions;* it cannot be imposed on the action or else the wrong equations of motion result.

This formulation is satisfactory for a classical treatment but it is not simple to impose the constraint in the quantum theory. This will not be important for our purposes, and we leave further discussion to the references. Our main interest in this action is a certain $SL(2, \mathbf{R})$ symmetry. Let

$$G_{E\mu\nu} = e^{-\Phi/2} G_{\mu\nu} \, , \quad \tau = C_0 + ie^{-\Phi} \, , \tag{12.1.30a}$$

$$\mathcal{M}_{ij} = \frac{1}{\operatorname{Im}\tau} \begin{bmatrix} |\tau|^2 & -\operatorname{Re}\tau \\ -\operatorname{Re}\tau & 1 \end{bmatrix} \, , \quad F_3^i = \begin{bmatrix} H_3 \\ F_3 \end{bmatrix} \, . \tag{12.1.30b}$$

Then

$$S_{\text{IIB}} = \frac{1}{2\kappa_{10}^2} \int d^{10}x \, (-G_E)^{1/2} \left(R_E - \frac{\partial_\mu \bar{\tau} \partial^\mu \tau}{2(\operatorname{Im}\tau)^2} \right.$$
$$\left. - \frac{\mathcal{M}_{ij}}{2} F_3^i \cdot F_3^j - \frac{1}{4} |\tilde{F}_5|^2 \right) - \frac{\epsilon_{ij}}{8\kappa_{10}^2} \int C_4 \wedge F_3^i \wedge F_3^j \, ,$$
$$\tag{12.1.31}$$

the Einstein metric (12.1.30a) being used everywhere. This is invariant under the following $SL(2, \mathbf{R})$ symmetry:

$$\tau' = \frac{a\tau + b}{c\tau + d} \, , \tag{12.1.32a}$$

$$F_3^{i\prime} = \Lambda^i_j F_3^j \, , \quad \Lambda^i_j = \begin{bmatrix} d & c \\ b & a \end{bmatrix} \, , \tag{12.1.32b}$$

$$\tilde{F}_5' = \tilde{F}_5 \, , \quad G'_{E\mu\nu} = G_{E\mu\nu} \, , \tag{12.1.32c}$$

with a, b, c, and d real numbers such that $ad - bc = 1$. The $SL(2, \mathbf{R})$ invariance of the τ kinetic term is familiar, and that of the F_3 kinetic term follows from

$$\mathcal{M}' = (\Lambda^{-1})^T \mathcal{M} \Lambda^{-1} \, . \tag{12.1.33}$$

This $SL(2, \mathbf{R})$ invariance is as claimed in the second line of table B.3. Any given value τ is invariant under an $SO(2, \mathbf{R})$ subgroup so the moduli space is the coset $SL(2, \mathbf{R})/SO(2, \mathbf{R})$. If we now compactify on tori, the moduli and other fields fall into multiplets of the larger symmetries indicated in the table and the low energy action has the larger symmetry.

Observe that this $SL(2, \mathbf{R})$ mixes the two 2-form potentials. We know that the NS–NS form couples to the string and the R–R form does not. The $SL(2, \mathbf{R})$ might thus seem to be an accidental symmetry of the low energy theory, not relevant to the full string theory. Indeed, this was assumed for some time, but now we know better. As we will explain in chapter 14, the discrete subgroup $SL(2, \mathbf{Z})$ is an exact symmetry.

Type I superstring

To obtain the type I supergravity action requires three steps: set to zero the IIB fields C_0, B_2, and C_4 that are removed by the Ω projection; add the gauge fields, with appropriate dilaton dependence for an open string field; and, modify the F_3 field strength. This gives

$$S_\mathrm{I} = S_\mathrm{c} + S_\mathrm{o} \,, \tag{12.1.34a}$$

$$S_\mathrm{c} = \frac{1}{2\kappa_{10}^2} \int d^{10}x \, (-G)^{1/2} \left[e^{-2\Phi} \left(R + 4\partial_\mu \Phi \partial^\mu \Phi \right) - \frac{1}{2}|\tilde{F}_3|^2 \right] , \tag{12.1.34b}$$

$$S_\mathrm{o} = -\frac{1}{2g_{10}^2} \int d^{10}x (-G)^{1/2} e^{-\Phi} \mathrm{Tr_v}(\, |F_2|^2) \,. \tag{12.1.34c}$$

The open string $SO(32)$ potential and field strength are written as matrix-valued forms A_1 and F_2, which are in the vector representation as indicated by the subscript on the trace. Here

$$\tilde{F}_3 = dC_2 - \frac{\kappa_{10}^2}{g_{10}^2}\omega_3 \,, \tag{12.1.35}$$

and ω_3 is the *Chern–Simons 3-form*

$$\omega_3 = \mathrm{Tr_v}\left(A_1 \wedge dA_1 - \frac{2i}{3} A_1 \wedge A_1 \wedge A_1 \right) . \tag{12.1.36}$$

Again the modification of the field strength implies a modification of the gauge transformation. Under an ordinary gauge transformation $\delta A_1 = d\lambda - i[A_1, \lambda]$, the Chern-Simons form transforms as

$$\delta\omega_3 = d\mathrm{Tr_v}(\lambda dA_1). \tag{12.1.37}$$

Thus it must be that

$$\delta C_2 = \frac{\kappa_{10}^2}{g_{10}^2}\mathrm{Tr_v}(\lambda dA_1) \,. \tag{12.1.38}$$

The 2-form transformation $\delta C_2 = d\lambda_1$ is unaffected.

The action appears to contain two parameters, κ_{10} with units of L^4 and g_{10} with units of L^3. We can think of κ_{10} as setting the scale because it is dimensionful, but there is one dimensionless combination $\kappa_{10}g_{10}^{-4/3}$. However, under an additive shift $\Phi \to \Phi + \zeta$, the couplings change $\kappa_{10} \to e^{\zeta}\kappa_{10}$ and $g \to e^{\zeta/2}g$ and so this ratio can be set to any value by a change of the background. Thus the low energy theory reflects the familiar string property that the coupling is not a fixed parameter but depends on the dilaton. The form of the action (12.1.34) is fixed by supersymmetry, but when we consider this as the low energy limit of string theory there is a relation between the closed string coupling κ_{10}, the open string coupling g_{10}, and the type I α'. We will derive this in the next chapter, from a D-brane calculation, as we did for the corresponding relation in the bosonic string.

Heterotic strings

The heterotic strings have the same supersymmetry as the type I string and so we expect the same action. However, in the absence of open strings or R–R fields the dilaton dependence should be $e^{-2\Phi}$ throughout:

$$S_{\text{het}} = \frac{1}{2\kappa_{10}^2} \int d^{10}x \, (-G)^{1/2} e^{-2\Phi} \left[R + 4\partial_\mu \Phi \partial^\mu \Phi - \frac{1}{2}|\tilde{H}_3|^2 - \frac{\kappa_{10}^2}{g_{10}^2} \text{Tr}_v(|F_2|^2) \right].$$

$$\text{(12.1.39)}$$

Here

$$\tilde{H}_3 = dB_2 - \frac{\kappa_{10}^2}{g_{10}^2}\omega_3 \,, \quad \delta B_2 = \frac{\kappa_{10}^2}{g_{10}^2}\text{Tr}_v(\lambda \, dA_1) \qquad \text{(12.1.40)}$$

are the same as in the type I string, with the form renamed to reflect the fact that it is from the NS sector.

Because of the high degree of supersymmetry, the type I and heterotic actions can differ only by a field redefinition. Indeed the reader can check that with the type I and heterotic fields related by

$$G_{I\mu\nu} = e^{-\Phi_h}G_{h\mu\nu} \,, \quad \Phi_I = -\Phi_h \,, \qquad \text{(12.1.41a)}$$

$$\tilde{F}_{I3} = \tilde{H}_{h3} \,, \quad A_{I1} = A_{h1} \,, \qquad \text{(12.1.41b)}$$

the action (12.1.34) becomes the action (12.1.39). For the heterotic string, the relation among κ_{10}, g_{10}, and α' will be obtained later in this chapter, by two different methods; it is, of course, different from the relation in the type I theory.

For $E_8 \times E_8$ there is no vector representation, but it is convenient to use a normalization that is uniform with $SO(32)$. In place of $\text{Tr}_v(t^a t^b)$ in the action use $\frac{1}{30}\text{Tr}_a(t^a t^b)$. This has the property that for fields in any $SO(16) \times SO(16)$ subgroup it reduces to $\text{Tr}_v(t^a t^b)$.

12.2   Anomalies

It is an important phenomenon that some classical symmetries are anomalous, meaning that they are not preserved by quantization. We encountered this for the Weyl anomaly in chapter 3. We also saw there that if the left- and right-moving central charges were not equal there was an anomaly in two-dimensional coordinate invariance.

In general, anomalies in local symmetries make a theory inconsistent, as unphysical degrees of freedom no longer decouple. Anomalies in global symmetries are not harmful but imply that the symmetry is no longer exact. Both kinds of anomalies play a role in the Standard Model. Potential local anomalies in gauge and coordinate invariance cancel among the quarks and leptons of each generation. Anomalies in global chiral symmetries of the strong interaction are important in accounting for the π^0 decay rate and the η' mass.

In this section we consider potential anomalies in the spacetime gauge and coordinate invariances in the various string theories. If the theories we have constructed are consistent these anomalies must be absent, and in fact they are. Although this can be understood in purely string theoretic terms it can also be understood from analysis of the low energy field theory, and it is useful to take both points of view.

We can analyze anomalies from the purely field theoretic point of view because of the odd property that they are both short-distance and long-distance effects. They are short-distance in the sense that they arise because the measure cannot be defined — the theory cannot be regulated — in an invariant way. They are long-distance in the sense that this impossibility follows entirely from the nature of the massless spectrum.

Let us illustrate this with another two-dimensional example, which is also of interest in its own right. Suppose we have left- and right-moving current algebras with the same algebra g, with the coefficients of the Schwinger terms being $\hat{k}_{L,R}\delta^{ab}$. Couple a gauge field to the current,

$$S_{\text{int}} = \int d^2z \left(j^a_z A^a_{\bar{z}} + j^a_{\bar{z}} A^a_z \right) . \tag{12.2.1}$$

The OPE determines the jj expectation value, so to second order the path integral is

$$Z[A] = \frac{1}{2} \int d^2z_1 \, d^2z_2 \left[\frac{\hat{k}_L}{z_{12}^2} A^a_{\bar{z}}(z_1, \bar{z}_1) A^a_{\bar{z}}(z_2, \bar{z}_2) + \frac{\hat{k}_R}{\bar{z}_{12}^2} A^a_z(z_1, \bar{z}_1) A^a_z(z_2, \bar{z}_2) \right] . \tag{12.2.2}$$

Now make a gauge transformation, which to leading order is $\delta A^a_l = d\lambda^a$.

Integrate by parts and use $\partial_z(1/\bar{z}^2) = -2\pi\partial_{\bar{z}}\delta^2(z, \bar{z})$ to obtain

$$\delta Z[A] = -\pi \int d^2z\, \lambda^a(z, \bar{z})\left[\hat{k}_L \partial_z A_{\bar{z}}^a(z, \bar{z}) + \hat{k}_R \partial_{\bar{z}} A_z^a(z, \bar{z})\right] . \qquad (12.2.3)$$

Now, consider the case that $\hat{k}_L = \hat{k}_R = \hat{k}$, where

$$\delta Z[A] = \pi\hat{k}\,\delta \int d^2z\, A_z^a(z, \bar{z})A_{\bar{z}}^a(z, \bar{z}) . \qquad (12.2.4)$$

Then

$$\begin{aligned}
Z'[A] &= Z[A] - \pi\hat{k} \int d^2z\, A_z^a(z, \bar{z})A_{\bar{z}}^a(z, \bar{z}) \\
&= \frac{\hat{k}}{2} \int d^2z_1\, d^2z_2\, \ln|z_{12}^2|\, F_{z\bar{z}}^a(z_1, \bar{z}_1)F_{z\bar{z}}^a(z_2, \bar{z}_2) \qquad (12.2.5)
\end{aligned}$$

is gauge-invariant.

Let us run through the logic here. The path integral (12.2.2) is nonlocal, but its gauge variation is local. The latter is necessarily true because the variation can be thought of as arising from the regulator if we actually evaluate the path integral by brute force. Although the variation is local, it is not in general the variation of a local operator. When it is so, as is the case for $\hat{k}_L = \hat{k}_R$ here, one can subtract that local operator from the action to restore gauge invariance. In fact, with a gauge-invariant regulator the needed local term will be produced by the path integral automatically. The OPE is unambiguous only for nonzero separation, so the OPE calculation above doesn't determine the local terms — it doesn't know which regulator we choose to use.

The final form (12.2.5) is written in terms of the field strength. For an Abelian theory the full path integral is just the exponential of this. For a non-Abelian theory the higher order terms are more complicated, but the condition $\hat{k}_L = \hat{k}_R$ for the symmetry to be preserved is still necessary and sufficient.

The two-dimensional gravitational anomaly was similarly determined from the z^{-4} term in the TT OPE. Also, if there is a z^{-3} term in a Tj OPE then there is a *mixed anomaly*: the current has an anomaly proportional to the curvature and the coordinate invariance an anomaly proportional to the field strength.

Note that these anomalies are all odd under parity, being proportional to $\hat{k}_L - \hat{k}_R$ or $c_L - c_R$. Parity-symmetric theories can be defined invariantly using a Pauli–Villars regulator. Also, the anomalies are unaffected if we add additional massive degrees of freedom. This follows from a field theory decoupling argument. Massive degrees of freedom give a contribution to $Z[A]$ which at asymptotically long distance looks local (analytic in momentum). Any gauge variation of this can therefore be written as the variation of a local operator, and removed by a counterterm. For this reason the anomalies in superstring theory are determined by the massless spectrum, independent of the stringy details at short distance.

A single fermion of charge q coupled to a $U(1)$ gauge field contributes q^2 to the jj OPE. The anomaly cancellation conditions for free fermions coupled to such a field are

$$\text{gauge anomaly: } \sum_L q^2 - \sum_R q^2 = 0 , \qquad (12.2.6a)$$

$$\text{gravitational anomaly: } \sum_L 1 - \sum_R 1 = 0 , \qquad (12.2.6b)$$

$$\text{mixed anomaly: } \sum_L q - \sum_R q = 0 . \qquad (12.2.6c)$$

In four dimensions things are slightly different. For dimensional reasons the dangerous amplitudes have three currents and the anomaly is quadratic in the field strengths and curvatures. The antiparticle of a left-handed fermion of charge q is a right-handed fermion of charge $-q$, so the two terms in the anomaly are automatically equal for odd powers of q and opposite for even powers (including the purely gravitational anomaly), leaving the conditions:

$$\text{gauge anomaly: } \sum_L q^3 = 0 , \qquad (12.2.7a)$$

$$\text{mixed anomaly: } \sum_L q = 0 . \qquad (12.2.7b)$$

If there is more than one gauge group the necessary and sufficient condition for anomaly cancellation is that the above hold for every linear combination of generators.

The IIA theory is parity-symmetric and so automatically anomaly-free, while the others have potential anomalies. In ten dimensions the anomaly involves amplitudes with six currents (the hexagon graph) and is of fifth order in the field strengths and curvatures. The calculation has been done in detail in the literature; we will not repeat it here but just quote the result. First we must establish some notation. For the gravitational field, it is convenient to work in the tangent space (tetrad) formalism. In this formalism there are two local symmetries, coordinate invariance and local Lorentz transformations

$$e_\mu{}^p(x)' = e_\mu{}^q(x)\Theta_q{}^p(x) . \qquad (12.2.8)$$

Both are necessary for the decoupling of unphysical degrees of freedom, and in fact when there is a coordinate anomaly one can by adding counterterms to the action convert it to a Lorentz anomaly, which closely resembles a gauge anomaly. The Riemann tensor can be written $R_{\mu\nu}{}^p{}_q$, with mixed spacetime and tangent space indices, and in this way be regarded as a 2-form R_2 which is a $d \times d$ tangent space matrix. Similarly

$e_\mu{}^q$ is written as a one-form which is a column vector in tangent space, and the field strength is written as a matrix 2-form $F_2 = F_2^a t_r^a$; here r is the representation carried by the matter.

The anomaly can be written in compact form in terms of an *anomaly polynomial*, a formal $(d+2)$-form $\hat{I}_{d+2}(R_2, F_2)$. This has the property that it is the exterior derivative of a $(d+1)$-form, whose variation is the exterior derivative of a d-form:

$$\hat{I}_{d+2} = d\hat{I}_{d+1} , \quad \delta\hat{I}_{d+1} = d\hat{I}_d . \tag{12.2.9}$$

The anomalous variation of the path integral is then

$$\delta \ln Z = \frac{-i}{(2\pi)^5} \int \hat{I}_d(F_2, R_2) . \tag{12.2.10}$$

The anomaly cancellation condition is that the total anomaly polynomial vanish.

In the ten-dimensional supergravity theories there are three kinds of chiral field: the spinors **8** and **8'**, the gravitinos **56** and **56'**, and the field strengths [5]$_+$ and [5]$_-$ of the IIB theory. Parity interchanges the two fields in each pair so these make opposite contributions to the anomaly. The anomaly polynomials have been calculated. For the Majorana–Weyl **8**,

$$\begin{aligned}
\hat{I}_8(F_2, R_2) = &-\frac{\mathrm{Tr}(F_2^6)}{1440} \\
&+ \frac{\mathrm{Tr}(F_2^4)\mathrm{tr}(R_2^2)}{2304} - \frac{\mathrm{Tr}(F_2^2)\mathrm{tr}(R_2^4)}{23040} - \frac{\mathrm{Tr}(F_2^2)[\mathrm{tr}(R_2^2)]^2}{18432} \\
&+ \frac{n\,\mathrm{tr}(R_2^6)}{725760} + \frac{n\,\mathrm{tr}(R_2^4)\mathrm{tr}(R_2^2)}{552960} + \frac{n\,[\mathrm{tr}(R_2^2)]^3}{1327104} .
\end{aligned} \tag{12.2.11}$$

For the Majorana–Weyl **56**,

$$\hat{I}_{56}(F_2, R_2) = -495\frac{\mathrm{tr}(R_2^6)}{725760} + 225\frac{\mathrm{tr}(R_2^4)\mathrm{tr}(R_2^2)}{552960} - 63\frac{[\mathrm{tr}(R_2^2)]^3}{1327104} . \tag{12.2.12}$$

For the self-dual tensor,

$$\hat{I}_{\mathrm{SD}}(R_2) = 992\frac{\mathrm{tr}(R_2^6)}{725760} - 448\frac{\mathrm{tr}(R_2^4)\mathrm{tr}(R_2^2)}{552960} + 128\frac{[\mathrm{tr}(R_2^2)]^3}{1327104} . \tag{12.2.13}$$

The 'tr' denotes the trace on the tangent space indices p, q. In this section we will write products and powers of forms without the $\wedge$, to keep expressions compact. The 'Tr' denotes the trace of the field strength in the representation carried by the fermion. In particular, $n = \mathrm{Tr}(1)$ is the dimension of the representation. If the representation r is reducible, $r = r_1 + r_2 + \dots$, the corresponding traces add: $\mathrm{Tr}_r = \mathrm{Tr}_{r_1} + \mathrm{Tr}_{r_2} + \dots$.

Now let us consider the anomalies in the various chiral string theories.

Type IIB anomalies

In type IIB supergravity there are two $\mathbf{8}'$s, two $\mathbf{56}$s, and one $[5]_+$, giving the total anomaly polynomial

$$\hat{I}_{\text{IIB}}(R_2) = -2\hat{I}_{\mathbf{8}}(R_2) + 2\hat{I}_{\mathbf{56}}(R_2) + \hat{I}_{\text{SD}}(R_2) = 0 . \qquad (12.2.14)$$

There are no gauge fields so only the three purely gravitational terms enter, and the coefficients of these conspire to produce zero total anomaly. From the point of view of the low energy theory, this is somewhat miraculous. In fact, it seems accidental that there are any consistent chiral theories at all. There are three anomaly terms that must vanish and three free parameters — the net number of $\mathbf{8}$ minus $\mathbf{8}'$, of $\mathbf{56}$ minus $\mathbf{56}'$, and of $[5]_+$ minus $[5]_-$. Barring a numerical coincidence the only solution would be that all these differences vanish, a nonchiral theory. One can view string theory as explaining this numerical coincidence: the conditions for the internal consistency of string theory are reasonably straightforward, and having satisfied them, the low energy theory must be nonanomalous.

The existence of consistent chiral theories is a beautiful example of the consistency of string theory, and is also of some practical importance. The fermion content of the Standard Model is chiral — the weak interactions violate parity. This chiral property seems to be an important clue, and it has been a difficulty for many previous unifying ideas. Of course, in string theory we are still talking about the ten-dimensional spectrum, but we will see in later chapters that there is some connection between chirality in higher dimensions and in four.

Type I and heterotic anomalies

The type I and heterotic strings have the same low energy limits so we can discuss their anomalies together. There is an immediate problem. The only charged chiral field is the $\mathbf{8}'$, so there is apparently no possibility of cancellation of gauge and mixed anomalies. This is a paradox because we have claimed that these string theories were constructed to satisfy all the conditions for unitarity. Our arguments were perhaps heuristic in places, but it is not so hard to carry out an explicit string calculation at one loop and verify the decoupling of null states. This contradiction led Green and Schwarz to a careful study of the structure of the string amplitude, and they found a previously unknown, and canceling, contribution to the anomaly.

The assertion that the anomaly cannot be canceled by local counterterms takes into account only terms constructed from the gauge field and metric. Consider, however, the Chern–Simons interaction

$$S' = \int B_2 \, \text{Tr}(F_2^4) \qquad (12.2.15)$$

(in any representation r, for now). This is invariant under gauge transformations of the vector potential because it is constructed from the field strength, and under the 2-form transformation $\delta B_2 = d\lambda_1$ using integration by parts and the Bianchi identity for the field strength. However, we have seen that in the $N = 1$ supergravity theory the 2-form has a nontrivial gauge transformation $\delta B_2 \propto \text{Tr}(\lambda dA_1)$, eq. (12.1.40). Then

$$\delta S' \propto \int \text{Tr}(\lambda dA_1)\text{Tr}(F_2^4) \; . \qquad (12.2.16)$$

This is of the form (12.2.9) with

$$\hat{I}_d \propto \text{Tr}(\lambda dA_1)\text{Tr}(F_2^4) \; , \quad \hat{I}_{d+1} \propto \text{Tr}(AF_2)\text{Tr}(F_2^4) \; , \quad (12.2.17a)$$
$$\hat{I}_{d+2} \propto \text{Tr}(F_2^2)\text{Tr}(F_2^4) \; . \qquad (12.2.17b)$$

Thus it can cancel an anomaly of this form. Similarly, the variation of

$$S'' = \int B_2 \, [\text{Tr}(F_2^2)]^2 \qquad (12.2.18)$$

can cancel the anomaly polynomial $[\text{Tr}(F_2^2)]^3$.

The pure gauge anomaly polynomial has a different group-theoretic form $\text{Tr}_a(F_2^6)$, now in the adjoint representation because the charged fields are gauginos. However, for certain algebras there are relations between the different invariants. For $SO(n)$, it is convenient to convert all traces into the vector representation. The fermions of the supergravity theory are always in the adjoint; in terms of the vector traces these are

$$\text{Tr}_a(t^2) = (n-2)\text{Tr}_v(t^2) \; , \qquad (12.2.19a)$$
$$\text{Tr}_a(t^4) = (n-8)\text{Tr}_v(t^4) + 3\,\text{Tr}_v(t^2)\text{Tr}_v(t^2) \; , \qquad (12.2.19b)$$
$$\text{Tr}_a(t^6) = (n-32)\text{Tr}_v(t^6) + 15\,\text{Tr}_v(t^2)\text{Tr}_v(t^4) \; . \qquad (12.2.19c)$$

Here t is any linear combination of generators, but this implies the same relations for symmetrized products of different generators. Symmetrized products appear when the anomaly polynomial is expanded in sums over generators, because the 2-forms F_2^a and F_2^b commute.

The last of these identities implies that precisely for $SO(32)$ the gauge anomaly $\text{Tr}_a(F_2^6)$ is equal to a product of lower traces and so can be canceled by the variations of S' and S''. This is the *Green–Schwarz mechanism*. This is of course the same SO group that arises in the type I and heterotic strings, and not surprisingly the necessary interactions occur in these theories with the correct coefficients.

Also for the group E_8, the sixth order trace can be reduced to lower order traces,

$$\text{Tr}_a(t^4) = \frac{1}{100}[\text{Tr}_a(t^2)]^2 \, , \quad \text{Tr}_a(t^6) = \frac{1}{7200}[\text{Tr}_a(t^2)]^3 \, . \tag{12.2.20}$$

Using the relation $\text{Tr}_a(t^m) = \text{Tr}_{a1}(t^m) + \text{Tr}_{a2}(t^m)$, it follows that the sixth power trace can be reduced for $E_8 \times E_8$ as well (with only one factor of E_8 the *gravitational* anomaly does not cancel, as we will see).

Now let us consider the full anomaly, including mixed anomalies. Generalizing S' and S'' to

$$\int B_2 X_8(F_2, R_2) \, , \tag{12.2.21}$$

makes it possible to cancel an anomaly of the form $\text{Tr}(F_2^2)X_8(F_2, R_2)$ for arbitrary 8-form $X_8(F_2, R_2)$. In addition, the B_2 field strength includes also a gravitational Chern–Simons term:

$$\tilde{H}_3 = dB_2 - c\omega_{3Y} - c'\omega_{3L} \tag{12.2.22}$$

with c and c' constants. Here $\omega_{3Y} = A_1 dA_1 - i\frac{2}{3}A_1^3$ is the gauge Chern–Simons term as before and

$$\omega_{3L} = \omega_1 d\omega_1 + \frac{2}{3}\omega_1^3 \tag{12.2.23}$$

is the Lorentz Chern–Simons term, with $\omega_1 \equiv \omega_\mu{}^p{}_q dx^\mu$ the spin connection. This has the property

$$\delta\omega_{3L} = d\text{tr}(\Theta d\omega_1) \, . \tag{12.2.24}$$

The combined Lorentz and Yang–Mills transformation law must then be

$$\delta A_1 = d\lambda \, , \tag{12.2.25a}$$
$$\delta\omega_1 = d\Theta \, , \tag{12.2.25b}$$
$$\delta B_2 = c\text{Tr}(\lambda dA_1) + c'\text{tr}(\Theta d\omega_1) \, . \tag{12.2.25c}$$

Again, we only indicate the leading, Abelian, terms. With this transformation the interaction (12.2.21) cancels an anomaly of the form

$$[c\text{Tr}(F_2^2) + c'\text{Tr}(R_2^2)]X_8(F_2, R_2) \, . \tag{12.2.26}$$

The gravitational Chern–Simons term was not included in the earlier low energy effective action because it is a higher derivative effect. The spin connection ω_1 is proportional to the derivative of the tetrad, so the gravitational term in the field strength (12.2.22) contains three derivatives where the other terms contain one. However, its contribution is important in discussing the anomaly.

The chiral fields of $N = 1$ supergravity with gauge group g are the gravitino **56**, a neutral fermion **8′**, and an **8** gaugino in the adjoint representation, for total anomaly

$$\hat{I}_1 = \hat{I}_{56}(R_2) - \hat{I}_8(R_2) + \hat{I}_8(F_2, R_2)$$

$$= \frac{1}{1440}\left\{ -\text{Tr}_a(F_2^6) + \frac{1}{48}\text{Tr}_a(F_2^2)\text{Tr}_a(F_2^4) - \frac{[\text{Tr}_a(F_2^2)]^3}{14400} \right\}$$

$$+ (n - 496)\left\{ \frac{\text{tr}(R_2^6)}{725760} + \frac{\text{tr}(R_2^4)\text{tr}(R_2^2)}{552960} + \frac{[\text{tr}(R_2^2)]^3}{1327104} \right\} + \frac{Y_4 X_8}{768}.$$

$$(12.2.27)$$

Here

$$Y_4 = \text{tr}(R_2^2) - \frac{1}{30}\text{Tr}_a(F_2^2), \tag{12.2.28a}$$

$$X_8 = \text{tr}(R_2^4) + \frac{[\text{tr}(R_2^2)]^2}{4} - \frac{\text{Tr}_a(F_2^2)\text{tr}(R_2^2)}{30} + \frac{\text{Tr}_a(F_2^4)}{3} - \frac{[\text{Tr}_a(F_2^2)]^2}{900}. \tag{12.2.28b}$$

The anomaly has been organized into a sum of three terms. The third is of the factorized form that can be canceled by the Green–Schwarz mechanism but the first two cannot, and so for the theory to be anomaly-free the combination of traces on the first line must vanish for the adjoint representation, and the total number of gauge generators must be 496. For the groups $SO(32)$ and $E_8 \times E_8$, both properties hold.[1] The net anomaly is then

$$\frac{Y_4 X_8}{768}. \tag{12.2.29}$$

Of the various additional heterotic string theories constructed in the previous chapter, all but the diagonal theory are chiral, and in all cases the anomalies factorize.

In six-dimensional compactifications, some of which will be discussed in chapter 19, there can be multiple tensors. The Green–Schwarz mechanism can then cancel a sum of products $Y_4 X_4$. Also, the same mechanism generalizes to forms of other rank; for example, a scalar in place of B_2 can cancel an anomaly $Y_2 X_d$. For $d = 4$ this will arise in section 18.7.

Relation to string theory

From the low energy point of view, the cancellation of the anomaly involves several numerical accidents: the identity for the gauge traces, the correct number of generators, the factorized form (12.2.27). Again, these are explained by the existence of consistent string theories. In constructing new string theories, it is in principle not necessary to check the low

---

[1] They also hold for $E_8 \times U(1)^{248}$ and $U(1)^{496}$, but no corresponding string theories are known.

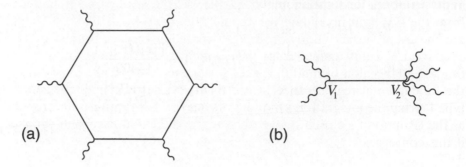

Fig. 12.1. Graphs contributing to the anomalies. One of the six external lines is a current and the others are gauge or gravitational fields: (a) hexagon graph; (b) canceling graph from exchange of $B_{\mu\nu}$ field.

energy anomaly, since this is guaranteed to vanish if the string consistency requirements have been satisfied. In practice, it is very useful as a check on the calculations and as a check that no subtle inconsistency has been overlooked.

In terms of Feynman graphs, the unphysical gauge and gravitational polarizations decouple by a cancellation between the two graphs of figure 12.1. The loop is the usual anomaly graph. The vertices of the tree graph come respectively from the H_3 kinetic term and the interaction (12.2.21). It is curious that a tree graph can cancel a loop, and it is interesting to look more closely at the coupling constant dependence. As discussed below eq. (12.1.11), in order to do the loop counting we need to write the R–R field as $C_2 = e^{-\Phi}C_2'$. Both vertices in figure 12.1(b) are then proportional to $e^{-\Phi}$ and so are 'half-loop' effects; they come from the disk amplitude. In the heterotic string no rescaling is needed. The vertex V_1 is proportional to $e^{-2\Phi}$ and so is a tree-level effect, while the vertex V_2 does not depend on the dilaton and so is actually a one-loop effect.

In each string theory, the hexagon loop and the tree graph arise from the same topology but different limits of moduli space. In the type I theory, the topology is the cylinder. The loop graph is from the short-cylinder limit and the tree graph from the long-cylinder limit. In the heterotic theory, the topology is the torus. The hexagon graph is from the limit $\tau_2 \to \infty$, while the tree graph is from the limit where two vertex operators approach one another.

In the heterotic string, the gauge group was determined by the requirement of modular invariance. In the type I string it was determined by cancellation of tadpole divergences. The relation with the field theory anomaly is as follows. One can prove the decoupling of null states for-

mally in either field theory or string theory; the issue is whether terms
from the UV limit invalidate the formal argument. In string theory these
are the usual surface terms on moduli space. In the heterotic string the
effective UV cutoff comes from the restriction of the integration to the
fundamental region of moduli space. Surface terms from the boundary of
the fundamental region cancel *if* the theory is modular-invariant. In the
type I string the integration is not cut off but the 'UV' limit is reinterpreted
as the IR limit of a closed string exchange, and the anomaly then vanishes
if this converges.

12.3 Superspace and superfields

To formulate superstring perturbation theory it is useful to give superconformal symmetry a more geometric interpretation. To do this we need a
supermanifold, a world-sheet with one ordinary complex coordinate z and
one anticommuting complex coordinate θ, with

$$\theta^2 = \bar{\theta}^2 = \{\theta, \bar{\theta}\} = 0 \,. \tag{12.3.1}$$

What do we mean by anticommuting coordinates? Because of the anticommuting property, the Taylor series for any function of θ and $\bar{\theta}$ terminates. We can then think of any function on a supermanifold as the
collection of ordinary functions appearing in the Taylor expansion. However, just as the operation '$\int d\theta$' has so many of the properties of ordinary
integration that it is useful to call it integration, θ behaves so much like a
coordinate that it is useful to think of a manifold with both ordinary and
anticommuting coordinates.

We can think about ordinary conformal transformations as follows.
Under a general change of world-sheet coordinates $z'(z, \bar{z})$ the derivative
transforms as

$$\partial_z = \frac{\partial z'}{\partial z}\partial_{z'} + \frac{\partial \bar{z}'}{\partial z}\partial_{\bar{z}'} \,. \tag{12.3.2}$$

The conformal transformations are precisely those that take ∂_z into a
multiple of itself.

Define the *superderivatives*,

$$D_\theta = \partial_\theta + \theta\partial_z \,, \quad D_{\bar{\theta}} = \partial_{\bar{\theta}} + \bar{\theta}\partial_{\bar{z}} \,, \tag{12.3.3}$$

which have the properties

$$D_\theta^2 = \partial_z \,, \quad D_{\bar{\theta}}^2 = \partial_{\bar{z}} \,, \quad \{D_\theta, D_{\bar{\theta}}\} = 0 \,. \tag{12.3.4}$$

A superconformal transformation $z'(z, \theta)$, $\theta'(z, \theta)$ is one that takes D_θ into
a multiple of itself. From

$$D_\theta = D_\theta\theta'\partial_{\theta'} + D_\theta z'\partial_{z'} + D_\theta\bar{\theta}'\partial_{\bar{\theta}'} + D_\theta\bar{z}'\partial_{\bar{z}'} \,, \tag{12.3.5}$$

it follows that a superconformal transformation satisfies

$$D_\theta \bar{\theta}' = D_\theta \bar{z}' = 0 , \quad D_\theta z' = \theta' D_\theta \theta' , \tag{12.3.6}$$

and so

$$D_\theta = (D_\theta \theta') D_{\theta'} . \tag{12.3.7}$$

Using $D_\theta^2 = \partial_z$, this also implies

$$\partial_{\bar{z}} z' = \partial_{\bar{\theta}} z' = \partial_{\bar{z}} \theta' = \partial_{\bar{\theta}} \theta' = 0 \tag{12.3.8}$$

and the conjugate relations. These conditions can be solved to express a general superconformal transformation in terms of a holomorphic function $f(z)$ and an anticommuting holomorphic function $g(z)$,

$$z'(z, \theta) = f(z) + \theta g(z) h(z) , \quad \theta'(z, \theta) = g(z) + \theta h(z) , \tag{12.3.9a}$$

$$h(z) = \pm \left[\partial_z f(z) + g(z) \partial_z g(z) \right]^{1/2} . \tag{12.3.9b}$$

Infinitesimally,

$$\delta z = \epsilon[v(z) - i\theta\eta(z)] , \quad \delta\theta = \epsilon[-i\eta(z) + \tfrac{1}{2}\theta\partial v(z)] \tag{12.3.10}$$

with ϵ and v commuting and η anticommuting. These satisfy the super-conformal algebra (10.1.11).

A *tensor superfield* of weight $(h, \tilde{h})$ transforms as

$$(D_\theta \theta')^{2h} (D_{\bar{\theta}} \bar{\theta}')^{2\tilde{h}} \phi'(z', \bar{z}') = \phi(z, \bar{z}) , \tag{12.3.11}$$

where z stands for (z, θ). This is analogous to the transformation (2.4.15) of a conformal tensor. Under an infinitesimal superconformal transformation $\delta\theta = \epsilon\eta(z)$,

$$\delta\phi(z, \bar{z}) = -\epsilon\left[2h\theta\partial\eta(z) + \eta(z)Q_\theta + 2\tilde{h}\bar{\theta}\bar{\partial}\bar{\eta}(\bar{z}) + \bar{\eta}(\bar{z})Q_{\bar{\theta}} \right]\phi(z, \bar{z}) , \tag{12.3.12}$$

where $Q_\theta = \partial_\theta - \theta\partial_z$ and $Q_{\bar{\theta}} = \partial_{\bar{\theta}} - \bar{\theta}\partial_{\bar{z}}$. Expand in powers of θ, and concentrate for simplicity on the holomorphic side,

$$\phi(z) = \mathcal{O}(z) + \theta\Psi(z) . \tag{12.3.13}$$

Then the infinitesimal transformation (12.3.12) is

$$\delta\mathcal{O} = -\epsilon\eta\Psi , \quad \delta\Psi = -\epsilon[2h\partial\eta\mathcal{O} + \eta\partial\mathcal{O}] . \tag{12.3.14}$$

In terms of the OPE coefficients (10.3.4) this is

$$G_{-1/2} \cdot \mathcal{O} = \Psi , \quad G_r \cdot \mathcal{O} = 0 , \, r \geq \tfrac{1}{2} , \tag{12.3.15a}$$

$$G_{-1/2} \cdot \Psi = \partial\mathcal{O} , \quad G_{1/2} \cdot \Psi = 2h\mathcal{O} , \quad G_r \cdot \Psi = 0 , \, r \geq \tfrac{3}{2} . \tag{12.3.15b}$$

Either by using the NS algebra, or by considering a purely conformal transformation $\delta z = \epsilon v(z)$, one finds that $\mathcal{O}$ is a tensor of weight h and Ψ a tensor of weight $h + \tfrac{1}{2}$, so that both are annihilated by all of the

Virasoro lowering generators. The lowest component $\mathcal{O}$ of the tensor superfield is a *superconformal primary field*, being annihilated by all the lowering generators of the NS algebra.

The analog of a rigid translation is a rigid world-sheet supersymmetry transformation, $\delta\theta = -\frac{1}{2}i\epsilon\eta$, $\delta z = -i\epsilon\theta\eta$. The Ward identity for T_F then gives the corresponding generator

$$G_{-1/2} \cdot \sim -iQ_\theta = -i(\partial_\theta - \theta\partial_z) \qquad (12.3.16)$$

acting on any superfield. This generalizes the relation $L_{-1} \cdot \sim \partial_z$ obtained in CFT.

Actions and backgrounds

The super-Jacobian (A.2.29) of the transformation (12.3.9) is

$$dz' \, d\theta' = dz \, d\theta \, D_\theta \theta' . \qquad (12.3.17)$$

To make a superconformally invariant action, the Lagrangian density must therefore be a weight $(\frac{1}{2}, \frac{1}{2})$ tensor superfield. The product of two tensor superfields is a superfield, with the weights additive, $(h, \tilde{h}) = (h_1, \tilde{h}_1) + (h_2, \tilde{h}_2)$. Also, the superderivative D_θ takes a $(0, \tilde{h})$ tensor superfield into a $(\frac{1}{2}, \tilde{h})$ tensor superfield, and $D_{\bar\theta}$ takes an $(h, 0)$ tensor superfield into an $(h, \frac{1}{2})$ tensor superfield.

These rules make it easy to write superconformal-invariant actions. A simple invariant action can be built from d weight $(0, 0)$ tensors $X^\mu(z, \bar{z})$:

$$S = \frac{1}{4\pi} \int d^2z \, d^2\theta \, D_{\bar\theta} X^\mu D_\theta X_\mu . \qquad (12.3.18)$$

The Taylor expansion in θ is

$$X^\mu(z, \bar{z}) = X^\mu + i\theta\psi^\mu + i\bar{\theta}\tilde{\psi}^\mu + \theta\bar{\theta}F^\mu . \qquad (12.3.19)$$

In this section we set $\alpha' = 2$ to make the structure clearer; the reader can restore dimensions by $X^\mu \to X^\mu(2/\alpha')^{1/2}$. The integral $d^2\theta = d\theta \, d\bar{\theta}$ in the action picks out the coefficient of $\bar{\theta}\theta$,

$$S = \frac{1}{4\pi} \int d^2z \left(\partial_{\bar{z}} X^\mu \partial_z X_\mu + \psi^\mu \partial_{\bar{z}} \psi_\mu + \tilde{\psi}^\mu \partial_z \tilde{\psi}_\mu + F^\mu F_\mu \right) . \qquad (12.3.20)$$

The field F^μ is an auxiliary field, meaning that it is completely determined by the equation of motion; in fact it vanishes here. The rest of the action is the same as the earlier (10.1.5), as are the superconformal transformations of the component fields.

Many of the earlier results can be recast in superfield form. The equation of motion is

$$D_\theta D_{\bar\theta} X^\mu(z, \bar{z}) = 0 . \qquad (12.3.21)$$

For the OPE, invariance under translations and rigid supersymmetry transformations implies that it is a function only of $z_1 - z_2 - \theta_1\theta_2$ and $\theta_1 - \theta_2$, and their conjugates. In this case,

$$X^\mu(z_1,\bar{z}_1)X^\nu(z_2,\bar{z}_2) \sim -\eta^{\mu\nu} \ln |z_1 - z_2 - \theta_1\theta_2|^2 \,, \qquad (12.3.22)$$

as one can verify by expanding both sides in the anticommuting variables.

The superconformal ghost action is constructed from $(\lambda - \frac{1}{2}, 0)$ and $(1 - \lambda, 0)$ tensor superfields B and C,

$$S_{BC} = \frac{1}{2\pi} \int d^2z \, d^2\theta \, BD_{\bar{\theta}}C \,. \qquad (12.3.23)$$

The equation of motion is

$$D_{\bar{\theta}}B = D_{\bar{\theta}}C = 0 \,. \qquad (12.3.24)$$

Acting on this equation with $D_{\bar{\theta}}$ gives $\partial_{\bar{z}}B = \partial_{\bar{z}}C = 0$, and so also $\partial_{\bar{\theta}}B = \partial_{\bar{\theta}}C = 0$. The equation of motion thus implies

$$B(z) = \beta(z) + \theta b(z) \,, \quad C(z) = c(z) + \theta\gamma(z) \,. \qquad (12.3.25)$$

This is the same as the theory (10.1.17). The OPE is

$$B(z_1)C(z_2) \sim \frac{\theta_1 - \theta_2}{z_1 - z_2 - \theta_1\theta_2} = \frac{\theta_1 - \theta_2}{z_1 - z_2} \,. \qquad (12.3.26)$$

The superfield form makes it easy to write down the nonlinear sigma model action

$$\begin{aligned}
S &= \frac{1}{4\pi} \int d^2z \, d^2\theta \, [G_{\mu\nu}(X) + B_{\mu\nu}(X)]D_{\bar{\theta}}X^\nu D_\theta X^\mu \\
&= \frac{1}{4\pi} \int d^2z \, \Big\{ [G_{\mu\nu}(X) + B_{\mu\nu}(X)]\partial_z X^\mu \partial_{\bar{z}} X^\nu \\
&\quad + G_{\mu\nu}(X)(\psi^\mu \mathscr{D}_{\bar{z}}\psi^\nu + \tilde{\psi}^\mu \mathscr{D}_z \tilde{\psi}^\nu) + \tfrac{1}{2} R_{\mu\nu\rho\sigma}(X)\psi^\mu \psi^\nu \tilde{\psi}^\rho \tilde{\psi}^\sigma \Big\} \,,
\end{aligned}$$
$$(12.3.27)$$

after eliminating the auxiliary field. The Christoffel connection and antisymmetric tensor field strength combine in the covariant derivative,

$$\mathscr{D}_{\bar{z}}\psi^\nu = \partial_{\bar{z}}\psi^\nu + \Big[\Gamma^\nu_{\rho\sigma}(X) + \tfrac{1}{2}H^\nu_{\rho\sigma}(X)\Big]\partial_{\bar{z}}X^\rho \psi^\sigma \,, \qquad (12.3.28a)$$

$$\mathscr{D}_z\tilde{\psi}^\nu = \partial_z\tilde{\psi}^\nu + \Big[\Gamma^\nu_{\rho\sigma}(X) - \tfrac{1}{2}H^\nu_{\rho\sigma}(X)\Big]\partial_z X^\rho \tilde{\psi}^\sigma \,. \qquad (12.3.28b)$$

This describes a general NS–NS background in either type II string theory. R–R backgrounds are hard to describe in this framework because the superconformal transformations have branch cuts at the operators. The dilaton does not appear in the flat world-sheet action but does appear in the superconformal generators. The reader should beware of a common convention in the literature, $B^{\text{here}}_{\mu\nu} = 2B^{\text{there}}_{\mu\nu}$.

All the above applies to the heterotic string, using only $\bar\theta$ and not θ. One needs the superfields

$$X^\mu = X^\mu + i\bar\theta\tilde\psi^\mu , \tag{12.3.29a}$$

$$\lambda^A = \lambda^A + \bar\theta G^A . \tag{12.3.29b}$$

The field G^A is auxiliary. The nonlinear sigma model is

$$
\begin{aligned}
S &= \frac{1}{4\pi}\int d^2z\, d\bar\theta \left\{ [G_{\mu\nu}(X) + B_{\mu\nu}(X)]\partial_z X^\mu D_{\bar\theta} X^\nu - \lambda^A \mathscr{D}_{\bar\theta}\lambda^A \right\} \\
&= \frac{1}{4\pi}\int d^2z \left\{ [G_{\mu\nu}(X) + B_{\mu\nu}(X)]\partial_z X^\mu \partial_{\bar z} X^\nu + G_{\mu\nu}(X)\tilde\psi^\mu \mathscr{D}_z \tilde\psi^\nu \right. \\
&\qquad\qquad \left. + \lambda^A \mathscr{D}_{\bar z}\lambda^A + \tfrac{1}{2}F_{\rho\sigma}^{AB}(X)\lambda^A\lambda^B\tilde\psi^\rho\tilde\psi^\sigma \right\} , \tag{12.3.30}
\end{aligned}
$$

where $\mathscr{D}_z\tilde\psi^\nu$ is as above and

$$\mathscr{D}_{\bar\theta}\lambda^A = D_{\bar\theta}\lambda^A - iA_\mu^{AB}(X)D_{\bar\theta}X^\mu\lambda^B , \tag{12.3.31a}$$

$$\mathscr{D}_{\bar z}\lambda^A = \partial_{\bar z}\lambda^A - iA_\mu^{AB}(X)\partial_{\bar z}X^\mu\lambda^B . \tag{12.3.31b}$$

It is worth noting that the modified gauge transformation of the 2-form potential, which played an important role in the cancellation of spacetime anomalies, has a simple origin in terms of a world-sheet anomaly. A spacetime gauge transformation

$$\delta A_\mu^{AB} = D_\mu\chi^{AB} , \qquad \delta\lambda^A = i\chi^{AB}\lambda^B \tag{12.3.32}$$

leaves the classical action invariant. However, this acts only on left-moving world-sheet fermions and so has an anomaly in the world-sheet path integral. We can use the result (12.2.3) with $\hat k_L = 1$, $\hat k_R = 0$, and

$$A_{\bar z}^{AB}(z,\bar z) = \frac{1}{2\pi}A_\mu^{AB}(X)\partial_{\bar z}X^\mu , \tag{12.3.33}$$

the factor of 2π correcting for the nonstandard normalization of the Noether current in CFT. Then after the addition of a counterterm,

$$\delta Z[A] = \frac{1}{8\pi}\int d^2z\, \mathrm{Tr}_{\mathrm{v}}[\chi(X)F_{\mu\nu}(X)]\partial_z X^\mu \partial_{\bar z}X^\nu . \tag{12.3.34}$$

This is precisely canceled if we also change the background,

$$\delta B_{\mu\nu} = \frac{1}{2}\mathrm{Tr}_{\mathrm{v}}(\chi F_{\mu\nu}) . \tag{12.3.35}$$

Comparing to the supergravity result (12.1.40) gives

$$\frac{\kappa_{10}^2}{g_{10}^2} = \frac{1}{2} \to \frac{\alpha'}{4} . \tag{12.3.36}$$

Noting that the left-hand side has units of L^2, we have restored α' by introducing one factor of $\alpha'/2$. This is the correct result for the relation between gravitational and gauge couplings in the heterotic string. For future reference, let us note that if we study a vacuum with a nonzero dilaton, the physical couplings differ from the parameters in the action by an additional e^{Φ}, so that also

$$\frac{\kappa^2}{g_{YM}^2} \equiv \frac{e^{2\Phi}\kappa_{10}^2}{e^{2\Phi}g_{10}^2} = \frac{\alpha'}{4} . \qquad (12.3.37)$$

(We will discuss slightly differing conventions for the gauge coupling in chapter 18.)

Vertex operators

Recall that the bosonic string vertex operators came in two forms. The state–operator mapping gave them as $\tilde{c}c$ times a $(1,1)$ matter tensor. In the gauge-fixed Polyakov path integral this was the appropriate form for a vertex operator whose coordinate had been fixed. For an integrated vertex operator the $\tilde{c}c$ was omitted, replaced by a d^2z. The vertex operators of the superstring have a similar variety of forms, or *pictures*. We will derive this idea here by analogy to the bosonic string, and explain it in a more geometric way in section 12.5.

The state–operator mapping in chapter 10 gave the NS–NS vertex operators as

$$\delta(\gamma)\delta(\tilde{\gamma}) = e^{-\phi-\tilde{\phi}} \qquad (12.3.38)$$

times a $(\frac{1}{2}, \frac{1}{2})$ superconformal tensor. These are the analog of the fixed bosonic vertex operators. We have seen that the superconformal tensors are the lowest components of superfields, which do indeed correspond to the value of the superfield when θ and $\bar{\theta}$ are fixed at 0. Calling this tensor $\mathcal{O}$, eq. (12.3.15) gives the vertex operator integrated over θ and $\bar{\theta}$ as

$$\mathscr{V}^{0,0} = G_{-1/2}\tilde{G}_{-1/2} \cdot \mathcal{O} . \qquad (12.3.39)$$

This operator appears without the $\delta(\gamma)\delta(\tilde{\gamma})$. The nonlinear sigma model action has just this form, the $d^2\theta$ integral of a $(\frac{1}{2}, \frac{1}{2})$ superfield. It is conventional to label vertex operators by their ϕ and $\tilde{\phi}$ charges as here, so that an operator of charges $(q, \tilde{q})$ is said to be in the $(q, \tilde{q})$ *picture*. The θ-integrated operator (12.3.39) is in the (0,0) picture and the fixed operators

$$\mathscr{V}^{-1,-1} = e^{-\phi-\tilde{\phi}}\mathcal{O} \qquad (12.3.40)$$

are in the $(-1,-1)$ picture. Of course, all of this extends to the open and heterotic cases with only one copy of the superconformal algebra, so we would have there the -1 and 0 pictures.

Let us consider as an example the massless states

$$\psi^\mu_{-1/2}\tilde\psi^\nu_{-1/2}|0;k\rangle_{\text{NS}} ,\tag{12.3.41}$$

with vertex operators

$$\mathcal{V}^{-1,-1} = g_c e^{-\phi-\tilde\phi}\psi^\mu\tilde\psi^\nu e^{ik\cdot X} .\tag{12.3.42}$$

The bosonic coordinates can be integrated or fixed independently of the fermionic ones, so for convenience we treated them as integrated. From

$$
\begin{aligned}
&G_{-1/2}\tilde G_{-1/2}\psi^\mu_{-1/2}\tilde\psi^\nu_{-1/2}|0;k\rangle_{\text{NS}}\\
&= -(\alpha^\mu_{-1} + \alpha_0\cdot\psi_{-1/2}\psi^\mu_{-1/2})(\tilde\alpha^\nu_{-1} + \tilde\alpha_0\cdot\tilde\psi_{-1/2}\tilde\psi^\nu_{-1/2})|0;k\rangle_{\text{NS}} ,
\end{aligned}\tag{12.3.43}
$$

we obtain the integrated vertex operators

$$\mathcal{V}^{0,0} = -\frac{2g_c}{\alpha'}(i\partial_z X^\mu + \tfrac12\alpha' k\cdot\psi\,\psi^\mu)(i\partial_{\bar z} X^\nu + \tfrac12\alpha' k\cdot\tilde\psi\,\tilde\psi^\nu)e^{ik\cdot X} ,\tag{12.3.44}$$

with α' again restored. Note the resemblance to the massless bosonic vertex operators, with additional fermionic terms. These additional terms correspond to the connection and curvature pieces in the nonlinear sigma models. For massless open string vectors,

$$\mathcal{V}^{-1} = g_o t^a \psi^\mu e^{ik\cdot X} ,\tag{12.3.45a}$$
$$\mathcal{V}^0 = g_o(2\alpha')^{-1/2} t^a(i\dot X^\mu + 2\alpha' k\cdot\psi\,\psi^\mu)e^{ik\cdot X} ,\tag{12.3.45b}$$

where t^a is the Chan–Paton factor. For heterotic string vectors,

$$\mathcal{V}^{-1} = g_c \hat k^{-1/2} j^a \tilde\psi^\mu e^{ik\cdot X} ,\tag{12.3.46a}$$
$$\mathcal{V}^0 = g_c(2/\alpha')^{1/2}\hat k^{-1/2} j^a(i\bar\partial X^\mu + \tfrac12\alpha' k\cdot\tilde\psi\,\tilde\psi^\mu)e^{ik\cdot X} .\tag{12.3.46b}$$

For convenient reference, we give the relations between the vertex operator normalizations and the various couplings in the low energy actions of section 12.1:

$$\text{type I:}\quad g_o = g_{\text{YM}}(2\alpha')^{1/2} ;\quad g_{\text{YM}}\equiv g_{10}e^{\Phi/2} ,\tag{12.3.47a}$$

$$\text{heterotic:}\quad g_c = \frac{\kappa}{2\pi} = \frac{\alpha'^{1/2}g_{\text{YM}}}{4\pi} ;\quad \kappa\equiv\kappa_{10}e^\Phi ,\quad g_{\text{YM}}\equiv g_{10}e^\Phi ,\tag{12.3.47b}$$

$$\text{type I/II:}\quad g_c = \frac{\kappa}{2\pi} ;\quad \kappa\equiv\kappa_{10}e^\Phi .\tag{12.3.47c}$$

These can be obtained by comparing the calculations of the next section with the field theory amplitudes. Note that the amplitudes depend on the background value of the dilaton in combination with the parameters κ_{10} and g_{10} from the action.

12.4 Tree-level amplitudes

It is now straightforward to guess the form of the tree-level amplitudes. In the next section we will justify this from a more geometric point of view.

We want the expectation value on the sphere or disk of the product of vertex operators with an appropriate number of bosonic and fermionic coordinates fixed. In the bosonic string it was necessary to fix three vertex operators on the sphere because of the existence of three c and three $\tilde{c}$ zero modes. There are two γ and two $\tilde{\gamma}$ zero modes on the sphere, namely $1, z$ and $1, \bar{z}$: these are holomorphic at infinity for a weight $-\frac{1}{2}$ field. We need this many factors of $\delta(\gamma)$ and $\delta(\tilde{\gamma})$, else the zero-mode integrals diverge. Thus we should fix the θ, $\bar{\theta}$ coordinates of two vertex operators. Similarly on the disk, we must fix the θ coordinates of two open string vertex operators.

We can also see this in the bosonized form. The anomaly in the ϕ current requires a total ϕ charge of -2 and a total $\tilde{\phi}$ charge of -2. Thus we need two vertex operators in the $(-1, -1)$ picture and the rest in the $(0, 0)$ picture. For open strings on the disk (or heterotic strings on the sphere) we need two in the -1 picture and the rest in the 0 picture.

The R sector vertex operators have ϕ charge $-\frac{1}{2}$ from the ghost ground state (10.4.24). This is midway between the fixed and integrated pictures and does not have such a simple interpretation. Nevertheless, conservation of ϕ charge tells us that the sum of the ϕ charges must be -2. Thus for an amplitude with two fermions and any number of bosons we can use the pictures $-\frac{1}{2}$ for the fermions, -1 for one boson, and 0 for the rest. For four fermions and any number of bosons we can use the pictures $-\frac{1}{2}$ for the fermions and 0 for all the bosons. This is enough for all the cases we will treat in this section. To go to six or more fermions we clearly need to understand things better, as we will do in the next section.

Three-point amplitudes

Type I disk amplitudes: According to the discussion above, the type I three-boson amplitude is

$$\frac{1}{\alpha' g_o^2} \left\langle c\mathcal{V}_1^{-1}(x_1) c\mathcal{V}_2^{-1}(x_2) c\mathcal{V}_3^0(x_3) \right\rangle + (\mathcal{V}_1 \leftrightarrow \mathcal{V}_2), \qquad (12.4.1)$$

where we take $x_1 > x_2 > x_3$. The relevant expectation values for massless amplitudes are

$$\langle c(x_1)c(x_2)c(x_3)\rangle = x_{12}x_{13}x_{23} \,, \tag{12.4.2a}$$

$$\left\langle e^{-\phi}(x_1)e^{-\phi}(x_2)\right\rangle = x_{12}^{-1} \,, \tag{12.4.2b}$$

$$\langle \psi^\mu(x_1)\psi^\nu(x_2)\rangle = \eta^{\mu\nu}x_{12}^{-1} \,, \tag{12.4.2c}$$

in the bc, $\beta\gamma$, and ψ CFTs, and

$$\left\langle \psi^\mu e^{ik_1\cdot X}(x_1)\,\psi^\nu e^{ik_2\cdot X}(x_2)\,(i\dot{X}^\rho + 2\alpha'k_3\cdot\psi\,\psi^\rho)e^{ik_3\cdot X}(x_3)\right\rangle$$

$$= 2i\alpha'(2\pi)^{10}\delta^{10}(\textstyle\sum_i k_i)\left(-\frac{\eta^{\mu\nu}k_1^\rho}{x_{12}x_{13}} - \frac{\eta^{\mu\nu}k_2^\rho}{x_{12}x_{23}} + \frac{\eta^{\mu\rho}k_3^\nu - \eta^{\nu\rho}k_3^\mu}{x_{13}x_{23}}\right) \tag{12.4.3}$$

in the combined $X\psi$ CFT. We have given the expectation value within each CFT a simple normalization and included an overall normalization factor $1/g_o^2\alpha'$, equal to the one (6.4.14) found in the bosonic theory. One can verify this normalization by a unitarity calculation as in the bosonic string, with the convention (12.4.1) that we sum separately over the reversed-cyclic orientation (which is always equal in this unoriented theory). That is, an n-particle amplitude is a sum of $(n-1)!$ orderings which are equal in pairs.

Combining these, using momentum conservation and transversality, and including the factor $g_o^3(2\alpha')^{-1/2}$ from the vertex operators, we obtain the type I three-gauge-boson amplitude

$$ig_{\mathrm{YM}}(2\pi)^{10}\delta^{10}(\textstyle\sum_i k_i)\,e_{1\mu}e_{2\nu}e_{3\rho}V^{\mu\nu\rho}\,\mathrm{Tr}_{\mathrm{v}}([t^{a_1},t^{a_2}]t^{a_3}) \,, \tag{12.4.4}$$

where

$$V^{\mu\nu\rho} = \eta^{\mu\nu}k_{12}^\rho + \eta^{\nu\rho}k_{23}^\mu + \eta^{\rho\mu}k_{31}^\nu \,, \tag{12.4.5}$$

and $k_{ij} = k_i - k_j$. This is the ordinary Yang–Mills amplitude, with g_{YM} related to g_o as in eq. (12.3.47a) so as to agree with the definition in the low energy action. Unlike the bosonic open string amplitude (6.5.15) there is no k^3 term and so no F^3 term in the low energy effective action. Indeed, it is known that such a term is not allowed by the $d = 10$, $N = 1$ supersymmetry.

Now consider amplitudes with two fermions and a boson. The CFT amplitudes are

$$\left\langle e^{-\phi/2}(x_1)e^{-\phi/2}(x_2)e^{-\phi}(x_3)\right\rangle = x_{12}^{-1/4}x_{13}^{-1/2}x_{23}^{-1/2} \,, \tag{12.4.6a}$$

$$\langle \Theta_\alpha(x_1)\Theta_\beta(x_2)\rangle = x_{12}^{-5/4}C_{\alpha\beta} \,, \tag{12.4.6b}$$

$$\langle \Theta_\alpha(x_1)\Theta_\beta(x_2)\psi^\mu(x_3)\rangle = 2^{-1/2}(C\Gamma^\mu)_{\alpha\beta}\,x_{12}^{-3/4}x_{13}^{-1/2}x_{23}^{-1/2} \,. \tag{12.4.6c}$$

The ghost amplitude is a free-field calculation, and in principle the matter part can be done in this way as well using bosonization. However, bosonization requires grouping the fermions in pairs and so spoils manifest Lorentz invariance. For explicit calculations it is often easier to use Lorentz and conformal invariance. The two-point amplitude (12.4.6b) is determined up to normalization by these symmetries. Note that $C_{\alpha\beta}$ is the charge conjugation matrix (section B.1), and that only for spinors of opposite chirality is it nonzero: in ten dimensions the product of like-chirality spinors does not include an invariant. The three-point amplitude (12.4.6c) is then deduced by using the OPE

$$\psi^\mu(x)\Theta_\alpha(0) = (2x)^{-1/2}\Theta_\beta(0)\Gamma^\mu_{\beta\alpha} + O(x^{1/2}) \qquad (12.4.7)$$

to determine the x_3 dependence. This amplitude is nonvanishing only for spinors of like chirality. The gaugino-gaugino-gauge-boson amplitude, with respective polarizations $u_{1,2}$ and e^μ, is then[2]

$$ig_{\rm YM}(2\pi)^{10}\delta^{10}(\textstyle\sum_i k_i)\,e_\mu\,\bar{u}_1\Gamma^\mu u_2\,{\rm Tr}_{\rm v}([t^{a_1},t^{a_2}]t^{a_3})\,. \qquad (12.4.8)$$

We have used $u_1^T C\Gamma^\mu u_2 = \bar{u}_1\Gamma^\mu u_2$, from the Majorana condition.[3]

Heterotic sphere amplitudes: The closed string three-point amplitudes are the products of open-string amplitudes. For the heterotic string we need the expectation values of two and three currents. The OPE gives

$$\left\langle\, j^a(z_1)j^b(z_2)\,\right\rangle = \frac{\hat{k}\delta^{ab}}{z_{12}^2} \qquad (12.4.9a)$$

$$\left\langle\, j^a(z_1)j^b(z_2)j^c(z_3)\,\right\rangle = \frac{i\hat{k}f^{abc}}{z_{12}z_{13}z_{23}}\,, \qquad (12.4.9b)$$

where the expectation value without insertions is normalized to unity. Each vertex operator thus needs a factor of $\hat{k}^{-1/2}$ to normalize the two-point function (as discussed in section 9.1). For the ten-dimensional heterotic string $k = 1$. In order to make contact with the discussion in the rest of this chapter, we will use the trace in the vector representation as the inner product, and then it follows from the discussion below eq. (11.5.13) that $\psi^2 = 1$ and $\hat{k} = \frac{1}{2}$. Including these factors, the normalization of the current algebra three-point function is

$$i\hat{k}^{-1/2}f^{abc} = 2^{1/2}{\rm Tr}_{\rm v}([t^a,t^b]t^c)\,. \qquad (12.4.10)$$

The result can also be obtained from the free-fermion form $j^a = 2^{-1/2}it^a_{AB}\lambda^A\lambda^B$, or from the free-boson form. Another necessary expec-

---

[2] In order that the gauge couplings of the gauge boson and gaugino agree — an indirect application of unitarity — we have normalized the fermion vertex operator as $g_{\rm o}\alpha'^{1/4}e^{-\phi/2}\Theta_\alpha e^{ik\cdot X}$.

[3] We are using standard field theory conventions, but to compare with much of the string literature one needs $g_{\rm YM}^{\rm here} = \frac{1}{2}g^{\rm there}$ and $u_i^{\rm here} = 2^{1/2}u_i^{\rm there}$.

tation value is

$$\left\langle \prod_{i=1}^{3} ie_i \cdot \partial X e^{ik_i \cdot X}(z_i, \bar{z}_i) \right\rangle = \frac{\alpha'^2 e_{1\mu} e_{2\nu} e_{3\rho} T^{\mu\nu\rho}}{8iz_{12}z_{13}z_{23}} , \qquad (12.4.11)$$

where

$$T^{\mu\nu\rho} = k_{23}^\mu \eta^{\nu\rho} + k_{31}^\nu \eta^{\rho\mu} + k_{12}^\rho \eta^{\mu\nu} + \frac{\alpha'}{8} k_{23}^\mu k_{31}^\nu k_{12}^\rho . \qquad (12.4.12)$$

This is the same as for the bosonic string, section 6.6, where we have used the mass-shell condition $k_i^2 = 0$ and transversality $e_i \cdot k_i = 0$.

Now we can write all the massless three-point amplitudes. Including an overall factor $8\pi/\alpha'g_c^2$ which is the same as in the bosonic string, the heterotic string three-gauge-boson amplitude is

$$4\pi g_c \alpha'^{-1/2}(2\pi)^{10}\delta^{10}(\textstyle\sum_i k_i)e_{1\mu}e_{2\nu}e_{3\rho}V^{\mu\nu\rho}\text{Tr}_\text{v}([t^a, t^b]t^c) . \qquad (12.4.13)$$

Up to the definition of the coupling this is the same as the open string amplitude (12.4.4). In particular there is no k^3 correction, again consistent with supersymmetry. Note that the vector part of this amplitude comes from the right-moving supersymmetric side. The heterotic amplitude for three massless neutral bosons (graviton, dilaton, or antisymmetric tensor) is

$$\pi ig_c(2\pi)^{10}\delta^{10}(\textstyle\sum_i k_i)e_{1\mu\sigma}e_{2\nu\omega}e_{3\rho\lambda}T^{\mu\nu\rho}V^{\sigma\omega\lambda} . \qquad (12.4.14)$$

One can relate the coupling g_c to the constants appearing in the heterotic string low energy action as in eq. (12.3.47b). In particular, the relation between g_{YM} and κ is in agreement with the anomaly result (12.3.37). The heterotic amplitude for two gauge bosons and one neutral boson is

$$\pi ig_c(2\pi)^{10}\delta^{10}(\textstyle\sum_i k_i) \, e_{1\mu\nu}e_{2\rho}e_{3\sigma}k_{23}^\nu V^{\mu\rho\sigma}\delta^{ab} . \qquad (12.4.15)$$

The antisymmetric part contains a Chern–Simons interaction, with ω_{3Y}.

Type I/II sphere amplitudes: In any type I or II theory, the amplitude for three massless NS–NS bosons on the sphere is

$$\pi ig_c(2\pi)^{10}\delta^{10}(\textstyle\sum_i k_i)e_{1\mu\sigma}e_{2\nu\omega}e_{3\rho\lambda}V^{\mu\nu\rho}V^{\sigma\omega\lambda} . \qquad (12.4.16)$$

The normalization factor $8\pi/g_c^2\alpha'$ and the relation $\kappa = 2\pi g_c$ are the same as in other closed string theories.

The tensor structure is simpler than in the corresponding heterotic amplitude (12.4.14), with terms only of order k^2. The bosonic side of the heterotic string makes a more complicated contribution and the amplitude has terms of order k^2 and k^4. An R^2 correction to the action would give a three-point amplitude of order k^4, and an R^3 correction would give an amplitude of order k^6. Here 'R' is shorthand for the whole Riemann tensor, not just the Ricci scalar. The type I/II amplitude (12.4.16) implies no R^2 or R^3 corrections. In the heterotic string there is a correction of order R^2 but none of order R^3. The absence of R^2 and R^3 corrections in the type II theories is a consequence of the greater supersymmetry (32 generators rather than 16).

By taking two polarizations symmetric and one antisymmetric, there is in the heterotic string an order k^4 interaction of two gravitons and an antisymmetric tensor. An effective interaction built out of field strengths and curvatures would have *five* derivatives. The interaction we have found must therefore be the gravitational Chern–Simons interaction $H_3 \wedge *\omega_{3L}$, which figured in the heterotic anomaly cancellation. No such term was expected in the type II theories and none has appeared. We do need such a term in the type I theory, which has the same massless spectrum as the heterotic string and so needs the same Green–Schwarz cancellation. However, as explained at the end of section 12.2, in the type I theory this will come from the disk rather than the sphere. We can also understand this from the field redefinition (12.1.41). An R^2 interaction which is a tree-level heterotic effect maps

$$(-G_h)^{1/2}e^{-2\Phi_h}R_h^2 \to (-G_I)^{1/2}e^{-\Phi_I}R_I^2 \, , \qquad (12.4.17)$$

which is the correct dilaton dependence for a disk or projective plane amplitude.

The various other three-point amplitudes are left as exercises.

Four-point amplitudes

All the four-point amplitudes of massless fields have been calculated. Many of the calculations are a bit tedious, though for supersymmetric strings the results tend to simplify. We will do a few simple calculations and quote some characteristic results, leaving the rest to the references.

Let us begin with the type I four-gaugino amplitude, each vertex operator being $g_o(\alpha')^{1/4}t^a \mathcal{V}_\alpha e^{ik\cdot X}u_\alpha$. We need the expectation value of four $\mathcal{V}$s (of the same chirality). The OPE

$$\mathcal{V}_\alpha(z)\mathcal{V}_\beta(0) \sim \frac{(C\Gamma^\mu)_{\alpha\beta}}{2^{1/2}z}e^{-\phi}\psi_\mu \, , \qquad (12.4.18)$$

follows from the three-point function (12.4.6c). Then

$$\langle \mathcal{V}_\alpha(z_1)\mathcal{V}_\beta(z_2)\mathcal{V}_\gamma(z_3)\mathcal{V}_\gamma(z_4) \rangle$$
$$= \frac{(C\Gamma^\mu)_{\alpha\beta}(C\Gamma_\mu)_{\gamma\delta}}{2z_{12}z_{23}z_{24}z_{34}} + \frac{(C\Gamma^\mu)_{\alpha\gamma}(C\Gamma_\mu)_{\delta\beta}}{2z_{13}z_{34}z_{32}z_{42}} + \frac{(C\Gamma^\mu)_{\alpha\delta}(C\Gamma_\mu)_{\beta\gamma}}{2z_{14}z_{42}z_{43}z_{23}} \, , \quad (12.4.19)$$

from consideration of the singularities in z_1. An additional holomorphic term is forbidden because the expectation value (12.4.19) must fall as z_1^{-2}

at infinity. Cancellation of the z_1^{-1} term further requires that

$$\Gamma^\mu_{\alpha\beta}\Gamma_{\mu\gamma\delta} + \Gamma^\mu_{\alpha\gamma}\Gamma_{\mu\delta\beta} + \Gamma^\mu_{\alpha\delta}\Gamma_{\mu\beta\gamma} = 0 \,. \tag{12.4.20}$$

This is indeed an identity, and plays an important role in ten-dimensional spacetime supersymmetry.

It is then straightforward to evaluate the rest of the amplitude. For the cyclic ordering 1234, let the vertex operators lie on the real axis and fix $x_1 = 0$, $x_3 = 1$, $x_4 \to \infty$ as usual to obtain

$$\frac{i}{2}g_o^2(2\pi)^{10}\delta^{10}(\textstyle\sum_i k_i)\text{Tr}_v(t^{a_1}t^{a_4}t^{a_2}t^{a_3}) \int_0^1 dx_2\, x^{-\alpha's-1}(1-x)^{-\alpha'u-1}$$

$$\times(\bar{u}_1\Gamma^\mu u_2\,\bar{u}_3\Gamma^\mu u_4 + x\,\bar{u}_1\Gamma^\mu u_3\,\bar{u}_2\Gamma^\mu u_4) \,. \tag{12.4.21}$$

Evaluating the integral and summing over cyclic orderings gives the final result

$$-16ig_{\text{YM}}^2\alpha'^2(2\pi)^{10}\delta^{10}(\textstyle\sum_i k_i)K(u_1,u_2,u_3,u_4)$$

$$\times\left[\text{Tr}_v(t^{a_1}t^{a_2}t^{a_3}t^{a_4})\frac{\Gamma(-\alpha's)\Gamma(-\alpha'u)}{\Gamma(1-\alpha's-\alpha'u)} + 2 \text{ permutations}\right] \,. \tag{12.4.22}$$

The kinematic factor

$$K(u_1,u_2,u_3,u_4) = \frac{1}{8}(u\,\bar{u}_1\Gamma^\mu u_2\,\bar{u}_3\Gamma_\mu u_4 - s\,\bar{u}_1\Gamma^\mu u_4\,\bar{u}_3\Gamma_\mu u_2) \tag{12.4.23}$$

is fully antisymmetric in the spinors. We recall the definitions

$$s = -(k_1+k_2)^2 \,, \quad t = -(k_1+k_3)^2 \,, \quad u = -(k_1+k_4)^2 \,. \tag{12.4.24}$$

Replacing some of the gauginos with gauge bosons leads to the same form (12.4.22), with only the factor K altered. For four gauge bosons,

$$K(e_1,e_2,e_3,e_4) = \frac{1}{8}\left(4M_{\mu\nu}^1 M_{\nu\sigma}^2 M_{\sigma\rho}^3 M_{\rho\mu}^4 - M_{\mu\nu}^1 M_{\nu\mu}^2 M_{\sigma\rho}^3 M_{\rho\sigma}^4\right)$$

$$+2 \text{ permutations}$$

$$\equiv t^{\mu\nu\sigma\rho\alpha\beta\gamma\delta}k_{1\mu}e_{1\nu}k_{2\sigma}e_{2\rho}k_{3\alpha}e_{3\beta}k_{4\gamma}e_{4\delta} \,, \tag{12.4.25}$$

where $M_{\mu\nu}^i = k_{i\mu}e_{i\nu} - e_{i\mu}k_{i\nu}$. The permutations replace the cyclic order 1234 with 1342 and 1423. The tensor t is antisymmetric within each $\mu_i\nu_i$ pair and symmetric under the interchange of two pairs, $\mu_i\nu_i$ with $\mu_j\nu_j$. This determines it to be a sum of the indicated two tensor structures. The result can also be written out

$$K(e_1,e_2,e_3,e_4) = -\frac{1}{4}\left(st\,e_1\cdot e_4\,e_2\cdot e_3 + 2 \text{ permutations}\right)$$

$$+\frac{1}{2}\left(s\,e_1\cdot k_4\,e_3\cdot k_2\,e_2\cdot e_4 + 11 \text{ permutations}\right) \,. \tag{12.4.26}$$

Each sum runs over all inequivalent terms obtained by permuting the four external lines.

It is interesting to consider the low energy limit of the bosonic amplitude. The expansion of the $\Gamma\Gamma/\Gamma$ factor begins

$$\frac{1}{\alpha'^2 su} - \frac{\pi^2}{6} + O(\alpha') , \qquad (12.4.27)$$

the $O(\alpha'^{-1})$ term vanishing. We have used $\Gamma''(1) - \Gamma'(1)^2 = \zeta(2) = \pi^2/6$, where the zeta function is defined below. The leading term represents the Yang–Mills interaction in the low energy theory. Combined with the kinematic factor K it gives a sum of single poles, corresponding to exchange of massless gauge bosons, as well as the local quartic gauge interaction. The $O(\alpha'^0)$ terms correspond to a higher-derivative low energy interaction. To convert the scattering amplitude to a Lagrangian density replace $k_{[\mu}e_{\nu]} \cong -iF_{\mu\nu}/2g_{\mathrm{YM}}$ (so that the kinetic term has canonical normalization $\frac{1}{2}k^2 e^\mu e_\mu = \frac{1}{2}k^2$) and include a factor of $1/4!$ for the identical fields to obtain

$$\frac{\pi^2 \alpha'^2}{2 \times 4! \, g_{\mathrm{YM}}^2} t^{\mu\nu\sigma\rho\alpha\beta\gamma\delta} \mathrm{Tr}_{\mathrm{v}}(F_{\mu\nu} F_{\sigma\rho} F_{\alpha\beta} F_{\gamma\delta}) . \qquad (12.4.28)$$

The net g_{YM}^{-2} is as expected for a tree-level string effect. The additional factor of α'^2 reflects the fact that this is a string correction to the low energy effective action, suppressed by the fourth power of the string length. The absence of an F^3 term is in agreement with the three-point amplitude.

The relation (6.6.23) between open and closed string tree amplitudes continues to hold in the superstring,

$$A_{\mathrm{c}}(s,t,u,\alpha',g_{\mathrm{c}}) = -\frac{\pi i g_{\mathrm{c}}^2 \alpha'}{g_{\mathrm{o}}^4} A_{\mathrm{o}}(s,t,\tfrac{1}{4}\alpha',g_{\mathrm{o}}) A_{\mathrm{o}}(t,u,\tfrac{1}{4}\alpha',g_{\mathrm{o}})^* \sin\frac{\pi\alpha't}{4} , \qquad (12.4.29)$$

where the open string amplitudes represent just one of the six cyclic orderings, and the factors $(2\pi)^{10}\delta^{10}(\sum_i k_i)$ are omitted in $A_{\mathrm{c,o}}$. The type II amplitude with four massless NS–NS bosons is then

$$-\frac{i\kappa^2 \alpha'^3}{4} \frac{\Gamma(-\tfrac{1}{4}\alpha's)\Gamma(-\tfrac{1}{4}\alpha't)\Gamma(-\tfrac{1}{4}\alpha'u)}{\Gamma(1+\tfrac{1}{4}\alpha's)\Gamma(1+\tfrac{1}{4}\alpha't)\Gamma(1+\tfrac{1}{4}\alpha'u)} K_{\mathrm{c}}(e_1,e_2,e_3,e_4) . \qquad (12.4.30)$$

Here,

$$K_{\mathrm{c}}(e_1,e_2,e_3,e_4) = t^{\mu_1\nu_1\dots\mu_4\nu_4} t^{\rho_1\sigma_1\dots\rho_4\sigma_4} \prod_{j=1}^{4} e_{j\mu_j\rho_j} k_{j\nu_j} k_{j\sigma_j} . \qquad (12.4.31)$$

The expansion of the ratio of gamma functions is

$$-\frac{64}{\alpha'^3 stu} - 2\zeta(3) + O(\alpha') \qquad (12.4.32)$$

where the zeta function is

$$\zeta(k) = \sum_{m=1}^{\infty} \frac{1}{m^k} . \tag{12.4.33}$$

The first term is the low energy gravitational interaction; note that it is proportional to κ^2 with no α' dependence. From the normalization of the gravitational kinetic term, $e_{\mu\rho}k_\nu k_\sigma$ contracted into t becomes $R_{\mu\nu\sigma\rho}/4\kappa$; including a symmetry factor $1/4!$, the second term corresponds to an interaction

$$\frac{\zeta(3)\alpha'^3}{2^9 \times 4! \, \kappa^2} t^{\mu_1 \nu_1 \cdots \mu_4 \nu_4} t^{\rho_1 \sigma_1 \cdots \rho_4 \sigma_4} R_{\mu_1 \nu_1 \rho_1 \sigma_1} R_{\mu_2 \nu_2 \rho_2 \sigma_2} R_{\mu_3 \nu_3 \rho_3 \sigma_3} R_{\mu_4 \nu_4 \rho_4 \sigma_4} . \tag{12.4.34}$$

This interaction, which is often identified by its distinctive coefficient $\zeta(3)$, has several interesting consequences; we will mention one in section 19.6. The absence of R^2 and R^3 corrections is again as expected from the three-point amplitude. For the heterotic string, the smaller supersymmetry allows more corrections.

We close with a few brief remarks about the heterotic amplitude with four gauginos or gauge bosons. The current algebra part of the amplitude is

$$\hat{k}^{-2} \langle \, j^{a_1}(z_1) j^{a_2}(z_2) j^{a_3}(z_3) j^{a_4}(z_4) \, \rangle = \frac{\delta^{a_1 a_2} \delta^{a_3 a_4}}{z_{12}^2 z_{34}^2} - \frac{f^{a_1 a_2 b} f^{b a_3 a_4}}{\hat{k} z_{12} z_{23} z_{24} z_{34}}$$
$$+ (2 \leftrightarrow 3) + (2 \leftrightarrow 4) . \tag{12.4.35}$$

This is obtained by using the OPE to find the singularities in z_1. An additional holomorphic term is forbidden by the behavior at infinity. In fact, the $(1,0)$ current must fall off as z_1^{-2}, and the three asymptotics of order z_1^{-1} do sum to zero by the Jacobi identity. Let us note further that $\delta^{a_1 a_2} = \text{Tr}_v(t^{a_1} t^{a_2})$ and that

$$-\hat{k}^{-1} f^{a_1 a_2 b} f^{b a_3 a_4} = 2 \text{Tr}_v([t^{a_1}, t^{a_2}] t^b) \text{Tr}_v(t^b [t^{a_3}, t^{a_4}])$$
$$= 2 \text{Tr}_v([t^{a_1}, t^{a_2}][t^{a_3}, t^{a_4}]) , \tag{12.4.36}$$

where the last equality holds for $SO(32)$ (or for states in an $SO(16) \times SO(16)$ subgroup of $E_8 \times E_8$) by completeness.

The remaining pieces of the amplitudes were obtained above, so it is straightforward to carry the calculation through. The amplitudes have the same factorized form (12.4.22) as in the type I theory, but with a more complicated group theory factor. In particular, the terms with two traces include effects from the exchange of massless supergravity states, which are of higher order in the type I theory.

All other three- and four-point massless amplitudes can be found in the references. We should mention that all of these were obtained first in the light-cone gauge, before the development of covariant methods. In fact, while we have emphasized the covariant approach, for actual calculation the two methods are roughly comparable. The advantage of covariance is offset by the complication of the ghosts, and the realization of spacetime supersymmetry is more complicated.

12.5 General amplitudes

Pictures

Amplitudes should not depend on which vertex operators have their θ coordinates fixed. We demonstrate this in two different formalisms. The first, operator, method is particularly common in the older literature. The second leans more heavily on the BRST symmetry.

Let the two θ-fixed vertex operators also be $z, \bar{z}$-fixed, and use an $SL(2, \mathbf{C})$ transformation to bring them to 0 and ∞. In operator form, the amplitude becomes

$$\int d^2 z_4 \ldots d^2 z_n \langle\!\langle \mathscr{V}_1^{-1} | \mathrm{T}[\mathscr{V}_3^0 \mathscr{V}_4^0 \ldots \mathscr{V}_n^0] | \mathscr{V}_2^{-1} \rangle_{\mathrm{matter}} \,. \qquad (12.5.1)$$

We are working in the old covariant formalism, where the ghosts appear in a definite way. They then contribute only an overall factor to the amplitude, so we need only consider the matter part, as indicated. Then

$$|\mathscr{V}_2^{-1}\rangle = 2L_0^{\mathrm{m}}|\mathscr{V}_2^{-1}\rangle = \{G_{1/2}^{\mathrm{m}}, G_{-1/2}^{\mathrm{m}}\}|\mathscr{V}_2^{-1}\rangle = G_{1/2}^{\mathrm{m}} G_{-1/2}^{\mathrm{m}}|\mathscr{V}_2^{-1}\rangle \,, \quad (12.5.2)$$

using the physical state conditions. The $G_{-1/2}^{\mathrm{m}}$ converts $|\mathscr{V}_2^{-1}\rangle$ into $|\mathscr{V}_2^0\rangle$. The $G_{1/2}^{\mathrm{m}}$ can be moved to the left, the commutators making no contribution because of the superconformal invariance of the vertex operators, where it converts $\langle\!\langle \mathscr{V}_1^{-1}|$ to $\langle\!\langle \mathscr{V}_1^0|$. The final form

$$\int d^2 z_4 \ldots d^2 z_n \langle\!\langle \mathscr{V}_1^0 | \mathrm{T}[\mathscr{V}_3^0 \mathscr{V}_4^0 \ldots \mathscr{V}_n^0] | \mathscr{V}_2^0 \rangle_{\mathrm{matter}} \qquad (12.5.3)$$

has all matter vertex operators in the 0 picture.

The BRST argument starts by considering the *picture-changing operator (PCO)*

$$X(z) \equiv Q_{\mathrm{B}} \cdot \xi(z) = T_F(z)\delta(\beta(z)) - \partial b(z)\delta'(\beta(z)) \,, \qquad (12.5.4)$$

where ξ is from bosonization of the superconformal ghosts. The calculation of $Q_{\mathrm{B}} \cdot \xi$ can be done in two ways. The first is to bosonize the BRST operator, expressing it in terms of ϕ, ξ, and η, calculate the OPE, and convert back. We will use a less direct but more instructive method. First, we claim that

$$\delta(\beta) \cong e^\phi \,. \qquad (12.5.5)$$

The logic is exactly the same as that of $\delta(\gamma) \cong e^{-\phi}$. Now, it is generally true that

$$\gamma(z)f(\beta(0),\gamma(0)) \sim \frac{1}{z}\partial_\beta f(\beta(0),\gamma(0)) \,, \tag{12.5.6}$$

from all ways of contracting γ with a β in f. Now, we claim that the step function bosonizes as

$$\theta(\beta) \cong \xi \,. \tag{12.5.7}$$

Taking the OPE with $\gamma = e^\phi\eta$, this is consistent with the previous two equations, and this determines the left-hand side up to a function of γ alone; this function must vanish because both sides have a nonsingular product with $\beta = e^{-\phi}\partial\xi$. The explicit form (10.5.21) of the BRST current then gives

$$j_\mathrm{B}(z)\theta(\beta(0)) \sim -\frac{1}{z^2}b(z)\delta'(\beta(0)) + \frac{1}{z}T_F(z)\delta(\beta(0)) \,. \tag{12.5.8}$$

The two terms come from two or one $\gamma\beta$ contractions respectively. Integrating the current on a contour around the origin gives the result (12.5.4).

To understand the role of the PCO we need to examine an unusual feature of the $\beta\gamma$ bosonization. The $(0,0)$ ξ field has one zero mode on the sphere, while the $(1,0)$ η field has none. One factor of ξ is then needed to give a nonvanishing path integral. However, the only ghost factors in the vertex operators are $e^{-\phi}$ and $e^{-\phi/2}$. The correct rule is that the $\beta\gamma$ path integral is equal to the $\phi\eta\xi$ path integral with the various operators bosonized *and* with one additional $\xi(z)$ in the path integral. The position of the ξ insertion is irrelevant because the expectation value is proportional to the zero mode, which is constant. We can simply normalize

$$\langle\, \xi(z) \,\rangle = 1 \,. \tag{12.5.9}$$

To verify the decoupling of a null state we need to pull the BRST contour off the sphere. The ξ insertion would seem to be an obstruction, because the contour integral of the BRST charge around ξ is nonzero: it is just the definition (12.5.4) of the PCO. However, when the ξ insertion is replaced by X in this way, the path integral vanishes because of the ξ zero mode, and so there is no problem.

Now consider the path integral with one PCO and with the ξ insertion, as well as additional BRST-invariant operators. Then

$$X(z_1)\xi(z_2) = Q_\mathrm{B}\cdot\xi(z_1)\,\xi(z_2) = \xi(z_1)\,Q_\mathrm{B}\cdot\xi(z_2) = \xi(z_1)X(z_2) \,. \tag{12.5.10}$$

In the middle step we have pulled the BRST contour from $\xi(z_1)$ to $\xi(z_2)$ as in figure 12.2. There are two signs, from changing the order of Q_B and $\xi(z_1)$, and from changing the direction of the contour. Although $X(z)$ is formally null, its expectation value does not vanish because of the contour

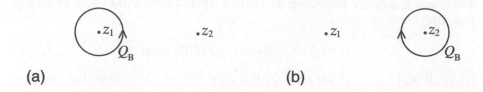

(a)   (b)

Fig. 12.2.   Moving the PCO. The contour around z_1 in (a) is pulled around the sphere until it becomes a contour around z_2 in (b).

integral of Q_B around the ξ insertion. Unlike the same contribution in the previous paragraph, this does not vanish because the $\xi(z_1)$ remains to saturate the zero-mode integral. We already know that the path integral is independent of the position of the ξ insertion, so eq. (12.5.10) shows that it is also independent of the position of the PCO.

Consider now

$$\lim_{z \to 0} X(z)\mathscr{V}^{-1}(0) , \qquad (12.5.11)$$

where for convenience we concentrate on the holomorphic side. The -1 picture vertex operator is $e^{-\phi}\mathcal{O}$ with $\mathcal{O}$ a matter superconformal primary. Consider now the term in $X(z)$ that involves the matter fields,

$$e^{\phi}T_F^{\mathrm{m}}(z)e^{-\phi}\mathcal{O}(0) = zT_F^{\mathrm{m}}(z)\mathcal{O}(0) + O(z^2) . \qquad (12.5.12)$$

The $z \to 0$ limit picks out the coefficient of the z^{-1} in the matter OPE, which is precisely $G_{-1/2} \cdot \mathcal{O} = \mathscr{V}^0$, the 0 picture vertex operator. The purely ghost terms in X vanish as $z \to 0$, so that

$$\lim_{z \to 0} X(z)\mathscr{V}^{-1}(0) = \mathscr{V}^0(0) . \qquad (12.5.13)$$

In the bosonic n-point amplitude with two -1 picture operators and $(n-2)$ 0 picture operators, we can pull a PCO out of each of the latter to be left with $(n-2)$ PCOs and n vertex operators, all of which are in the -1 picture. This is the 'natural' picture, the one given by the state–operator mapping. This also shows how to define a general tree-level amplitude, with n_B bosons and n_F (which must be even) fermions. Put all the bosons in the natural -1 picture, all the fermions in the natural $-\frac{1}{2}$ picture, and include $(n_B + \frac{1}{2}n_F - 2)$ PCOs. By taking some of the PCOs coincident with vertex operators, possibly more than one PCO at the same vertex operator, one obtains a representation with the vertex operators in higher pictures.

Finally, let us tie up a loose end. The operator product (12.4.18) is just the product of two spacetime supersymmetry currents, $\mathscr{V}_\alpha \equiv j_\alpha$. By the Ward identity and the supersymmetry algebra, we would expect the z^{-1}

term to be the translation current. Instead it is $e^{-\phi}\psi^\mu$. However, this is the zero-momentum vector vertex operator in the -1 picture; if we move a PCO to the operator we get the 0 picture ∂X^μ which is indeed the translation current. So the algebra is correct. The $(1,0)$ operator $e^{-\phi}\psi^\mu$ is the translation current; which picture it appears in has no effect on the physics.

Super-Riemann surfaces

The preceding discussion suggests a natural generalization to all orders of perturbation theory. That is, string amplitudes are given by an integral over moduli space and the ghost plus matter path integral with the following insertions: the appropriate vertex operator for each incoming or outgoing string in the natural -1 or $-\frac{1}{2}$ picture, the b-ghosts for the measure on moduli space as in the bosonic string, plus the appropriate number of PCOs to give a sensible path integral. At genus g, the Riemann–Roch theorem gives the number of beta zero modes minus the number of gamma zero modes as $2g - 2$. Equivalently, the total ϕ charge of the insertions must be $2g - 2$. To obtain this, the total number of PCOs must be

$$n_X = 2g - 2 + n_B + \frac{n_F}{2} , \qquad (12.5.14)$$

at arbitrary points; this is for the open string or one side of the closed string. The same formal arguments as in the case of the bosonic string show that this defines a consistent unitary theory. In particular, the PCOs are BRST-invariant and do not affect the decoupling of null states.

This prescription is sufficient for all the calculations we will carry out. However, in the remainder of this section we will develop superstring perturbation theory from a more general and geometric point of view. One reason for this is that the picture-changing prescription is rather *ad hoc* and it would be satisfying to see it derived in some way. Another is that this prescription actually has a subtle ambiguity at higher genus, which is best resolved from the more geometric point of view.

The needed idea is *supermoduli space*, the space of *super-Riemann surfaces (SRSs)*. These are defined by analogy to Riemann surfaces. Cover the surface with overlapping coordinate patches. The mth has coordinates z_m, θ_m. Patches are glued together with superconformal transformations. That is, if patches m and n overlap, identify points such that

$$z_m = f_{mn}(z_n) + \theta_n g_{mn}(z_n) h_{mn}(z_n) , \qquad (12.5.15a)$$
$$\theta_m = g_{mn}(z_n) + \theta_n h_{mn}(z_n) , \qquad (12.5.15b)$$
$$h_{mn}^2(z_n) = \partial_z f_{mn}(z_n) + g_{mn}(z_n) \partial_z g_{mn}(z_n) . \qquad (12.5.15c)$$

The holomorphic functions f_{mn} and the anticommuting holomorphic functions g_{mn} define the SRS. Two SRSs are equivalent if there is a one-to-one mapping between them such that the respective coordinates are related by a superconformal transformation. Tensor fields are defined by analogy to tensors on an ordinary manifold, as functions in each patch with appropriate transformations between patches. Supermoduli space is the set of equivalence classes of super-Riemann surfaces. The coordinates on supermoduli space are the bosonic (even) moduli t_j and the anticommuting (odd) moduli v_a. The Riemann–Roch theorem gives the number of odd moduli minus the number of globally defined odd superconformal transformations as $2g - 2$.

Again one can define all of this by Taylor expanding all functions in the anticommuting variables θ and v_a. The term in f_{mn} of order v_a^0 defines an ordinary (not super-) Riemann surface, and everything is expressed in terms of functions on this surface with the component form of the superconformal transformation between patches. Incidentally, z and $\bar{z}$ are no longer formally conjugates of one another on a SRS, particularly in the heterotic string where $\bar{z}$ transforms as the conjugate of eq. (12.5.15) while z transforms as on a 'bosonic' Riemann surface. However, if one defines everything by the Taylor expansion then z and $\bar{z}$ are again conjugates on the resulting ordinary Riemann surface.

For any SRS, setting the v_a to zero makes the anticommuting g_{mn} vanish and leaves

$$z_m = f_{mn}(z_n) , \tag{12.5.16a}$$
$$\theta_m = \theta_n h_{mn}(z_n) , \quad h_{mn}^2(z_n) = \partial_z f_{mn}(z_n) . \tag{12.5.16b}$$

The transformation of z defines a Riemann surface, but that of θ requires the additional choice of which square root to take in each h_{mn}. This choice is known as a *spin structure*; it is the same data one would need to put a spin-$\frac{1}{2}$ field on the surface. The signs are not all independent. If three patches overlap then the transition functions must satisfy the *cocycle condition*

$$h_{mn} h_{np} h_{pm} = 1 . \tag{12.5.17}$$

Also, a coordinate change $\theta_p \to -\theta_p$ in the patch p_0 changes the signs of all the h_{pn}. The net result is that there is one meaningful sign for each nontrivial closed path on the surface, $2g$ for a genus g surface. These define 2^{2g} different spin structures, topologically distinct ways to put a spinor field on the surface.

Any sphere is equivalent to the one with two patches (z, θ), (u, ϕ) and transition functions

$$u = 1/z , \quad \phi = i\theta/z . \tag{12.5.18}$$

Clearly there are no spin structures. The index theorem implies superconformal Killing transformations. One can look for infinitesimal transformations as in the bosonic case, with the result that δf must be at most
quadratic in z and δg linear in z. The general finite transformation is then
of the superconformal form with

$$f(z) = \frac{\alpha z + \beta}{\gamma z + \delta}, \quad g(z) = \epsilon_1 + \epsilon_2 z \tag{12.5.19}$$

with $\alpha\delta - \beta\gamma = 1$. In particular there are two odd transformations, ϵ_1 and
ϵ_2, consistent with the Riemann–Roch theorem. These can be used to fix
the odd coordinates of two NS vertex operators to zero.

A torus can be described as the (z, θ) plane modded by a group of rigid
superconformal transformations,

$$(z, \theta) \cong (z + 2\pi, \eta_1\theta) \cong (z + 2\pi\tau, \eta_2\theta) . \tag{12.5.20}$$

The η_1 and η_2 are each ± 1, defining the four spin structures. When θ
changes sign around a loop, the bosonic and fermionic components of
any superfield will have opposite periodicities, and in particular T_F will
be antiperiodic. We thus denote the spin structures (P,P), (P,A), (A,P), and
(A,A), giving the $z \to z + 2\pi$ periodicity first. The periodicities on the
right-moving side have the same form, with $\bar{\tau}$ the conjugate of τ but with
independent $\tilde{\eta}_1$ and $\tilde{\eta}_2$.

On a torus the only holomorphic functions are the constants, so β and
γ zero modes are possible only in the (P,P) case, in which case there is
one of each. There is then an odd supermodulus ν, giving rise to the more
general periodicity[4]

$$(z, \theta) \cong (z + 2\pi, \theta) \cong (z + 2\pi\tau + \theta\nu, \theta + \nu) . \tag{12.5.21}$$

There is also the *superconformal Killing vector (SCKV)*

$$(z, \theta) \to (z + \theta\epsilon, \theta + \epsilon) . \tag{12.5.22}$$

The number of odd moduli minus the number of SCKVs is zero in all
sectors, being $1 - 1$ for the (P,P) spin structure and $0 - 0$ for the others.
The modular group and the fundamental region for τ are the same as in
the bosonic string.

Returning to a general SRS, if the positions of n vertex operators
are singled out then there is an additional nontrivial closed curve circling
each, giving 2^{2g+n} spin structures altogether. The additional spin structures
come from the choice of R or NS boundary conditions of the external

---

[4] We could introduce a second odd parameter into the $z + 2\pi$ periodicity, but one of the two
parameters can be removed by a linear redefinition of (z, θ). Also, it might appear that a similar
generalization is possible in the antiperiodic case, but a coordinate redefinition returns the
periodicity to the form (12.5.20).

strings. To describe the supermoduli space of SRSs with n_B NS vertex operators and n_F R vertex operators, it is useful to extend the approach used in section 5.4. First in the bosonic case, consider a specific patching together of a Riemann surface, with n marked points. We will define another Riemann surface as equivalent to this one if there is a one-to-one holomorphic mapping between the two which leaves the coordinates of the points invariant. That is, $f(z) - z$ must vanish linearly at the vertex operators. For simplicity we take each operator to be at $z = 0$ in its own tiny patch. Since we are modding by a smaller group, with two real conditions for each vertex operator, we obtain a correspondingly larger coset space, with two additional moduli for each vertex operator. This is similar to the treatment of vertex operator positions in section 5.4, but more abstract. In the superconformal case, we mod out the superconformal transformations for which $f(z) - z$ and $g(z)$ vanish linearly at each NS vertex operator. At each R vertex operator, $g(z)$ has a branch cut, and so it is appropriate to require $f(z) - z$ to vanish linearly z and $g(z)$ to vanish as $z^{1/2}$. The NS vertex reduces the odd coordinate degrees of freedom by one and so increases the number of inequivalent surfaces: the number of odd moduli increases by one, which we can take to be the θ coordinate of the operator. The condition for the R vertex operator is essentially half as restrictive, so that there is an additional odd modulus for each *pair* of R vertex operators. This has no simple interpretation as a vertex operator position; an R vertex operator produces a branch cut in θ, so there can be no well-defined θ coordinate for the operator. The total number of odd moduli is

$$n_v = 2g - 2 + n_B + \frac{n_F}{2} \,. \tag{12.5.23}$$

The measure on supermoduli space

The expression (5.4.19) for the bosonic string S-matrix now generalizes in a natural way,

$$S(1;\ldots;n) = \sum_{\chi,\gamma} \frac{e^{-\lambda\chi}}{n_R} \int_{\chi,\gamma} d^{n_e}t d^{n_o}v \left\langle \prod_{j=1}^{n_e} B_j \prod_{a=1}^{n_o} \delta(B_a) \prod_{i=1}^{n} \hat{\mathcal{V}}_i \right\rangle \,. \tag{12.5.24}$$

The sum is over topologies χ and spin structures γ. The integral runs over the corresponding supermoduli space. There are n_e even moduli, n_o odd moduli, and n external strings. The quantity B_j in the ghost insertions is

$$B_j = \sum_{(mn)} \int_{C_{mn}} \frac{dz_m d\theta_m}{2\pi i} B(z_m,\theta_m) \left[\frac{\partial z_m}{\partial t_j} - \frac{\partial \theta_m}{\partial t_j} \theta_m \right]_{z_n,\theta_n} \,, \tag{12.5.25}$$

plus a right-moving piece of the same form; B_a is given by an identical expression with v_a replacing t_j. The sum again runs over all pairs of overlapping patches m and n, clockwise as seen from m, with the z_m integration along a contour between the two patches and the θ_m integral of the usual Berezin form. The B ghost superfield is as in eq. (12.3.25), $B = \beta + \theta b$.

The logic of this expression is the same as for the earlier bosonic expression. First, the number of commuting and anticommuting ghost insertions is correct for a well-defined path integral. Second, the path integral depends only on the superconformal structure and not on the particular choice of patches and transition functions. In particular it is unchanged if we make a superconformal transformation within a single coordinate patch. The combination $\partial z_m + (\partial \theta_m)\theta_m$ transforms as a $(-1, 0)$ tensor superfield, so the integrand is a $(\frac{1}{2}, 0)$ tensor superfield and the integral is invariant. Third, under a change of coordinates in supermoduli space, the product $\prod_{j=1} B_j \prod_a \delta(B_a)$ transforms as a density, inversely to the measure on supermoduli space. Finally, the commutator of the BRST charge with $B_{j,a}$ is $T_{j,a}$, defined in the same way but with B replaced by

$$Q_B \cdot B(z) = T(z) = T_F(z) + \theta T_B(z) \,. \tag{12.5.26}$$

The insertion of $T_{j,a}$ generates a relative coordinate transformation of adjacent patches, which is just the derivative with respect to the super-modulus of the world-sheet.

It is interesting to work out the form of the amplitude more explicitly for a special choice of patches and transition functions. Namely, let patch 1 be contained entirely within patch 2, so that the overlap is an annulus. Let the 1-2 transition functions depend only on a single odd modulus v, as follows:

$$f_{12}(z_2) = z_2 \,, \quad g_{12}(z_2) = v\alpha(z_2) \,, \tag{12.5.27}$$

for some holomorphic function $\alpha(z)$. The ghost factor (12.5.25) is proportional to

$$B[\alpha] = \oint \frac{dz_1}{2\pi i} \alpha(z_1)\beta(z, \theta) \,. \tag{12.5.28}$$

Similarly the path integral depends on v only through the insertion

$$v T[\alpha] \,, \tag{12.5.29}$$

where β is replaced by T_F. We can then perform the integration over v, so that the net effect of the supermodulus is the insertion in the path integral of

$$T[\alpha]\delta(B[\alpha]) = Q_B \cdot \theta(B[\alpha]) \,. \tag{12.5.30}$$

The function $\alpha(z_1)$ is holomorphic in the annular overlap of the patches, but in general cannot be extended holomorphically into the full inner patch z_1. If it can, the contour integrals $B[\alpha]$ and $T[\alpha]$ vanish. In this case v is not a modulus at all because it can be transformed away. A nontrivial case is

$$\alpha(z_1) = \frac{1}{z_1 - z_0}\,, \tag{12.5.31}$$

for which $B[\alpha] = \beta(z_0)$ and the insertion (12.5.30) just becomes the PCO $X(z_0)$ (the second term in X is from normal ordering). Thus, the PCO is the result of integrating out an odd modulus in this special parameterization of the SRSs. Note that the number (12.5.23) of odd moduli is the same as the number (12.5.14) of PCOs needed in the *ad hoc* approach. This provides the desired geometric derivation of the picture-changing prescription.

The parameterization (12.5.27) is always possible locally on supermoduli space. It can also be used globally, with careful treatment of the modular identification and the limits of moduli space. There is a literature on the 'ambiguity of superstring perturbation theory,' which arose from parameterizations that did not precisely cover supermoduli space. It appears that superstring perturbation theory to arbitrary order is understood in principle, and certain special amplitudes have been calculated at higher orders of perturbation theory. However, the subject is somewhat unfinished — a fully explicit proof of the perturbative consistency of the theory seems to be lacking. With the immense progress in nonperturbative string theory, filling this technical gap does not seem to be a key issue.

We derived the bosonic version (5.4.19) of the measure (12.5.24) by starting with a path integral over the world-sheet metric, whereas in the present case we have written it down directly. One can partly work backwards to an analogous description as follows. Although $\alpha(z_1)$ cannot be extended holomorphically into patch 1 it can be extended smoothly. It can then be removed by a change of variables in the path integral, but not one that leaves the action invariant. The odd modulus v appears in the final action, multiplying T_F and a function that can be regarded as the world-sheet gravitino field. In particular, the PCO can be regarded as coming from a pointlike gravitino, a gauge where the gravitino has delta-function support.

12.6　　One-loop amplitudes

We will illustrate one-loop superstring calculations with two examples where the low energy limit can be obtained in closed form.

The first is the heterotic string amplitude with four gauge bosons and one antisymmetric tensor. The Green–Schwarz anomaly cancellation requires a one-loop Chern–Simons term

$$\int B_2 \mathrm{Tr_v}(F_2^4) \, . \tag{12.6.1}$$

We would like to confirm the appearance of this term by an explicit string calculation.

Note first that this can only arise from the (P,P) path integral. This is because it is odd under spacetime parity: written out in components, it involves the ten-dimensional ϵ-tensor. The heterotic string world-sheet action and constraints are invariant under parity. The parity asymmetry of the theory, the fact that the massless fermions are in a **16** and not a **16$'$**, comes about from the GSO projection in the right-moving R sector, the choice of $\exp(\pi i F)$ to be $+1$ or -1. The (P,P) path integral produces this term in the projection operator. The path integral is then

$$\left(\frac{2}{\alpha'}\right)^{5/2} g_c^5 \int_F \frac{d\tau d\bar\tau}{8\tau_2} \left[\prod_{i=1}^{5} \int d^2 w_i\right] \Bigg\langle b(0)\tilde b(0)\tilde c(0)c(0)X(0) $$

$$\times \left[\prod_{i=1}^{4} \hat k^{-1/2} j^{a_i}(i e_i\cdot\bar\partial X + \tfrac{1}{2}\alpha' k_i\cdot\tilde\psi\, e_i\cdot\tilde\psi)e^{ik_i\cdot X}(w_i,\bar w_i)\right]$$

$$\times i e_{5\mu\nu}\partial X^\mu \delta(\tilde\gamma)\tilde\psi^\nu e^{ik_5\cdot X}(w_5,\bar w_5)\Bigg\rangle_{(P,P)} \tag{12.6.2}$$

The bc ghosts and corresponding measure are the same as in the bosonic string, with an extra $\frac{1}{2}$ from the GSO projection operator. For the (P,P) spin structure there is one PCO and one -1 picture vertex operator.

We will consider in order the $\tilde\psi^\mu$, X^μ, bc, $\beta\gamma$, and j^a path integrals. In the vacuum amplitude the $\tilde\psi^\mu$ path integral vanishes in the (P,P) sector. In terms of a trace, this is due to a cancellation between the R sector ground states. In terms of a path integral it is due to the Berezin integration over the zero mode of ψ^μ (which exists only for this spin structure). In the latter form it is clear that we need at least ten factors of $\tilde\psi$ to obtain a nonzero path integral. In fact, the path integral (12.6.2) has a maximum of ten $\tilde\psi$s, including one from the term

$$\delta(\tilde\beta)i(2/\alpha')^{1/2}\tilde\psi^\rho\bar\partial X_\rho \tag{12.6.3}$$

in the PCO. The relevant path integral is easily obtained from a trace, giving

$$\left\langle \prod_{i=1}^{10} \tilde\psi^{\mu_i} \right\rangle_{\tilde\psi(P,P)} = \epsilon^{\mu_1\cdots\mu_{10}}\bar q^{10/24}\prod_{n=1}^{\infty}(1-\bar q^n)^{10}$$

$$= \epsilon^{\mu_1\cdots\mu_{10}}[\eta(\tau)^{10}]^* \, . \tag{12.6.4}$$

The X^μ path integral is then reduced to

$$\left\langle \partial X^\mu(w_5)\bar\partial X^\rho(0) \prod_{i=1}^{5} e^{ik_i \cdot X(w_i,\bar w_i)} \right\rangle_X , \qquad (12.6.5)$$

the gradients coming from the tensor vertex operator and the PCO. To make things simple we now take the $k_i \to 0$ limit. Contractions between the gradients and exponentials, and among the exponentials, are then suppressed. Only the contraction between the gradients survives, $-\alpha'/8\pi\tau_2$ from the background charge term $\alpha'(\mathrm{Im}\, w_{ij})^2/4\pi\tau_2$ in the Green's function (7.2.3). The leading term in the expectation value (12.6.5) is then

$$- i(2\pi)^{10}\delta^{10}(\textstyle\sum_i k_i)\frac{\eta^{\mu\rho}\alpha'}{8\pi\tau_2(4\pi^2\alpha'\tau_2)^5|\eta(\tau)|^{20}} . \qquad (12.6.6)$$

The bc path integral is

$$\left\langle b(0)\tilde b(0)\tilde c(0)c(0) \right\rangle_{bc} = |\eta(\tau)|^4 , \qquad (12.6.7)$$

just as in the bosonic string. The $\beta\gamma$ path integral is the reciprocal of the right-moving part of this,

$$\left\langle \delta(\tilde\beta(0))\delta(\tilde\gamma(\bar w_5)) \right\rangle_{\beta\gamma} = [\eta(\tau)^{-2}]^* . \qquad (12.6.8)$$

Finally for the current algebra, we need

$$\hat k^{-2} \left\langle j^{a_1}(w_1)j^{a_2}(w_2)j^{a_3}(w_3)j^{a_4}(w_4) \right\rangle_g . \qquad (12.6.9)$$

We continue to use the convention $\hat k = \frac12$ for the rest of the chapter. Note first that all other expectation values are independent of w_i. The integrations over w_i thus have the effect of averaging over $\mathrm{Re}(w_i)$ and so we can replace each current with the corresponding charge, Q^{a_i}. We can then evaluate the expectation value as a trace. However, a careful treatment of the $k \to 0$ limit shows that an additional contact term is needed when two vertex operators coincide,

$$j^a(w)j^b(0) \to \mathrm{T}\!\left[\hat Q^a(w)\hat Q^b(0)\right] - \pi\delta^2(w,\bar w)\delta^{ab} . \qquad (12.6.10)$$

To see this, integrate both sides over the region of world-sheet $-\delta < \sigma^2 < \delta$. On the left we have

$$\frac{\delta^{ab}}{2} \int_{|\sigma^2|<\delta} d^2w \,\frac{1}{w^2}(w\bar w)^{k\cdot k'} , \qquad (12.6.11)$$

where we have introduced small k and k'. The $(w\bar w)^{k\cdot k'}$ factor from the X^μ path integral then regulates the integral at the origin. We have kept the leading term in the OPE because this is the only one that contributes at small δ. Writing the integrand as

$$(-1+k\cdot k')^{-1}\partial_w(w^{-1+k\cdot k'}\bar w^{k\cdot k'}) , \qquad (12.6.12)$$

we can integrate by parts to convert this to a surface integral, which is then easily evaluated at $k, k' \to 0$ to give -2π. On the right, the first term has no singularity that would allow a nonzero limit as $\delta \to 0$ ($\hat{Q}^a$ is conserved, so it is a constant except for the time ordering), and so the delta function is needed.

For the product of two currents we would then have

$$\langle j^{a_1}(w_1)j^{a_2}(w_2)\rangle \to \mathrm{Tr}\left\{\exp(2\pi i\tau H)\mathrm{T}\left[\hat{Q}^{a_1}(w_1)\hat{Q}^{a_2}(w_2)\right]\right\}$$
$$- \delta^{ab}\pi\delta^2(w_{12}, \bar{w}_{12})$$
$$\to \mathrm{Tr}\left\{\exp(2\pi i\tau H)\hat{Q}^{(a_1}\hat{Q}^{a_2)}\right\} - \frac{\delta^{ab}}{8\pi\tau_2}. \quad (12.6.13)$$

In the second line we have used the fact that all other expectation values are independent of the w_i, so that the integrations will have the effect of averaging over w_i. In the first term this symmetrizes the operators as indicated; in the second it allows us to replace the delta function with its average over the torus. For four currents the combinatorics are conveniently summarized in terms of the generating function

$$f(q, z) \equiv \langle \exp(z \cdot \bar{j})\rangle$$
$$= \exp\left(-\frac{z \cdot z}{16\pi\tau_2}\right)\mathrm{Tr}\left[\exp(2\pi i\tau H)\exp(z \cdot \hat{Q})\right], \quad (12.6.14)$$

where $\bar{j}^a$ is the average over the torus and the dot denotes a sum on a. The needed expectation value is the fourth derivative with respect to z^a. The trace is most easily carried out in the bosonic form, where it becomes an oscillator sum plus sum over the $SO(32)$ or $E_8 \times E_8$ lattice:

$$f(q, z) = \eta(\tau)^{-16}\exp\left(-\frac{z \cdot z}{16\pi\tau_2}\right)\sum_{l\in\Gamma}q^{l^2/2}\exp(2^{-1/2}z \cdot l). \quad (12.6.15)$$

Gathering all factors, including $(8\pi\tau_2)^5$ from integrating over the w_i, the amplitude becomes

$$-\frac{ig_c^5}{\pi\alpha'^3}(2\pi)^{10}\delta^{10}(\textstyle\sum_i k_i)\epsilon^{\mu_1...\mu_{10}}k_{1\mu_1}e_{1\mu_2}...k_{4\mu_7}e_{4\mu_8}e_{5\mu_9\mu_{10}}$$
$$\times \int_F \frac{d^2\tau}{\tau_2^2}\left.\frac{\partial^4\hat{f}(q, z)}{\partial z^{a_1}...\partial z^{a_4}}\right|_{z=0}, \quad (12.6.16)$$

where

$$\hat{f}(q, z) = \eta(\tau)^{-8}f(q, z). \quad (12.6.17)$$

We leave it as an exercise to show that this is modular-invariant. Substituting $e_{\mu\nu} \to B_{\mu\nu}/2\kappa$ and $k_{[\mu}e_{\nu]} \to -iF_{\mu\nu}/2g_{\mathrm{YM}}$ from the normalization of the kinetic terms, including a factor 2^5 from converting the ϵ-tensor to form notation, and using the relation between the couplings, the amplitude can be concisely summarized by the effective action

$$-\frac{1}{2^9\pi^6\alpha'}\int B_2\int_F \frac{d^2\tau}{\tau_2^2}\,\hat{f}(q,F_2)\,. \qquad (12.6.18)$$

The form integration picks out the term of order F_2^4. The integral can be written as a surface term and given in closed form, using

$$\frac{\hat{f}(q,F_2)}{\tau_2^2}=-\frac{32\pi i}{F_2\cdot F_2}\frac{\partial \hat{f}(q,F_2)}{\partial\bar{\tau}}\,. \qquad (12.6.19)$$

Due to modular invariance, only the limit $\tau_2\to\infty$ contributes. The effective action becomes

$$-\frac{1}{2^4\pi^5\alpha'}\int \frac{B_2}{F_2\cdot F_2}\hat{f}(q,F_2)\Big|_{q^0\ \text{term}}\,. \qquad (12.6.20)$$

Only the lattice momenta with $l^2=2$ contribute to the q^0 term. These form the adjoint of the gauge group, so the lattice sum reduces to a trace in the adjoint representation and the effective interaction is

$$-\frac{1}{2^4\pi^56!\,\alpha'}\int \frac{B_2\,\mathrm{Tr}_a(F_2^6)}{F_2\cdot F_2}=-\frac{1}{2^4\pi^54!\,\alpha'}\int B_2\frac{\mathrm{Tr}_a(F_2^6)}{\mathrm{Tr}_a(F_2^2)}\,. \qquad (12.6.21)$$

Ordinarily dividing by a form would make no sense, but we know from the discussion of anomalies that $\mathrm{Tr}_a(F_2^6)\propto F_2\cdot F_2\,X_8(F_2)$, so the effective interaction is proportional to $\int B_2 X_8$ as required by anomaly cancellation (and with the correct coefficient). For $SO(32)$ the ratio of forms is $\frac{1}{2}\mathrm{Tr}_v(F_2^4)$; for $E_8\times E_8$ it is

$$\frac{1}{7200}\Big\{[\mathrm{Tr}_{a1}(F_2^2)]^2+[\mathrm{Tr}_{a2}(F_2^2)]^2-\mathrm{Tr}_{a1}(F_2^2)\mathrm{Tr}_{a2}(F_2^2)\Big\}\,. \qquad (12.6.22)$$

With somewhat more effort one can also find the required curvature terms. That we were able with modest effort to bring this string loop amplitude to a closed form is not too surprising, since this is a very special amplitude whose coefficient is determined by symmetry (anomaly cancellation). However, many of the physically interesting corrections to the low energy effective action can be obtained in a closed form.

Next we consider the heterotic string amplitude with four gauge bosons but without the antisymmetric tensor. In contrast to the previous amplitude, which came only from the (P,P) spin structure, the present one comes only from the other three spin structures: with one fewer vertex operator there are not enough insertions of $\tilde{\psi}$ to saturate the zero modes of the

(P,P) path integral. The amplitude is then

$$\frac{4g_c^4}{\alpha'^2} \int_F \frac{d\tau d\bar\tau}{8\tau_2} \left[\prod_{i=1}^4 \int d^2 w_i\right] \sum_{\gamma \neq (\text{P,P})} \left\langle b(0)\tilde b(0)\tilde c(0)c(0) \right.$$

$$\times \left. \left[\prod_{i=1}^4 \hat k^{-1/2} j^{a_i}(ie_i \cdot \bar\partial X + \tfrac{1}{2}\alpha' k_i \cdot \tilde\psi \, e_i \cdot \tilde\psi)e^{ik_i \cdot X}(w_i, \bar w_i)\right]\right\rangle_\gamma . \quad (12.6.23)$$

For these spin structures, all vertex operators are in the 0 picture and there is no PCO.

We could proceed to calculate in a straightforward way, but the final result simplifies substantially and it would be better to simplify at the start. In fact, this is one amplitude that is much more easily obtained in the light-cone superstring formalism, and so we will effectively convert the calculation to that form.

First, analytically continue the momenta so that $k^0 = k^1 = 0$. To be consistent with the mass-shell condition the momenta must become complex but this will not be a problem. Also, take the polarizations to vanish in the longitudinal directions. The longitudinal degrees of freedom then do not appear in the vertex operators, and so in the (P,A), (A,P), and (A,A) sectors the longitudinal path integrals just give determinants that cancel against the corresponding ghost path integrals. In particular, the combined longitudinal and ghost path integrals for these three sectors are independent of the spin structure, so the net spin structure dependence comes only from the eight transverse $\tilde\psi^i$.

Now, we will temporarily change the problem and also add in the (P,P) spin structure for the $\tilde\psi^i$, even though in the real amplitude this is multiplied by zero from the $\tilde\psi^{0,1}$ zero modes. The sum over four spin structures gives a GSO projection in the transverse $\tilde\psi^i$ CFT by itself. Consider the vertex operator $\tilde\Theta_\alpha$ for the R ground state, and bosonize:

$$\tilde\Theta_\alpha \to \left\{ \begin{array}{ll} \exp[\tfrac{1}{2}i(\tilde H_1 + \tilde H_2 + \tilde H_3 + \tilde H_4)] &= \exp(i\tilde H_1') \\ \exp[\tfrac{1}{2}i(\tilde H_1 + \tilde H_2 - \tilde H_3 - \tilde H_4)] &= \exp(i\tilde H_2') \\ \exp[\tfrac{1}{2}i(\tilde H_1 - \tilde H_2 + \tilde H_3 - \tilde H_4)] &= \exp(i\tilde H_3') \\ \exp[\tfrac{1}{2}i(\tilde H_1 - \tilde H_2 - \tilde H_3 + \tilde H_4)] &= \exp(i\tilde H_4') \end{array} \right\} \to \tilde\theta_\alpha . \quad (12.6.24)$$

Precisely for eight $\tilde\psi^i$, the linear combinations of scalars appearing in the spin field are themselves scalars of canonical normalization

$$\tilde H_i'(\bar z)\tilde H_j'(0) \sim -\delta_{ij} \ln \bar z . \quad (12.6.25)$$

Thus after bosonizing and going to a new basis for the scalars we can refermionize in terms of free $(0, \tfrac{1}{2})$ fields $\tilde\theta_\alpha(\bar z)$. Note that only for eight $\tilde\psi^i$ does the spin field have weight $\tfrac{1}{2}$. Thus we turn the $\tilde\psi^i$ path integral into a $\tilde\theta_\alpha$ path integral. Moreover, we claim that the spin structures are related as follows:

$$\frac{1}{2}\sum_\gamma \langle \quad \rangle_{\tilde\psi,\gamma} = \langle \quad \rangle_{\tilde\theta(P,P)} \; . \tag{12.6.26}$$

This follows because the spin field $\tilde\Theta_\alpha$ survives the GSO projection, implying that it is single-valued on the torus. The periodicity in the time direction implies the insertion of a factor $\exp(\pi i \tilde F_\theta)$ in the sum over states, which is appropriate because $\tilde\theta_\alpha$ is a spacetime spinor.

Now we come to the payoff: for this spin structure the $\tilde\theta_\alpha$ have eight zero modes, and so there must be eight factors of $\tilde\theta_\alpha$ to get a nonzero result. We need to refermionize the vertex operators, but this is easy. The spinors appear only in the combination

$$k_i e_j \tilde\psi^{[i} \tilde\psi^{j]} \; , \tag{12.6.27}$$

where we can antisymmetrize because $e \cdot k = 0$. The product of fermions is just an $SO(8)$ rotation current, so we can immediately write

$$\tilde\psi^{[i} \tilde\psi^{j]} \to \frac{1}{4} \tilde\theta^T \Gamma^{ij} \tilde\theta \; . \tag{12.6.28}$$

One can check this by taking the OPE of the two sides with $\tilde\Theta_\alpha$ and $\tilde\theta_\alpha$ respectively. The fermionic terms in the vertex operators then provide precisely the eight $\tilde\theta$s needed to saturate the zero modes, with the result

$$\left\langle \prod_{a=1}^{4} \frac{1}{4} \tilde\theta^T \Gamma^{i_a j_a} \tilde\theta \right\rangle_{\tilde\theta(P,P)} = \frac{1}{2^8} \epsilon^{\alpha_1 \dots \alpha_8} \Gamma^{i_1 j_1}_{\alpha_1 \alpha_2} \cdots \Gamma^{i_4 j_4}_{\alpha_7 \alpha_8}$$

$$= t^{i_1 j_1 \dots i_4 j_4} + \epsilon^{i_1 j_1 \dots i_4 j_4} \; . \tag{12.6.29}$$

Here t is the same tensor (12.4.25) that appears in the tree-level amplitudes.

It remains to separate out the unwanted (P,P) sector of the $\tilde\psi$ path integral, but this is easy. It is the only one that is odd under a reflection of one of the transverse directions, so it is responsible for the term $\epsilon^{i_1 j_1 \dots i_4 j_4}$. Thus we omit this term, which in any case does not contribute because the momenta with which it contracts are not linearly independent. Further, the tensor t has a unique covariant extension.

The remaining factors are much as in the previous amplitude, leading for $SO(32)$ to the effective interaction

$$\frac{1}{2^8 \pi^5 4! \, \alpha'} t^{\mu\nu\sigma\rho\alpha\beta\gamma\delta} \mathrm{Tr}_{\mathrm{v}}(F_{\mu\nu} F_{\sigma\rho} F_{\alpha\beta} F_{\gamma\delta}) \; . \tag{12.6.30}$$

For $E_8 \times E_8$ one has instead the group theory structure (12.6.22). Given the similarity of the F^4 amplitude to the BF^4 amplitude, the reader may not be surprised that they are in fact related by supersymmetry.

By this same method several other amplitudes can be obtained, including the type I cylinder with four open string gauge bosons and the type II torus with four gravitons. In the latter case both the ψ^i and the $\tilde{\psi}^i$ are refermionized, and the tensor structure is the same as for the tree-level amplitude (12.4.34), with two t tensors.

The $\tilde{\theta}_\alpha$ are the first free fields carrying a spacetime spinor index that we have encountered. One might have expected these to arise at some earlier stage. In fact the covariant Green–Schwarz superstring theory, with manifest spacetime supersymmetry, has such fields. It is equivalent to the RNS superstring: after putting each theory in the light cone they are related by the refermionization above. However, the constraints and gauge fixing in the Green–Schwarz description are rather more complicated, and so we have chosen not to emphasize this subject.

Nonrenormalization theorems

It follows from the preceding calculations that any amplitude with three or fewer massless particles vanishes because there are too few factors of θ_α to saturate the zero-mode integrations. One consequence is that there is no renormalization of Newton's constant, which can be measured in the three-graviton amplitude.

It also follows that all amplitudes vanish at least as k^4 when $k \to 0$, from the explicit momentum factors in the vertex operators.[5] This has the important physical consequence that the constant background

$$G_{\mu\nu}(x) = \eta_{\mu\nu} , \quad \Phi(x) = \Phi_0 \tag{12.6.31}$$

around which we are expanding remains a solution of the field equations to one-loop order. No interaction

$$\int d^{10}x \, (-G)^{1/2} V(\Phi) \tag{12.6.32}$$

is generated. Actually we already knew this from the calculation of the one-loop vacuum amplitude in chapter 10, which vanished by cancellation between bosons and fermions. These nonrenormalization theorems have been argued to extend to all orders of string perturbation theory; the details are left to the references.

Nonrenormalization can also be understood from a spacetime point of view. The tree-level action has local supersymmetry. Therefore the loop corrections must respect this symmetry or else the unphysical polarizations of the gravitino will not decouple. However, no interaction of the

---

[5] This kind of argument is subtle because one can obtain offsetting poles from $\int d^2w \, (w\bar{w})^{-1+k\cdot k'}$, but the necessary singularity in w does not appear here because the zero modes are independent of $\bar{w}$.

form (12.6.32) is allowed by $d = 10$, $N = 1$ or $N = 2$ supersymmetry.[6] This argument applies to all orders of perturbation theory and in fact it is a nonperturbative (exact) result. The latter fact is very striking, because with the string technology developed so far we have no direct way to understand strings beyond perturbation theory. It should be noted that the leap from 'all orders of perturbation theory' to 'exact' is quite non-trivial, because in theories with less symmetry there are many examples of corrections that arise only from nonperturbative effects. We will see some of these later.

Exercises

12.1 Derive the $SO(n)$ trace identities (12.2.19). You can assume a basis in which the generator t is a linear combination of the commuting generators H^i.

12.2 Obtain the trace relations (12.2.20) for E_8, and show that $\text{Tr}_a(t^6)$ can be reduced to lower order traces for $E_8 \times E_8$.

12.3 Show that the anomaly factorizes for the massless spectrum of the $SO(16) \times SO(16)$ nonsupersymmetric heterotic string.

12.4 Show that the superfield forms for the superconformal transformation δX^μ, the OPEs (12.3.22) and (12.3.26), and the action (12.3.23) reduce to the correct component forms.

12.5 Show that the superfield form of the sigma model action reduces to components as shown in eq. (12.3.27).

12.6 Using the contour method from sections 6.2 and 6.3, show that the sphere amplitude must have total ϕ charge -2 as discussed at the beginning of section 12.4.

12.7 (a) Calculate the tree-level heterotic string amplitude with two gauginos and a gauge boson.
(b) Calculate the tree-level heterotic string amplitude with two gauginos and a massless tensor.

12.8 Calculate the tree-level type II amplitude with one NS–NS boson and two R–R bosons.

12.9 Calculate the tree-level heterotic string amplitude with four gauginos. You can either do this directly, or by first calculating an appropriate bosonic open string amplitude and then using the open–closed relation (12.4.29).

---

[6] The massive IIA supergravity theory (12.1.24) effectively has such a term after setting the fields M and F_{10} to fixed background values, but the dilaton dependence is fixed and corresponds to a tree-level effect.

12.10 Calculate the $k_i \to 0$ limit of the type I cylinder amplitude with four gauge bosons, where two open string vertex operators are on each boundary. This is often referred to as the *nonplanar amplitude*.

12.11 Calculate the same amplitude as in the previous problem but with all four vertex operators on one boundary. This *planar amplitude* has a divergence; calculate the canceling Möbius amplitude.

13

D-branes

In chapter 8 we found that a number of new phenomena, unique to string theory, emerged when the theory was toroidally compactified. Most notable were the T-duality of the closed oriented theory and the appearance of D-branes in the $R \to 0$ limit of the open string theory. These subjects become richer still with the introduction of supersymmetry. We will see that the D-branes are BPS states and carry R–R charges. We will argue that the type I, IIA, and IIB string theories are actually different states in a single theory, which also includes states containing general configurations of D-branes. Whereas previously we considered only parallel D-branes all of the same dimension, we now wish to study more general configurations. We will be concerned with the breaking of supersymmetry, the spectrum and effective action of strings stretched between different D-branes, and scattering and bound states of D-branes. In the present chapter we are still in the realm of string perturbation theory, but many of the results will be used in the next chapter to understand the strongly coupled theory.

13.1 T-duality of type II strings

Even in the closed oriented type II theories T-duality has an interesting new effect. Compactify a single coordinate X^9 in either type II theory and take the $R \to 0$ limit. This is equivalent to the $R \to \infty$ limit in the dual coordinate, whose right-moving part is reflected

$$X_R'^9(\bar{z}) = -X_R^9(\bar{z}) \tag{13.1.1}$$

just as in the bosonic string. By superconformal invariance we must also reflect $\tilde{\psi}^9(\bar{z})$,

$$\tilde{\psi}'^9(\bar{z}) = -\tilde{\psi}^9(\bar{z}) . \tag{13.1.2}$$

136

However, this implies that the chirality of the right-moving R sector ground state is reversed: the raising and lowering operators $\tilde{\psi}_0^8 \pm i\tilde{\psi}_0^9$ are interchanged. Simply put, T-duality is a spacetime parity operation on just one side of the world-sheet, and so reverses the relative chiralities of the right- and left-moving ground states. If we begin with the IIA theory and take the compactification radius to be small, we obtain the IIB theory at large radius, and vice versa. The same is true if we T-dualize — that is, carry out the change of variables (13.1.1) and (13.1.2) — on any odd number of dimensions, while T-dualizing on an even number returns one to the type II theory with which one began. Thus the two type II theories are related in the same way as the two heterotic theories: in each case the two noncompact theories are different limits of a single space of compactified theories. The type II relation is even simpler than the heterotic relation, in that one takes the radius to zero without having also to include a Wilson line.

Since the IIA and IIB theories have different R–R fields, T-duality must transform one set into the other. Again focus on T-duality in just the 9-direction. In order to preserve the OPE between $\tilde{\psi}^\mu$ and the spin field, this must act as

$$\mathcal{V}'_\alpha(z) = \mathcal{V}_\alpha(z) , \quad \tilde{\mathcal{V}}'_\alpha(\bar{z}) = \beta_{\alpha\beta}^9 \tilde{\mathcal{V}}_\beta(\bar{z}) , \tag{13.1.3}$$

where β^9 is the parity transformation (9-reflection) on the spinors. It anticommutes with Γ^9 and commutes with the remaining Γ^μ, so $\beta^9 = \Gamma^9\Gamma$. Now consider the effect on the R–R vertex operators

$$\overline{\mathcal{V}}\Gamma^{\mu_1\cdots\mu_p}\tilde{\mathcal{V}} . \tag{13.1.4}$$

The T-duality multiplies the product of Γ matrices by $\Gamma^9\Gamma$ on the right. The Γ just gives ± 1 because the R ground states have definite chirality. The Γ^9 adds a '9' to the set $\mu_1 \ldots \mu_p$ if none is present, or removes one if it is present via $(\Gamma^9)^2 = 1$. This is how T-duality acts on the R–R field strengths and potentials, adding or subtracting the index for the dualized dimensions. Thus, if we start from the IIA string we get the IIB R–R fields as follows (up to signs),

$$C_9 \to C , \tag{13.1.5a}$$
$$C_\mu, C_{\mu\nu 9} \to C_{\mu 9}, C_{\mu\nu} , \tag{13.1.5b}$$
$$C_{\mu\nu\lambda} \to C_{\mu\nu\lambda 9} , \tag{13.1.5c}$$

where here μ stands for a nondualized dimension. We could go on, getting $C_{\mu\nu\lambda\omega}$ from $C_{\mu\nu\lambda\omega 9}$ and so on, but these are not independent fields, and give rather the Poincaré dual description of the fields listed.

For T-duality on multiple dimensions replace β^9 with

$$\prod_m \beta^m \, , \tag{13.1.6}$$

where $\beta^m = \Gamma\Gamma^m$ and the product runs over the dualized directions. There are some signs which should be noted but should not distract attention from the main physical point. Since $\beta^m\beta^n = -\beta^n\beta^m$ for $m \neq n$, T-dualities in different directions do not quite commute but differ by a sign in the right-moving R sector. We can write this as

$$\beta^m\beta^n = \exp(\pi i\tilde{\mathbf{F}})\beta^n\beta^m \, , \tag{13.1.7}$$

where $\tilde{\mathbf{F}}$ is the spacetime fermion number of the right-moving state of the string; this is a symmetry that flips the sign of all right-moving R states. Also, we have defined β^m so as to preserve the Hermiticity of $\tilde{\mathcal{V}}_\alpha$ (that is, it is real in a Majorana basis), but then $\beta^m\beta^m = -1$ and so acting twice with T-duality gives $\exp(\pi i\tilde{\mathbf{F}})$.

13.2 T-duality of type I strings

Taking the $R \to 0$ limit of the open and unoriented type I $SO(32)$ theory leads to D-branes and orientifold planes by the same arguments as for the bosonic string in chapter 8, which the reader should review. In particular, taking the T-dual on a single dimension leads to a space with 16 D8-branes between two orientifold hyperplanes.

Let us first consider the bulk physics of the T-dual theory, obtained by taking $R \to 0$ and concentrating on a region of the dual spacetime that is far away from the fixed planes and D-branes, as illustrated in figure 13.1. The local physics is that of a closed oriented superstring theory: closed because the open strings live far away on the D-branes; oriented because the orientation projection relates the state of any string to that of its image behind the fixed plane, but does not locally constrain the space of states. Thus the local physics must be that of a type II theory. In particular there are two gravitinos, and any closed string scattering process will be invariant under the 32 supersymmetries of the type II theory. Since the type I theory with which we started has equal left- and right-moving chiralities, taking the T-dual in one dimension makes them opposite: the local physics is the IIA superstring. Taking the T-dual on any odd number of dimensions has the same effect; taking the T-dual on any even number of dimensions gives the IIB theory in the bulk.

Now take the $R \to 0$ limit while concentrating on the neighborhood of one D-brane in the T-dual theory, adjusting the Wilson lines so that again the fixed plane and other D-branes move away in the T-dual spacetime. The low energy degrees of freedom on the D-brane are the massless open

Fig. 13.1. A D-brane, with one attached open string and one closed string moving in the bulk. The physics away from the D-brane is described by a type II string theory, so the string theory with the D-brane has the physical properties of a state of the *type II* theory containing an extended object.

string states

$$\psi^{\mu}_{-1/2}|k\rangle_{\mathrm{NS}} , \quad \psi^{9}_{-1/2}|k\rangle_{\mathrm{NS}} , \quad |\alpha;k\rangle_{\mathrm{R}} . \tag{13.2.1}$$

As in the bosonic theory, the bosonic states are a gauge field living on the D-brane and the collective coordinates for the D-brane. The fermionic states are the superpartners of these.

Consider now a process where closed strings scatter from the D-brane; this necessarily involves a world-sheet with boundary. Now, the open string boundary conditions are invariant only under $d = 10$, $N = 1$ supersymmetry. In the original type I theory, the left-moving world-sheet current for spacetime supersymmetry $\mathcal{V}_{\alpha}(z)$ flows into the boundary and the right-moving current $\tilde{\mathcal{V}}_{\alpha}(\bar{z})$ flows out, so only the total charge $Q_{\alpha} + \tilde{Q}_{\alpha}$ of the left- and right-movers is conserved. Under T-duality this becomes

$$Q'_{\alpha} + (\beta^{9}\tilde{Q}')_{\alpha} . \tag{13.2.2}$$

The scattering amplitudes of closed strings from the D-brane are invariant only under these 16 supersymmetries.

To see the significance of this, consider first the conservation of momentum. There is a nonzero amplitude for a closed string to reflect backwards from the D-brane, which clearly does not conserve momentum in the direction orthogonal to the D-brane. This occurs because the Dirichlet boundary conditions explicitly break translational invariance. However, from the spacetime point of view the breaking is spontaneous: we are expanding around a D-brane in some definite location, but there are degenerate states with the D-brane translated by any amount.[1] For a spontaneously broken symmetry the consequences are more subtle than for an unbroken symmetry: the apparent violation of the conservation law is related to the amplitude to emit a long-wavelength Goldstone boson. For the D-brane, as for any extended object, the Goldstone bosons are the collective coordinates for its motion. In fact, the nonconservation of momentum is measured by the integral of the corresponding current over the world-sheet boundary,

$$ \frac{1}{2\pi\alpha'} \int_{\partial M} ds\, \partial_n X'^9 \, , \tag{13.2.3} $$

which up to normalization is just the (0 picture) vertex operator for the collective coordinate, with zero momentum in the Neumann directions.

We conclude by analogy that the D-brane also spontaneously breaks 16 of the 32 spacetime supersymmetries, the ones that are explicitly broken by the open string boundary conditions. The integrals

$$ \int_{\partial M} ds\, \mathcal{V}'_\alpha = - \int_{\partial M} ds\, (\beta^9 \tilde{\mathcal{V}}')_\alpha \, , \tag{13.2.4} $$

which measure the breaking of supersymmetry, are just the vertex operators for the fermionic open string state (13.2.1). Thus this state is a *goldstino*, the Goldstone state associated with spontaneously broken supersymmetry.

It is not surprising that the D-brane breaks some supersymmetry. The only state invariant under all supersymmetries is the vacuum. Rather, what is striking is that it leaves half the supersymmetries unbroken: *it is a BPS state*. This same argument holds for any number of dualized dimensions, and so for Dp-branes for all p. The unbroken supersymmetry is

$$ Q'_\alpha + (\beta^\perp \tilde{Q}')_\alpha \, , \tag{13.2.5} $$

where

$$ \beta^\perp = \prod_m \beta^m \, , \tag{13.2.6} $$

the product running over all the dimensions perpendicular to the D-brane.

---

[1] The Mermin–Wagner–Coleman theorem from quantum field theory implies that if the D-brane has two or more noncompact directions there will indeed be an infinite number of degenerate states. If it has one or zero noncompact directions, quantum fluctuations force it into a unique translationally invariant state. The latter effect shows up in perturbation theory through IR divergences. For a spontaneously broken *supersymmetry* the fluctuations are less effective: supersymmetry can be broken even by a zero-dimensional object.

BPS states, which are discussed in section B.2, must carry conserved charges. In the present case there is a natural set of charges with the correct Lorentz properties, namely the antisymmetric R–R charges. The world-volume of a p-brane naturally couples to a $(p+1)$-form potential

$$\int C_{p+1} , \tag{13.2.7}$$

the integral running over the D-brane world-volume. By T-duality we can reach the IIA theory with a Dp-brane of any even p. Thus we need 1-, 3-, 5-, 7-, and 9-form potentials. Indeed, the 1-form and 3-form are in the IIA theory and the 5-form and 7-form give equivalent descriptions of the same physics. The 9-form potential we have discussed in section 12.1 in the context of massive IIA supergravity. Although it is not associated with propagating states, and so was not detected in the quantization of the IIA string, the existence of D8-branes shows that it must be included.

By analogy with electromagnetism in four dimensions, where the 1-form electric potential can be replaced with a 1-form magnetic potential, a Dirichlet p-brane and $(6-p)$-brane are like electric and magnetic sources for the same field strength. For example, the free field equation and Bianchi identity for a 2-form field strength, $d*F_2 = dF_2 = 0$, are symmetric between F_2 and $(*F)_8$, and can be written either in terms of a 1-form or 7-form potential:

$$F_2 = dC_1 , \quad d \wedge *dC_1 = 0 , \tag{13.2.8a}$$
$$*F_2 = (*F)_8 = dC_7 , \quad d \wedge *dC_7 = 0 . \tag{13.2.8b}$$

At an electric source, which would be a D0-brane for C_1 or a D6-brane for C_7, the field equation has a source term. At a magnetic source, a D6-brane for C_1 or a D0-brane for C_7, the Bianchi identity breaks down, and the potential cannot be globally defined: one must introduce a Dirac string, or use different potentials in different patches.[2]

For the IIB theory we need 2-, 4-, 6-, 8-, and 10-form potentials. The first four arise in either the electric or magnetic description of the propagating R–R states. The existence of the 10-form was deduced in section 10.8, from the study of type I divergences. Indeed, we argued there for the coupling (13.2.7) for the 10-form, where the integral runs over all spacetime. This fits with a point made in chapter 8, that it is natural to interpret each Chan–Paton degree of freedom in the fully Neumann theory as a 9-brane filling spacetime. All the other R–R couplings follow from this one by T-duality, since each time we T-dualize in an additional

---

[2] It should be mentioned that there is no local covariant action for a system with both electric and magnetic charges, even though the physics is covariant and presumably satisfies the axioms of local quantum field theory.

direction the dimension of the p-branes goes down by one and the R–R form loses an index.

The IIB theory also has a 0-form potential C_0, the R–R scalar. This should couple to a '(-1)-brane.' Indeed, there is a natural interpretation to this: it is defined by Dirichlet boundary conditions in all directions, time as well as space, so its world-sheet is zero-dimensional and the integral (13.2.7) reduces to the value of C_0 at that point. An object that is localized in time as well as space is an instanton. Instantons in Euclidean path integrals correspond to tunneling events, and we will argue shortly that these must be present in string theory.

We will verify the R–R couplings of D-branes in the next section; for the remainder of this section we will discuss some of the consequences. The discovery that D-branes carry R–R charges neatly ties together two loose ends. On the one hand, it was argued in section 12.1 that the ordinary string states do not have R–R charges, but now we see that string theory does have a source for every gauge field.[3] This extends the result from chapter 8, that the gauge field from compactification of the antisymmetric tensor (under which all states in quantum field theory are neutral) couples to winding strings. On the other hand, the existence of so many different kinds of extended object, Dp-branes for every p, might have seemed excessive, but we now see that these are in one-to-one correspondence with the R–R potentials of the respective type II theories.

The divergence of the type I theory for groups other than $SO(32)$ arose from the R–R 10-form field equation. This divergence is unaffected by toroidal compactification and again cancels only for $SO(32)$. It would have been surprising if toroidal compactification made a consistent theory inconsistent, or the reverse, and it is not hard to verify explicitly that this does not happen. The effect of toroidal compactification is to add world-sheets that wrap around the periodic directions of spacetime. These correspond to exchange of closed strings with winding number, which are massive and so do not have dangerous tadpoles.

The spacetime interpretation of the divergence in the T-dual picture with D-branes is again an inconsistency in the R–R field equations. One can picture field lines emerging from each D-brane, orthogonal to the noncompact dimensions, and these field lines must end somewhere. Further, all D-branes must have the same sign of the charge: the full set of D-branes is still a BPS state, being T-dual to the type I theory, and the total mass is linear in the total charge for a BPS state. We know that the disk tadpole is canceled by the unoriented cross-cap. In the T-dual spacetime the cross-cap must be localized near one of the orientifold

---

[3] In the next chapter we will discuss a seemingly different kind of R–R source, the *black p-brane*.

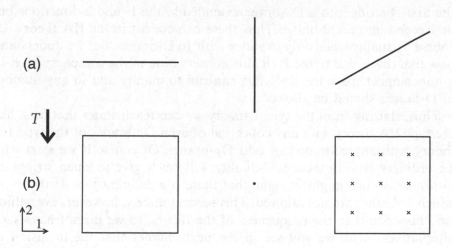

Fig. 13.2. Effect of a T-duality in the 2-direction on D1-branes at various angles in the (1,2) plane: (a) before T-duality; (b) after T-duality. The ×s indicate a magnetic field on the D2-brane.

planes, because the string theory in the bulk is oriented. Thus we deduce that the orientifold planes are sinks for R–R charge. If we T-dualize on k dimensions there are 2^k orientifold planes but still 16 D-branes, so the charge of an orientifold plane must be -2^{4-k} times that of a D-brane of the same dimension.

New connections between string theories

Starting from the toroidally compactified type I theory, we can reach either $d = 10$ type II theory. Simply take an odd or even number of radii to zero, while moving the D-branes and fixed planes off to infinity as the dual spacetime expands. Thus, just as for the two heterotic theories, these should be thought of as limits of a single theory. The theory has many other states as well: we can take the limit while keeping some of the D-branes in fixed positions, so that we obtain the compact theory in a state with D-branes. The simple T-duality leads only to parallel D-branes of equal dimension, but since the D-branes are dynamical we can continuously vary their configurations. We can then reach states with p-branes of different dimension as follows. Consider two D1-branes (D-strings) in the IIB theory, from dualizing in eight directions. Let one be along the 1-direction and the other be rotated to lie along the 2-direction. As illustrated in figure 13.2, a further T-duality in the 2-direction reverses Dirichlet and Neumann boundary conditions in this direction and so turns the first D-string into a D2-brane extended in the 1- and 2-directions but the second into a D0-brane. Thus these can coexist in the IIA theory. Of course T-duality leads only to states with 16 D-branes, but we understand now that this is due to the R–R flux conservation in the compact space. In a noncompact space the R–R flux can run to infinity and so any number of D-branes should be allowed.

Thus, starting from the type I theory we can reach states that look like the type IIA theory with any collection of even Dp-branes or the type IIB theory with any collection of odd Dp-branes. Of course if we start with an ordinary type II theory, T-duality will never give us open strings or D-branes, so one might imagine that there is a different type II theory in which D-branes are not allowed. This seems unlikely, however: everything we know points to the uniqueness of the theory, so we do not have such alternatives. Also, we will see in the next chapter that the inclusion of D-branes leads to a much more elegant and symmetric theory.

In summary, we are now considering a single theory, which has a state that contains no D-branes and looks like the ordinary IIA theory, a second state (T-dual to the first) that contains no D-branes and looks like the ordinary IIB theory, and a third state that contains 16 D9-branes (and an orientifold 9-plane) that looks like the type I theory. It also contains an infinite number of other states with very general configurations of D-branes.

We can now write down the supersymmetry algebra for this theory:

$$\{Q_\alpha, \bar{Q}_\beta\} = -2\Big[P_M + (2\pi\alpha')^{-1}Q_M^{\mathrm{NS}}\Big]\Gamma_{\alpha\beta}^M\,, \tag{13.2.9a}$$

$$\{\tilde{Q}_\alpha, \tilde{\bar{Q}}_\beta\} = -2\Big[P_M - (2\pi\alpha')^{-1}Q_M^{\mathrm{NS}}\Big]\Gamma_{\alpha\beta}^M\,, \tag{13.2.9b}$$

$$\{Q_\alpha, \tilde{\bar{Q}}_\beta\} = -2\sum_p \frac{\tau_p}{p!}Q_{M_1\ldots M_p}^{\mathrm{R}}(\beta^{M_1}\cdots\beta^{M_p})_{\alpha\beta}\,. \tag{13.2.9c}$$

The spacetime supersymmetries Q_α and $\tilde{Q}_\alpha$ act respectively on the right- and left-movers.

The anticommutator (13.2.9b) of two right-moving supersymmetries is the same as the heterotic string anticommutator (11.6.32), containing the charge that couples to the NS–NS 2-form. The argument for the appearance of this term is the same as before: the $\tilde{\mathcal{V}}_\alpha\tilde{\mathcal{V}}_\beta$ OPE contains the right-moving momentum $\bar{\partial}X^\mu$, which involves both ordinary momentum and winding number. Similarly, the $\mathcal{V}_\alpha\mathcal{V}_\beta$ OPE contains the left-moving momentum ∂X^μ, so the NS–NS charge appears in the left-moving anticommutator with the opposite sign. We have added a superscript NS to distinguish this charge from the charges Q^{R} that couple to R–R forms. Also, we have changed conventions so that all charges are now normalized to one per unit world-volume of the respective extended object, and so the string tension $(2\pi\alpha')^{-1}$ appears explicitly. As discussed in section 11.6,

Q_M^{NS} is the charge carried by the *fundamental string,* meaning the original quantized string. Henceforth we use this term, or *F-string,* to distinguish it from the D1-brane.

From the argument that D-branes are BPS states, we expect the R–R charges to appear in the algebra as well, and the natural place for an R–R charge to appear is in the anticommutator of a left- and a right-moving supersymmetry. The sum on p runs over even values in the IIA theory and odd values in the IIB theory. By analogy with the NS–NS case we have included the D-brane tensions τ_p, whose values will be obtained in the next section; the factor of $1/p!$ offsets the sum over permutations of indices. To see that the algebra is correct, focus on a state that contains a single static Dp-brane. The nonzero charge is $Q_{\mu_1\cdots\mu_p}$, where the indices run over the directions tangent to the Dp-brane. Note that

$$\beta^{\mu_1}\cdots\beta^{\mu_p}=\beta^\perp\Gamma^0\,,\tag{13.2.10}$$

up to a possible overall sign that can be reabsorbed in the definition of $\tilde{Q}$; $\beta^\perp$ is the same as in eq. (13.2.5). It then follows that the anticommutator of $Q+\beta^\perp\tilde{Q}$ with any supercharge vanishes in this state, as required by the BPS property. (In eq. (13.2.5) we included primes on the supercharges to indicate that we were working in a T-dual description to the type I theory with which we began. In writing the algebra (13.2.9) we are considering an arbitrary state with D-branes, without necessarily obtaining it from T-duality, so there are no primes.)

Incidentally, the central charge (13.2.9) is still not complete: the *magnetic NS–NS* charge is missing. This is not carried by D-branes or F-strings. We will discuss this further in the next chapter.

Finally, let us also explain the necessity of D-instantons, localized in time. We could try to use T-duality in the time direction, but it is not clear whether this is meaningful. Rather, consider D0-branes, whose world-lines are one-dimensional, in a space with one compact spatial dimension. For an ordinary quantized particle in a path sum description we would have to include closed paths that wind around the compact direction. Such paths are responsible for Casimir energies and other effects of compactification. Presumably we must do the same for the D0-branes as well. The shortest such winding path is a straight line in the compact spatial direction. This is localized in time and so is essentially an instanton: Casimir energies, in the path sum description, are essentially instanton effects. Further, by a T-duality in the compact dimension we obtain a D-instanton that is localized in all directions. We know from chapter 8 that the D-brane action depends on the closed string coupling as $O(1/g)$, so the D-instanton amplitude is of order $e^{-O(1/g)}$. Thus we have found an example of the enhanced nonperturbative effects that were inferred in section 9.7 from the high order behavior of string perturbation theory.

On the heterotic world-sheet there are no boundary conditions that preserve the world-sheet gauge symmetries, and there is no indication that D-branes exist. Nevertheless, we will see in the next chapter that the D-branes of the type I/II theory enable us to learn a great deal about the heterotic string as well.

13.3 The D-brane charge and action

There is no force between static BPS objects of like charge. The multi-object state is still supersymmetric and so its energy is determined only by its charge and is independent of the separations. For parallel Dp-branes, the unbroken supersymmetry (13.2.5) is the same as for a single Dp-brane.

The vanishing of the force comes about from a cancellation between attraction due to the graviton and dilaton and repulsion due to the R–R tensor. We can calculate these forces explicitly from the usual cylinder vacuum amplitude. The exchange of light NS–NS closed strings was isolated in eq. (10.8.4). Modify this expression by removing the factors for the momentum integrations in the Dirichlet directions and introducing a term for the tension of a string stretched over a separation y^μ:

$$
\begin{aligned}
\mathscr{A}_{\text{NS–NS}} &\approx \frac{iV_{p+1}4 \times 16}{8\pi(8\pi^2\alpha')^5} \int_0^\infty \frac{\pi dt}{t^2}(8\pi^2\alpha' t)^{(9-p)/2} \exp\left(-\frac{ty^2}{2\pi\alpha'}\right) \\
&= iV_{p+1}2\pi(4\pi^2\alpha')^{3-p}G_{9-p}(y)
\end{aligned}
\tag{13.3.1}
$$

with $G_d(y) = 2^{-2}\pi^{-d/2}\Gamma(\frac{1}{2}d - 1)y^{2-d}$ the scalar Green's function. The Chan–Paton weight is 2 here, from the two orientations of the open string, and there is no factor of $\frac{1}{2}$ from the orientation projection because the physics is locally oriented. Due to supersymmetric cancellation in the trace, the R–R exchange amplitude is

$$
\mathscr{A}_{\text{R–R}} = -\mathscr{A}_{\text{NS–NS}}
\tag{13.3.2}
$$

and so the total force vanishes as expected.

The field theory calculation (8.7.25) for the dilaton–graviton potential changes only by the substitution $6 = (D - 2)/4 \to 2$, and so is

$$
2i\kappa^2\tau_p^2 G_{9-p}(y) \, .
\tag{13.3.3}
$$

Thus

$$
\tau_p^2 = \frac{\pi}{\kappa^2}(4\pi^2\alpha')^{3-p} \, .
\tag{13.3.4}
$$

This satisfies the same T-duality relation as in the bosonic string. For the R–R exchange, the low-energy action is

$$
-\frac{1}{4\kappa_{10}^2} \int d^{10}x \, (-G)^{1/2}|F_{p+2}|^2 + \mu_p \int C_{p+1} \, .
\tag{13.3.5}
$$

The kinetic term is canonically normalized, so the propagator for any given component (such as the one parallel to the D-brane) is $2\kappa_{10}^2 i/k^2$, and the field theory amplitude is

$$- 2\kappa_{10}^2 i\mu_p^2 G_{9-p}(y) . \tag{13.3.6}$$

Hence

$$\mu_p^2 = \frac{\pi}{\kappa_{10}^2}(4\pi^2\alpha')^{3-p} = e^{2\Phi_0}\tau_p^2 = T_p^2 . \tag{13.3.7}$$

The reader can carry out a similar calculation of the force between a D-brane and an orientifold plane and show that it has an additional $-(2^{5-k})$. We deduced from the cancellation of divergences that the charge of the orientifold plane should have a factor of $-(2^{4-k})$; the extra factor of 2 in the force arises because the orientifold geometry squeezes the flux lines into half the solid angle.

The calculation of the interaction confirms our earlier deduction that D-branes carry the R–R charges. It is interesting to see how this is consistent with our earlier discussion of string vertex operators. The R–R vertex operator (12.1.14) is in the $(-\frac{1}{2}, -\frac{1}{2})$ picture, which can be used in almost all processes. On the disk, however, the total left- plus right-moving ghost number must be -2. With two or more R–R vertex operators, all can be in the $(-\frac{1}{2}, -\frac{1}{2})$ picture (with PCOs included as well), but a single vertex operator must be in either the $(-\frac{3}{2}, -\frac{1}{2})$ or the $(-\frac{1}{2}, -\frac{3}{2})$ picture. The $(-\frac{1}{2}, -\frac{1}{2})$ vertex operator is essentially $e^{-\phi}G_0$ times the $(-\frac{3}{2}, -\frac{1}{2})$ operator, so besides the shift in the ghost number the latter has one less power of momentum and one less Γ-matrix. The missing factor of momentum turns F into C, and the missing Γ-matrix gives the correct Lorentz representations for the potential rather than the field strength.

Dirac quantization condition

There is an important consistency check on the value of the R–R charge, which generalizes the Dirac quantization condition for magnetic monopole charge. Let us review the Dirac condition, shown in figure 13.3. Consider a magnetic charge μ_m at the origin. The integrated flux is

$$\int_{S_2} F_2 = \mu_m . \tag{13.3.8}$$

Because the integral over a closed surface is nonzero, we cannot write $F_2 = dA_1$ for any vector potential. However, we can write $F_2 = dA_1$ except along a Dirac string ending on the monopole. Now consider an electric charge μ_e moving in this field. Its coupling to the field produces a

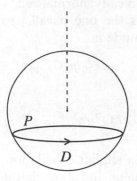

Fig. 13.3. Sphere surrounding monopole, with a Dirac string running upward. The particle path P is bounded by the lower cap D.

phase

$$\exp\left(i\mu_e \oint_P A_1\right) = \exp\left(i\mu_e \int_D F_2\right) \quad (13.3.9)$$

when the particle moves on a closed path P. The surface D spans P and does not intersect the Dirac string. Now consider the limit as the path is contracted to a small circle around the Dirac string. The phase becomes

$$\exp\left(i\mu_e \int_{S_2} F_2\right) = \exp(i\mu_e\mu_m) . \quad (13.3.10)$$

The Dirac string must be invisible, so this phase must be 1. Equivalently, this is the condition that the phase (13.3.9) is unchanged if we instead take the surface $D' = S_2 - D$ spanning P in the upper hemisphere. The result is the Dirac quantization condition,

$$\mu_e\mu_m = 2\pi n \quad (13.3.11)$$

for some integer n.

A p-brane and $(6-p)$-brane are sources for F_{p+2} and F_{8-p} respectively. These two field strengths are Poincaré dual to one another, so again there is a Dirac quantization condition that must be satisfied by the product of their charges. Let us think about F_{p+2} as the field strength, so that the p-brane is an electric source and the $(6-p)$-brane a magnetic source. In nine dimensions a $(6-p)$-dimensional object is surrounded by a $(p+2)$-sphere, so by analogy to the magnetic flux (13.3.8),

$$\int_{S_{p+2}} F_{p+2} = 2\kappa_{10}^2 \mu_{6-p} . \quad (13.3.12)$$

One can then repeat the same argument. For example, let the p-brane be extended in the directions $4 \le \mu \le p+3$ and the $(6-p)$-brane in the directions $p + 4 \leq \mu \leq 9$. The system essentially reduces to the three-dimensional situation of figure 13.3 in the directions $\mu = 1, 2, 3$, and the charges must satisfy

$$\mu_p \mu_{6-p} = 4\kappa_{10}^2 \pi n \ . \tag{13.3.13}$$

Remarkably, the charges (13.3.7), arrived at in an entirely different way, satisfy this relation with the minimum quantum $n = 1$.

D-brane actions

The coupling of a D-brane to NS–NS closed string fields is the same Dirac–Born–Infeld action as in the bosonic string,

$$S_{Dp} = -\mu_p \int d^{p+1}\xi \, \mathrm{Tr}\Big\{ e^{-\Phi}[-\det(G_{ab} + B_{ab} + 2\pi\alpha' F_{ab})]^{1/2} \Big\} \ , \tag{13.3.14}$$

where G_{ab} and B_{ab} are the components of the spacetime NS–NS fields parallel to the brane and F_{ab} is the gauge field living on the brane. The argument leading to this form is exactly as in the bosonic case, section 8.7. Recall that for n D-branes at small separation, where the strings stretched between them are light enough to be included in the low energy action, the collective coordinates $X^\mu(\xi)$, gauge fields $A_a(\xi)$, and their fermionic partners $\lambda(\xi)$ all become $n \times n$ matrices. The trace in the action is in this $n \times n$ space. In addition there is a term

$$O([X^m, X^n]^2) \tag{13.3.15}$$

in the potential. As discussed in chapter 8, the effect of this potential is that in the flat directions the collective coordinates become diagonal. They can then be interpreted as n ordinary collective coordinates for n objects. At small separation the full matrix dynamics is crucial, as we will see.

The coupling to the R–R background also includes corrections involving the gauge field on the brane. Like the Born–Infeld action, these can be deduced via T-duality. Consider, as an example, a 1-brane in the (1,2) plane. The action is

$$\int C_2 = \int dx^0(dx^1 C_{01} + dx^2 C_{02}) = \int dx^0 \, dx^1 \Big(C_{01} + \partial_1 X^2 C_{02}\Big) \ . \tag{13.3.16}$$

Under a T-duality in the 2-direction this becomes

$$\int dx^0 \, dx^1 \ (C_{012} + 2\pi\alpha' F_{12} C_0) \ . \tag{13.3.17}$$

We have used the T-transformation of the C fields, eq. (13.1.5). A D-brane at an angle is T-dual to one with a magnetic field, as in figure 13.2. We are not keeping track of the normalization but one could, with the result $\mu_p = \mu_{p-1}/2\pi\alpha'^{1/2}$ in agreement with the explicit calculation. The generalization of (13.3.17) to an arbitrary configuration, and to multiple D-branes, gives the Chern–Simons-like result

$$i\mu_p \int_{p+1} \text{Tr}\left[\exp(2\pi\alpha' F_2 + B_2) \wedge \sum_q C_q\right] . \tag{13.3.18}$$

The expansion of the integrand (13.3.18) involves forms of various rank; the integral picks out precisely the terms that are proportional to the volume form of the p-brane world-volume. There are similar couplings with the spacetime curvature in addition to the field strength; these can be obtained from a string calculation.

Thus far we have given only the action for the bosonic fields on the brane. For the leading fluctuations around a flat D-brane in flat spacetime the fermionic action is of the usual Dirac form

$$-i \int d^{p+1}\xi \, \text{Tr}(\bar{\lambda}\Gamma^a D_a \lambda) . \tag{13.3.19}$$

The full nonlinear supersymmetric form is left to the references.

Coupling constants

The ratio of the F-string tension to the D-string tension is

$$\frac{\tau_{F1}}{\tau_{D1}} = \frac{1}{2\pi\alpha'}\frac{\kappa}{4\pi^{5/2}\alpha'} = \frac{\kappa}{8\pi^{7/2}\alpha'^2} . \tag{13.3.20}$$

Up to now there has been no natural convention for defining the additive normalization of the dilaton field or the multiplicative normalization of the closed string coupling $g = e^\Phi$. The dimensionless ratio (13.3.20) is proportional to the closed string coupling, and it turns out to be very convenient to take it as the definition of the coupling,

$$g = \frac{\tau_{F1}}{\tau_{D1}} . \tag{13.3.21}$$

Then the gravitational coupling is

$$\kappa^2 = \tfrac{1}{2}(2\pi)^7 g^2 \alpha'^4 \tag{13.3.22}$$

and the D-brane tension is

$$\tau_p = \frac{1}{g(2\pi)^p \alpha'^{(p+1)/2}} = (2\kappa^2)^{-1/2}(2\pi)^{(7-2p)/2}\alpha'^{(3-p)/2} . \tag{13.3.23}$$

Also, the constant appearing in the low energy actions of section 12.1 is

$$\kappa_{10}^2 = \tfrac{1}{2}(2\pi)^7 \alpha'^4 ; \tag{13.3.24}$$

this differs from the physically measured κ because the latter depends on the dilaton background.

Expanding the action (13.3.14) gives the coupling of the Yang–Mills theory on the Dp-brane,

$$g_{\mathrm{D}p}^2 = \frac{1}{(2\pi\alpha')^2\tau_p} = (2\pi)^{p-2}g\alpha'^{(p-3)/2} . \qquad (13.3.25)$$

Notice that for $p = 3$ this coupling is dimensionless, as expected in a $(3 + 1)$-dimensional gauge theory. For $p < 3$ the coupling has units of length to a negative power, and for $p > 3$ length to a positive power.

We now wish to obtain the relation among κ, g_{YM}, and α' in the type I theory. We cannot quite identify $g_{\mathrm{D}p}$ for $p = 9$ with g_{YM}, because the former has been obtained in a locally oriented theory and there are some additional factors of 2 in the type I case. Rather than repeat the string calculation we will make a more roundabout but possibly instructive argument using T-duality.

First, we should note that the coupling (13.3.25) is for the $U(n)$ gauge theory of coincident branes in the oriented theory: it appears in the form

$$\frac{1}{4g_{\mathrm{D}p}^2}\mathrm{Tr}_{\mathrm{f}} , \qquad (13.3.26)$$

where the trace is in the $n \times n$ fundamental representation. Now let us consider moving the branes to an orientifold plane so that the gauge symmetry is enlarged to $SO(2n)$. An $SU(n)$ generator t is embedded in $SO(2n)$ as

$$\tilde{t} = \begin{bmatrix} t & 0 \\ 0 & -t^T \end{bmatrix} , \qquad (13.3.27)$$

because the orientation projection reverses the order of the Chan–Paton factors and the sign of the gauge field. Comparing the low energy actions gives

$$\frac{1}{4g_{\mathrm{D}p}^2}\mathrm{Tr}_{\mathrm{f}}(t^2) = \frac{1}{4g_{\mathrm{D}p,SO(2n)}^2}\mathrm{Tr}_{\mathrm{v}}(\tilde{t}^2) \qquad (13.3.28)$$

and so $g_{\mathrm{D}p,SO(2n)}^2 = 2g_{\mathrm{D}p}^2$.

Now consider the type I theory compactified on a k-torus with all radii equal to R. The couplings in the lower-dimensional $SO(32)$ theory are related to those in the type I theory by

$$\kappa_{10-k}^2 = (2\pi R)^{-k}\kappa^2 \text{ (type I)} , \quad g_{10-k,\mathrm{YM}}^2 = (2\pi R)^{-k}g_{\mathrm{YM}}^2 \text{ (type I)} . \qquad (13.3.29)$$

In the T-dual picture, the bulk theory is of type II and the gauge fields live on a D$(p - k)$-brane, and

$$\kappa_{10-k}^2 = 2(2\pi R')^{-k}\kappa^2 , \quad g_{10-k,\mathrm{YM}}^2 = g_{\mathrm{D}(9-k),SO(32)}^2 . \qquad (13.3.30)$$

The dimensional reduction for κ_{10-k}^2 has an extra factor of 2 because the compact space is an orientifold, its volume halved. The gauge coupling is independent of the volume because the fields are localized on the D-brane. Combining these results with the relations (13.3.22) and (13.3.25) gives, independent of k, the type I relation

$$\frac{g_{YM}^2}{\kappa} = 2(2\pi)^{7/2}\alpha' \quad \text{(type I)}. \tag{13.3.31}$$

As one final remark, the Born–Infeld form for the gauge action applies by T-duality to the type I theory:

$$S = -\frac{1}{(2\pi\alpha')^2 g_{YM}^2} \int d^{10}x \, \text{Tr}\{[-\det(\eta_{\mu\nu} + 2\pi\alpha' F_{\mu\nu})]^{1/2}\}, \tag{13.3.32}$$

whose normalization is fixed by the quadratic term in F. In the previous chapter we obtained the tree-level string correction (12.4.28) to the type I effective action. If the gauge field lies in an Abelian subgroup, the tensor structure simplifies to

$$\frac{(2\pi\alpha')^2}{32 g_{YM}^2} \text{Tr}_v\left(4F_{\mu\nu}F^{\nu\sigma}F_{\sigma\rho}F^{\rho\mu} - F_{\mu\nu}F^{\nu\mu}F_{\sigma\rho}F^{\rho\sigma}\right). \tag{13.3.33}$$

This is indeed the quartic term in the expansion of the Born–Infeld action, as one finds by using

$$\det^{1/2}(1 + M) = \exp\left[\frac{1}{2}\text{tr}\left(M - \frac{1}{2}M^2 + \frac{1}{3}M^3 - \frac{1}{4}M^4 + \dots\right)\right] \tag{13.3.34}$$

with $M_\mu^{\ \nu} = 2\pi\alpha' F_{\mu\sigma}\eta^{\sigma\nu}$. The trace here is on the Lorentz indices, and $\text{tr}(x^{2k+1}) = 0$ for antisymmetric x. Note that only when the gauge field can be diagonalized can we give a geometric interpretation to the T-dual configuration and so derive the Born–Infeld form.

13.4   D-brane interactions: statics

Many interesting new issues arise with D-branes that are not parallel, or are of different dimensions. In this section we focus on static questions. The first of these concerns the breaking of supersymmetry. Let us consider a Dp-brane and a Dp'-brane, which we take first to be aligned along the coordinate axes. That is, we can partition the spacetime directions μ into two sets S_D and S_N according to whether the coordinate X^μ has Dirichlet or Neumann boundary conditions on the first D-brane, and similarly into two sets S_D' and S_N' depending on the alignment of the second D-brane. The *DD coordinates* are $S_D \cap S_D'$, the *ND coordinates* are $S_N \cap S_D'$, and so on.

The first D-brane leaves unbroken the supersymmetries

$$Q_\alpha + (\beta^\perp \tilde{Q})_\alpha, \quad \beta^\perp = \prod_{m \in S_D} \beta^m. \tag{13.4.1}$$

Similarly the second D-brane leaves unbroken

$$Q_\alpha + (\beta^{\perp\prime}\tilde{Q})_\alpha = Q_\alpha + [\beta^\perp(\beta^{\perp-1}\beta^{\perp\prime})\tilde{Q}]_\alpha \,, \quad \beta^{\perp\prime} = \prod_{m \in S'_D} \beta^m \,. \tag{13.4.2}$$

The complete state is invariant only under supersymmetries that are of both forms (13.4.1) and (13.4.2). These are in one-to-one correspondence with spinors left invariant by $\beta^{\perp-1}\beta^{\perp\prime}$. The operator $\beta^{\perp-1}\beta^{\perp\prime}$ is a reflection in the DN and ND directions. Let us denote the total number of DN and ND directions $\#_{\text{ND}}$. Since $p - p'$ is always even the number $\#_{\text{ND}} = 2j$ is also even. We can then pair these dimensions and write $(\beta^\perp)^{-1}\beta^{\perp\prime}$ as a product of rotations by π,

$$\beta \equiv (\beta^\perp)^{-1}\beta^{\perp\prime} = \exp[\pi i(J_1 + \ldots + J_j)] \,. \tag{13.4.3}$$

In a spinor representation, each $\exp(i\pi J)$ has eigenvalues $\pm i$, so there will be unbroken supersymmetries only if j is even and so $\#_{\text{ND}}$ is a multiple of 4. In this case there are 8 unbroken supersymmetries, one quarter of the original 32. Note that T-duality switches NN↔DD and ND↔DN and so leaves $\#_{\text{ND}}$ invariant. When $\#_{\text{ND}} = 0$, then $(\beta^\perp)^{-1}\beta^{\perp\prime} = 1$ identically and there are 16 unbroken supersymmetries. This is the same as for the original type I theory, to which it is T-dual.

An open string can have both ends on the same D-brane or one on each. The p-p and p'-p' spectra are the same as obtained before by T-duality from the type I string, but the p-p' strings are new. Each of the four possible boundary conditions can be written with the doubling trick

$$X^\mu(w, \bar{w}) = X^\mu(w) + \tilde{X}^\mu(\bar{w}) \tag{13.4.4}$$

in terms of one of two mode expansions,

$$X^\mu(w) = x^\mu + \left(\frac{\alpha'}{2}\right)^{1/2}\left[-\alpha_0^\mu w + i \sum_{\substack{m \in \mathbf{Z} \\ m \neq 0}} \frac{\alpha_m^\mu}{m} \exp(imw)\right] \,, \tag{13.4.5a}$$

$$X^\mu(w) = i\left(\frac{\alpha'}{2}\right)^{1/2} \sum_{r \in \mathbf{Z}+1/2} \frac{\alpha_r^\mu}{r} \exp(irw) \,. \tag{13.4.5b}$$

The periodic expansion (13.4.5a) describes NN strings for

$$\tilde{X}^\mu(\bar{w}) = X^\mu(2\pi - \bar{w}) \tag{13.4.6}$$

and DD strings for

$$\tilde{X}^\mu(\bar{w}) = -X^\mu(2\pi - \bar{w}) \,. \tag{13.4.7}$$

The antiperiodic expansion (13.4.5b) similarly defines DN and ND strings, with $\tilde{X}^\mu(\bar{w}) = \pm X^\mu(2\pi - w)$. For ψ^μ, the periodicity in the R sector is the same as for X^μ because T_F is periodic. In the NS sector it is the opposite.

The string zero-point energy is zero in the R sector as always, because bosons and fermions with the same periodicity cancel. In the NS sector it is

$$(8 - \#_{ND})\left(-\frac{1}{24} - \frac{1}{48}\right) + \#_{ND}\left(\frac{1}{48} + \frac{1}{24}\right) = -\frac{1}{2} + \frac{\#_{ND}}{8} . \quad (13.4.8)$$

The oscillators can raise the level in half-integer units, so only for $\#_{ND}$ a multiple of 4 is degeneracy between the R and NS sectors possible. This agrees with the analysis above: the $\#_{ND} = 2$ and $\#_{ND} = 6$ systems cannot be supersymmetric. Later we will see that there are supersymmetric *bound states* when $\#_{ND} = 2$, but their description is rather different.

Branes at general angles

It is interesting to consider the case of D-branes at general angles to one another. To be specific consider two D4-branes. Let both initially be extended in the (2,4,6,8)-directions, and separated by some distance y_1 in the 1-direction. Now rotate one of them by an angle ϕ_1 in the $(2, 3)$ plane, ϕ_2 in the $(4, 5)$ plane, and so on; call this rotation ρ. The supersymmetry unbroken by the rotated 4-brane is

$$Q_\alpha + (\rho^{-1}\beta^\perp \rho \tilde{Q})_\alpha . \quad (13.4.9)$$

Supersymmetries left unbroken by both branes then correspond to spinors left invariant by

$$(\beta^\perp)^{-1}\rho^{-1}\beta^\perp\rho = (\beta^\perp)^{-1}\beta^\perp\rho^2 = \rho^2 . \quad (13.4.10)$$

In the usual s-basis the eigenvalues of ρ^2 are

$$\exp\left(2i\sum_{a=1}^{4}s_a\phi_a\right) . \quad (13.4.11)$$

In the **16** the $(2s_1, 2s_2, 2s_3, 2s_4)$ run over all 16 combinations of ± 1s; each combination such that the phase (13.4.11) is 1 gives an unbroken supersymmetry. There are many possibilities — for example:

- For generic ϕ_a there are no unbroken supersymmetries.

- For angles $\phi_1 + \phi_2 + \phi_3 + \phi_4 = 0$ mod 2π (but otherwise generic) there are two unbroken supersymmetries, namely those with $s_1 = s_2 = s_3 = s_4$. The rotated D4-brane breaks seven-eighths of the supersymmetry of the first.

- For $\phi_1 + \phi_2 + \phi_3 = \phi_4 = 0$ mod 2π there are four unbroken supersymmetries.

- For $\phi_1 + \phi_2 = \phi_3 + \phi_4 = 0$ mod 2π there are four unbroken supersymmetries.

- For $\phi_1 + \phi_2 = \phi_3 = \phi_4 = 0$ mod 2π there are eight unbroken supersymmetries.

Also, when k angles are $\pi/2$ and the rest are zero this reduces to the earlier analysis with $\#_{\mathrm{ND}} = 2k$.

For later reference let us also describe these results as follows. Join the coordinates into complex pairs, $Z^1 = X^2 + iX^3$ and so on, with the conjugate $\overline{Z^a}$ denoted $Z^{\bar{a}}$. Then ρ takes Z^a to $\exp(i\phi_a)Z^a$. The $SO(8)$ rotation group on the transverse dimensions has a $U(4)$ subgroup that preserves the complex structure. That is, it rotates $Z'^a = U^{ab}Z^b$, whereas a general $SO(8)$ rotation would mix in $Z^{\bar{b}}$ as well. The rotation ρ in particular is the $U(4)$ matrix

$$\mathrm{diag}\Big[\exp(i\phi_1), \exp(i\phi_2), \exp(i\phi_3), \exp(i\phi_4)\Big] \, . \qquad (13.4.12)$$

When $\phi_1 + \phi_2 + \phi_3 + \phi_4 = 0$ mod 2π, which is the condition for two supersymmetries to be unbroken, the determinant of ρ is 1 and so it actually lies in the $SU(4)$ subgroup of $U(4)$. Then we can summarize the above by saying that a general $U(4)$ rotation breaks all the supersymmetry, an $SU(4)$ rotation breaks seven-eighths, an $SU(3)$ or $SU(2) \times SU(2)$ rotation breaks three-quarters, and an $SU(2)$ rotation half. Further, if we consider several branes, so that in general the rotations ρ_i cannot be simultaneously diagonalized, then as long as all of them lie within a given subgroup the number of unbroken supersymmetries is as above.

Now let us calculate the force between these rotated branes. The cylinder graph involves traces over the p-p' strings, so we need to generalize the mode expansion to the rotated case. Letting the $\sigma^1 = 0$ endpoint be on the unrotated brane and the $\sigma^1 = \pi$ endpoint on the rotated brane, it follows that the boundary conditions are

$$\sigma_1 = 0 : \quad \partial_1 \mathrm{Re}(Z^a) = \mathrm{Im}(Z^a) = 0 \, , \qquad (13.4.13a)$$

$$\sigma_1 = \pi : \quad \partial_1 \mathrm{Re}[\exp(i\phi_a)Z^a] = \mathrm{Im}[\exp(i\phi_a)Z^a] = 0 \, . \quad (13.4.13b)$$

These are satisfied by

$$Z^a(w, \bar{w}) = \mathscr{Z}^a(w) + \overline{\mathscr{Z}^a}(-\bar{w}) \, ,$$

$$= \exp(-2i\phi_a)\mathscr{Z}^a(w + 2\pi) + \overline{\mathscr{Z}^a}(-\bar{w}) \, , \qquad (13.4.14)$$

where $w = \sigma^1 + i\sigma^2$. This implies the mode expansion

$$\mathscr{Z}^a(w) = i\left(\frac{\alpha'}{2}\right)^{1/2} \sum_{r \in \mathbf{Z}+v_a} \frac{\alpha_r^a}{r} \exp(irw) \, , \qquad (13.4.15)$$

with $v_a = \phi_a/\pi$. The modes $(\alpha_r^a)^\dagger$ are linearly independent.

The partition function for one such complex scalar is

$$q^{E_0} \prod_{m=0}^{\infty} \left[1 - q^{m+(\phi/\pi)}\right]^{-1} \left[1 - q^{m+1-(\phi/\pi)}\right]^{-1} = -i\frac{\exp(\phi^2 t/\pi)\eta(it)}{\vartheta_{11}(i\phi t/\pi, it)} \quad (13.4.16)$$

with $q = \exp(-2\pi t), 0 < \phi < \pi$ (else subtract the integer part of ϕ/π), and

$$E_0 = \frac{1}{24} - \frac{1}{2}\left(\frac{\phi}{\pi} - \frac{1}{2}\right)^2 . \quad (13.4.17)$$

The definitions and properties of theta functions are collected in section 7.2, but we reproduce here the results that will be most useful:

$$\vartheta_{11}(v, it) = -2q^{1/8} \sin \pi v \prod_{m=1}^{\infty} (1 - q^m)(1 - zq^m)(1 - z^{-1}q^m) ,$$
$$(13.4.18a)$$

$$\vartheta_{11}(-iv/t, i/t) = -it^{1/2} \exp(\pi v^2/t)\vartheta_{11}(v, it) , \quad (13.4.18b)$$

where $z = \exp(2\pi iv)$. Similarly in each of the sectors of the fermionic path integral one replaces the $Z^\alpha_\beta(it)$ that appears for parallel D-branes with[4]

$$Z^\alpha_\beta(\phi, it) = \frac{\vartheta_{\alpha\beta}(i\phi t/\pi, it)}{\exp(\phi^2 t/\pi)\eta(it)} . \quad (13.4.19)$$

The full fermionic partition function is

$$\frac{1}{2}\left[\prod_{a=1}^{4} Z^0_0(\phi_a, it) - \prod_{a=1}^{4} Z^0_1(\phi_a, it) - \prod_{a=1}^{4} Z^1_0(\phi_a, it) - \prod_{a=1}^{4} Z^1_1(\phi_a, it)\right] ,$$
$$(13.4.20)$$

generalizing the earlier $Z^+_\psi(it)$. By a generalization of the abstruse identity (7.2.41), the fermionic partition function can be rewritten

$$\prod_{a=1}^{4} Z^1_1(\phi'_a, it) , \quad (13.4.21)$$

where

$$\phi'_1 = \frac{1}{2}(\phi_1 + \phi_2 + \phi_3 + \phi_4) , \quad \phi'_2 = \frac{1}{2}(\phi_1 + \phi_2 - \phi_3 - \phi_4) , \quad (13.4.22a)$$

$$\phi'_3 = \frac{1}{2}(\phi_1 - \phi_2 + \phi_3 - \phi_4) , \quad \phi'_4 = \frac{1}{2}(\phi_1 - \phi_2 - \phi_3 + \phi_4) . \quad (13.4.22b)$$

This identity has a simple physical origin. If we refermionize, writing the theory in terms of the free fields θ_α as in eq. (12.6.24), we get the

---

[4] If one applies the formalism of the previous chapter, in the (P,P) spin structure there are two $\beta\gamma$ zero modes and two longitudinal ψ zero modes for a net $0^2/0^2$. One can define this by a more careful gauge fixing, or equivalently by adding a graviton vertex operator (which allows all the zero modes to be soaked up) and relating the zero-momentum graviton coupling to the potential. However, we simply rely on the physical input of the Coleman–Weinberg formula.

form (13.4.21) directly. In particular, the $\exp(\pm i\phi'_a)$ are the eigenvalues of ρ in the spinor **8** of $SO(8)$.

Collecting all factors, the potential is

$$V = -\int_0^\infty \frac{dt}{t}(8\pi^2\alpha't)^{-1/2}\exp\left(-\frac{ty_1^2}{2\pi\alpha'}\right)\prod_{a=1}^4 \frac{\vartheta_{11}(i\phi'_a t/\pi, it)}{\vartheta_{11}(i\phi_a t/\pi, it)} \ . \qquad (13.4.23)$$

Note that for nonzero angles the stretched strings are confined near the point of closest approach of the two 4-branes. The function $\vartheta_{11}(v, it)$ is odd in v and so vanishes when $v = 0$. If any of the ϕ_a vanish the denominator has a zero. This is because the 4-branes become parallel in one direction and the strings are then free to move in that direction. One must replace

$$\vartheta_{11}(i\phi_a t/\pi, it)^{-1} \rightarrow iL\eta(it)^{-3}(8\pi^2\alpha't)^{-1/2} \ . \qquad (13.4.24)$$

This gives the usual factors for a noncompact direction, L being the length of the spatial box. Taking $\phi_4 \rightarrow 0$ so the 4-branes both run in the 8-direction, one can T-dualize in this direction to get a pair of 3-branes with relative rotations in three planes. The fermionic partition function is unaffected, while the factors (13.4.24) are instead replaced by

$$\eta(it)^{-3}\exp\left[-\frac{t(y_8^2 + y_9^2)}{2\pi\alpha'}\right] \ , \qquad (13.4.25)$$

allowing for the possibility of a separation in the (8,9) plane. Taking the T-dual in the 9-direction instead one obtains 5-branes that are separated in the 1-direction, extended in the (8,9)-directions, and with relative rotations in the other three planes. The effect is an additional factor of $L_9(8\pi^2\alpha't)^{-1/2}$. The extension to other p is straightforward.

If instead any of the ϕ'_a vanishes, the potential is zero. The reason is that there is unbroken supersymmetry: the phases (13.4.11) include $\exp(\pm 2i\phi'_a)$. Curiously this covers only eight of the sixteen phases (13.4.11), so that if some phases (13.4.11) are unity but not those of the form $\exp(\pm 2i\phi'_a)$, then supersymmetry is unbroken but the potential is nonzero. This is an exception to the usual rule that the vacuum loop amplitudes vanish by Bose–Fermi cancellation. The rotated D-branes leave only two supersymmetries unbroken, so that BPS multiplets of open strings contain a single bosonic or fermionic state.

The potential is a complicated function of position, but at long distance it simplifies. The exponential factor in the integral (13.4.23) forces t to be small, and then the ϑ-functions simplify,

$$\prod_{a=1}^4 \frac{\vartheta_{11}(i\phi'_a t/\pi, it)}{\vartheta_{11}(i\phi_a t/\pi, it)} \rightarrow \prod_{a=1}^4 \frac{\sin\phi'_a}{\sin\phi_a} \ , \qquad (13.4.26)$$

by using the modular transformation of ϑ_{11}. The t-integral then gives a power of the separation y_1. The result agrees with the low energy field

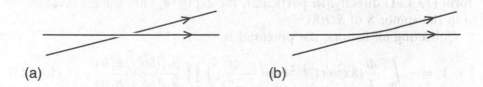

(a)   (b)

Fig. 13.4.   (a) D-branes at relative angle. (b) Lower energy configuration.

theory calculation, including the angular factor. For 4-branes with all ϕ_a nonzero the potential grows linearly with y_1 at large distance, for 3-branes with all ϕ_a nonzero it falls as $1/y_1$, and so on.

In nonsupersymmetric configurations a tachyon can appear. For simplicity let only ϕ_1 be nonzero, with $0 \leq \phi_1 \leq \pi$. The NS ground state energy is $-(1/2) + (\phi_1/2\pi)$, and the first excited state $\psi_{-(1/2)+(\phi_1/\pi)}|0\rangle_{NS}$, which survives the GSO projection, has weight $-\phi_1/2\pi$. Including the energy from tension, the lightest state has

$$m^2 = \frac{y_1^2}{4\pi^2\alpha'^2} - \frac{\phi_1}{2\pi\alpha'}, \qquad 0 \leq \phi_1 \leq \pi. \tag{13.4.27}$$

This is negative if the separation is small enough. A special case is $\phi_1 = \pi$, when the 4-branes are antiparallel rather than parallel. The NS–NS and R–R exchanges are then both attractive, and below the critical separation $y_1^2 = 8\pi^2\alpha'$ the cylinder amplitude diverges as $t \to \infty$. This is where the tachyon appears — evidently it represents D4-brane/anti-D4-brane annihilation. Even when the D-branes are nearly parallel they can lower their energy by reconnecting as in figure 13.4(b), and this is the origin of the instability. This is one example where the tachyon has a simple physical interpretation and we can see that the decay has no end: the reconnected strings move apart indefinitely. On the other hand, for the same instability but with the strings wound on a two-torus there is a lower bound to the energy.

13.5   D-brane interactions: dynamics

D-brane scattering

For parallel static D-branes the potential energy is zero, but if they are in relative motion all supersymmetry is broken and there is a velocity-dependent force. This can be obtained by an analytic continuation of the static potential for rotated branes. Consider the case that only ϕ_1 is nonzero, so the rotated brane satisfies $X^3 = X^2 \tan \phi_1$. Analytically continue $X^2 \to iX'^0$ and let $\phi_1 = -iu$, with $u > 0$. Then

$$X^3 = X'^0 \tanh u , \qquad (13.5.1)$$

which describes a D-brane moving with constant velocity. Continue also $X^0 \to -iX'^2$ to eliminate the spurious extra time coordinate. The interaction amplitude (13.4.23) between the D-branes becomes

$$\mathscr{A} = -iV_p \int_0^\infty \frac{dt}{t} (8\pi^2\alpha' t)^{-p/2} \exp\left(-\frac{ty^2}{2\pi\alpha'}\right) \frac{\vartheta_{11}(ut/2\pi, it)^4}{\eta(it)^9 \vartheta_{11}(ut/\pi, it)} , \qquad (13.5.2)$$

where we have extended the result to general p by using T-duality.[5] It is also useful to give the modular transformation

$$\mathscr{A} = \frac{V_p}{(8\pi^2\alpha')^{p/2}} \int_0^\infty \frac{dt}{t} t^{(6-p)/2} \exp\left(-\frac{ty^2}{2\pi\alpha'}\right) \frac{\vartheta_{11}(iu/2\pi, i/t)^4}{\eta(i/t)^9 \vartheta_{11}(iu/\pi, i/t)} . \qquad (13.5.3)$$

We can write this as an integral over the world-line,

$$\mathscr{A} = -i \int_{-\infty}^\infty d\tau \, V(r(\tau), v) , \qquad (13.5.4)$$

where

$$r(\tau)^2 = y^2 + v^2\tau^2 , \quad v = \tanh u , \qquad (13.5.5)$$

and

$$V(r, v) = i \frac{2V_p}{(8\pi^2\alpha')^{(p+1)/2}} \int_0^\infty dt \, t^{(5-p)/2}$$

$$\times \exp\left(-\frac{tr^2}{2\pi\alpha'}\right) \frac{(\tanh u)\vartheta_{11}(iu/2\pi, i/t)^4}{\eta(i/t)^9 \vartheta_{11}(iu/\pi, i/t)} . \qquad (13.5.6)$$

The interaction has a number of interesting properties. The first is that as $v \to 0$ (so that $u \to 0$), it vanishes as v^4 from the zeros of the theta functions. We expect only even powers of v by time-reversal invariance. The vanishing of the v^2 interaction, like the vanishing of the static interaction, is a consequence of supersymmetry. The low energy field theory of the D-branes is a $U(1) \times U(1)$ supersymmetric gauge theory with 16 supersymmetries. What we are calculating is a correction to the effective action from integrating out massive states, strings stretched between the D-branes. The vanishing of the v^2 term is then consistent with the assertion in section B.6 that with 16 supersymmetries corrections to the kinetic term are forbidden — the moduli space is flat. If we had instead taken $\phi_3 = \phi_4 = \pi/2$ so that $\#_{\text{ND}} = 4$, there would only be two zeros in the numerator and thus a v^2 interaction. This is consistent with

---

[5] We find it difficult to keep track of the sign during the continuation, but it is easily checked by looking at the contribution of NS–NS exchange in the static limit. Note that the ϑ_{11} are negative for small positive u.

the result that corrections to the kinetic term are allowed when there are eight unbroken supersymmetries.

The interaction (13.5.6) is in general a complicated function of the separation, but in an expansion in powers of the velocity the leading $O(v^4)$ term is simple,

$$V(r,v) = -v^4 \frac{V_p}{(8\pi^2\alpha')^{(p+1)/2}} \int_0^\infty dt \, t^{(5-p)/2} \exp\left(-\frac{tr^2}{2\pi\alpha'}\right) + O(v^6)$$

$$= -\frac{v^4}{r^{7-p}} \frac{V_p}{\alpha'^{p-3}} 2^{2-2p} \pi^{(5-3p)/2} \Gamma\left(\frac{7-p}{2}\right) + O(v^6) . \qquad (13.5.7)$$

At long distances this is in agreement with low energy supergravity. It is also the leading behavior if we expand in powers of $1/r$ rather than v.

In general the behavior of $V(r,v)$ as $r \to 0$ is quite different from the behavior as $r \to \infty$. The r-dependence of the integral (13.5.6) arises from the factor $\exp(-tr^2/2\pi\alpha')$, so that $t \approx 2\pi\alpha'/r^2$ governs the behavior at given r. Large r corresponds to small t, where the asymptotic behavior is given by tree-level exchange of light closed strings — hence the agreement with classical supergravity. Small r corresponds to large t, where the asymptotic behavior is given by a loop of the light open strings. The cross-over is at $r^2 \sim 2\pi\alpha'$. This is as we expect: string theory modifies gravity at distances below the string scale.

This simple r-dependence of the v^4 term is another consequence of supersymmetry. The fact that this term is singular as $r \to 0$ might seem to conflict with the assertion that string theory provides a short-distance cutoff. However, one must look more carefully. To obtain the small-r behavior of the scattering amplitude (13.5.6), take the large-t limit without expanding in v to obtain

$$V(r,v) \approx -2V_{p+1} \int_0^\infty \frac{dt}{(8\pi^2\alpha't)^{(p+1)/2}} \exp\left(-\frac{tr^2}{2\pi\alpha'}\right) \frac{\tanh u \, \sin^4 ut/2}{\sin ut} .$$

$$(13.5.8)$$

Since $t \approx 2\pi\alpha'/r^2$ and $v \approx u$, the arguments of the sines are $ut \approx 2\pi\alpha'v/r^2$. No matter how small v is the v^4 term will cease to dominate at small enough r. The oscillations of the integrand then smooth the small-r behavior on a scale $ut \approx 1$. The effective scale probed by the scattering is

$$r \approx \alpha'^{1/2}v^{1/2} . \qquad (13.5.9)$$

A small-velocity D-brane probe is thus sensitive to distances shorter than the string scale. This is in contrast to the behavior we have seen in string scattering at weak coupling, but fits nicely with the understanding of strongly coupled strings in the next chapter.

Let us expand on this result. A slower D-brane probes shorter distances, but the scattering process takes longer, $\delta t \approx r/v$. Then

$$\delta x\, \delta t \gtrsim \alpha' \ . \tag{13.5.10}$$

This is a suggestion for a new uncertainty relation involving only the coordinates. It is another indication of 'noncommutative geometry,' perhaps connected with the promotion of D-brane collective coordinates to matrices.

For a pointlike D0-brane probe there is a minimum distance that can be measured by scattering. The wavepacket in which it is prepared satisfies

$$\delta x \gtrsim \frac{1}{m\delta v} = \frac{g\alpha'^{1/2}}{\delta v} \ . \tag{13.5.11}$$

The combined uncertainties (13.5.9) and (13.5.11) are minimized by $v \approx g^{2/3}$, for which

$$\delta x \gtrsim g^{1/3} \alpha'^{1/2} \ . \tag{13.5.12}$$

We will see the significance of this scale in the next chapter.

D0-brane quantum mechanics

The nonrelativistic effective Lagrangian for n D0-branes is

$$L = \text{Tr}\left\{ \frac{1}{2g\alpha'^{1/2}} D_0 X^i D_0 X^i + \frac{1}{4g\alpha'^{1/2}(2\pi\alpha')^2} [X^i, X^j]^2 \right.$$
$$\left. - \frac{i}{2}\lambda D_0 \lambda + \frac{1}{4\pi\alpha'} \lambda \Gamma^0 \Gamma^i [X^i, \lambda] \right\} \ . \tag{13.5.13}$$

The first term is the usual nonrelativistic kinetic energy with $m = \tau_0 = 1/g\alpha'^{1/2}$, dropping the constant rest mass $n\tau_0$. The coefficients of the other terms are most easily obtained by T-duality from the ten-dimensional super-Yang–Mills action (B.6.13), with $A^i \to X^i/2\pi\alpha'$. We have taken a basis in which the fermionic field λ is Hermitean, and rescaled λ to obtain a canonical kinetic term. The index i runs over the nine spatial dimensions. The gauge field A^0 has no kinetic term but remains in the covariant derivatives. It couples to the $U(n)$ charges, so its equation of motion amounts to the constraint that only $U(n)$-invariant states are allowed. Only terms with at most two powers of the velocity have been kept, not the full Born–Infeld action.

The Hamiltonian is

$$H = \text{Tr}\left\{ \frac{g\alpha'^{1/2}}{2} p_i p_i - \frac{1}{16\pi^2 g\alpha'^{5/2}} [X^i, X^j]^2 - \frac{1}{4\pi\alpha'} \lambda \Gamma^0 \Gamma^i [X^i, \lambda] \right\} \ . \tag{13.5.14}$$

Note that the potential is positive because $[X^i, X^j]$ is anti-Hermitean. The canonical momentum, like the coordinate, is a matrix,

$$[p_{iab}, X^j_{cd}] = -i\delta^j_i \delta_{ad} \delta_{bc} \ . \tag{13.5.15}$$

Now we define

$$X^i = g^{1/3}\alpha'^{1/2}Y^i , \tag{13.5.16}$$

so that also $p_i = p_{Yi}/g^{1/3}\alpha'^{1/2}$. The Hamiltonian becomes

$$H = \frac{g^{1/3}}{\alpha'^{1/2}}\text{Tr}\left\{\frac{1}{2}p_{Yi}p_{Yi} - \frac{1}{16\pi^2}[Y^i, Y^j]^2 - \frac{1}{4\pi}\lambda\Gamma^0\Gamma^i[Y^i, \lambda]\right\} . \tag{13.5.17}$$

The parameters g and α' now appear only in the overall normalization. It follows that the wavefunctions are independent of the parameters when expressed in terms of the variables Y^i. In terms of the original coordinates X^i their characteristic size scales as $g^{1/3}\alpha'^{1/2}$, the same scale (13.5.12) found above. The energies scale as $g^{1/3}/\alpha'^{1/2}$ from the overall normalization of H, and the characteristic time scale as the inverse of this, so we find again the relation (13.5.10).

Recall from the discussion of D-brane scattering that at distances less than the string scale only the lightest open string states (those which become massless when the D-branes are coincident) contribute. In this regime the cylinder amplitude reduces to a loop amplitude in the low energy field theory (13.5.13).

The #ND = 4 system

Another low energy action with many applications is that for a Dp-brane and Dp'-brane with relative $\#_{\text{ND}} = 4$. There are three kinds of light strings: p-p, p-p', and p'-p', with ends on the respective D-branes. We will consider explicitly the case $p = 5$ and $p' = 9$, where we can take advantage of the $SO(5, 1) \times SO(4)$ spacetime symmetry; all other cases are related to this by T-duality.

The 5-5 and 9-9 strings are the same as those that arise on a single D-brane. The new feature is the 5-9 strings; let us study their massless spectrum. The NS zero-point energy is zero. The moding of the fermions differs from that of the bosons by $\frac{1}{2}$, so there are four periodic world-sheet fermions ψ^m, namely those in the ND directions $m = 6, 7, 8, 9$. The four zero modes then generate $2^{4/2} = 4$ degenerate ground states, which we label by their spins in the (6,7) and (8,9) planes,

$$|s_3, s_4\rangle_{\text{NS}} , \tag{13.5.18}$$

with s_3, s_4 taking values $\pm\frac{1}{2}$. Now we need to impose the GSO projection. This was defined in eq. (10.2.22) in terms of s_a, so that with the extra sign from the ghosts it is

$$-\exp[\pi i(s_3 + s_4)] = +1 \Rightarrow s_3 = s_4 . \tag{13.5.19}$$

In terms of the symmetries, the four states (13.5.18) are invariant under $SO(5,1)$ and form spinors $\mathbf{2} + \mathbf{2}'$ of the 'internal' $SO(4)$, and only the $\mathbf{2}$ survives the GSO projection. In the R sector, of the transverse fermions ψ^i only those with $i = 2, 3, 4, 5$ are periodic, so there are again four ground states

$$|s_1, s_2\rangle_{\mathrm{R}} .$$ (13.5.20)

The GSO projection does not have a extra sign in the R sector so it requires $s_1 = -s_2$. The surviving spinors are invariant under the internal $SO(4)$ and form a $\mathbf{2}'$ of the $SO(4)$ little group of a massless particle.

The system has six-dimensional Lorentz invariance and eight unbroken supersymmetries, so we can classify it by $d = 6$, $N = 1$ supersymmetry (section B.7). The massless content of the 5-9 spectrum amounts to half of a hypermultiplet. The other half comes from strings of opposite orientation, 9-5. The action is fully determined by supersymmetry and the charges; we write the bosonic part:

$$S = -\frac{1}{4g_{\mathrm{D}9}^2} \int d^{10}x \, F_{MN} F^{MN} - \frac{1}{4g_{\mathrm{D}5}^2} \int d^6x \, F'_{MN} F'^{MN}$$

$$- \int d^6x \left[D_\mu \chi^\dagger D^\mu \chi + \frac{g_{\mathrm{D}5}^2}{2} \sum_{A=1}^{3} (\chi_i^\dagger \sigma_{ij}^A \chi_j)^2 \right] .$$ (13.5.21)

The integrals run respectively over the 9-brane and the 5-brane, with $M = 0, \ldots, 9$, $\mu = 0, \ldots, 5$, and $m = 6, \ldots, 9$. The covariant derivative is $D_\mu = \partial_\mu + iA_\mu - iA'_\mu$ with A_μ and A'_μ the 9-brane and 5-brane gauge fields. The field χ_i is a doublet describing the hypermultiplet scalars. The 5-9 strings have one endpoint on each D-brane so χ carries charges $+1$ and -1 under the respective symmetries. The gauge couplings $g_{\mathrm{D}p}$ were given in eq. (13.3.25). We are using a condensed notation,

$$A'_M \rightarrow A'_\mu , \quad X'_m/2\pi\alpha' .$$ (13.5.22)

The massless 5-5 (and also 9-9) strings separate into $d = 6$, $N = 1$ vector and hypermultiplets. The final potential term is the 5-5 D-term required by the supersymmetry. One might have expected a 9-9 D-term as well by T-duality, but this is inversely proportional to the volume of the D9-brane in the (6,7,8,9)-directions, which we have taken to be infinite.

Under T-dualities in any of the ND directions, one obtains $(p, p') = (8, 6)$, $(7, 7)$, $(6, 8)$, or $(5, 9)$, but the intersection of the branes remains $(5 + 1)$-dimensional and the p-p' strings live on the intersection with action (13.5.21). T-dualities in r NN directions give $(p, p') = (9 - r, 5 - r)$. The vector components in the dualized directions become collective coordinates as usual,

$$A_i \rightarrow X_i/2\pi\alpha' , \quad A'_i \rightarrow X'_i/2\pi\alpha' .$$ (13.5.23)

The term $D_i\chi^\dagger D^i\chi$ then becomes

$$\left(\frac{X_i - X_i'}{2\pi\alpha'}\right)^2 \chi^\dagger\chi \,. \tag{13.5.24}$$

This just reflects the fact that when the $(9 - r)$-brane and $(5 - r)$-brane are separated, the strings stretched between them become massive.

The action for several branes of each type is given by the non-Abelian extension.

13.6   D-brane interactions: bound states

Bound states of D-branes with strings and with each other, and supersymmetric bound states in particular, present a number of interesting dynamical problems. Further, these bound states will play an important role in the next chapter in our attempts to deduce the strongly coupled behavior of string theory.

FD bound states

The first case we consider is a state with p F-strings and q D-strings in the IIB theory, all at rest and aligned along the 1-direction. For a state with these charges, the supersymmetry algebra (13.2.9) becomes

$$\frac{1}{2}\left\{\begin{bmatrix} Q_\alpha \\ \tilde{Q}_\alpha \end{bmatrix}, \begin{bmatrix} Q_\beta^\dagger & \tilde{Q}_\beta^\dagger \end{bmatrix}\right\} = M\delta_{\alpha\beta}\begin{bmatrix} 1 & 0 \\ 0 & 1 \end{bmatrix} + \frac{L_1}{2\pi\alpha'}(\Gamma^0\Gamma^1)_{\alpha\beta}\begin{bmatrix} p & q/g \\ q/g & -p \end{bmatrix}, \tag{13.6.1}$$

where L_1 is the length of the system. The eigenvalues of $\Gamma^0\Gamma^1$ are ± 1, so those of the right-hand side are

$$M \pm L_1 \frac{(p^2 + q^2/g^2)^{1/2}}{2\pi\alpha'} \,. \tag{13.6.2}$$

The left-hand side of the algebra is positive — its expectation value in any state is a matrix with positive eigenvalues. This implies a BPS bound on the total energy per unit length,

$$\frac{M}{L_1} \geq \frac{(p^2 + q^2/g^2)^{1/2}}{2\pi\alpha'} \,. \tag{13.6.3}$$

This inequality is saturated by the F-string, which has $(p, q) = (1, 0)$, and by the D-string, with $(p, q) = (0, 1)$.

For one F-string and one D-string, the total energy per unit length is

$$\tau_{(0,1)} + \tau_{(1,0)} = \frac{g^{-1} + 1}{2\pi\alpha'} \,. \tag{13.6.4}$$

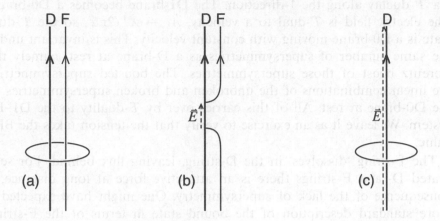

Fig. 13.5. (a) Parallel D-string and F-string. The loop signifies a 7-sphere surrounding the strings. (b) The F-string breaks, its ends attaching to the D-string. (c) Final state: D-string with flux.

This exceeds the BPS bound

$$\tau_{(1,1)} = \frac{(g^{-2} + 1)^{1/2}}{2\pi\alpha'} , \tag{13.6.5}$$

and so this configuration is not supersymmetric. One can also see this directly. The F-string is invariant under supersymmetries satisfying

$$\text{left-moving: } \Gamma^0\Gamma^1 Q = Q , \quad \text{right-moving: } \Gamma^0\Gamma^1 \tilde{Q} = -\tilde{Q} , \tag{13.6.6}$$

and no linear combination of these is of the form $Q_\alpha + (\beta^\perp \tilde{Q})_\alpha$ preserved by the D-string (note that $\beta^\perp \Gamma^0 \Gamma^1 = \Gamma^0 \Gamma^1 \beta^\perp$).

However, the system can lower its energy as shown in figure 13.5. The F-string breaks, with its endpoints attached to the D-string. The endpoints can then move off to infinity, leaving only the D-string behind. This cannot be the whole story because the F-string carries the NS–NS 2-form charge, as measured by the integral of $*H$ over the 7-sphere in the figure: this flux must still be nonzero in the final configuration. This comes about because the F-string endpoints are charged under the D-string gauge field, so an electric flux runs between them. This flux remains in the end. Further, from the D-string action

$$S_1 = -T_1 \int d^2\xi \, e^{-\Phi}[-\det(G_{ab} + B_{ab} + 2\pi\alpha' F_{ab})]^{1/2} , \tag{13.6.7}$$

one sees that $B_{\mu\nu}$ has a source proportional to the invariant electric flux $\mathscr{F}_{ab} = F_{ab} + B_{ab}/2\pi\alpha'$ on the D-string.

The simplest way to see that the resulting state is supersymmetric is via T-duality along the 1-direction. The D1-brane becomes a D0-brane. The electric field is T-dual to a velocity, $\dot{A}_1 \to \dot{X}^1/2\pi\alpha'$, so the T-dual state is a D0-brane moving with constant velocity. This is invariant under the same number of supersymmetries as a D-brane at rest, namely the Lorentz boost of those supersymmetries. The boosted supersymmetries are linear combinations of the unbroken and broken supersymmetries of the D0-brane at rest. All of this carries over by T-duality to the D1–F1 system. We leave it as an exercise to verify that the tension takes the BPS value.

The F-string 'dissolves' in the D-string, leaving flux behind. For separated D- and F-strings there is an attractive force at long distance, a consequence of the lack of supersymmetry. One might have expected a more standard description of the bound state in terms of the F-string moving in this potential well. However, this description breaks down at short distance; happily, the D-brane effective theory gives a simple alternative description. Note that the bound state is quite deep: the binding tension

$$\tau_{(1,0)} + \tau_{(0,1)} - \tau_{(1,1)} = \frac{1 - O(g)}{2\pi\alpha'} \tag{13.6.8}$$

is almost the total tension of the F-string.

String theory with a constant open string field strength has a simple world-sheet description. The variation of the world-sheet action includes a surface term

$$\oint_{\partial M} ds \, \delta X^\mu \left(\frac{1}{2\pi\alpha'} \partial_n X_\mu + iF_{\mu\nu} \partial_t X^\nu \right) , \tag{13.6.9}$$

implying the linear boundary condition

$$\partial_n X_\mu + 2\pi\alpha' iF_{\mu\nu} \partial_t X^\nu = 0 . \tag{13.6.10}$$

This can also be seen from the T-dual relation to the moving D-brane.

All of the above extends immediately to p F-strings and one D-string forming a supersymmetric $(p,1)$ bound state. The general case of p F-strings and q D-strings is more complicated because the gauge dynamics on the D-strings is non-Abelian. A two-dimensional gauge coupling has units of inverse length-squared; we found the precise value $g_{D1}^2 = g/2\pi\alpha'$ in eq. (13.3.25). For dynamics on length scale l the effective dimensionless coupling is $gl^2/2\pi\alpha'$. No matter how weak the underlying string coupling g, the D-string dynamics at long distances is strongly coupled — this is a *relevant* coupling. The theory cannot then be solved directly, but it has been shown by indirect means that there is a bound string saturating the BPS bound for all (p,q) such that p and q are relatively prime. We will sketch the argument and leave the details to the references.

Focus for example on two D-strings and one F-string. There is a state with a separated $(1, 1)$ bound state and $(0, 1)$ D-string. The tension

$$\tau_{(1,1)} + \tau_{(0,1)} = \frac{(g^{-2} + 1)^{1/2} + g^{-1}}{2\pi\alpha'} = \frac{2g^{-1} + g/2 + O(g^3)}{2\pi\alpha'} \quad (13.6.11)$$

exceeds the BPS bound

$$\tau_{(1,2)} = \frac{(4g^{-2} + 1)^{1/2}}{2\pi\alpha'} = \frac{2g^{-1} + g/4 + O(g^3)}{2\pi\alpha'} . \quad (13.6.12)$$

The electric flux is on the first D-brane, so as a $U(2)$ matrix this is proportional to

$$\begin{bmatrix} 1 & 0 \\ 0 & 0 \end{bmatrix} = \frac{1}{2}\begin{bmatrix} 1 & 0 \\ 0 & 1 \end{bmatrix} + \frac{1}{2}\begin{bmatrix} 1 & 0 \\ 0 & -1 \end{bmatrix} . \quad (13.6.13)$$

We have separated this into $U(1)$ and $SU(2)$ pieces. When we bring the two D-strings together, the $SU(2)$ field becomes strongly coupled as we have explained but the $U(1)$ part remains free. The $U(1)$ flux is then unaffected by the dynamics, and in particular there are no charged fields that might screen it. However, if the $SU(2)$ part is screened by the massless fields on the D-strings, then the total energy in the flux (which is proportional to the trace of the square of the matrix) is reduced by a factor of 2, from (13.6.11) to the BPS value (13.6.12).

That this does happen has been shown as follows. Focus on four of the 16 supersymmetries, forming the equivalent of $d = 4$, $N = 1$ supersymmetry. The six scalars $X^{4,\ldots,9}$ can be written as three chiral superfields Φ_i, with the potential coming from a superpotential $\text{Tr}(\Phi_1[\Phi_2, \Phi_3])$. Now change the problem, adding to the superpotential a mass term,

$$W(\Phi) = \text{Tr}(\Phi_1[\Phi_2, \Phi_3]) + m\,\text{Tr}(\Phi_i\Phi_i) . \quad (13.6.14)$$

This is an example of a general strategy for finding supersymmetric bound states: the D-string is a BPS state even under the reduced supersymmetry algebra. Its mass is then determined by the algebra and cannot depend on the parameter m. By now increasing m we can reduce the effective dimensionless coupling $g/2\pi\alpha'm^2$ to a value where the system becomes weakly coupled. It can then be shown that the $SU(2)$ system has a supersymmetric ground state.

The same argument goes through for all relatively prime p and q. When these are not relatively prime, $(p, q) = (kp, kq)$ and the system is only marginally unstable against falling apart into k subsystems. The dynamics is then quite different, and there is believed to be no bound string in this case. The bound string formed from p F-strings and q D-strings is called a (p, q)-string (as opposed to a p-p' string, which is an open string whose endpoints move on Dp- and Dp'-branes).

D0–Dp BPS bound

For a system with the charges of a D0-brane and a Dp-brane extended in the $(1, \ldots, p)$-directions, the supersymmetry algebra becomes

$$\frac{1}{2}\left\{\left[\begin{array}{c} Q_\alpha \\ \tilde{Q}_\alpha \end{array}\right], \left[Q_\beta^\dagger \ \tilde{Q}_\beta^\dagger\right]\right\} = M\left[\begin{array}{cc} 1 & 0 \\ 0 & 1 \end{array}\right]\delta_{\alpha\beta} + \left[\begin{array}{cc} 0 & Z \\ -Z^\dagger & 0 \end{array}\right]\Gamma_{\alpha\beta}^0 , \quad (13.6.15)$$

where

$$Z = \tau_0 + \tau_p V_p \beta , \quad \beta = \beta^1 \cdots \beta^p . \quad (13.6.16)$$

We have wrapped the Dp-brane on a torus of volume V_p so that its mass will be finite. The positivity of the left-hand side implies that

$$M^2\left[\begin{array}{cc} 1 & 0 \\ 0 & 1 \end{array}\right] \geq \left[\begin{array}{cc} 0 & Z \\ -Z^\dagger & 0 \end{array}\right]\Gamma^0\left[\begin{array}{cc} 0 & Z \\ -Z^\dagger & 0 \end{array}\right]\Gamma^0 = \left[\begin{array}{cc} ZZ^\dagger & 0 \\ 0 & Z^\dagger Z \end{array}\right] ,$$

$$\hspace{11cm} (13.6.17a)$$

$$ZZ^\dagger = \tau_0^2 + \tau_p V_p(\beta + \beta^\dagger) + \tau_p^2 V_p^2 \beta\beta^\dagger . \quad (13.6.17b)$$

For p a multiple of 4, β is Hermitean and $\beta^2 = 1$ by the same argument as in eq. (13.4.3). The BPS bound is then

$$M \geq \tau_0 + \tau_p V_p . \quad (13.6.18)$$

For $p = 4k + 2$, β is anti-Hermitean, $\beta^2 = -1$, and the BPS bound is

$$M \geq (\tau_0^2 + \tau_p^2 V_p^2)^{1/2} . \quad (13.6.19)$$

These bounds are consistent with our earlier results on supersymmetry breaking, noting that $\#_{\text{ND}} = p$. For $p = 4k$, a separated 0-brane and p-brane saturate the BPS bound (13.6.18), agreeing with the earlier conclusion that they leave some supersymmetry unbroken. For $p = 4k + 2$ they do not saturate the bound and so cannot be in a BPS state, as found before. The reader can extend the analysis of the BPS bound to general values of p and p'.

D0–D0 bound states

The BPS bound for the quantum numbers of two 0-branes is $2\tau_0$, so any bound state will be at the lower edge of the continuous spectrum of two-body states. Nevertheless there is a well-defined, and as it turns out very important, question as to whether a *normalizable* state of energy $2\tau_0$ exists.

Let us first look at an easier problem. Compactify the 9-direction and add one unit of compact momentum, $p_9 = 1/R$. In a two-body state this momentum must be carried by one 0-brane or the other for minimum total energy

$$\tau_0 + \left(\tau_0 + \frac{p_9^2}{2\tau_0} \right) . \tag{13.6.20}$$

For a bound state of mass $2\tau_0$, on the other hand, the minimum energy is

$$2\tau_0 + \frac{p_9^2}{4\tau_0} , \tag{13.6.21}$$

a finite distance below the continuum states. The reader may note some resemblance between these energies and the earlier (13.6.11) and (13.6.12). In fact the two systems are T-dual to one another. Taking the T-dual along the 9-direction, the D0-branes become D1-branes and the unit of momentum becomes a unit of fundamental string winding to give the $(1, 2)$ system, now at finite radius $R' = \alpha'/R$. Quantizing the $(1, 2)$ string wrapped on a circle gives the 2^8 states of an ultrashort BPS multiplet. In terms of the previous analysis, the $SU(2)$ part has a unique ground state in finite volume while the zero modes of the 16 components of the $U(1)$ gaugino generate 2^8 states. The earlier analysis is valid for the T-dual radius R' large, but having found an ultrashort multiplet we know that it must saturate the BPS bound exactly — its mass is determined by its charges and cannot depend on R. Similarly for n D-branes with m units of compact momentum, when m and n are relatively prime there is an ultrashort multiplet of bound states.

Now let us try to take $R \to \infty$ in order to return to the earlier problem. Having found that a bound state exists at any finite radius, it is natural to suppose that it persists in the limit. Since for any n we can choose a relatively prime m, it appears that there is one ultrashort bound state multiplet for any number of D0-branes. However, it is a logical possibility that the size of these states grows with R such that the states becomes nonnormalizable in the limit. To show that the bound states actually exist requires a difficult analysis, which has been carried out fully only for $n = 2$.

D0–D2 bound states

Here the BPS bound (13.6.19) puts any bound state discretely below the continuum. One can see hints of a bound state: the long-distance force is attractive, and for a coincident 0-brane and 2-brane the NS 0-2 string has a negative zero-point energy (13.4.8) and so a tachyon (which survives the GSO projection), indicating instability towards something. We cannot follow the tachyonic instability directly, but there is a simple alternative description of where it must end up.

Let us compactify the 1- and 2-directions and take the T-dual only in the first, so that the 0-brane becomes a D-string wrapped in the 1-direction and the 2-brane becomes a D-string wrapped in the 2-direction. Now there is an obvious state with the same charges and lower energy, a single D-string running at an angle to wrap once in each direction. A single wrapped D-string is a BPS state (an ultrashort multiplet to be precise). Now use T-duality to return to the original description. As in figure 13.2, this will be a D2-brane with a nonzero magnetic field, such that

$$\int_{D2} F_2 = 2\pi \; . \tag{13.6.22}$$

We can also check that this state has the correct R–R charges. Expanding out the Chern–Simons action (13.3.18) gives

$$i\mu_2 \int (C_3 + 2\pi\alpha' F_2 \wedge C_1) \; . \tag{13.6.23}$$

Thus the magnetic field induces a D0-brane charge on the D2-branes, and the normalizations are consistent with $\mu_0 = 4\pi^2\alpha'\mu_2$.

The D0-brane dissolves in the D2-brane, turning into flux. The reader may note several parallels with the discussion of a D-string and an F-string, and wonder whether the systems are equivalent. In fact, they are not related to one other by T-duality or any other symmetry visible in string perturbation theory, but we will see in the next chapter that they are related by nonperturbative dualities.

The analysis extends directly to n D2-branes and m D0-branes: there is a single ultrashort multiplet of bound states.

D0–D4 bound states

As with the D0–D0 case, the BPS bound (13.6.18) implies that any bound state is marginally stable. We can proceed as before, first compactifying another dimension and adding a unit of momentum so that the bound state lies below the continuum. The low energy D0–D4 action is as discussed at the end of the previous section. Again it is an interacting theory, with a coupling that becomes large at low energy, but again the existence of supersymmetric bound states can be established by deforming the Hamiltonian; the details are left to the references. A difference from the D0–D0 case is that these bound states are invariant only under one-quarter of the original supersymmetries, the intersection of the supersymmetries of the 0-brane and of the 4-brane. The bound states then lie in a short (but not ultrashort) multiplet of 2^{12} states. It is useful to imagine that the D4-brane is wound on a finite but large torus. In this limit the massless 4-4 strings are essentially decoupled from the 0-4 and 0-0 strings. The 16 zero modes of the massless 4-4 fermion then generate 2^8 ground states delocalized on the D4-brane. The fermion in the 0-4 hypermultiplet has eight real components (the smallest spinor in six dimensions) and their zero modes generate 2^4 ground states localized on the D0-brane. The tensor product gives the 2^{12} states.

For two D0-branes and one D4-brane, one gets the correct count as follows. We can have the two D0-branes bound to the D4-brane independently of one another; for a large D4-brane their interactions can be neglected. Each D0-brane has 2^4 states as noted above, eight bosonic and eight fermionic. Now count the number of ways two D0-branes can be put into these states: there are eight states with both D0-branes in the same (bosonic) state and $\frac{1}{2} \times 8 \times 7$ states with the D-branes in different bosonic states, for a total of $\frac{1}{2} \times 8 \times 9$ states. There are also $\frac{1}{2} \times 8 \times 7$ states with the D0-branes in different fermionic states and 8×8 with one in a bosonic state and one a fermionic state. Summing and tensoring with the 2^8 D4-brane states gives 2^{15} states. However, we could also imagine the two D0-branes first forming a D0–D0 bound state. The $SU(2)$ dynamics decouples and the resulting $U(1)$ dynamics is essentially the same as that of a single D0-brane. This bound state can then bind to the D4-brane, giving 2^{4+8} states as for a single D0-brane. The total number is 9×2^{12}.

This counting extends to n D0-branes and one D4-brane. The degeneracy D_n is given by the generating function

$$\sum_{n=0}^{\infty} q^n D_n = 2^8 \prod_{k=1}^{\infty} \left(\frac{1+q^k}{1-q^k} \right)^8 . \tag{13.6.24}$$

The term k in the product comes from bound states of k D0-branes which are then bound to the D4-brane. For each k there are eight bosonic states and eight fermionic states, and the expression (13.6.24) is then the product of the partition functions for all species. The coefficient of q^2 in its expansion is indeed 9×2^{12}. This proliferation of bound states is in contrast to the single ultrashort multiplet for n D0-branes and one D2-brane. The difference is that all the latter states are spread over the D2-brane, whereas the D0–D4 bound states are localized.

By T-duality the above system is converted into one D0-brane and n D4-branes, so the number of bound states of the latter is the same D_n. For m D0-branes and n D4-branes one gets the correct answer by the following argument. The equality of the degeneracy for one D0-brane and n D4-branes with that for n D0-branes and one D4-brane suggests that the systems are really the same — that in the former case we can somehow picture the D0-brane bound to n D4-branes as separating into n 'fractional branes,' each of which can then bind to each other in all combinations as in the earlier case. Then m D0-branes separate into mn fractional branes.

The degeneracy would then be D_{mn}, defined as in eq. (13.6.24). This is apparently correct, but the justification is not simple.

D-branes as instantons

The D0–D4 system is interesting in other ways. Consider its scalar potential

$$\frac{g_{D0}^2}{4} \sum_{A=1}^{3} (\chi_i^\dagger \sigma_{ij}^A \chi_j)^2 + \sum_{i=1}^{5} \frac{(X_i - X_i')^2}{2\pi\alpha'} \chi^\dagger \chi , \qquad (13.6.25)$$

as at the end of the previous section. The second term by itself has two branches of zeros,

$$X_i - X_i' = 0 , \quad \chi \neq 0 \qquad (13.6.26)$$

and

$$X_i - X_i' \neq 0 , \quad \chi = 0 . \qquad (13.6.27)$$

The first of these, where the hypermultiplet scalars are nonzero, is known as a Higgs branch. The second, where the vector multiplet scalars are nonzero, is known as a Coulomb branch. In the present case the first term in the potential, the D-term, eliminates the Higgs branch. The condition

$$D^A \equiv \chi_i^\dagger \sigma_{ij}^A \chi_j = 0 \qquad (13.6.28)$$

implies that $\chi = 0$. For example, if there were a nonzero solution we could by an $SU(2)$ rotation make only the upper component nonzero, and then D^3 is nonzero. However, for *two* D4-branes χ acquires a D4-brane index $a = 1, 2$ and the D-term condition is

$$\chi_{ia}^\dagger \sigma_{ij}^A \chi_{ja} = 0 . \qquad (13.6.29)$$

This now is solved by

$$\chi_{ia} = v\delta_{ia} \qquad (13.6.30)$$

for any v, or more generally

$$\chi_{ia} = vU_{ia} \qquad (13.6.31)$$

for any constant v and unitary U. Further, U can be taken to lie in $SU(2)$ by absorbing its phase into v, and the latter can then be made real by a 4–4 $U(1)$ gauge rotation.

The Coulomb branch has an obvious physical interpretation, corresponding to the separation of the D0- and D4-brane in the directions transverse to the latter. But what of the Higgs branch?

Recall that non-Abelian gauge theories in four Euclidean dimensions have classical solutions, instantons, that are localized in all four dimensions. Their distinguishing property is that the field strength is self-dual or anti-self-dual,

$$*F_2 = \pm F_2 \, , \tag{13.6.32}$$

so that the Bianchi identity implies the field equations. Because the classical theory is scale-invariant, the characteristic size of the configuration is undetermined — there is a family of solutions parameterized by scale size ρ. The $U(n)$ gauge theory on coincident D4-branes is five-dimensional, so a configuration that looks like an instanton in the four spatial dimensions and is independent of time is a static classical solution, a soliton.

This soliton has many properties in common with the D0-brane bound to the D4-branes. First, it is a BPS state, breaking half of the supersymmetries of the D4-branes. The supersymmetry variation of the gaugino is

$$\delta\lambda \propto F_{MN}\Gamma^{MN}\zeta \, . \tag{13.6.33}$$

Here the nonzero terms involve the components of Γ^{MN} in the spatial directions of the D4-brane. These are then generators of the $SO(4) = SU(2) \times SU(2)$ rotation group. The self-duality relation (13.6.32) amounts to the statement that only the generators of the first or second $SU(2)$ appear in the variation. The ten-dimensional spinor ζ decomposes into

$$(\mathbf{4,2,1}) + (\mathbf{\bar{4},1,2}) \tag{13.6.34}$$

under $SO(5,1) \times SU(2) \times SU(2)$, so half the components are invariant under each $SU(2)$ and half the supersymmetry variations (13.6.33) are zero. Second, it carries the same R–R charge as the D0-brane. Expanding the Chern–Simons action (13.3.18) gives the term

$$\frac{1}{2}(2\pi\alpha')^2\mu_4 \int C_1 \wedge \mathrm{Tr}(F_2 \wedge F_2) \, . \tag{13.6.35}$$

The topological charge of the instanton is

$$\int_{D4} \mathrm{Tr}(F_2 \wedge F_2) = 8\pi^2 \, , \tag{13.6.36}$$

so the total coupling to a constant C_1 is $(4\pi^2\alpha')^2\mu_4 = \mu_0$, exactly the charge of the D0-brane. Finally, the moduli (13.6.31) for the $SU(2)$ Higgs branch just match those of the $SU(2)$ instanton, v to the scale size ρ and U to the orientation of the instanton in the gauge group.[6] Let us check the counting of the moduli, as follows. There are eight real hypermultiplet scalars in χ. The three D-term conditions and the gauge rotation each remove one

---

[6] For a single instanton the latter are not regarded as moduli because they can be changed by a global gauge transformation, but with more than one instanton there are moduli for the relative orientation. The same is true of the D0-branes.

to leave four moduli. There are also four additional 0-0 moduli for the position of the particle within the D4-branes.

The precise connection between the D0-brane and the instanton is this. When the scale size ρ is large compared to the string scale, the low energy effective field theory on the D4-branes should give a good description of the instanton. However, as ρ is reduced below the string length, this description is no longer accurate. Happily, the D0-brane picture provides a description that is accurate in the opposite limit: the point $v = 0$ where the Higgs and Coulomb branches meet is the zero-size instanton, and turning on the Higgs moduli expands the instanton: as in the D0–D2 case, the D0-brane is dissolving into flux. This picture also accounts for the absence of a Higgs branch for a single D4-brane because there are no instantons for $U(1)$.

The gauge field of the small instanton can be measured directly. Recall that a slow D0-brane probe is sensitive to distances below the string scale. One can consider the D0–D4 bound system with an additional probe D0-brane. This has been studied in a slightly different form, taking first the T-dual to the D5–D9 system and using a D1-brane probe. As discussed earlier, only the effective field theory of the light open string states enters, though this is still rather involved because each open string endpoint can lie on a D1-, D5-, or D9-brane. However, after integrating out the massive fields (which get mass because they stretch from the probe to the other D-branes), the effective theory on moduli space displays the instanton gauge field. This provides a physical realization of the so-called Atiyah–Drinfeld–Hitchin–Manin (ADHM) construction of the general instanton solution.

Note the following curious phenomenon. Start with a large instanton, an object made out of the gauge fields that live on the D4-branes. Contract it to zero size, where the branches meet, and now pull it off the D4-branes along the Coulomb branch. The 'instanton' can no longer be interpreted as being made of the gauge fields, because these exist only on the D4-branes.

It should be noted that because the Higgs moduli are 0-4 fields their vertex operators are rather complicated: the different boundary conditions on the two endpoints mean that the world-sheet boundary conditions on the two sides of the vertex operator are different. They are similar to orbifold twisted state vertex operators — in fact, using the doubling trick, they are essentially half of the latter. It is therefore difficult to discuss in string theory a background with nonzero values for these fields, so the D0-brane picture is really an expansion in ρ, whereas the low energy field theory is an expansion in $1/\rho$.

Returning to the bound state problem, the system with m D0-branes bound to n D4-branes is equivalent to quantum mechanics on the moduli space of m $SU(n)$ instantons. The number of supersymmetric states is related to the topology of this space, and the answer has been argued to be D_{mn} as asserted before. (The connection with the fractional-brane picture is complicated and the latter is perhaps unphysical.)

D0–D6 bound states

The relevant bound is (13.6.19) and again any bound state would be below the continuum. This is as in the D0–D2 case, but the situation is different. The long-distance force is repulsive and the zero-point energy of coincident 0-6 NS strings is positive, so there is no sign of instability toward a supersymmetric state. One can give 0-brane charge to the 6-brane by turning on flux, but there is no configuration that has only these two charges and saturates the BPS bound. So it appears that there are *no* supersymmetric bound states.

D0–D8 bound states

This system is complicated in a number of ways and we will not pursue it. As one example of the complication, the R–R fields of the D8-brane do not fall off with distance (it has codimension 1, like a planar source in $3+1$ dimensions). The total energy is then infinite, and when the couplings to the dilaton and metric are taken into account the dilaton diverges a finite distance from the D8-brane. Thus the D8-brane cannot exist as an independent object, but only in connection with orientifold planes such as arise in the T-dual of the type I theory.

Exercises

13.1 (a) For the various massless fields of each of the type II string theories, write out the relation between the field at (x^μ, x^m) and at the orientifold image point $(x^\mu, -x^m)$. The analogous relation for the bosonic string was given as eq. (8.8.3).
(b) At the eight-dimensional orientifold plane (obtained from type I by T-duality on a single axis), which massless type IIA fields satisfy Dirichlet boundary conditions and which Neumann ones?

13.2 Find the scattering amplitude involving four bosonic open string states attached to a Dp-brane. [Hint: this should be very little work.]

13.3 (a) Consider three D4-branes that are extended along the (6,7,8,9)-, (4,5,8,9)-, and (4,5,6,7)-directions respectively. What are the unbroken supersymmetries?
(b) Add a D0-brane to the previous configuration. Now what are the unbroken supersymmetries?
(c) Call this configuration $(p_1, p_2, p_3, p_4) = (4, 4, 4, 0)$. By T-dualities, what other configurations of D-branes can be reached?

13.4 (a) Calculate the static potential between a D2-brane and a D0-brane from the cylinder amplitude by applying T-duality to the result (13.4.23).
(b) Do the same calculation in low energy field theory and compare the result with part (a) at distances long compared to the string scale.
(c) Extend parts (a) and (b) to a Dp-brane and D($p + 2$)-brane oriented such that #ND = 2.

13.5 Repeat parts (a) and (b) of the previous exercise for a D0-brane and D6-brane.

13.6 (a) Find the velocity-dependent interaction between a D4-brane and D0-brane due to the cylinder. You can do this by analytic continuation of the potential (13.4.23), with appropriate choice of angles.
(b) Expand the interaction in powers of v and find the explicit r-dependence at $O(v^2)$.
(c) Compare the interaction at distances long compared to the string scale with that obtained from the low energy field theory. One way to do this is to determine the long-range fields of the D4-brane by solving the linearized field equations with a D4-brane source, insert these into the D0-brane action, and expand in the velocity.

13.7 For the D4-brane and D0-brane, determine the interaction at distances short compared to the string scale as follows. Truncate the low energy action given at the end of section 13.5 to the massless 0-0 strings and the lightest 0-4 strings. The D0–D4 interaction arises as a loop correction to the effective action of the 0-0 collective coordinate, essentially a propagator correction for the field we called X_i'. Calculate this Feynman graph and compare with part (b) of the previous exercise at short distance. This is a bit easier than the corresponding D0–D0 calculation because the 0-4 strings do not include gauge fields. You need the Lagrangian for the 0-4 fermions; this is the dimensional reduction of the $(5 + 1)$-dimensional fermionic Lagrangian density $-i\overline{\psi}\Gamma^\mu D_\mu\psi$.

13.8 (a) Continuing the previous two exercises, obtain the full v-dependence at large r from the cylinder amplitude. Compare the result with the low energy supergravity (graviton–dilaton–R–R) exchange.
(b) Obtain the full v-dependence at small r and compare with the same from the open string loop.

13.9 Find a configuration of an infinite F-string and infinite D3-brane that leaves some supersymmetry unbroken.

13.10 From the D-string action, calculate the tension with q units of electric flux and compare with the BPS bound (13.6.3) for a $(q, 1)$ string.

13.11 Carry out in detail the counting that leads to the bound state degeneracy (13.6.24).

13.12 Consider one of the points in figure 13.5(b) at which the F-string attaches to the D-string. At this point a $(1,0)$ and a $(0,1)$ string join to form a $(1,1)$ string; alternatively, if we count positive orientation as being inward, it is a junction of $(1,0)$, $(0,1)$, and $(-1,-1)$ strings. Consider the junction of three semi-infinite straight strings of general (p_i, q_i), with vanishing total p and q. Find the conditions on the angles such that the system is mechanically stable. Show that, with these angles, one-quarter of the original supersymmetries leave all three strings invariant.

14
Strings at strong coupling

Thus far we have understood string interactions only in terms of perturbation theory — small numbers of strings interacting weakly. We know from quantum field theory that there are many important phenomena, such as quark confinement, the Higgs mechanism, and dynamical symmetry breaking, that arise from having many degrees of freedom and/or strong interactions. These phenomena play an essential role in the physics of the Standard Model. If one did not understand them, one would conclude that the Standard Model incorrectly predicts that the weak and strong interactions are both long-ranged like electromagnetism; this is the famous criticism of Yang–Mills theory by Wolfgang Pauli.

Of course string theory contains quantum field theory, so all of these phenomena occur in string theory as well. In addition, it likely has new nonperturbative phenomena of its own, which must be understood before we can connect it with nature. Perhaps even more seriously, the perturbation series does not even define the theory. It is at best asymptotic, not convergent, and so gives the correct qualitative and quantitative behavior at sufficiently small coupling but becomes useless as the coupling grows.

In quantum field theory we have other tools. One can define the theory (at least in the absence of gravity) by means of a nonperturbative lattice cutoff on the path integral. There are a variety of numerical methods and analytic approximations available, as well as exactly solvable models in low dimensions. The situation in string theory was, until recently, much more limited.

In the past few years, new methods based on supersymmetry have revolutionized the understanding both of quantum field theory and of string theory. In the preceding chapters we have assembled the tools needed to study this. We now consider each of the five string theories and deduce the physics of its strongly coupled limit. We will see that all are limits of a single theory, which most surprisingly has a limit in which spacetime becomes *eleven-dimensional*. We examine one proposal, matrix theory, for a formulation of this unified theory. We conclude with a discussion of related progress on one of the central problems of quantum gravity, the quantum mechanics of black holes.

We will use extensively the properties of D-brane states in mapping out the physics of strongly coupled strings. This allows a natural connection with our previous perturbative discussion. We should note, however, that most of these results were deduced by other methods before the role of D-branes was understood. Many properties of the R–R states were guessed (subject to many consistency checks) before the explicit D-brane picture was known.

14.1 Type IIB string and $SL(2,Z)$ duality

In the IIB theory, consider an infinite D-string stretched in the 1-direction. Let us determine its massless excitations, which come from the attached strings. The gauge field has no dynamics in two dimensions, so the only bosonic excitations are the transverse fluctuations. The Dirac equation for the massless R sector states

$$(\Gamma^0 \partial_0 + \Gamma^1 \partial_1)u = 0 \qquad (14.1.1)$$

implies that $\Gamma^0\Gamma^1 u = \pm u$ for the left- and right-movers respectively, or that the boost eigenvalue $s_0 = \pm\frac{1}{2}$. The open string R sector ground state decomposes as

$$\mathbf{16} \to (\tfrac{1}{2}, \mathbf{8}) + (-\tfrac{1}{2}, \mathbf{8}') \qquad (14.1.2)$$

under $SO(9,1) \to SO(1,1) \times SO(8)$, so the left-moving fermionic open strings on the D-string are in an $\mathbf{8}$ of $SO(8)$ and the right-movers are in an $\mathbf{8}'$.

Now consider an infinite fundamental string in the same theory. The massless bosonic fluctuations are again the transverse fluctuations. The massless fermionic fluctuations are superficially different, being the spacetime vectors ψ^μ and $\tilde{\psi}^\mu$. However, these are not entirely physical — the GSO projection forbids a single excitation of these fields. To identify the physical fermionic fluctuations, recall from the discussion in section 13.2 that these can be thought of as the Goldstone fermions of the supersymmetries broken by the string. The supersymmetry algebra for a state containing a long string was given in eq. (13.6.1), where $(p,q) = (1,0)$ for the fundamental string. The broken supersymmetries are those whose anticommutators do not vanish when acting on the BPS state; for the IIB F-string these are the Q_α with $\Gamma^0\Gamma^1 = +1$ and the $\tilde{Q}_\alpha$ with $\Gamma^0\Gamma^1 = -1$. The decomposition (14.1.2) then shows that the Goldstone fermions on the IIB F-string have the same quantum numbers as on the IIB D-string; for the IIA F-string, on the other hand, the Goldstone fermions moving in both directions are **8**s. The relation between these excitations and the ψ^μ is just the refermionization used in section 12.6.

The D-string and F-string have the same massless excitations but they are not the same object. Their tensions are different,

$$\frac{\tau_{F1}}{\tau_{D1}} = g = e^\Phi \, . \tag{14.1.3}$$

This relation is a consequence of supersymmetry and so is exact. The field dependence of the central charge is connected by supersymmetry to the field dependence of the moduli space metric, and this receives no corrections for 16 or more supersymmetries. At weak coupling the F-string is much lighter than the D-string, but consider what happens as the coupling is adiabatically increased. Quantum mechanics does not allow the D-string states to simply disappear from the spectrum, and they must continue to saturate the BPS bound because their multiplet is smaller than the non-BPS multiplet. Thus at very strong coupling the D-string is still in the spectrum but it is much *lighter* than the F-string. It is tempting to conclude that the theory with coupling $1/g$ is the same as the theory with coupling g, but with the two strings reversing roles.

Let us amplify this as follows. Consider also a third scale, the gravitational length

$$l_0 = (4\pi^3)^{-1/8}\kappa^{1/4} \, , \tag{14.1.4}$$

where the important feature is the dependence on κ; the numerical constants are just included to simplify later equations. The relevant length scales are in the ratios

$$\tau_{F1}^{-1/2} : l_0 : \tau_{D1}^{-1/2} = g^{-1/4} : 1 : g^{1/4} \, . \tag{14.1.5}$$

At $g \ll 1$, if we start at long distance and consider the physics at progressively shorter scales, before reaching the scale where gravity would become strong we encounter the fundamental string scale and all the excited states of the fundamental string. At $g \gg 1$, we again encounter another scale before reaching the scale where gravity is strong, namely the D-string scale. We cannot be certain that the physics is the same as at weak coupling, but we do know that gravity is weak at this scale, and we can reproduce much of the same spectrum — the long straight string is a BPS state, as are states with arbitrary left- *or* right-moving excitations, so we can identify these. States with both left- and right-moving excitations are not BPS states, but at low energy the interactions are weak and we can identify them approximately.

Of course we have no nonperturbative definition of string theory and anything can happen. For example there could be very light non-BPS states below the D-string scale in the strongly coupled string theory, with no analogs in the weakly coupled theory. However, given that we can identify many similarities in the g and $1/g$ theories with F- and D-strings reversed, the simplest explanation is that there is a symmetry that relates them. Furthermore we will see that in every string theory there is a unique natural candidate for its strongly coupled dual, that the various tests we can make on the basis of BPS states work, and that this conjecture fits well with observations about the symmetries of low energy supergravity and in some cases with detailed calculations of low energy amplitudes.

One might have imagined that at strong string coupling one would encounter a phase with strongly coupled gravity and so with exotic space-time physics, but what happens instead seems to be the same physics as at weak coupling. Of course for $g \approx 1$, neither theory is weakly coupled and there is no quantitative understanding of the theory, but the fact that we have the $g \approx 1$ theory 'surrounded' surely limits how exotic it can be. Such weak–strong dualities have been known in low-dimensional quantum field theories for some time. They were conjectured to occur in some four-dimensional theories, notably $N = 4$ non-Abelian gauge theory. There is now very strong evidence that this is true. It should be noted though that even in field theory, where we have a nonperturbative definition of the theory, weak–strong duality has not been shown directly. This seems to require new ideas, which are likely to come from string theory.

The D-string has many massive string excitations as well. These have no analog in the heterotic string, but this is not relevant. They are not supersymmetric and decay to massless excitations at a rate of order g^2. As g becomes strong they become broader and broader 'resonances' and disappear into multi-particle states of the massless spectrum.

As a further test, the effective low energy IIB action (12.1.26), known exactly from supersymmetry, must be invariant. Since the coupling is determined by the value of the dilaton, this must take $\Phi \to -\Phi$. Setting the R–R scalar C_0 to zero for simplicity, the reader can check that the action is invariant under

$$\Phi' = -\Phi \,, \quad G'_{\mu\nu} = e^{-\Phi} G_{\mu\nu} \,, \tag{14.1.6a}$$

$$B'_2 = C_2 \,, \quad C'_2 = -B_2 \,, \tag{14.1.6b}$$

$$C'_4 = C_4 \,. \tag{14.1.6c}$$

The Einstein metric, defined to have a dilaton-independent action, is

$$G_{\mathrm{E}\mu\nu} = e^{-\Phi/2} G_{\mu\nu} = e^{-\Phi'/2} G'_{\mu\nu} \tag{14.1.7}$$

and so is invariant.

SL(2,Z) duality

The transformation (14.1.6) is one of the $SL(2, \mathbf{R})$ symmetries (12.1.32) of the low energy theory, with $\tau' = -1/\tau$. Consider the action of a general element on the 2-form coupling of the fundamental string,

$$\int_M B_2' = \int_M (B_2 d + C_2 c) . \tag{14.1.8}$$

For general real c and d there is no state with this coupling, but for the integer subgroup $SL(2, \mathbf{Z})$, the condition $ad - bc = 1$ implies that d and c are relatively prime. In this case we know from the previous chapter that there is a supersymmetric (d, c)-string with these quantum numbers. It is described at weak coupling as a bound state of c D-strings and d F-strings, and its existence at strong coupling follows from the continuation argument used above. This is a strong indication that this integer subgroup is an exact symmetry of the theory, with the weak–strong duality as one consequence. The BPS bound can be written in $SL(2, \mathbf{Z})$-invariant form as

$$\tau_{(p,q)}^2 = l_0^{-4} (\mathcal{M}^{-1})^{ij} q_i q_j = l_0^{-4} \left[e^{\Phi} (p + C_0 q)^2 + e^{-\Phi} q^2 \right] . \tag{14.1.9}$$

Note that a subgroup of $SL(2, \mathbf{R})$, with $a = d = 1$ and $c = 0$, is visible in perturbation theory. This leaves the dilaton invariant and shifts

$$C_0 \to C_0 + b . \tag{14.1.10}$$

This shift is a symmetry of perturbation theory because the R–R scalar C_0 appears only through its field strength (gradient). The coupling to D-strings then breaks this down to integer shifts. This is evident from the bound (14.1.9), which is invariant under $C_0 \to C_0 + 1$ with $(p, q) \to (p - q, q)$. The integer shift takes τ to $\tau + 1$, and the full $SL(2, \mathbf{Z})$ is generated by this symmetry plus the weak–strong duality.

The IIB NS5-brane

Let us consider how the weak–strong duality acts on the various extended objects in the theory. We know that it takes the F- and D-strings into one another. It leaves the potential C_4 invariant and so should take the D3-brane into itself. The D5-brane is a magnetic source for the R–R 2-form charge: the integral of F_3 over a 3-sphere surrounding it is nonzero. This must be transformed into a magnetic source for the NS–NS 2-form charge. We have not encountered such an object before — it is neither a string nor a D-brane. Rather, it is a soliton, a localized classical solution to the field equations.

Consider the action for a graviton, dilaton, and q-form field strength in d dimensions, of the general form

$$\int d^{10}x \,(-G)^{1/2} e^{-2\Phi} (R + 4\partial_\mu \Phi \partial^\mu \Phi) - \frac{1}{2} \int e^{2\alpha\Phi} |F_q|^2 \,, \qquad (14.1.11)$$

where α is -1 for an NS–NS field and 0 for an R–R field. We can look for a solution which is spherically symmetric in $q + 1$ directions and independent of the other $8 - q$ spatial dimensions and of time, and which has a fixed 'magnetic' charge

$$\int_{S_q} F_q = Q \,. \qquad (14.1.12)$$

Here the q-sphere is centered on the origin in the $q + 1$ spherically symmetric dimensions. This would be an $(8 - q)$-brane. The field equation

$$d * (e^{2\alpha\Phi} F_q) = 0 \qquad (14.1.13)$$

is automatic as a consequence of the spherical symmetry. The dual field strength is $F_{10-q} = *e^{2\alpha\Phi} F_q$, for which eq. (14.1.13) becomes the Bianchi identity. An 'electric' solution with

$$\int_{S_{10-q}} *e^{2\alpha\Phi} F_q = Q' \qquad (14.1.14)$$

would be a $(q - 2)$-brane.

A generalization of Birkhoff's theorem from general relativity guarantees a unique solution for given mass M and charge Q. For M/Q greater than a critical value $(M/Q)_c$ the solution is a black hole, with a singularity behind a horizon. More precisely, the solution is a *black p-brane*, meaning that it is extended in p spatial dimensions and has a black hole geometry in the other $9 - p$. Essentially the source for the field strength is hidden in the singularity. For $M/Q < (M/Q)_c$, there is a naked singularity. The solution with $M/Q = (M/Q)_c$ is called extremal, and in most cases it is a supersymmetric solution, saturating the BPS bound. The naked singularities would then be excluded by the bound.

For the NS5-brane, the extremal solution is supersymmetric and takes the form

$$G_{mn} = e^{2\Phi} \delta_{mn} \,, \qquad G_{\mu\nu} = \eta_{\mu\nu} \,, \qquad (14.1.15a)$$
$$H_{mnp} = -\epsilon_{mnp}{}^q \partial_q \Phi \,, \qquad (14.1.15b)$$
$$e^{2\Phi} = e^{2\Phi(\infty)} + \frac{Q}{2\pi^2 r^2} \,. \qquad (14.1.15c)$$

Here the x^m are transverse to the 5-brane, the x^μ are tangent to it, and $r^2 = x^m x^m$. This is the magnetically charged object required by string duality. The product $\tau_{D1}\tau_{D5} = \pi/\kappa^2$ should equal $\tau_{F1}\tau_{NS5}$ by the Dirac

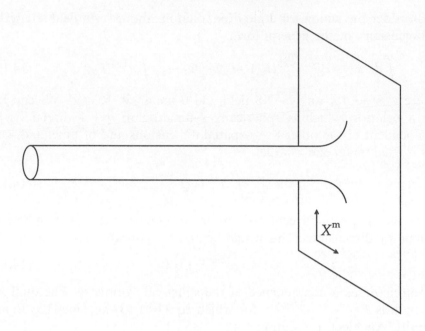

Fig. 14.1.   Infinite throat of an NS5-brane, with asymptotically flat spacetime on the right. The x^μ-directions, in which the 5-brane is extended, are not shown.

quantization condition (which determines the product of the charges) combined with the BPS condition (which relates the charges to the tensions). This gives

$$\tau_{\rm NS5} = \frac{2\pi^2\alpha'}{\kappa^2} = \frac{1}{(2\pi)^5 g^2 \alpha'^3} \cdot \qquad (14.1.16)$$

There must also be bound states of this with the D5-brane, which are presumably described by adding R–R flux to the above solution.

The geometry of the metric (14.1.15), shown in figure 14.1, is interesting. There is an infinite throat. The point $x^m = 0$ is at infinite distance, and as one approaches it the radius of the angular 3-spheres does not shrink to zero but approaches an asymptotic value $(Q/2\pi^2)^{1/2}$. The dilaton grows in the throat of the 5-brane, diverging at infinite distance. String perturbation theory thus breaks down some distance down the throat, and the effective length is probably finite. Because of the strong coupling one cannot describe this object quite as explicitly as the fundamental strings and D-brane, but one can look at fluctuations of the fields around the classical solution. There are normalizable massless fluctuations corresponding to translations and also ones which transform as a vector on the 5-brane, and the fermionic partners of these. These are the same as for the D5-brane, as should be true by duality.

It should be noted that the description above is in terms of the string metric, which is what appears in the string world-sheet action and is relevant for the dynamics of the string. The geometry is rather different in the Einstein metric $G_{E\mu\nu} = e^{-\Phi/2}G_{\mu\nu}$. In the string metric the radial distance is $ds \propto x^{-1}dx$, while in the Einstein metric it is $ds \propto x^{-3/4}dx$. The latter is singular but integrable, so the singularity is at finite distance. Thus, different probes can see a very different geometry.

Let us make a few more comments on this solution. For an NS field strength, a shift of the dilaton just multiplies the classical action by a constant. The solution is then independent of the dilaton, and its size can depend only on α' and the charge Q. The charge is quantized, $Q = nQ_0$, by the Dirac condition. The radius is then of order $\alpha'^{1/2}$ times a function of n, which in fact is $n^{1/2}$. For small n the characteristic scale of the solution is the string scale, so the low energy theory used to find the solution is not really valid. However, there are nonrenormalization theorems, which have been argued to show that the solution does not receive corrections. There is also a description of the throat region that is exact at string tree level — it does not use sigma model perturbation theory but is an exact CFT. The geometry of the throat is $S_3 \times R_1 \times$ six-dimensional Minkowski space. The CFT similarly factorizes. The six dimensions parallel to the brane world-volume are the usual free fields. The CFT of the radial coordinate is the linear dilaton theory that we have met before, with the dilaton diverging at infinite distance. The CFT of the angular directions is an $SU(2) \times SU(2)$ current algebra at level n, in a form that we will discuss in the next chapter.

This construction might seem to leave us with an embarrassment of riches, for we can similarly construct NS–NS electrically charged solutions and R–R charged solutions, for which we already have the F-string and D-branes as sources. In fact, the NS–NS electrically charged solution has a pointlike singularity and the fields satisfy the field equations with a δ-function source at the singularity. Thus this solution just gives the external fields produced by the F-string. The R–R charged solutions are black p-branes. Their relation to the D-branes will be considered at the end of this chapter.

A fundamental string can end on a D5-brane. It follows by weak–strong duality that a D-string should be able to end on an NS5-brane. A plausible picture is that it extends down the infinite throat. Its energy is finite in spite of the infinite length because of the position-dependence of the dilaton. Similarly a D-string should be able to end on a D3-brane. There is a nontrivial aspect to the termination of one object 'A' on a second

'B', which we saw in figure 13.5. Since A carries a conserved charge, the coupling between spacetime forms and world-brane fields of B must be such as to allow it to carry the charge of A.

The solution for any number of parallel NS5-branes is simply given by substituting

$$e^{2\Phi} = e^{2\Phi(\infty)} + \frac{1}{2\pi^2} \sum_{i=1}^{N} \frac{Q_i}{(x - x_i)^2} \qquad (14.1.17)$$

into the earlier solution (14.1.15). A D-string can run from one 5-brane to another, going down the throat of each. The ground state of this D-string is a BPS state. It is related by string duality to a ground state F-string stretched between two D5-branes, which is related by T-duality to a massless open string in the original type I theory. The mass of the D-string is given by the classical D-string action in the background (14.1.17), and agrees with string duality. In particular it vanishes as the NS5-branes become coincident, so like D-branes these have a non-Abelian symmetry in this limit. The limiting geometry is a single throat with twice the charge; in the limit, the non-Abelian degrees of freedom are in the strong coupling region down the throat and cannot be seen explicitly.

D3-branes and Montonen–Olive duality

Consider a system of n D3-branes. The dynamics on the D-branes is a $d = 4$, $N = 4$ $U(n)$ gauge theory, with the gauge coupling (13.3.25) equal to

$$g_{D3}^2 = 2\pi g \ . \qquad (14.1.18)$$

In particular this is dimensionless, as it should be for a gauge theory in four dimensions. At energies far below the Planck scale, the couplings of the closed strings to the D-brane excitations become weak and we can consider the D-brane gauge theory separately.

The $SL(2, \mathbf{Z})$ duality of the IIB string takes this system into itself, at a different coupling. In particular the weak–strong duality $g \to 1/g$ takes

$$g_{D3}^2 \to \frac{4\pi^2}{g_{D3}^2} \ . \qquad (14.1.19)$$

This is a weak–strong duality transformation within the gauge theory itself. Thus, the self-duality of the IIB string implies a similar duality within $d = 4$, $N = 4$ gauge theory. Such a duality was conjectured by Montonen and Olive in 1979. The evidence for it is of the same type as for string duality: duality of BPS masses and degeneracies and of the low energy effective action. Nevertheless the reaction to this conjecture was for a long time skeptical, until the development of supersymmetric gauge theory in the past few years placed it in a broader and more systematic context.

To understand the full $SL(2, \mathbf{Z})$ symmetry we need also to include the coupling (13.3.18) to the R–R scalar,

$$\frac{1}{4\pi} \int C_0 \operatorname{Tr}(F_2 \wedge F_2) .\qquad (14.1.20)$$

This is the Pontrjagin (instanton winding number) term, with $C_0 = \theta/2\pi$. The full gauge theory action, in a constant C_0 background, is

$$-\frac{1}{2g_{D3}^2} \int d^4x \operatorname{Tr}(|F_2|^2) + \frac{\theta}{8\pi^2} \int \operatorname{Tr}(F_2 \wedge F_2) .\qquad (14.1.21)$$

The duality $C_0 \to C_0 + 1$ is then the shift $\theta \to \theta + 2\pi$, corresponding to quantization of instanton charge. This and the weak–strong duality generate the full $SL(2, \mathbf{Z})$.

Let the D3-branes be parallel but slightly separated, corresponding to spontaneous breaking of $U(n)$ to $U(1)^n$. The ground state of an F-string stretched between D3-branes is BPS, and corresponds to a vector multiplet that has gotten mass from spontaneous breaking. The weak–strong dual is a D1-string stretched between D3-branes. To be precise, this is what it looks like when the separation of the D3-branes is large compared to the string scale. When the separation is small there is an alternative picture of this state as an 't Hooft-Polyakov magnetic monopole in the gauge theory. The size of the monopole varies inversely with the energy scale of gauge symmetry breaking and so inversely with the separation. This is similar to the story of the instanton in section 13.6, which has a D-brane description when small and a gauge theory description when large.

The relation between the IIB and Montonen–Olive dualities is one example of the interplay between the spacetime dynamics of various branes and the nonperturbative dynamics of the gauge theories that live on them. This is a very rich subject, and one which at this time is developing rapidly.

14.2   U-duality

The effect of toroidal compactification is interesting. The symmetry group of the low energy supergravity theory grows with the number k of compactified dimensions, listed as G in table B.3. We are familiar with two subgroups of each of these groups. The first is the $SL(2, \mathbf{R})$ symmetry of the uncompactified IIB theory. The second is the perturbative $O(k, k, \mathbf{R})$ symmetry of compactification of strings on T^k, which we encountered in the discussion of Narain compactification in chapter 8. In each case the actual symmetry of the full theory is the integer subgroup, the $O(k, k, \mathbf{Z})$

T-duality group and the $SL(2, \mathbf{Z})$ of the ten-dimensional IIB theory. The continuous $O(k, k, \mathbf{R})$ is reduced to the discrete $O(k, k, \mathbf{Z})$ by the discrete spectrum of (p_L, p_R) charges, and the continuous $SL(2, \mathbf{R})$ to the discrete $SL(2, \mathbf{Z})$ by the discrete spectrum of (p, q)-strings. In the massless limit the charged states do not appear and the symmetry appears to be continuous.

The natural conjecture is that in each case the maximal integer subgroup of the low energy symmetry is actually a symmetry of the full theory. This subgroup has been given the name *U-duality*. In perturbation theory we only see symmetries that act linearly on g and so are symmetries of each term in the perturbation series — these are the T-dualities plus shifts of the R–R fields. The other symmetries take small g to large and so require some understanding of the exact theory. The principal tools here are the constraints of supersymmetry on the low energy theory, already used in writing table B.3, and the spectrum of BPS states, which can be determined at weak coupling and continued to strong.

Let us look at the example of the IIB string on T^5, which by T-duality is the same as the IIA string on T^5. This is chosen because it is the setting for the simplest black hole state counting, and also because the necessary group theory is somewhat familiar from grand unification.

Let us first count the gauge fields. From the NS–NS sector there are five Kaluza–Klein gauge bosons and five gauge bosons from the antisymmetric tensor. There are also 16 gauge bosons from the dimensional reduction of the various R–R forms: five from $C_{\mu n}$, ten from $C_{\mu npq}$ and one from $C_{\mu npqrs}$. The index μ is in the noncompact dimensions, and in each case one sums over all antisymmetric ways of assigning the compact dimensions to the roman indices. Finally, in five noncompact dimensions the 2-form $B_{\mu\nu}$ is equivalent by Poincaré duality to a vector field, giving 27 gauge bosons in all.

Let us see how T-duality acts on these. This group is $O(5, 5, \mathbf{Z})$, generated by T-dualities on the various axes, linear redefinitions of the axes, and discrete shifts of the antisymmetric tensor. This mixes the first ten NS–NS gauge fields among themselves, and the 16 R–R gauge fields among themselves, and leaves the final NS–NS field invariant. Now, a representation of $O(10, \mathbf{R})$ automatically gives a representation of $O(5, 5, \mathbf{R})$ by analytic continuation, and so in turn a representation of the subgroup $O(5, 5, \mathbf{Z})$. The group $O(10, \mathbf{R})$ has a vector representation $\mathbf{10}$, spinor representations $\mathbf{16}$ and $\mathbf{16}'$, and of course a singlet $\mathbf{1}$. The gauge fields evidently transform in these representations; which spinor occurs depends on whether we start with the IIA or IIB theory, which differ by a parity transformation on $O(5, 5, \mathbf{Z})$.

According to table B.3, the low energy supergravity theory for this compactification has a continuous symmetry $E_{6(6)}$, which is a noncompact version of E_6. The maximal discrete subgroup is denoted $E_{6(6)}(\mathbf{Z})$. The group E_6 has a representation **27**, and a subgroup $SO(10)$ under which

$$27 \rightarrow 10 + 16 + 1 \ . \tag{14.2.1}$$

This may be familiar to readers who have studied grand unification; some of the relevant group theory was summarized in section 11.4 and exercise 11.5. Evidently the gauge bosons transform as this **27**.

Now let us identify the states carrying the various charges. The charges **10** are carried by the Kaluza–Klein and winding strings. Then U-duality also requires states in the **16**. These are just the various wrapped D-branes. Finally, the state carrying the **1** charge is the NS5-brane, fully wrapped around the T^5 so that it is localized in the noncompact dimensions.

U-duality and bound states

It is interesting to see how some of the bound state results from the previous chapter fit the predictions of U-duality in detail. We will generate U transformations as a combination of $T_{mn\cdots p}$, which is a T-duality in the indicated directions, and S, the IIB weak–strong transformation. The former switches between Neumann and Dirichlet boundary conditions and between momentum and winding number in the indicated directions. The latter interchanges the NS–NS and R–R 2-forms but leaves the R–R 4-form invariant, and acts correspondingly on the solitons carrying these charges. We denote by $D_{mn\cdots p}$ a D-brane extended in the indicated directions, and similarly for F_m a fundamental string extended in the given direction and p_m a momentum-carrying BPS state.

The first duality chain is

$$(D_9, F_9) \stackrel{T_{78}}{\rightarrow} (D_{789}, F_9) \stackrel{S}{\rightarrow} (D_{789}, D_9) \stackrel{T_9}{\rightarrow} (D_{78}, D_\emptyset) \ . \tag{14.2.2}$$

Thus the D-string/F-string bound state is U-dual to the D0–D2 bound state. The constructions of these bound states were similar, but the precise relation goes through the nonperturbative step S. In each case there is one short multiplet of BPS states.

The second chain is

$$(D_{6789}, D_\emptyset) \stackrel{T_6}{\rightarrow} (D_{789}, D_6) \stackrel{S}{\rightarrow} (D_{789}, F_6) \stackrel{T_{6789}}{\rightarrow} (D_6, p_6) \stackrel{S}{\rightarrow} (F_6, p_6) \ . \tag{14.2.3}$$

The bound states of n D0-branes and m D4-branes are thus U-dual to fundamental string states with momentum n and winding number m in one direction. Let us compare the degeneracy of BPS states in the two cases. For the winding string, the same argument as led to eq. (11.6.28) for the heterotic string shows that the BPS strings satisfy

$$(N, \tilde{N}) = \begin{cases} (nm, 0) \ , & nm > 0 \ , \\ (0, -nm) \ , & nm < 0 \ . \end{cases} \tag{14.2.4}$$

Here N and $\tilde{N}$ are the number of excitations above the massless ground state. We see that BPS states have only left-moving or only right-moving excitations. The generating function for the number of BPS states is the usual string partition function,

$$\operatorname{Tr} q^N = 2^8 \prod_{k=1}^{\infty} \left(\frac{1+q^k}{1-q^k} \right)^8 , \tag{14.2.5}$$

or the same with $\tilde{N}$. Note that we are counting the states of one string with winding number m, not of a bound state of m strings of winding number 1. The latter does not exist at small g — except insofar as one can think of the multiply wound string in this way. The counting (14.2.5) is most easily done with the refermionized θ_α. In terms of the ψ^μ the GSO projection gives several terms, which simplify using the abstruse identity. The string degeneracy (14.2.5) precisely matches the degeneracy D_{nm} of D0–D4 bound states in section 13.6.

14.3 $SO(32)$ **type I–heterotic duality**

In the type I theory, the only R–R fields surviving the Ω projection are the 2-form, which couples electrically to the D1-brane and magnetically to the D5-brane, and the nondynamical 10-form which couples to the D9-brane. This is consistent with the requirement for unbroken supersymmetry — the D1- and D5-branes both have $\#_{\mathrm{ND}} = 4k$ relative to the D9-brane.[1]

Consider again an infinite D-string stretched in the 1-direction. The type I D-string differs from that of the IIB theory in two ways. The first is the projection onto oriented states. The $U(1)$ gauge field, with vertex operator $\partial_t X^\mu$, is removed. The collective coordinates, with vertex operators $\partial_n X^\mu$, remain in the spectrum because the normal derivative is even under reversal of the orientation of the boundary. That is, in terms of its action on the X oscillators Ω has an additional -1 for the $m = 2, \ldots, 9$ directions, as compared to the action on the usual 9-9 strings. By superconformal symmetry this must extend to the ψ^μ, so that in particular on the ground states Ω is no longer -1 but acts as

$$ -\beta = -\exp[\pi i(s_1 + s_2 + s_3 + s_4)] , \tag{14.3.1}$$

with an additional rotation by π in the four planes transverse to the string. From the fermionic 1-1 strings of the IIB D-string, this removes the left-moving **8** and leaves the right-moving **8′**.

---

[1] It is conceivable that the D3- and D7-branes exist as non-BPS states. However, they would be expected to decay rapidly; also, there is some difficulty at the world-sheet level in defining them, as explained later.

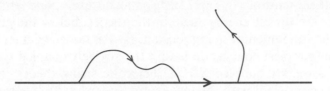

Fig. 14.2. D-string in type I theory with attached 1-1 and 1-9 strings.

The second modification is the inclusion of 1-9 strings, strings with one end on the D1-brane and one on a D9-brane. The end on the D9-brane carries the type I Chan–Paton index, so these are vectors of $SO(32)$. These strings have $\#_{\text{ND}} = 8$ so that the NS zero-point energy (13.4.8) is positive, and there are no massless states in the NS sector. The R ground states are massless as always. Only ψ^0 and ψ^1 are periodic in the R sector, so their zero modes generate two states

$$|s_0; i\rangle \,, \tag{14.3.2}$$

where $s_0 = \pm\frac{1}{2}$ and i is the Chan–Paton index for the 9-brane end. One of these two states is removed by the GSO projection; our convention has been

$$\exp(\pi i F) = -i \exp[2\pi i(s_0 + \ldots + s_4)] \,, \tag{14.3.3}$$

so that the state with $s_0 = +\frac{1}{2}$ would survive. We now impose the G_0 condition, which as usual (e.g. eq. (14.1.1)) reduces to a Dirac equation and then to the condition $s_0 = +\frac{1}{2}$ for the left-movers and $s_0 = -\frac{1}{2}$ for the right-movers. The right-moving 1-9 strings are thus removed from the spectrum by the combination of the GSO projection and G_0 condition. Finally we must impose the Ω projection; this determines the 9-1 state in terms of the 1-9 state, but otherwise makes no constraint.

To summarize, the massless bosonic excitations are the usual collective coordinates. The massless fermionic excitations are right-movers in the $\mathbf{8'}$ of the transverse $SO(8)$ and left-movers that are invariant under $SO(8)$ and are vectors under the $SO(32)$ gauge group. This is the same as the excitation spectrum of a long $SO(32)$ heterotic string. Incidentally, this explains how it can be consistent with supersymmetry that the 1-9 strings have massless R states and no massless NS states: the supersymmetry acts only on the right-movers. This is also a check that our conventions above were consistent — supersymmetry requires the 1-9 fermions to move in the opposite direction to the 1-1 fermions. From a world-sheet point of view, this is necessary in order that the gravitino OPE be consistent.

The D-string tension $\tau_{D1} = 1/2\pi\alpha' g$ is again exact, and at strong coupling this is the lowest energy scale in the theory, below the gravitational scale and the fundamental string tension. By the same arguments as in the IIB case, the simplest conclusion is that the strongly coupled type I theory is actually a weakly coupled $SO(32)$ heterotic string theory. As a check, this must be consistent with the low energy supergravity theories. We have already noted that these must be the same up to field redefinition, because the supersymmetry algebras are the same. It is important, though, that the redefinition (12.1.41),

$$G_{I\mu\nu} = e^{-\Phi_h} G_{h\mu\nu} , \quad \Phi_I = -\Phi_h , \tag{14.3.4a}$$

$$\widetilde{F}_{I3} = \widetilde{H}_{h3} , \quad A_{I1} = A_{h1} , \tag{14.3.4b}$$

includes a reversal of the sign of the dilaton.

The conclusion is that there is a single theory, which looks like a weakly coupled type I theory when $e^{\Phi_I} \ll 1$ and like a weakly coupled $SO(32)$ heterotic theory when $e^{\Phi_I} \gg 1$. The type I supergravity theory is a good description of the *low energy* physics throughout. Even if the dimensionless string coupling is of order 1, the couplings in the low energy theory are all irrelevant in ten dimensions (and remain irrelevant as long as there are at least five noncompact dimensions) and so are weak at low energy.

As a bonus we have determined the strong-coupling physics of the $SO(32)$ heterotic string, namely the type I string. It would have been harder to do this directly. The strategy we have used so far, which would require finding the type I string as an excitation of the heterotic theory, would not work because a long type I string is not a BPS state. The NS–NS 2-form, whose charge is carried by most fundamental strings, is not present in the type I theory. The R–R 2-form remains, but its charge is carried by the type I D-string, not the F-string. That the long type I F-string is not a BPS state is also evident from the fact that it can break and decay. As the type I coupling increases, this becomes rapid and the type I string disappears as a recognizable excitation.

The strings of the type I theory carry only symmetric and antisymmetric tensor representations of the gauge group, while the strings of the heterotic theory can appear in many representations. We see that the corresponding states appear in the type I theory as D-strings, where one gets large representations of the gauge group by exciting many 1-9 strings. Note in particular that type I D-strings can carry the spinor representation of $SO(32)$; this representation is carried by fundamental heterotic strings but cannot be obtained in the product of tensor representations. Consider a long D-string wrapped around a periodic dimension of length L. The massless 1-9 strings are associated with fermionic fields Λ^i living on the

D-string, with i the $SO(32)$ vector index. The zero modes of these,

$$\Lambda_0^i = L^{-1/2} \int_0^L dx^1 \, \Lambda^i(x^1) \,, \tag{14.3.5}$$

satisfy a Clifford algebra

$$\{\Lambda_0^i, \Lambda_0^j\} = \delta^{ij} \,. \tag{14.3.6}$$

The quantization now proceeds just as for the *fundamental* heterotic string, giving spinors $2^{15} + 2^{15'}$ of $SO(32)$. Again, the Λ^i are fields that create light *strings*, but they play the same role here as the λ^i that create excitations *on* the heterotic string.

The heterotic string automatically comes out in fermionic form, and so a GSO projection is needed. We can think of this as gauging a discrete symmetry that acts as -1 on every D-string endpoint (the idea of gauging a discrete group was explained in section 8.5). This adds in the NS sectors for the fields Λ^i and removes one of the two spinor representations. Recall that in the IIB D-string there is a continuous $U(1)$ gauge symmetry acting on the F-string endpoints. The part of this that commutes with the Ω projection and so remains on the type I D-string is just the discrete gauge symmetry that we need to give the current algebra GSO projection.

Quantitative tests

Consider the tension of the heterotic D-string,

$$\tau_{D1} \, (\text{type I}) = \frac{\pi^{1/2}}{2^{1/2}\kappa}(4\pi^2\alpha') = \frac{g_{YM}^2}{8\pi\kappa^2} \,. \tag{14.3.7}$$

We have used the type I relation (13.3.31) to express the result in terms of the low energy gauge and gravitational couplings, which are directly measurable in scattering experiments. It should be noted that the type I cylinder amplitude for the D-brane interaction has an extra $\frac{1}{2}$ from the orientation projection as compared to the type II amplitude, so the D-brane tension is multiplied by $2^{-1/2}$. The result (14.3.7), obtained at weak type I coupling, is exact as a consequence of the BPS property. Hence it should continue to hold at strong type I coupling, and therefore agree with the relation between the *heterotic* string tension and the low energy couplings at weak heterotic coupling. Indeed, this is precisely eq. (12.3.37).

As another example, consider the $F_{\mu\nu}^4$ interaction (12.4.28) found in type I theory from the disk amplitude,

$$\frac{\pi^2\alpha'^2}{2 \times 4! \, g_{YM}^2}(tF^4) = \frac{g_{YM}^2}{2^{10}\pi^5 4! \, \kappa^2}(tF^4) \,, \tag{14.3.8}$$

and the same interaction found in the $SO(32)$ heterotic theory from the torus,

$$\frac{1}{2^8\pi^5 4!\,\alpha'}(tF^4) = \frac{g_{YM}^2}{2^{10}\pi^5 4!\,\kappa^2}(tF^4)\,. \tag{14.3.9}$$

Here (tF^4) is an abbreviation for the common Lorentz and gauge structure of the two amplitudes. In each theory we have expressed α' appropriately in terms of the low energy couplings. The agreement between the numerical coefficients of the respective interactions is not an accident but is required by type I–heterotic duality. To explain this, first we must assert without proof the fact that supersymmetry completely determines the dilaton dependence of the $F^4_{\mu\nu}$ interaction in a theory with 16 supersymmetries.[2] Hence we can calculate the coefficient when Φ_I is large and negative and the type I calculation is valid, and it must agree with the result at large positive Φ_I where the heterotic calculation is valid.

Actually, this particular agreement is not an independent test of duality, but is a consequence of the consistency of each string theory separately. The (tF^4) interaction is related by supersymmetry to the $B_2F_2^4$ interaction, and the coefficient of the latter is fixed in terms of the low energy spectrum by anomaly cancellation. However, this example illustrates the fact that weak–strong dualities in general can relate calculable amplitudes in the dual theories, and not only incalculable strong-coupling effects. In more complicated examples, such as compactified theories, there are many such successful relations that are not preordained by anomaly cancellation. As in this example, a tree-level amplitude on one side can be related to a loop amplitude on the other, or to an instanton calculation.

Type I D5-branes

The type I D5-brane has some interesting features. The D5–D9 system is related by T-duality to the D0–D4 system. We argued that in the latter case the D0-brane was in fact the zero-size limit of an instanton constructed from the D4-brane gauge fields. The same is true here. The type I theory has gauge field solutions in which the fields are independent of five spatial dimensions and are a localized Yang–Mills instanton configuration in the other four: this is a 5-brane. It has collective coordinates for its shape, and also for the size and gauge orientation of the instanton. In the zero-size

---

[2] Notice that there are dilaton dependences hidden in the couplings in (14.3.8) and (14.3.9), which moreover are superficially different because of the different dilaton dependences of g_{YM} in the two string theories. However, the dilaton dependences are related by the field redefinition (14.3.4), and are correlated with the fact that the lower order terms in the action also have different dilaton dependences.

limit, the D5-brane description is accurate. As in the discussion of D0–D4 bound states, there are flat directions for the 5-9 fields. Again these have the interpretation of blowing the D5-brane up into a 9-9 gauge field configuration whose cross-section is the $SO(32)$ instanton and which is independent of the other six dimensions.

The heterotic dual of the type I D5-brane is simple to deduce. The blown-up instanton is an ordinary field configuration. The transformation (14.3.4) between the type I and heterotic fields leaves the gauge field invariant, so this just becomes an instanton in the heterotic theory. The transformation of the metric has an interesting effect. What looks in the type I theory like a small instanton becomes in the heterotic theory an instanton at the end of a long but finite throat; in the zero-size limit the throat becomes infinite as in figure 14.1.

There is one difference from the earlier discussion of D-branes. It turns out to be necessary to assume that the type I D5-brane carries an $SU(2)$ symmetry — that is, a two-valued Chan–Paton index. More specifically, it is necessary on the D5-branes to take a symplectic rather than orthogonal projection. We will first work out the consequences of this projection, and then discuss why it must be so.

The bosonic excitation spectrum consists of

$$\psi^{\mu}_{-1/2}|0,k;ij\rangle\lambda_{ij}\ , \quad \psi^{m}_{-1/2}|0,k;ij\rangle\lambda'_{ij}\ , \tag{14.3.10}$$

which are the D5-brane gauge field and collective coordinate respectively; i and j are assumed to be two-valued. The symplectic Ω projection gives

$$M\lambda M^{-1} = -\lambda^T\ , \quad M\lambda' M^{-1} = \lambda'^T\ , \tag{14.3.11}$$

with M the antisymmetric 2×2 matrix. The general solutions are

$$\lambda = \sigma^a\ , \quad \lambda' = I\ . \tag{14.3.12}$$

In particular the Chan–Paton wavefunction for the collective coordinate is the identity, so 'both' D5-branes move together. We should really then refer to *one* D5-brane, with a two-valued Chan–Paton index. This is similar to the T-dual of the type I string, where there are 32 Chan–Paton indices but 16 D-branes, each D-brane index being doubled to account for the orientifold image. The world-brane vectors have Chan–Paton wavefunctions σ^a_{ij} so the gauge group is $Sp(1) = SU(2)$, unlike the IIB D5-brane whose gauge group is $U(1)$. For k coincident D5-branes the group is $Sp(k)$.

The need for a two-valued Chan–Paton index can be seen in four independent ways. The first is that it is needed in order to get the correct instanton moduli space, the instanton gauge group now being $SO(n)$ rather than $SU(n)$. We will not work out the details of this, but in fact this is how the $SU(2)$ symmetry was first deduced. Note that starting from a large instanton, it is rather surprising that in the zero-size limit a new internal gauge symmetry appears. The appearance of new gauge symmetries at special points in moduli space is now known to occur in many contexts. The non-Abelian gauge symmetry of coincident IIB NS5-branes, pointed out in section 14.1, was another such surprise. The enhanced gauge symmetry of the toroidally compactified string is a perturbative example.

The second argument for a symplectic projection is based on the fact that in the type I theory the force between D1-branes, and between D5-branes, is half of what was calculated in section 13.3 due to the orientation projection. The tension and charge are then each reduced by a factor $2^{-1/2}$, so the product of the charges of a single D1-brane and single (one-valued) D5-brane would then be only half a Dirac unit. However, since the D5-branes with a symplectic projection always move in pairs, the quantization condition is respected. The third argument is based on the spectrum of 5-9 strings. For each value of the Chan–Paton indices there are two bosonic states, as in eq. (13.5.19). The D5–D9 system has eight supersymmetries, and these two bosons form half of a hypermultiplet (section B.7). In an oriented theory the 9-5 strings are the other half, but in this unoriented theory these are not independent. A half-hypermultiplet is possible only for pseudoreal representations, like the **2** of $SU(2)$ — hence the need for the $SU(2)$ on the D5-brane.

The final argument is perhaps the most systematic, but also the most technical. Return to the discussion of the orientation projection in section 6.5. The general projection was of the form

$$\hat{\Omega}|\psi; ij\rangle = \gamma_{jj'}|\Omega\psi; j'i'\rangle\gamma_{i'i}^{-1} \, . \qquad (14.3.13)$$

We can carry over this formalism to the present case, where now the Chan–Paton index in general runs over 1-, 5-, and 9-branes. In order for this to be a symmetry the matrix $\gamma_{jj'}$ must connect D-branes that are of the same dimension and coincident.[3] In chapter 6 we argued that $\hat{\Omega}^2 = 1$ and therefore that γ was either purely symmetric or purely antisymmetric. The first argument still holds, but the second rested on an assumption that is not true in general: that the operator Ω, the part of $\hat{\Omega}$ that acts on the fields, squares to the identity, $\Omega^2 = 1$. More generally, it may in fact be a phase.

Working out the phase of Ω is a bit technical. It is determined by the requirement that the symmetry be conserved by the operator product of the corresponding vertex operators. In the 5-5 sector, the massless

---

[3] This formalism also applies to the more general orientifold projection, where $\hat{\Omega}$ is combined with a spacetime symmetry. The matrix γ then connects each D-brane with its image under the spacetime symmetry.

vertex operator is $\partial_t X^\mu$ ($\Omega = -1$) for μ parallel to the 5-brane, and $\partial_n X^\mu$ ($\Omega = +1$) for μ perpendicular. On these states, $\Omega^2 = 1$, and the same is true for the rest of the 9-9 and 5-5 Hilbert spaces. To see this, use the fact that Ω multiplies any mode operator ψ_r by $\pm\exp(i\pi r)$. The mode expansions were given in section 13.4. In the NS sector this is $\pm i$, but the GSO projection requires that these mode operators act in pairs (the OPE is single-valued only for GSO-projected vertex operators). So $\Omega = \pm 1$, and this holds in the R sector as well by supersymmetry.

Now consider the NS 5-9 sector. The four X^μ with mixed Neumann–Dirichlet boundary conditions, say $\mu = 6, 7, 8, 9$, have a half-integer-mode expansion. Their superconformal partners ψ^μ then have an integer-mode expansion and the ground state is a representation of the corresponding zero-mode algebra. The vertex operator is thus a spin field: the periodic ψ^μ contribute a factor

$$V = e^{i(H_3 + H_4)/2} , \tag{14.3.14}$$

where $H_{3,4}$ are from the bosonization of the four periodic $\psi^{6,7,8,9}$. We need only consider this part of the vertex operator, as the rest is the same as in the 9-9 string and so has $\Omega^2 = +1$. Now, the operator product of V with itself (which is in the 5-5 or 9-9 sector) involves $e^{i(H_3+H_4)}$, which is the bosonization of $(\psi^6 + i\psi^7)(\psi^8 + i\psi^9)$. This in turn is the vertex operator for the state

$$(\psi^6 + i\psi^7)_{-1/2}(\psi^8 + i\psi^9)_{-1/2}|0\rangle . \tag{14.3.15}$$

Finally we can deduce the Ω eigenvalue. For $|0\rangle$ it is $+1$, because its vertex operator is the identity, while each $\psi_{-1/2}$ contributes either $-i$ (for a 9-9 string) or $+i$ (for a 5-5 string), giving an overall -1. That is, the Ω eigenvalue of VV is -1, and so therefore is the Ω^2 eigenvalue of V.

In the 5-9 sector $\Omega^2 = -1$. Separate γ into a block γ_9 that acts on the D9-branes and a block γ_5 that acts on the D5-branes. Then repeating the argument in section 6.5 gives

$$\gamma_9^T \gamma_9^{-1} = \Omega_{5\text{-}9}^2 \gamma_5^T \gamma_5^{-1} . \tag{14.3.16}$$

We still have $\gamma_9^T = +\gamma_9$ from tadpole cancellation, so we need $\gamma_5^T = -\gamma_5$, giving symplectic groups on the D5-brane. The minimum dimension for the symplectic projection is 2, so we need a two-valued Chan–Paton state. This argument seems roundabout, but it is faithful to the logic that the actions of Ω in the 5-5 and 9-9 sectors are related because they are both contained in the 5-9 $\times$ 9-5 product. Further, there does not appear to be any arbitrariness in the result. It also seems to be impossible to define the D3- or D7-brane consistently, as $\Omega^2 = \pm i$.

14.4   Type IIA string and M-theory

The type IIA string does not have D-strings but does have D0-branes, so let us consider the behavior of these at strong coupling. We focus on the D-brane of smallest dimension for the following reason. The D-brane tension $\tau_p = O(g^{-1}\alpha'^{-(p+1)/2})$ translates into a mass scale

$$(\tau_p)^{1/(p+1)} \approx g^{-1/(p+1)}\alpha'^{-1/2} \qquad (14.4.1)$$

so that at strong coupling the smallest p gives the lowest scale. Thus we need to find an effective field theory describing these degrees of freedom.

The D0-brane mass is

$$\tau_0 = \frac{1}{g\alpha'^{1/2}} \ . \qquad (14.4.2)$$

This is heavy at weak coupling but becomes light at strong coupling. We also expect that for any number n of D0-branes there is an ultrashort multiplet of bound states with mass

$$n\tau_0 = \frac{n}{g\alpha'^{1/2}} \ . \qquad (14.4.3)$$

This is exact, so as the coupling becomes large all these states become light and the spectrum approaches a continuum. Such a continuous spectrum of particle states is characteristic of a system that is becoming noncompact. In particular, the evenly spaced spectrum (14.4.3) matches the spectrum of momentum (Kaluza–Klein) states for a periodic dimension of radius

$$R_{10} = g\alpha'^{1/2} \ . \qquad (14.4.4)$$

Thus, as $g \to \infty$ an *eleventh* spacetime dimension appears. This is one of the greatest surprises in this subject, because perturbative superstring theory is so firmly rooted in ten dimensions.

From the point of view of supergravity all this is quite natural. Eleven-dimensional supergravity is the supersymmetric field theory with the largest possible Poincaré invariance. Beyond this, spinors have at least 64 components, and this would lead to massless fields with spins greater than 2; such fields do not have consistent interactions. We have used dimensional reduction of eleven-dimensional supergravity as a crutch to write down ten-dimensional supergravity, but now we see that it was more than a crutch: dimensional reduction keeps only the $p_{10} = 0$ states, but string theory has also states of $p_{10} \neq 0$ in the form of D0-branes and their bound states. Recall that in the reduction of the eleven-dimensional theory to IIA string theory, the Kaluza–Klein gauge boson which couples to p_{10} became the R–R gauge boson which couples to D0-branes. The eleventh dimension is invisible in string perturbation theory because this is an expansion around the zero-radius limit for the extra dimension, as is evident from eq. (14.4.4).

The eleven-dimensional gravitational coupling is given by dimensional reduction as

$$\kappa_{11}^2 = 2\pi R_{10} \kappa^2 = \frac{1}{2}(2\pi)^8 g^3 \alpha'^{9/2} \,. \tag{14.4.5}$$

The numerical factors here are inconvenient so we will define instead an eleven-dimensional Planck mass

$$M_{11} = g^{-1/3} \alpha'^{-1/2} \,, \tag{14.4.6}$$

in terms of which $2\kappa_{11}^2 = (2\pi)^8 M_{11}^{-9}$. The two parameters of the IIA theory, g and α', are related to the eleven-dimensional Planck mass and the radius of compactification by eqs. (14.4.4) and (14.4.6). Inverting these,

$$g = (M_{11} R_{10})^{3/2} \,, \quad \alpha' = M_{11}^{-3} R_{10}^{-1} \,. \tag{14.4.7}$$

The reader should be be alert to possible differences in convention in the definition of M_{11}, by powers of 2π; the choice here makes the conversion between string and M-theory parameters simple.

We know little about the eleven-dimensional theory. Its low energy physics must be described by $d = 11$ supergravity, but it has no dimensionless parameter in which to make a perturbation expansion. At energies of order M_{11} neither supergravity nor string theory is a useful description. It is hard to name a theory when one does not know what it is; it has been given the tentative and deliberately ambiguous name *M-theory*. Later in the chapter we will discuss a promising idea as to the nature of this theory.

U-duality and F-theory

Since we earlier deduced the strongly coupled behavior of the IIB string, and this is T-dual to the IIA string, we can also understand the strongly coupled IIA string in this way. Periodically identify the 9-direction. The IIB weak–strong duality S interchanges a D-string wound in the 9-direction with an F-string wound in the 9-direction. Under T-duality, the D-string becomes a D0-brane and the wound F-string becomes a string with nonzero p_9. So TST takes D0-brane charge into p_9 and vice versa. Thus we should be able to interpret D0-brane charge as momentum in a dual theory, as indeed we argued above. The existence of states with R–R charge and of the eleventh dimension was inferred in this way — as were the various other dualities — before the role of D-branes was understood.

It is notable that while the IIA and IIB strings are quite similar in perturbation theory, their strongly coupled behaviors are very different. The strongly coupled dual of the IIB theory is itself, while that of the

IIA theory is a new theory with an additional spacetime dimension. Nevertheless, we see that these results are consistent with the equivalence of the IIA and IIB theories under T-duality. The full set of dualities forms a rich interlocking web.

For the type II theory on a circle, the noncompact symmetry of the low energy theory is $SL(2,\mathbf{R}) \times SO(1,1,\mathbf{R})$ (table B.3) and the discrete U-duality subgroup is

$$d = 9: \quad U = SL(2,\mathbf{Z}) \,. \tag{14.4.8}$$

Regarded as a compactification of the IIB string, this is just the $SL(2,\mathbf{Z})$ symmetry of the ten-dimensional theory. Regarded as a compactification of the IIA string on a circle and therefore of M-theory on T^2, it is a geometric symmetry, the modular transformations of the spacetime T^2.

For the type II theory on T^2, the noncompact symmetry of the low energy theory is $SL(3,\mathbf{R}) \times SL(2,\mathbf{R})$ and the discrete U-duality subgroup is

$$d = 8: \quad U = SL(3,\mathbf{Z}) \times SL(2,\mathbf{Z}) \,. \tag{14.4.9}$$

In section 8.4 we studied compactification of strings on T^2 and found that the T-duality group was $SL(2,\mathbf{Z}) \times SL(2,\mathbf{Z})$, one factor being the geometric symmetry of the 2-torus and one factor being stringy. In the U-duality group the geometric factor is enlarged to the $SL(3,\mathbf{Z})$ of the M-theory T^3.

Under compactification of more dimensions, it is harder to find a geometric interpretation of the U-duality group. The type II string on T^4, which is M-theory on T^5, has the U-duality symmetry

$$d = 6: \quad U = SO(5,5,\mathbf{Z}) \,. \tag{14.4.10}$$

This is the same as the T-duality of *string theory* on T^5. This is suggestive, but this identity holds only for T^5 so the connection if any will be intricate. For compactification of M-theory on T^k for $k \geq 6$, the U-duality symmetry is a discrete exceptional group, which has no simple geometric interpretation. A good interpretation of these symmetries would likely be an important step in understanding the nature of M-theory.

Returning to the IIB string in ten dimensions, it has been suggested that the $SL(2,\mathbf{Z})$ duality has a geometric interpretation in terms of two additional toroidal dimensions. This construction was christened *F-theory*. It is clear that these dimensions are not on the same footing as the eleventh dimension of M-theory, in that there is no limit of the parameters in which the spectrum becomes that of twelve noncompact dimensions. However, there may be some sense in which it is useful to begin with twelve dimensions and 'gauge away' one or two of them. Independent of this, F-theory has been a useful technique for finding solutions to the field equations with nontrivial behavior of the dilaton and R–R scalar. As in eq. (12.1.30), these fields are joined in a complex parameter $\tau = C_0 + ie^{-\Phi}$ characterizing the complex structure of the additional 2-torus. Ten-dimensional solutions are then usefully written in terms of *twelve-dimensional* geometries.

IIA branes from eleven dimensions

The IIA theory has a rich spectrum of extended objects. It is interesting to see how each of these originates from compactification of M-theory on a circle. Let us first consider the extended objects of the eleven-dimensional theory. There is one tensor gauge field, the 3-form $A_{\mu\nu\rho}$. The corresponding electrically charged object is a 2-brane; in the literature the term *membrane* is used specifically for 2-branes. The magnetically charged object is a 5-brane. Of course the designations *electric* and *magnetic* interchange if we use instead a 6-form potential. However, $d = 11$ supergravity is one case in which one of the two Poincaré dual forms seems to be preferred (the 3-form) because the Chern–Simons term in the action cannot be written with a 6-form.

As in the discussion of the IIB NS5-brane, but with the dilaton omitted, we can always find a supersymmetric solution to the field equations having the appropriate charges. The M2- and M5-brane solutions are black p-branes, as described below eq. (14.1.14).

0-branes: The D0-branes of the IIA string are the BPS states of nonzero p_{10}. In M-theory these are the states of the massless graviton multiplet, an ultrashort multiplet of 2^8 states for each value of p_{10}.

1-branes: The 1-brane of the IIA theory is the fundamental IIA string. Its natural origin is as an M-theory supermembrane wrapped on the hidden dimension. As a check, such a membrane would couple to $A_{\mu\nu 10}$; this reduces to the NS–NS $B_{\mu\nu}$ field which couples to the IIA string. It was noted some time ago that the classical action of a wrapped M2-brane reduces to that of the IIA string.

2-branes: The obvious origin of the IIA D2-brane is as a transverse (rather than wrapped) M2-brane. The former couples to the R–R $C_{\mu\nu\rho}$, which is the reduction of the $d = 11$ $A_{\mu\nu\rho}$ to which the latter couples. Note that when written in terms of M-theory parameters, the D2-brane tension

$$\tau_{D2} = \frac{1}{(2\pi)^2 g\alpha'^{3/2}} = \frac{M_{11}^3}{(2\pi)^2} \tag{14.4.11}$$

depends only on the fundamental scale M_{11} and not on R_{10}, as necessary for an object that exists in the eleven-dimensional limit. On the other hand, the F-string tension

$$\tau_{F1} = \frac{1}{2\pi\alpha'} = 2\pi R_{10}\tau_{D2} \qquad (14.4.12)$$

is linear in R_{10}, as should be the case for a wrapped object.

The D2-brane is perpendicular to the newly discovered 10-direction, and so should have a collective coordinate for fluctuations in that direction. This is puzzling, because D-branes in general have collective coordinates only for their motion in the ten-dimensional spacetime of perturbative string theory. However, the D2-brane is special, because in 2+1 dimensions a vector describes the same physics, by Poincaré duality, as a scalar. It is interesting to see this in detail. The bosonic action for a D2-brane in flat spacetime is

$$S[F, \lambda, X] = -\tau_2 \int d^3x \left\{ [-\det(\eta_{\mu\nu} + \partial_\mu X^m \partial_\nu X^m + 2\pi\alpha' F_{\mu\nu})]^{1/2} \right.$$

$$\left. + \frac{\epsilon^{\mu\nu\rho}}{2} \lambda \partial_\mu F_{\nu\rho} \right\} . \qquad (14.4.13)$$

We are treating $F_{\mu\nu}$ as the independent field and so include a Lagrange multiplier λ to enforce the Bianchi identity. In this form $F_{\mu\nu}$ is an auxiliary field (its equation of motion determines it completely as a local function of the other fields) and it can be eliminated with the result

$$S[\lambda, X] = -\tau_2 \int d^3x \left\{ -\det[\eta_{\mu\nu} + \partial_\mu X^m \partial_\nu X^m + (2\pi\alpha')^{-2} \partial_\mu \lambda \partial_\nu \lambda] \right\}^{1/2} .$$

$$(14.4.14)$$

The algebra is left as an exercise. Defining $\lambda = 2\pi\alpha' X^{10}$, this is the action for a membrane in eleven dimensions. Somewhat surprisingly, it displays the full eleven-dimensional Lorentz invariance, even though this is broken by the compactification of X^{10}. This can be extended to the fermionic terms, and to membranes moving in background fields.

4-branes: These are wrapped M5-branes.

5-branes: The IIA theory, like the IIB theory, has a 5-brane solution carrying the magnetic NS–NS $B_{\mu\nu}$ charge. The solution is the same as in the IIA theory, because the actions for the NS–NS fields are the same. However, there is an interesting difference. Recall that a D1-brane can end on the IIB NS 5-brane. Under T-duality in a direction parallel to the 5-brane, we obtain a D2-brane ending on a IIA NS5-brane. From the point of view of the $(5 + 1)$-dimensional theory on the 5-brane, the end of a D1-brane in the IIB theory is a point, and is a source for the $U(1)$

gauge field living on the 5-brane. This is necessary so that the 5-brane can, through a Chern–Simons interaction, carry the R–R charge of the D1-brane. Similarly the end of the D2-brane in the IIA theory is a string in the 5-brane, and so should couple to a *2-form* field living on the IIA NS5-brane.

We were not surprised to find a $U(1)$ gauge field living on the IIB NS 5-brane, because it is related by S-duality to the IIB D5-brane which we know to have such a field. We cannot use this argument for the IIA NS 5-brane. However, in both cases the fields living on the world-sheet can be seen directly by looking at small fluctuations around the soliton solution. We do not have space here to develop in detail the soliton solutions and their properties, but we summarize the results. Modes that are normalizable in the directions transverse to the 5-brane correspond to degrees of freedom living on the 5-brane. These include the collective coordinates for its motion and in each case some R–R modes, which do indeed form a vector in the IIB case and a 2-form in the IIA case. The field strength of the 2-form is self-dual.

It is also interesting to look at this in terms of the unbroken supersymmetry algebras in the 5-brane world-volumes. Again we have space only to give a sketch. The supersymmetry variations of the gravitinos in a general background are

$$\delta\psi_M = D_M^-\zeta \,, \quad \delta\tilde\psi_M = D_M^+\zeta \,. \tag{14.4.15}$$

Here $D_M^\pm$ is a covariant derivative where the spin connection ω is replaced with $\omega^\pm = \omega \pm \frac{1}{2}H$ with H the NS–NS 3-form field strength. We have already encountered $\omega^\pm$ in the world-sheet action (12.3.28). The difference of sign on the two sides occurs because H is odd under world-sheet parity. Under

$$SO(9,1) \to SO(5,1) \times SO(4) \,, \tag{14.4.16}$$

the ten-dimensional spinors decompose

$$\mathbf{16} \to (\mathbf{4},\mathbf{2}) + (\mathbf{4}',\mathbf{2}') \,, \tag{14.4.17a}$$

$$\mathbf{16}' \to (\mathbf{4},\mathbf{2}') + (\mathbf{4}',\mathbf{2}) \,. \tag{14.4.17b}$$

The nonzero components of the connection for the 5-brane solution lie in the transverse $SO(4) = SU(2) \times SU(2)$, and for the NS5-branes ω^+ and ω^- have the property that they lie entirely in the first or second $SU(2)$ respectively. A constant spinor carrying the second $SU(2)$ (that is, a $\mathbf{2}'$ of $SO(4)$) is then annihilated by D_M^+, and one carrying the first (a $\mathbf{2}$) by D_M^-; these correspond to unbroken supersymmetries. The left-moving supersymmetries transforming as a $\mathbf{2}$ of $SO(4)$ are thus unbroken — these are a $\mathbf{4}$ in both the IIA and IIB theories. Also unbroken are the right-moving supersymmetries transforming as a $\mathbf{2}'$ of $SO(4)$, which for the IIA theory are a **4** and for the IIB theory a **4′**. In other words, the unbroken supersymmetry of the IIA NS5-brane is $d = 6$ $(2,0)$ supersymmetry, and the unbroken supersymmetry of the IIB NS5-brane is $d = 6$ $(1,1)$ supersymmetry. These supersymmetries are reviewed in section B.6. Curiously the nonchiral IIA theory has a chiral 5-brane, and the chiral IIB theory a nonchiral 5-brane.

These results fit with the fluctuation spectra. For the IIB NS5-brane the collective coordinates plus vector add up to a vector multiplet of $d = 6$ $(1,1)$ supersymmetry. For the IIA NS5-brane, the only low-spin multiplet is the tensor, which contains the self-dual tensor argued for above and five scalars.

The obvious interpretation of the IIA NS5-brane is as an M-theory 5-brane that is transverse to the eleventh dimension. As in the discussion of the 2-brane, it should then have a collective coordinate for motion in this direction. Four of the scalars in the tensor multiplet are from the NS–NS sector and are collective coordinates for the directions perpendicular to the 5-brane that are visible in string perturbation theory. The fifth scalar, from the R–R sector, must be the collective coordinate for the eleventh dimension. It is remarkable that the 2-brane and the 5-brane of the IIA theory know that they secretly live in eleven dimensions.

The tension of the IIA NS5-brane is the same as that of the IIB NS5-brane,

$$\tau_{NS5} = \frac{1}{(2\pi)^5 g^2 \alpha'^3} = \frac{\tau_{D2}^2}{2\pi} = \frac{M_{11}^6}{(2\pi)^5} . \qquad (14.4.18)$$

Like the tension of the D2-brane this is independent of R_{10}, as it must be for the eleven-dimensional interpretation,

$$\tau_{D2} = \tau_{M2} , \quad \tau_{NS5} = \tau_{M5} . \qquad (14.4.19)$$

This also fits with the interpretation of the D4-brane,

$$\tau_{D4} = 2\pi R_{10}\tau_{M5} . \qquad (14.4.20)$$

Since the IIA NS5-brane and D2-brane are both localized in the eleventh dimension, the configuration of a D2-brane ending on an NS5-brane lifts to an eleven-dimensional configuration of an M-theory 2-brane ending on an M-theory 5-brane. It is interesting to consider two nearby 5-branes with a 2-brane stretched between them, either in the IIA or M-theory context. The 2-brane is still extended in one direction and so behaves as a string. The tension is proportional to the distance r between the two 5-branes,

$$\tau_1 = r\tau_{M2} . \qquad (14.4.21)$$

In the IIB theory, the $r \to 0$ limit was a point of non-Abelian gauge symmetry. Here it is something new, a *tensionless string theory*. For small r the lightest scale in the theory is set by the tension of these strings. They are entirely different from the strings we have studied: they live in six dimensions, they are not associated with gravity, and they have no adjustable coupling constant — their interactions in fact are of order 1. Of all the new phases of gauge and string theories that have been discovered this is perhaps the most mysterious, and may be a key to understanding many other things.

6-branes: The D6-brane field strength is dual to that of the D0-brane. Since the D0-brane carries Kaluza–Klein electric charge, the D6-brane must be a Kaluza–Klein magnetic monopole. Such an object exists as a soliton, where the Kaluza–Klein direction is not independent of the noncompact directions but is combined with them in a smooth and topologically nontrivial way. This is a local object in three noncompact spatial dimensions and so becomes a 6-brane in nine noncompact spatial dimensions.

8-branes: The eleven-dimensional origin of the D8-brane will be seen in the next section.

14.5 The $E_8 \times E_8$ heterotic string

The final ten-dimensional string theory is the $E_8 \times E_8$ heterotic string. We should be able to figure out its strongly coupled behavior, since it is T-dual to the $SO(32)$ heterotic string whose strongly coupled limit is known. We will need to trace through a series of T- and S-dualities before we come to a weakly coupled description. In order to do this we will keep track of how the moduli — the dilaton and the various components of the metric — are related at each step.

Recall that in each string theory the natural metric to use is the one that appears in the F-string world-sheet action. The various dualities interchange F-strings with other kinds of string, and the 'string metrics' in the different descriptions differ, as one sees explicitly in the IIB transformation (14.1.6) and the type I–heterotic transformation (14.3.4). After composing a series of dualities, one is interested in how the final dilaton and metric vary as the original dilaton becomes large. We seek to reach a description in which the final dilaton becomes small (or at least stays fixed), and in which the final radii grow (or at least stay fixed). A description in which the dilaton becomes small and also the radii become small is not useful, because the effective coupling in a small-radius theory is increased by the contributions of light winding states. To get an accurate estimate of the coupling one must take the T-dual to a large-radius description.

T: *Heterotic $E_8 \times E_8$ on S_1 to heterotic $SO(32)$ on S_1.* Compactify the heterotic $E_8 \times E_8$ theory on a circle of large radius R_9 and turn on the Wilson line that breaks $E_8 \times E_8$ to $SO(16) \times SO(16)$. We will eventually take $R_9 \to \infty$ to get back to the ten-dimensional theory of interest, and then the Wilson line will be irrelevant. As discussed in section 11.6 this theory is T-dual to the $SO(32)$ heterotic string, again with a Wilson line breaking the group to $SO(16) \times SO(16)$. The couplings and radii are related

$$R_9' \propto R_9^{-1} , \quad g' \propto g R_9^{-1} . \tag{14.5.1}$$

Here primed quantities are for the $SO(32)$ theory and unprimed for the $E_8 \times E_8$ theory. We are only keeping track of the field dependence on each side, $R_9 \propto G_{99}^{1/2}$ and $g \propto e^{\Phi}$. The transformation of g follows by requiring that the two theories give the same answer for scattering of low energy gravitons. The low energy actions are proportional to

$$\frac{1}{g^2} \int d^{10}x = \frac{2\pi R_9}{g^2} \int d^9x \tag{14.5.2}$$

and so $R_9/g^2 = R_9'/g'^2$.

S: *Heterotic $SO(32)$ on S_1 to type I on S_1.* Now use type I–heterotic duality to write this as a type I theory with

$$g_{\mathrm{I}} \propto g'^{-1} \propto g^{-1} R_9 , \quad R_{9\mathrm{I}} \propto g'^{-1/2} R_9' \propto g^{-1/2} R_9^{-1/2} . \tag{14.5.3}$$

The transformation of G_{99} follows from the field redefinition (14.3.4). We are interested in the limit in which g and R_9 are both large. It appears that we can make g_{I} small by an appropriate order of limits. However, the radius of the type I theory is becoming very small and so we must go to the T-dual description as warned above.

T: *Type I on S_1 to type IIA on $S_1/\mathbf{Z}_2$.* Consider a T-duality in the 9-direction of the type I theory. The compact dimension becomes a segment of length $\pi \alpha'/R_{9\mathrm{I}}$ with eight D8-branes at each end, and

$$g_{\mathrm{I}'} \propto g_{\mathrm{I}} R_{9\mathrm{I}}^{-1} \propto g^{-1/2} R_9^{3/2} , \quad R_{9\mathrm{I}'} \propto R_{9\mathrm{I}}^{-1} \propto g^{1/2} R_9^{1/2} . \tag{14.5.4}$$

If we are taking $g \to \infty$ at fixed R_9 then we have reached a good description. However, our real interest is the ten-dimensional theory at fixed large coupling. The coupling $g_{\mathrm{I}'}$ then becomes large, but one final duality brings us to a good description. The theory that we have reached is often called the type I' theory. In the bulk, between the orientifold planes, it is the IIA theory, so we can also think of it as the IIA theory on the segment $S_1/\mathbf{Z}_2$. The coset must be an orientifold because the only spacetime parity symmetry of the IIA theory also includes a world-sheet parity transformation.

S: Type IIA on $S_1/\mathbf{Z}_2$ to M-theory on $S_1 \times S_1/\mathbf{Z}_2$. The IIA theory is becoming strongly coupled, so the physics between the orientifold planes is described in terms of a new periodic dimension. The necessary transformations (12.1.9) were obtained from the dimensional reduction of $d = 11$ supergravity, giving

$$R_{10M} \propto g_{I'}^{2/3} \propto g^{-1/3} R_9 \; , \quad R_{9M} \propto g_{I'}^{-1/3} R_{9I'} \propto g^{2/3} \; . \tag{14.5.5}$$

As the original R_9 is taken to infinity, the new R_{10} diverges linearly. Evidently we should identify the original 9-direction with the final 10-direction. Hence at the last step we also rename $(9, 10) \rightarrow (10', 9')$. The final dual for the strongly coupled $E_8 \times E_8$ theory in ten dimensions is M-theory, with ten noncompact dimensions and the $10'$-direction compactified. This is the same as the strongly coupled IIA theory. The difference is that here the $10'$-direction is not a circle but a segment, with boundaries at the orientifold planes. M-theory on S_1 is the strongly coupled IIA theory. M-theory on $S_1/\mathbf{Z}_2$ is the strongly coupled $E_8 \times E_8$ heterotic theory. At each end are the orientifold plane and eight D8-branes, but now both are nine-dimensional as they bound a ten-dimensional space. The gauge degrees of freedom thus live in these walls, one E_8 in each wall.

The full sequence of dualities is

$$\text{heterotic } E_8 \times E_8 \xrightarrow{T_9} \text{heterotic } SO(32) \xrightarrow{S} \text{type I} \xrightarrow{T_9} \text{type I}' \xrightarrow{S} \text{M-theory} \; . \tag{14.5.6}$$

A heterotic string running in the 8-direction becomes

$$\text{F}_8 \xrightarrow{T_9} \text{F}_8 \xrightarrow{S} \text{D}_8 \xrightarrow{T_9} \text{D}_{89} \xrightarrow{S} \text{M}_{8,10'} \; . \tag{14.5.7}$$

That is, it is a membrane running between the boundaries, as in figure 14.3. This whole picture is highly constrained by anomalies, and this in fact is how it was originally discovered. The $d = 11$ supergravity theory in a space with boundaries has anomalies unless the boundaries carry precisely E_8 degrees of freedom. Note also that

$$\text{p}_9 \xrightarrow{T_9} \text{F}_9 \xrightarrow{S} \text{D}_9 \xrightarrow{T_9} \text{D}_\emptyset \xrightarrow{S} \text{p}_{10} = \text{p}_{9'} \; . \tag{14.5.8}$$

This confirms the identification of the original 9-direction and final 10-direction.

Let us comment on the D8-branes. In string theory the D8-brane is a source for the dilaton. To first order the result is a constant gradient for the dilaton (since the D8-brane has codimension one), but the full nonlinear supergravity equations for the dilaton, metric, and R–R 9-form imply that

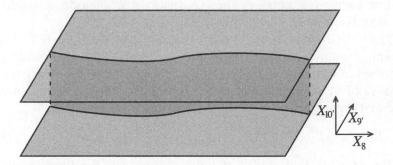

Fig. 14.3.   Strongly coupled limit of $E_8 \times E_8$ heterotic string theory, with one heterotic string shown. The shaded upper and lower faces are boundaries. In the strongly coupled IIA string the upper and lower faces would be periodically identified.

the dilaton diverges a finite distance from the D8-brane. To cure this, one must run into a boundary (orientifold plane) before the divergence: this sets a maximum distance between the D8-brane and the boundary. As one goes to the strongly coupled limit, the initial value for the dilaton is greater and so this distance is shorter. In the strongly coupled limit the D8-branes disappear into the boundary, and in the eleven-dimensional theory there is no way to pull them out. The moduli for their positions just become Wilson lines for the gauge theory in the boundary.

We have now determined the strongly coupled behaviors of all of the ten-dimensional string theories. One can apply the same methods to the compactified theories, and we will do this in detail for toroidal compactifications of the heterotic string in section 19.9. Almost all of that section can be read now; we defer it because to complete the discussion we will need some understanding of strings moving on the smooth manifold K3.

14.6   What is string theory?

What we have learned is shown in figure 14.4. There is a single theory, and all known string theories arise as limits of the parameter space, as does M-theory with 11 noncompact spacetime dimensions. For example, if one starts with the type I theory on T^2, then by varying the two radii, the string coupling, and the Wilson line in one of the compact directions, one can reach the noncompact weakly coupled limit of any of the other string theories, or the noncompact limit of M-theory. Figure 14.4 shows a two-dimensional slice through this four-parameter space. This is only

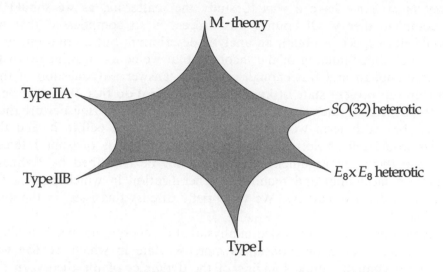

Fig. 14.4. All string theories, and M-theory, as limits of one theory.

one of many branches of the moduli space, and one with a fairly large number of unbroken supersymmetries, 16.

The question is, what is the theory of which all these things are limits? On the one hand we know a lot about it, in that we are able to put together this picture of its moduli space. On the other hand, over most of moduli space, including the M-theory limit, we have only the low energy effective field theory. In the various weakly coupled string limits, we have a description that is presumably valid at all energies but only as an asymptotic expansion in the coupling. This is very far from a complete understanding.

As an example of a question that we do not know how to answer, consider graviton–graviton scattering with center-of-mass energy E. Let us suppose that in moduli space we are near one of the weakly coupled string descriptions, at some small but finite coupling g. The ten-dimensional gravitational constant is of order $G_N \sim g^2 \alpha'^4$. The Schwarzchild radius of the system is of order $R \sim (G_N E)^{1/7}$. One would expect that a black hole would be produced provided that R is large compared to the Compton wavelength E^{-1} and also to the string length $\alpha'^{1/2}$. The latter condition is more stringent, giving

$$E \gtrsim \frac{1}{g^2 \alpha'^{1/2}} . \tag{14.6.1}$$

At this scale, these considerations show that the interactions are strong and string perturbation theory has broken down. Moreover, we do not even *in principle* have a way to study the scattering, as we should in a complete theory. Of course, this process is so complicated that we would not expect to obtain an analytic description, but a criterion for a complete understanding of the theory is that we could in principle, with a large enough and fast enough computer, answer any question of this sort. In our present state of knowledge we cannot do this. We could only instruct the computer to calculate many terms in the string perturbation series, but each term would be larger than the one before it, and the series would tell us nothing. This particular process is of some interest, because there are arguments that it cannot be described by ordinary quantum mechanics and requires a generalization in which pure states evolve to density matrices. We will briefly discuss this issue in the next section.

Even if one is only interested in physics at accessible energies, it is likely that to understand the nonsupersymmetric state in which we live will require a complete understanding of the dynamics of the theory. In the case of quantum field theory, to satisfy Wilson's criterion of 'computability in principle' required an understanding of the renormalization group, and this in turn gave much more conceptual insight into the dynamics of the theory.

One possibility is that each of the string theories (or perhaps, just some of them) can be given a nonperturbative definition in the form of string field theory, so that each would give a good nonperturbative definition. The various dualities would then amount to changes of variables from one theory to another. However, there are various reasons to doubt this. The most prominent is simply that string field theory has not been successful — it has not allowed us to calculate anything we did not already know how to calculate using string perturbation theory. Notably, all the recent progress in understanding nonperturbative physics has taken place without the aid of string field theory, and no connection between the two has emerged. On the contrary, the entire style of argument in the recent developments has been that there are different effective descriptions, each with its own range of validity, and there is no indication that in general any description has a wider range of validity than it should. That is, a given string theory is a valid effective description only near the corresponding cusp of figure 14.4. And if strings are the wrong degrees of freedom for writing down the full Hamiltonian, no bookkeeping device like string field theory will give a satisfactory description. We should also note that even in quantum field theory, where we have a nonperturbative definition, this idea of understanding dualities as changes of variables seems to work only in simple low dimensional examples. Even in field theory the understanding of duality is likely to require new ideas.

However, there must be some exact definition of the theory, in terms of some set of variables, because the graviton scattering question must have an answer in principle. The term M-theory, originally applied to the eleven-dimensional limit, has now come to denote the complete theory.

14.7 Is M for matrix?

A notable feature of the recent progress has been the convergence of many lines of work, as the roles of such constructions as D-branes, string solitons, and $d = 11$ supergravity have been recognized. It is likely that the correct degrees of freedom for M-theory are already known, but their full significance not appreciated. Indeed, one promising proposal is that D-branes, specifically D0-branes, are those degrees of freedom.

According to our current picture, D-branes give a precise description of part of the spectrum, the R–R charged states, but only near the cusps where the type I, IIA, and IIB strings are weakly coupled — elsewhere their relevance comes only from the usual supersymmetric continuation argument. To extend this to a complete description covering the whole parameter space requires some cleverness. The remainder of this section gives a description of this idea, *matrix theory*.

Consider a state in the IIA theory and imagine boosting it to large momentum in the hidden X^{10} direction. Of course 'boosting' is a deceptive term because the compactification of this dimension breaks Lorentz invariance, but at least at large coupling (and so large R_{10}), we should be able to make sense of this. The energy of a particle with n units of compact momentum is

$$E = (p_{10}^2 + q^2 + m^2)^{1/2} \approx p_{10} + \frac{q^2 + m^2}{2p_{10}} = \frac{n}{R_{10}} + \frac{R_{10}}{2n}(q^2 + m^2) . \quad (14.7.1)$$

Here q is the momentum in the other nine spatial dimensions. Recalling the connection between p_{10} and D0-brane charge, this is a state of n D0-branes, and the first term in the action is the D0-brane rest mass. Large boost is large n/R_{10}. In this limit, the second term in the energy is quite small. States that have finite energy in the original frame have

$$E - n/R_{10} = O(R_{10}/n) \qquad (14.7.2)$$

in the boosted frame. There are very few string states with the property (14.7.2). For example, even adding massless closed strings would add an energy q, which does not go to zero with R_{10}/n. Excited open strings connected to the D0-branes also have too large an energy. Thus it seems that we can restrict to ground state open strings attached to the D0-branes.

The Hamiltonian for these was given in eq. (13.5.14), which we now write
in terms of the M-theory parameters M_{11} and R_{10}:

$$H = R_{10}\text{Tr}\left\{\frac{1}{2}p_i p_i - \frac{M_{11}^6}{16\pi^2}[X^i, X^j]^2 - \frac{M_{11}^3}{4\pi}\lambda\Gamma^0\Gamma^i[X^i, \lambda]\right\}. \qquad (14.7.3)$$

We have dropped higher powers of momentum coming from the Born–
Infeld term because all such corrections are suppressed by the boost, just
as the square root in the energy (14.7.1) simplifies. Also, we drop the
additive term n/R_{10} from H.

The Hamiltonian (14.7.3) is conjectured to be the complete description
of systems with $p_{10} = n/R_{10} \gg M_{11}$. Now take R_{10} and n/R_{10} to infinity,
to describe a highly boosted system in eleven noncompact dimensions.
By eleven-dimensional Lorentz invariance, we can put any system in this
frame, so this should be a complete description of the whole of M-theory!
This is the *matrix theory* proposal. We emphasize that this is a conjecture,
not a derivation: we can derive the Hamiltonian (14.7.3) only at weak
string coupling, where we know what the theory is. In effect, we are taking
a specific result derived at the IIA cusp of figure 14.4 and conjecturing
that it is valid over the whole moduli space.

This is a remarkably simple and explicit proposal: the nine $n \times n$ matrices
X_{ab}^i are all one needs. As one check, let us recall the observation from the
previous chapter that only one length scale appears in this Hamiltonian,
$g^{1/3}\alpha'^{1/2}$. This is the minimum distance that can be probed by D0-brane
scattering, and now in light of M-theory we see that this scale has another
interpretation — it is M_{11}^{-1}, the eleven-dimensional Planck length. This is
the fundamental length scale of M-theory, and so the only one that should
appear.

At first sight, the normalization of the Hamiltonian (14.7.3) seems
to involve another parameter, R_{10}. Recall, however, that the system is
boosted and so internal times are dilated. The boost factor is proportional
to p_{10}, so the time-scale should be divided by a dimensionless factor
$p_{10}/M_{11} = n/M_{11}R_{10}$, and again only the scale M_{11} appears.

The description of the eleven-dimensional spacetime in matrix theory
is rather asymmetric: time is the only explicit coordinate, nine spatial
dimensions emerge from matrix functions of time, and the last dimension
is the Fourier transform of n. This asymmetric picture is similar to the
light-cone gauge fixing of a covariant theory.

Now let us discuss some of the physics. As in the discussion of IIA–M-
theory duality, a graviton of momentum $p_{10} = n/R_{10}$ is a bound state of
n D0-branes. Again, the existence of these bound states is necessary for
M-theory to be correct, and has been shown in part. For a bound state of
total momentum q_i, the $SU(n)$ dynamics is responsible for the zero-energy

bound state, and the center-of-mass energy from the $U(1)$ part $p_i = q_i I_n/n$ is

$$E = \frac{R_{10}}{2} \mathrm{Tr}(p_i p_i) = \frac{q^2}{2p_{10}} , \qquad (14.7.4)$$

correctly reproducing the energy (14.7.1) for a massless state.

Now let us consider a simple interaction, graviton–graviton scattering. Let the gravitons have 10-momenta $p_{10} = n_{1,2}/R_{10}$ and be at well-separated positions $Y_{1,2}^i$. The total number of D0-branes is $n_1 + n_2$, and the coordinate matrices X^i are approximately block diagonal. Write X^i as

$$X^i = X_0^i + x^i \qquad (14.7.5a)$$
$$X_0^i = Y_1^i I_1 + Y_2^i I_2 , \quad x^i = x_{11}^i + x_{22}^i + x_{12}^i + x_{21}^i . \qquad (14.7.5b)$$

Here I_1 and I_2 are the identity matrices in the two blocks, which are respectively $n_1 \times n_1$ and $n_2 \times n_2$, and we have separated the fluctuation x^i into a piece in each block plus off-diagonal pieces. First setting the off-diagonal $x_{12,21}^i$ to zero, the blocks decouple because $[x_{11}^i, x_{22}^j] = 0$. The wavefunction is then a product of the corresponding bound state wavefunctions,

$$\psi(x_{11}, x_{22}) = \psi_0(x_{11}) \psi_0(x_{22}) . \qquad (14.7.6)$$

Now consider the off-diagonal block. These degrees of freedom are heavy: the commutator

$$[X_0^i, x_{12}^j] = (Y_1^i - Y_2^i) x_{12}^j \qquad (14.7.7)$$

gives them a mass proportional to the separation of the gravitons. Thus we can integrate them out to obtain the effective interaction between the gravitons.

We would like to use this to test the matrix theory proposal, to see that the effective interaction at long distance agrees with eleven-dimensional supergravity. In fact, we can do this without any further calculation: all the necessary results can be extracted from the cylinder amplitude (13.5.6). At distances small compared to the string scale, the cylinder is dominated by the lightest open strings stretched between the D0-branes, which are precisely the off-diagonal matrix theory degrees of freedom. At distances long compared to the string scale, the cylinder is dominated by the lightest closed string states and so goes over to the supergravity result. This is ten-dimensional supergravity, but it is equivalent to the answer from eleven-dimensional supergravity for the following reason. In the process we are studying, the sizes of the blocks stay fixed at n_1 and n_2, meaning that the values of p_{10} and p'_{10} do not change in the scattering and the p_{10} of the exchanged graviton is zero. This has the effect of averaging over x^{10} and so giving the dimensionally reduced answer.

Finally, we should keep only the leading velocity dependence from the cylinder, because the time dilation from the boost suppresses higher powers as in eq. (14.7.1). The result (13.5.7) for $p = 0$, multiplying by the number of D0-branes in each clump, is

$$L_{eff} = -V(r,v) = 4\pi^{5/2}\Gamma(7/2)\alpha'^3 n_1 n_2 \frac{v^4}{r^7}$$

$$= \frac{15\pi^3}{2} \frac{p_{10}p'_{10}}{M_{11}^9 R_{10}} \frac{v^4}{r^7} . \tag{14.7.8}$$

Because the functional form is the same at large and small r, the matrix theory correctly reproduces the supergravity amplitude (in the matrix theory literature, the standard convention is $M_{11} = (2\pi)^{-1/3} M_{11}$ (here), which removes all 2πs from the matrix theory Hamiltonian).

This is an interesting result, but its significance is not clear. Some higher order extensions do not appear to work, and it may be that one must take the large n limit to obtain agreement with supergravity. The loop expansion parameter in the quantum mechanics is then large, so perturbative calculations are not sufficient. Also, the process being studied here, where the p_{10} of the exchanged graviton vanishes, is quite special. When this is not the case, one has a very different process where the sizes of the blocks change, meaning that D0-branes move from one clump to the other; this appears to be much harder to study.

Matrix theory, if correct, satisfies the 'computability' criterion: we can in principle calculate graviton–graviton scattering numerically at any energy. The analytic understanding of the bound states is still limited, but in principle they could be determined numerically to any desired accuracy and then the wavefunction for the two-graviton state evolved forward in time. Of course any simulation is at finite n and R_{10}, and the matrix theory proposal requires that we take these to infinity; but if the proposal is correct then the limits exist and can be taken numerically. For now all this is just a statement in principle, as various difficulties make the numerical calculation impractical. Most notable among these is the difficulty of preserving to sufficient precision the supersymmetric cancellations that are needed for the theory to make sense — for example, along the flat directions of the potential.

The M-theory membrane

If the matrices X^i are a complete set of degrees of freedom, then it must be possible to identify all the known states of M-theory, in particular the membranes. We might have expected that these would require us to add explicit D2-brane degrees of freedom, but remarkably the membranes are already present as excitations of the D0-brane Hamiltonian.

To see this, define the $n \times n$ matrices

$$U = \begin{bmatrix} 1 & 0 & 0 & 0 \\ 0 & \alpha & 0 & 0 \\ 0 & 0 & \alpha^2 & 0 & \cdots \\ 0 & 0 & 0 & \alpha^3 \\ \vdots & & & & \ddots \end{bmatrix}, \quad V = \begin{bmatrix} 0 & 0 & 0 & & 1 \\ 1 & 0 & 0 & \cdots & 0 \\ 0 & 1 & 0 & & 0 \\ 0 & 0 & 1 & & 0 \\ \vdots & & & \ddots \end{bmatrix}, \quad (14.7.9)$$

where $\alpha = \exp(2\pi i/n)$. These have the properties

$$U^n = V^n = 1, \quad UV = \alpha VU, \quad (14.7.10)$$

and these properties determine U and V up to change of basis. The matrices $U^r V^s$ for $1 \leq r,s \leq n$ form a complete set, and so any matrix can be expanded in terms of them. For example,

$$X^i = \sum_{r,s=[1-n/2]}^{[n/2]} X^i_{rs} U^r V^s, \quad (14.7.11)$$

with [ ] denoting the integer part and similarly for the fermion λ. To each matrix we can then associate a periodic function of two variables,

$$X^i \to X^i(p,q) = \sum_{r,s=[1-n/2]}^{[n/2]} X^i_{rs} \exp(ipr + iqs). \quad (14.7.12)$$

If we focus on matrices which remain smooth functions of p and q as n becomes large (so that the typical r and s remain finite), then the commutator maps

$$[X^i, X^j] \to \frac{2\pi i}{n}(\partial_q X^i \partial_p X^j - \partial_p X^i \partial_q X^j) + O(n^{-2})$$

$$\equiv \frac{2\pi i}{n}\{X^i, X^j\}_{\text{PB}} + O(n^{-2}). \quad (14.7.13)$$

One can verify this by considering simple monomials $U^r V^s$. Notice the analogy to taking the classical limit of a quantum system, with the Poisson bracket appearing. One can also rewrite the trace as an integral,

$$\text{Tr} = n \int \frac{dq\,dp}{(2\pi)^2}. \quad (14.7.14)$$

The Hamiltonian becomes

$$R_{10} \int dq\,dp \left(\frac{n}{8\pi^2}\Pi_i\Pi_i + \frac{M^6_{11}}{16\pi^2 n}\{X^i, X^j\}^2_{\text{PB}} - i\frac{M^3_{11}}{8\pi^2}\lambda\Gamma^0\Gamma^i\{X^i, \lambda\}_{\text{PB}} \right). \quad (14.7.15)$$

Since $X^i(p,q)$ is a function of two variables, this Hamiltonian evidently describes the quantum mechanics of a membrane. In fact, it is identical to the Hamiltonian one gets from an eleven-dimensional supersymmetric membrane action in the light-cone gauge. We do not have space to develop this in detail, but as an example consider the static configuration

$$X^1 = aq , \quad X^2 = bp ; \tag{14.7.16}$$

since q and p are periodic we must also suppose $X^{1,2}$ to be as well. Then the energy is

$$\frac{M_{11}^6 R_{10} a^2 b^2}{2n} = \frac{M_{11}^6 A^2}{2(2\pi)^4 p_{10}} = \frac{\tau_{M2}^2 A^2}{2p_{10}} . \tag{14.7.17}$$

Here $A = 4\pi^2 ab$ is the area of the membrane. The product $\tau_{M2} A$ is the mass of an M-theory membrane of this area, so this agrees with the energy (14.7.1).

There was at one time an effort to define eleven-dimensional super-gravity as a theory of fundamental membranes; this was one of the roots of the name M-theory. This had many difficulties, the most immediate being that the world-volume theory is nonrenormalizable. However, it was noted that the light-cone Hamiltonian (14.7.15) was the large-n limit of dimensionally reduced $d = 10$, $N = 1$ gauge theory (14.7.3), so the finite-n theory could be thought of as regularizing the membrane. Matrix theory puts this idea in a new context. One of the difficulties of the original interpretation was that the potential has flat directions, for example

$$X^i = Y_1^i I_1 + Y_2^i I_2 \tag{14.7.18}$$

as in eq. (14.7.5). This implies a continuous spectrum, which is physically unsatisfactory given the original interpretation of the Hamiltonian as aris-ing from gauge-fixing the action for a single membrane. However, we now interpret the configuration (14.7.18) as a two-particle state. The continu-ous spectrum is not a problem because the matrix theory is supposed to describe states with arbitrary numbers of particles. We should emphasize that in focusing on matrices that map to smooth functions of p and q we have picked out just a piece of the matrix theory spectrum, namely states of a single membrane of toroidal topology. Other topologies, other branes, and graviton states are elsewhere.

Since matrix theory is supposed to be a complete formulation of M-theory, it must in particular reproduce all of string theory. It is surprising that it can do this starting with just nine matrices, but we now see how it is possible — it contains membranes, and strings are just wrapped membranes. The point is that one can hide a great deal in a large matrix! If we compactify one of the nine X^i dimensions, the membranes wrapped in this direction reproduce string theory; arguments have been given that the string interactions are correctly incorporated.

Finite n and compactification

In arguing for matrix theory we took n to be large. Let us also ask, does the finite-n matrix theory Hamiltonian have any physical relevance? In fact it describes M-theory compactified in a *lightlike* direction,

$$(x^0, x^{10}) \cong (x^0 - \pi \tilde{R}_{10}, x^{10} + \pi \tilde{R}_{10}) . \qquad (14.7.19)$$

To see this — in fact, to define it — let us reach this theory as the limit of spacelike compactification,

$$(x^0, x^{10}) \cong (x^0 - \pi \tilde{R}_{10}, x^{10} + \pi \tilde{R}_{10} + 2\pi \epsilon^2 \tilde{R}_{10}) \qquad (14.7.20)$$

with $\epsilon \to 0$. The invariant length of the compact dimension is $2\pi\epsilon\tilde{R}_{10} + O(\epsilon^2)$, so this is Lorentz-equivalent to the spacelike compactification

$$(x'^0, x'^{10}) \cong (x'^0, x'^{10} + 2\pi\epsilon\tilde{R}_{10}) , \qquad (14.7.21)$$

where

$$x'^0 \pm x'^{10} = \epsilon^{\mp 1}(x^0 \pm x^{10}) . \qquad (14.7.22)$$

Unlike the $n \to \infty$ conjecture, the finite-n conjecture can actually be derived from things that we already know. Because the invariant radius (14.7.21) is going to zero, we are in the regime of weakly coupled IIA string theory. Moreover, states that have finite energy in the original frame acquire

$$E', p'_{10} \propto O(\epsilon^{-1}) , \quad E' - p'_{10} \propto O(\epsilon) \qquad (14.7.23)$$

under the boost (14.7.22). These are the only states that we are to retain. However, we have already carried out this exercise at the beginning of this section: this is eq. (14.7.2) where

$$R_{10} = \epsilon \tilde{R}_{10} . \qquad (14.7.24)$$

The derivation of the matrix theory Hamiltonian then goes through just as before, and it is surely correct because we are in weakly coupled string theory. The lightlike theory is often described as the *discrete light-cone quantization (DLCQ)* of M-theory, meaning light-cone quantization with a discrete spectrum of p_-. This idea has been developed in field theory, but one must be careful because the definition there is generally not equivalent to the lightlike limit.

Of course, the physics in a spacetime with lightlike compactification may be rather exotic, so this result does not directly enable us to understand the eleven-dimensional theory which is supposed to be recovered in the large-n limit. However, it has been very valuable in understanding how the matrix theory conjecture is to be extended to the case that some of the additional dimensions are compactified. Let us consider, for example, the case that k dimensions are periodic. Working in the frame (14.7.21), we are instructed to take R_{10} to zero holding fixed M_{11} and all momenta and distances in the transverse directions (those other than x^{10}). We then keep only states whose energy is $O(R_{10})$ above the BPS minimum. These clearly include the gravitons (and their superpartners) with nonzero p_{10}, which are just the D0-branes. In addition, let us consider M2-branes that are wrapped around the 10-direction and around one of the transverse directions x^m. From the IIA point of view these are F-strings winding in the x^m-direction. They have mass equal to $\tau_{M2}A = M_{11}^3 R_m R_{10}$ and so are candidates to survive in the limit. However, for M2-branes with vanishing p_{10}, $E = (q^2 + m^2)^{1/2}$ and we also need that they have zero momentum in the noncompact directions — this is a point of measure zero. The only membrane states that survive are M2-branes with nonzero p_{10}, which are F-strings that end on D0-branes in the IIA description. These F-strings must be in their ground states, but they can wind any number of times around the transverse compact directions. The lightlike limit now has many more degrees of freedom than in the noncompact case, because there is an additional winding quantum number for each compact dimension. In fact, it is simpler to use the T-dual description, where the D0-branes become Dk-branes and the winding number becomes momentum: the lightlike limit of matrix theory then includes the full $(k + 1)$-dimensional $U(n)$ Yang–Mills theory on the branes.

It is notable that the number of degrees of freedom goes up drastically with compactification of each additional dimension, as the dimension of the effective gauge theory increases. A difficulty is that for $k > 3$ the gauge theory on the brane is nonrenormalizable. However, for $k > 3$ our discussion of the lightlike limit is incomplete. In the first place, we have not considered all the degrees of freedom. For $k \geq 4$, an M5-brane that wraps the x^{10}-direction and four of the transverse directions also survives the limit. Moreover, the coupling of the T-dual string theory,

$$\frac{R_{10}}{\alpha'^{1/2}} \prod_m \frac{\alpha'^{1/2}}{R_m} = R_{10}^{(3-k)/2}(M_{11}^3/2\pi)^{(1-k)/2} \prod_m R_m^{-1} , \qquad (14.7.25)$$

diverges as $R_{10} \to 0$ for $k \geq 4$. The lightlike limit is then no longer a weakly coupled string theory, and it is necessary to perform further dualities. The various cases $k \geq 4$ are quite interesting, but we must leave the details to the references.

In summary, the various compactifications of matrix theory suggest a deep relation between large-n gauge theory and string theory. Such a relation has arisen from various other points of view, and may lead to a better understanding of gauge theories as well as string theory.

14.8 Black hole quantum mechanics

In the early 1970s it was found that classical black holes obey laws directly analogous to the laws of thermodynamics. This analogy was made sharper by Hawking's discovery that black holes radiate as black bodies at the corresponding temperature. Under this analogy, the entropy of a black hole is equal to the area of its event horizon divided by $\kappa^2/2\pi$. The analogy is so sharp that it has long been a goal to find a statistical mechanical theory associated with this thermodynamics, and in particular to associate the entropy with the density of states of the black hole. Many arguments have been put forward in this direction but until recently there was no example where the states of a black hole could be counted in a controlled way.

This has now been done for some string theory black holes. To see the idea, let us return to the relation between a D-brane and an R–R charged black p-brane. The thermodynamic and other issues are the same for black p-branes as for black holes. The explicit solution for a black p-brane with Q units of R–R charge is (for $p \leq 6$)

$$ds^2 = Z(r)^{-1/2}\eta_{\mu\nu}dx^\mu dx^\nu + Z(r)^{1/2}dx^m dx^m \, , \qquad (14.8.1a)$$

$$e^{2\Phi} = Z(r)^{(3-p)/2} \, . \qquad (14.8.1b)$$

Here x^μ is tangent to the p-brane, x^m is transverse, and

$$Z(r) = 1 + \frac{\rho^{7-p}}{r^{7-p}} \, , \quad r^2 = x^m x^m \, , \qquad (14.8.2a)$$

$$\rho^{7-p} = \alpha'^{(7-p)/2}gQ(4\pi)^{(5-p)/2}\Gamma\left(\frac{7-p}{2}\right) \, . \qquad (14.8.2b)$$

The numerical constant, which is not relevant to the immediate discussion, is obtained in exercise 14.6. The characteristic length ρ is shorter than the string scale when gQ is less than 1. In this case, the effective low energy field theory that we have used to derive the solution (14.8.1) is not valid. When gQ is greater than 1 the geometry is smooth on the string scale and the low energy field theory should be a good description.

Consider now string perturbation theory in the presence of Q coincident D-branes. The expansion parameter is gQ: each additional world-sheet boundary brings a factor of the string coupling g but also a factor of Q from the sum over Chan–Paton factors. When gQ is small, string perturbation theory is good, but when it is large it breaks down. Thus the situation appears to be very much as with the instanton in section 13.6: in one range of parameters the low energy field theory description is good, and in another range the D-brane description is good.

In the instanton case we can continue from one regime to the other by varying the instanton scale factor. In the black p-brane case we can do the same by varying the string coupling, as we have often done in the analysis of strongly coupled strings. We can use this to count supersymmetric (BPS) black hole states: we can do the counting at small gQ, where the weakly coupled D-brane description is good, and continue to large gQ, where the black hole description is accurate.

The particular solution (14.8.1) is not useful for a test of the black hole entropy formula because the event horizon, at $r = 0$, is singular. It can be made nonsingular by adding energy to give a nonsupersymmetric black hole, but in the supersymmetric (extremal) limit the area goes to zero. To obtain a supersymmetric black p-brane with a smooth horizon of nonzero area requires at least three nonzero charges. A simple example combines Q_1 D1-branes in the 5-direction with Q_5 D5-branes in the (5,6,7,8,9)-directions. To make the energy finite the (6,7,8,9)-directions are compactified on a T^4 of volume V_4. We also take the 5-direction to be finite, but it is useful to keep its length L large. The third charge is momentum p_5. The solution is

$$ds^2 = Z_1^{-1/2} Z_5^{-1/2} \left[\eta_{\mu\nu} dx^\mu dx^\nu + (Z_n - 1)(dt + dx_5)^2 \right]$$
$$+ Z_1^{1/2} Z_5^{1/2} dx^i dx^i + Z_1^{1/2} Z_5^{-1/2} dx^m dx^m , \quad (14.8.3a)$$

$$e^{-2\Phi} = Z_5/Z_1 . \quad (14.8.3b)$$

Here μ, ν run over the (0,5)-directions tangent to all the branes, i runs over the (1,2,3,4)-directions transverse to all branes, and m runs over the (6,7,8,9)-directions tangent to the D5-branes and transverse to the D1-branes. We have defined

$$Z_1 = 1 + \frac{r_1^2}{r^2} , \quad r_1^2 = \frac{(2\pi)^4 g Q_1 \alpha'^3}{V_4} , \quad (14.8.4a)$$

$$Z_5 = 1 + \frac{r_5^2}{r^2} , \quad r_5^2 = g Q_5 \alpha' , \quad (14.8.4b)$$

$$Z_n = 1 + \frac{r_n^2}{r^2} , \quad r_n^2 = \frac{(2\pi)^5 g^2 p_5 \alpha'^4}{L V_4} , \quad (14.8.4c)$$

with $r^2 = x^i x^i$. The event horizon is at $r = 0$; the interior of the black hole is not included in this coordinate system. The integers Q_1, Q_5, and $n_5 = p_5 L / 2\pi$ are all taken to be large so that this describes a classical black hole, with horizon much larger than the Planck scale.

The solution (14.8.3) is in terms of the string metric. The black hole area law applies to the Einstein metric $G_{E\mu\nu} = e^{-\Phi/2} G_{\mu\nu}$, whose action is field-independent. This is

$$ds_E^2 = Z_1^{-3/4} Z_5^{-1/4} \left[\eta_{\mu\nu} dx^\mu dx^\nu + (Z_n - 1)(dt + dx_5)^2 \right]$$
$$+ Z_1^{1/4} Z_5^{3/4} dx^i dx^i + Z_1^{1/4} Z_5^{-1/4} dx^m dx^m . \quad (14.8.5)$$

Now let us determine the horizon area. The eight-dimensional horizon is a 3-sphere in the transverse dimensions and is extended in the (5,6,7,8,9)-directions. Near the origin the angular metric is

$$\left(\frac{r_1^2}{r^2}\right)^{1/4}\left(\frac{r_5^2}{r^2}\right)^{3/4} r^2 d\Omega^2 = r_1^{1/2} r_5^{3/2} d\Omega^2 , \qquad (14.8.6)$$

with total area $2\pi^2 r_1^{3/4} r_5^{9/4}$. From $G_{55} = Z_1^{-3/4} Z_5^{-1/4} Z_n$ it follows that the invariant length of the horizon in the 5-direction is $r_1^{-3/4} r_5^{-1/4} r_n L$. Similarly the invariant volume in the toroidal directions is $r_1 r_5^{-1} V_4$. The area is the product

$$A = 2\pi^2 L V_4 r_1 r_5 r_n = 2^6 \pi^7 g^2 \alpha'^4 (Q_1 Q_5 n_5)^{1/2} = \kappa^2 (Q_1 Q_5 n_5)^{1/2} . \qquad (14.8.7)$$

This gives for the black hole entropy

$$S = \frac{2\pi A}{\kappa^2} = 2\pi (Q_1 Q_5 n_5)^{1/2} . \qquad (14.8.8)$$

The final result is quite simple, depending only on the integer charges and not on any of the moduli g, L, or V_4. This is a reflection of the classical black hole area law: under adiabatic changes in the moduli the horizon area cannot change.

Now let us consider the same black hole in the regime where the D-brane picture is valid. The dynamics of the $\#_{\mathrm{ND}} = 4$ system was discussed in chapter 13, and in particular the potential is

$$V = \frac{1}{(2\pi\alpha')^2}|X_i\chi - \chi Y_i|^2 + \frac{g_1^2}{4}D_1^A D_1^A + \frac{g_5^2}{4V_4}D_5^A D_5^A . \qquad (14.8.9)$$

This is generalized from the earlier (13.6.25) because there are multiple D1-branes and D5-branes. Thus in the first term the $Q_1 \times Q_1$ D1-brane collective coordinates X_i act on the left of the $Q_1 \times Q_5$ matrix χ, and the $Q_5 \times Q_5$ D5-brane collective coordinates Y_i act on the right. The black hole is a bound state of D1- and D5-branes, so the χ are nonzero. The first term in the potential then requires that

$$X^i = x^i I_{Q_1} , \qquad Y^i = x^i I_{Q_5} , \qquad (14.8.10)$$

and the center-of-mass x^i is the only light degree of freedom in the transverse directions. Also, the 1-1 X_m and 5-5 A_m are now charged under the $U(Q_1)$ and $U(Q_5)$ and so contribute to the D-terms in the general form (B.7.3).[4] What is important is the dimension of the moduli space, which can be determined by counting. The X_m contribute $4Q_1^2$ real scalars,

---

[4] The A^4 term is just a rewriting of the $[A_m, A_n]$ term from the dimensional reduction, and similarly for the X^4.

the A_m contribute $4Q_5^2$, and the χ contribute $4Q_1Q_5$. The vanishing of the D-terms imposes $3Q_1^2 + 3Q_5^2$ conditions; since the Qs are large we do not worry about the $U(1)$ parts, which are only $1/Q^2$ of the total. Also, the $U(Q_1)$ and $U(Q_5)$ gauge equivalences remove another $Q_1^2 + Q_5^2$ moduli, leaving $4Q_1Q_5$. This is a generalization of the counting that we did for the instanton in section 13.6.

These moduli are functions of x^5 and x^0. We are treating L as very large, but the counting extends to small L with some subtlety. So we have a two-dimensional field theory with $4Q_1Q_5$ real scalars and by supersymmetry $4Q_1Q_5$ Majorana fermions, and we need its density of states. This is a standard calculation, which in fact we have already done. For a CFT of central charge c, the general relation (7.2.30) between the central charge and the density of states gives

$$\text{Tr}[\exp(-\beta H)] \approx \exp(\pi c L/12\beta) . \tag{14.8.11}$$

We have effectively set $\tilde{c} = 0$ because only the left-movers are excited in the supersymmetric states. The earlier result was for a string of length 2π, so we have replaced $H \to LH/2\pi$ by dimensional analysis. The density of states is related to this by

$$\int_0^\infty dE \, n(E) \exp(-\beta E) = \text{Tr}[\exp(-\beta H)] , \tag{14.8.12}$$

giving in saddle point approximation

$$n(E) \approx \exp\left[(\pi c E L/3)^{1/2}\right] . \tag{14.8.13}$$

Finally, the central charge for our system is $c = 6Q_1Q_5$, while $E = 2\pi n_5/L$, and so

$$n(E) \approx \exp\left[2\pi(Q_1Q_5n_5)^{1/2}\right] , \tag{14.8.14}$$

in precise agreement with the exponential of the black hole entropy. This is a remarkable result, and another indication, beyond perturbative finiteness, that string theory defines a sensible theory of quantum gravity.

This result has been extended to other supersymmetric black holes, to the entropy of almost supersymmetric black holes, and to decay and absorption rates of almost supersymmetric black holes. In these cases the agreement is somewhat surprising, not obviously a consequence of supersymmetry. Subsequently the 'string' picture of the black holes has been extended to circumstances such as M-theory where there is no D-brane interpretation. These results are suggestive but the interpretation is not clear. We will discuss highly nonsupersymmetric black holes below.

Recalling the idea that D-branes can probe distances below the string scale, one might wonder whether the black p-brane metrics (14.8.1) and (14.8.3) can be seen even in the regime $gQ < 1$ in which the D-brane picture rather than low energy field theory is relevant. Indeed, in some cases they can; this is developed in exercise 14.7.

The metric simplifies very close to $r = 0$, where the terms 1 in Z_1, Z_5, and Z_n become negligible. Taking for simplicity the case $r_n = 0$, the metric becomes

$$ds^2 = \frac{r^2}{r_1 r_5} \eta_{\mu\nu} dx^\mu dx^\nu + \frac{r_1 r_5}{r^2} dr^2 + r_1 r_5 d\Omega^2 + \frac{r_1}{r_5} dx^m dx^m . \qquad (14.8.15)$$

This is a product space

$$AdS_3 \times S^3 \times T^4 . \qquad (14.8.16)$$

Here AdS_3 is three-dimensional anti-de-Sitter space, which is the geometry in the coordinates x^μ and r (to be precise, these coordinates cover only part of anti-de-Sitter space). In a similar way, the metric (14.8.1) near a black 3-brane is

$$AdS_5 \times S^5 . \qquad (14.8.17)$$

The case $p = 3$ is special because the dilaton remains finite at the horizon $r = 0$, as it does for the D1–D5 metric (14.8.3).

Very recently, a very powerful new duality proposal has emerged. Consider the IR dynamics of a system of N coincident Dp-branes. The bulk closed strings should decouple from the dynamics on the branes because gravity is an irrelevant interaction. The brane dynamics will then be described by the supersymmetric Yang–Mills theory on the brane, even for gN large. On the other hand, when gN is large the description of the system in terms of low energy supergravity should be valid as we have discussed. Thus we have two different descriptions which appear to have an overlapping range of validity. In the Yang–Mills description the effective expansion parameter gN is large, so perturbation theory is not valid. However, for g fixed, N is also very large. Noting that the gauge group on the branes is $U(N)$, this is the limit of a large number of 'colors,' the large-N limit. Field theories simplify in this limit, but it has been a long-standing unsolved problem to obtain any analytic understanding of Yang–Mills theories in this way. Now it appears, at least for theories with enough supersymmetry, that one can calculate amplitudes in the gauge theory by using the dual picture, where at low energy supergravity is essentially classical. If this idea is correct, it is a tremendous advance in the understanding of gauge field theories.

A correspondence principle

To make a precise entropy calculation we had to consider an extremal black hole with a specific set of charges. What of the familiar neutral

Schwarzschild black hole? Here too one can make a quantitative statement, but not at the level of precision of the supersymmetric case.

For a four-dimensional Schwarzschild black hole of mass M, the radius and entropy are

$$R \approx G_N M \,, \qquad\qquad (14.8.18a)$$

$$S_{bh} \approx \frac{R^2}{G_N} \approx G_N M^2 \,. \qquad\qquad (14.8.18b)$$

In this section we will systematically ignore numerical constants like 2 and π, for a reason to be explained below; hence the $\approx$. Let us consider what happens as we adiabatically change the dimensionless string coupling g. In four dimensions, dimensional analysis gives

$$G_N \approx g^2 \alpha' \,. \qquad\qquad (14.8.19)$$

As we vary g the dimensionless combination $G_N M^2$ stays fixed. The simplest way to see this is to appeal to the fact that the black hole entropy (14.8.18) has the same properties as the thermodynamic entropy, and so is invariant under adiabatic changes.

Now imagine making the coupling very weak. One might imagine that for sufficiently weak coupling the black hole will no longer be black. One can see where this should happen from the following argument. The preceding two equations imply that

$$\frac{R}{\alpha'^{1/2}} \approx g S_{bh}^{1/2} \,. \qquad\qquad (14.8.20)$$

We are imagining that S_{bh} is large so that the thermodynamic picture is good. Until g is very small, the Schwarzschild radius is then large compared to the string length and the gravitational dynamics should not be affected by stringy physics.

However, when g becomes small enough that $g S_{bh}^{1/2}$ is of order 1, stringy corrections to the action become important. If we try to extrapolate past this point, the black hole becomes smaller than a string! It is then unlikely that the field theory description of the black hole continues to be valid. Rather, the system should look like a state in weakly coupled string theory. This is how we can make the comparison: at this point, if the black hole entropy has a statistical interpretation, then the weakly coupled string theory should have the appropriate number of states of the given mass to account for this entropy. Since the entropy is assumed to be large we are interested in highly excited states. For a *single* highly excited string of mass M the density of states can be found as in section 9.8 and exercise 11.12,

$$\exp\left\{ \pi M [(c + \tilde{c})\alpha'/3]^{1/2} \right\} \,. \qquad\qquad (14.8.21)$$

In fact, one can show that with this exponential growth in their number, the single string states are a significant fraction of the total number of states of given energy. In particular, and in contrast to the R–R case, states with D-branes plus anti-D-branes would have a much lower entropy because of the energy locked in the D-brane rest mass. The entropy of weakly coupled string states is then the logarithm

$$S_s \approx M\alpha'^{1/2} \approx g^{-1}MG_N^{1/2} \,. \tag{14.8.22}$$

This entropy has a different parametric dependence than the black hole entropy (14.8.18). However, they are to be compared only at the point

$$gS_{bh}^{1/2} \approx 1 \,, \tag{14.8.23}$$

where the transition from one picture to the other occurs. Inserting this value for g, the string entropy (14.8.22) becomes $S_{bh}^{1/2}MG_N^{1/2} \approx S_{bh}$.

We see that the numerical coefficients cannot be determined in this approach, since we do not know the exact coupling where the transition occurs, and corrections are in any case becoming significant on each side. However, *a priori* the entropy could have failed to match by a power of the large dimensionless number in the problem, S_{bh}. One can show that the same agreement holds in any dimension (exercise 14.8) and for black holes with a variety of charges. This is further evidence for the statistical interpretation of the black hole entropy, and that string theory has the appropriate number of states to be a complete theory of quantum gravity.

The information paradox

A closely associated issue is the black hole information paradox. A black hole of given mass and charge can be formed in a very large number of ways. It will then evaporate, and the Hawking radiation is apparently independent of what went into the black hole. This is inconsistent with ordinary quantum mechanics, as it requires pure states to evolve into mixed states.

There are various schools of thought here. The proposal of Hawking is that this is just the way things are: the laws of quantum mechanics need to be changed. There is also strong skepticism about this view, partly because this modification of quantum mechanics is rather ugly and very possibly inconsistent. However, 20 years of investigation have only served to sharpen the paradox. The principal alternative, that the initial state is encoded in subtle correlations in the Hawking radiation, sounds plausible but in fact is even more radical.[5] The problem is that Hawking radiation

---

[5] A third major alternative is that the evaporation ends in a remnant, a Planck-mass object having an enormous number of internal states. This might be stable or might release its information over an exceedingly long time scale. This has its own problems of aesthetics and possibly consistency, and is generally regarded as less likely.

emerges from the region of the horizon, where the geometry is smooth and so ordinary low energy field theory should be valid. One can follow the Hawking radiation and see correlations develop between the fields inside and outside the black hole; the superposition principle then forbids the necessary correlations to exist strictly among the fields outside. To evade this requires that the locality principle in quantum field theory break down in some long-ranged but subtle way.

The recent progress in string duality suggests that black holes do obey the ordinary rules of quantum mechanics. The multiplets include black holes along with various nonsingular states, and we have continuously deformed a black hole into a system that obeys ordinary quantum mechanics. However, this is certainly not conclusive — we have two descriptions with different ranges of validity, and while the D-brane system has an explicit quantum mechanical description, one could imagine that as the coupling constant is increased a critical coupling is reached where the D-particles collapse to form a black hole. At this coupling there could be a discontinuous change (or a smooth crossover) from ordinary quantum behavior to information loss.

Certainly if matrix theory is correct, the ordinary laws of quantum mechanics are preserved and the information must escape (there are not enough states for the remnant idea). It should be noted that in matrix theory only locality in time is explicit, so the necessary nonlocality may be present. If so, it is important to see in detail how this happens. In particular it may give insight into the cosmological constant problem, which stands in the way of our understanding the vacuum and supersymmetry breaking. This is another place where the continued failure of mundane ideas suggests that we need something new and perhaps nonlocal.

Exercises

14.1 From the supersymmetry algebra (13.2.9), show that an infinite type II F-string with excitations moving in only one direction is a BPS state. Show the same for a D-string.

14.2 Using the multi-NS5-brane solution (14.1.15), (14.1.17) and the D-string action, calculate the mass of a D-string stretched between two NS5-branes. Using IIB S-duality, compare this with the mass of an F-string stretched between D5-branes.

14.3 For compactification of the type II string on T^4, where the U-duality group is $SO(5,5,\mathbf{Z})$ and the T-duality group is $SO(4,4,\mathbf{Z})$, repeat the discussion in section 14.2 of the representations carried by the vector fields.

14.4 (a) For the series of operations TST discussed in section 14.4, deduce the transformation of each gauge field and higher rank form.
(b) Deduce the transformation of each extended object (D-brane, F-string, or NS-brane, with the various possible orientations for each).
(c) In each case compare with the interpretation as a 90° rotation of M-theory compactified on T^2.

14.5 As discussed in section 14.6, consider the type I theory compactified on T^2. In terms of the two radii, the string coupling, and the Wilson line, determine the six limits of parameter space that give the six noncompact theories at the cusps of figure 14.4, with the coupling going to zero in the stringy limits.

14.6 Expand the black p-brane solution (14.8.1) to first order in gQ and compare with the field produced by a Dp-brane, calculated in the linearized low energy field theory as in section 8.7.

14.7 Consider a D1-brane aligned along the 1-direction. Evaluate the D1-brane action in the field (14.8.3) and expand to order v^2. For $n_5 = 0$, compare with the order v^2 interaction between a D1-brane and a collection of D1- and D5-branes as obtained from the annulus.

14.8 Extend the correspondence principle to Schwarzschild black holes in other dimensions. The necessary black hole properties can be obtained by dimensional analysis. The entropy is always equal to the horizon area (with units l^{d-2}) divided by G_N up to a numerical constant.

15
Advanced CFT

We have encountered a number of infinite-dimensional symmetry algebras on the world-sheet: conformal, superconformal, and current. While we have used these symmetries as needed to obtain specific physical results, in the present chapter we would like to take maximum advantage of them in determining the form of the world-sheet theory. An obvious goal, not yet reached, would be to construct the general conformal or superconformal field theory, corresponding to the general classical string background.

This subject is no longer as central as it once appeared to be, as spacetime rather than world-sheet symmetries have been the principal tools in recent times. However, it is a subject of some beauty in its own right, with various applications to string compactification and also to other areas of physics.

We first discuss the representations of the conformal algebra, and the constraints imposed by conformal invariance on correlation functions. We then study some examples, such as the minimal models, Sugawara and coset theories, where the symmetries do in fact determine the theory completely. We briefly summarize the representation theory of the $N = 1$ superconformal algebra. We then discuss a framework, rational conformal field theory, which incorporates all these CFTs. To conclude this chapter we present some important results about the relation between conformal field theories and nearby two-dimensional field theories that are *not* conformally invariant, and the application of CFT in statistical mechanics.

15.1 Representations of the Virasoro algebra

In section 3.7 we discussed the connection between classical string backgrounds and general CFTs. In particular, we observed that CFTs corresponding to compactification of the spatial dimensions are unitary and their spectra are discrete and bounded below. These additional conditions strongly restrict the world-sheet theory, and we will assume them throughout this chapter except for occasional asides.

Because the spectrum is bounded below, acting repeatedly with Virasoro lowering operators always produces a highest weight (primary) state $|h\rangle$, with properties

$$L_0|h\rangle = h|h\rangle , \qquad (15.1.1a)$$

$$L_m|h\rangle = 0, \quad m > 0 . \qquad (15.1.1b)$$

Starting from a highest weight state, we can form a representation of the Virasoro algebra

$$[L_m, L_n] = (m-n)L_{m+n} + \frac{c}{12}(m^3 - m)\delta_{m,-n} \qquad (15.1.2)$$

by taking $|h\rangle$ together with all the states obtained by acting on $|h\rangle$ with the Virasoro raising operators,

$$L_{-k_1}L_{-k_2}\ldots L_{-k_l}|h\rangle . \qquad (15.1.3)$$

We will denote this state $|h, \{k\}\rangle$ or $L_{-\{k\}}|h\rangle$ for short. The state (15.1.3) is known as a *descendant* of $|h\rangle$, or a *secondary*. A primary together with all of its descendants is also known as a *conformal family*. The integers $\{k\}$ may be put in the standard order $k_1 \geq k_2 \geq \ldots \geq k_l \geq 1$ by commuting the generators. This process terminates in a finite number of steps, because each nonzero commutator reduces the number of generators by one. To see that this is a representation, consider acting on $|h, \{k\}\rangle$ with any Virasoro generator L_n. For $n < 0$, commute L_n into its standard order; for $n \geq 0$, commute it to the right until it annihilates $|h\rangle$. In either case, the nonzero commutators are again of the form $|h, \{k'\}\rangle$. All coefficients are determined entirely in terms of the central charge c from the algebra and the weight h obtained when L_0 acts on $|h\rangle$; these two parameters completely define the highest weight representation.

It is a useful fact that for unitary CFTs all states lie in highest weight representations — not only can we always get from any state to a primary with lowering operators, but we can always get back again with raising operators. Suppose there were a state $|\phi\rangle$ that could not be expanded in terms of primaries and secondaries. Consider the lowest-dimension state with this property. By taking

$$|\phi\rangle \rightarrow |\phi\rangle - |i\rangle\langle i|\phi\rangle \qquad (15.1.4)$$

with $|i\rangle$ running over a complete orthonormal set of primaries and secondaries, we may assume $|\phi\rangle$ to be orthogonal to all primaries and secondaries. Now, $|\phi\rangle$ is not primary, so there is a nonzero state $L_n|\phi\rangle$ for some $n > 0$. Since the CFT is unitary this has a positive norm,

$$\langle \phi | L_{-n} L_n | \phi \rangle > 0 \ . \tag{15.1.5}$$

The state $L_n | \phi \rangle$ lies in a highest weight representation, since by assumption $| \phi \rangle$ is the lowest state that does not, and so therefore does $L_{-n} L_n | \phi \rangle$. Therefore it must be orthogonal to $| \phi \rangle$, in contradiction to eq. (15.1.5).

This need not hold in more general circumstances. Consider the operator ∂X of the linear dilaton theory. Lowering this gives the unit operator, $L_1 \cdot \partial X = -\alpha' V \cdot 1$, but $L_{-1} \cdot 1 = 0$ so we cannot raise this operator back to ∂X. The problem is the noncompactness of X combined with the position dependence of the dilaton, so that even $| 1 \rangle$ is not normalizable.

Now we would like to know what values of c and h are allowed in a unitary theory. The basic method was employed in section 2.9, using the Virasoro algebra to compute the inner product

$$\mathscr{M}^1 \equiv \langle h | L_1 L_{-1} | h \rangle = 2h \ , \tag{15.1.6}$$

implying $h \geq 0$. Consideration of another inner product gave $c \geq 0$. Now look more systematically, level by level. At the second level of the highest weight representation, the two states $L_{-1} L_{-1} | h \rangle$ and $L_{-2} | h \rangle$ have the matrix of inner products

$$\mathscr{M}^2 = \left[\begin{array}{c} \langle h | L_1^2 \\ \langle h | L_2 \end{array} \right] \left[\begin{array}{cc} L_{-1}^2 | h \rangle & L_{-2} | h \rangle \end{array} \right] \ . \tag{15.1.7}$$

Commuting the lowering operators to the right gives

$$\mathscr{M}^2 = \left(\begin{array}{cc} 8h^2 + 4h & 6h \\ 6h & 4h + c/2 \end{array} \right) \tag{15.1.8}$$

and

$$\det(\mathscr{M}^2) = 32h(h - h_+)(h - h_-) \ , \tag{15.1.9a}$$

$$16h_\pm = (5 - c) \pm [(1 - c)(25 - c)]^{1/2} \ . \tag{15.1.9b}$$

In a unitary theory the matrix of inner products, and in particular its determinant, cannot be negative. The determinant is nonnegative in the region $c \geq 1$, $h \geq 0$, but for $0 < c < 1$ a new region $h_- < h < h_+$ is excluded.

At level N, the matrix of inner products is

$$\mathscr{M}^N_{\{k\},\{k'\}}(c, h) = \langle h, \{k\} | h, \{k'\} \rangle \ , \qquad \sum_i k_i = N \ . \tag{15.1.10}$$

Its determinant has been found,

$$\det[\mathscr{M}^N(c, h)] = K_N \prod_{1 \leq rs \leq N} (h - h_{r,s})^{P(N - rs)} \tag{15.1.11}$$

with K_N a positive constant. This is the *Kac determinant*. The zeros of the determinant are at

$$h_{r,s} = \frac{c-1}{24} + \frac{1}{4}(r\alpha_+ + s\alpha_-)^2 , \qquad (15.1.12)$$

where

$$\alpha_\pm = (24)^{-1/2}\left[(1-c)^{1/2} \pm (25-c)^{1/2}\right] . \qquad (15.1.13)$$

The multiplicity $P(N - rs)$ of each root is the partition of $N - rs$, the number of ways that $N - rs$ can be written as a sum of positive integers (with $P(0) \equiv 1$):

$$\prod_{n=1}^{\infty} \frac{1}{1-q^n} = \sum_{k=0}^{\infty} P(k)q^k . \qquad (15.1.14)$$

At level 2, for example, the roots are $h_{1,1} = 0$, $h_{2,1} = h_+$, and $h_{1,2} = h_-$, each with multiplicity 1, as found above.

The calculation of the determinant (15.1.11) is too lengthy to repeat here. The basic strategy is to construct all of the null states, those corresponding to the zeros of the determinant, either by direct combinatoric means or using some tricks from CFT. The determinant is a polynomial in h and so is completely determined by its zeros, up to a normalization which can be obtained by looking at the $h \to \infty$ limit. The order of the polynomial is readily determined from the Virasoro algebra, so one can know when one has all the null states. Let us note one particular feature. At level 2, the null state corresponding to $h_{1,1}$ is $L_{-1}L_{-1}|h = 0\rangle$. This is a descendant of the level 1 null state $L_{-1}|h = 0\rangle$. In general, the zero $h_{r,s}$ appears first at level rs. At every higher level N are further null states obtained by acting with raising operators on the level rs state; the partition $P(N - rs)$ in the Kac determinant is the total number of ways to act with raising operators of total level $N - rs$.

A careful study of the determinant and its functional dependence on c and h shows (the analysis is again too lengthy to repeat here) that unitary representations are allowed only in the region $c \geq 1$, $h \geq 0$ and at a discrete set of points (c, h) in the region $0 \leq c < 1$:

$$c = 1 - \frac{6}{m(m+1)} , \qquad m = 2, 3, \dots ,$$

$$= 0, \frac{1}{2}, \frac{7}{10}, \frac{4}{5}, \frac{6}{7}, \dots , \qquad (15.1.15a)$$

$$h = h_{r,s} = \frac{[r(m+1) - sm]^2 - 1}{4m(m+1)} , \qquad (15.1.15b)$$

where $1 \leq r \leq m - 1$ and $1 \leq s \leq m$. The discrete representations are of great interest, and we will return to them in section 15.3.

For a unitary representation, the Kac determinant also determines whether the states $|h, \{k\}\rangle$ are linearly independent. If it is positive, they are; if it vanishes, some linear combination(s) are orthogonal to all states and so by unitarity must vanish. The representation is then said to be *degenerate*. Of the unitary representations, all the discrete representations (15.1.15) are degenerate, as are the representations $c = 1$, $h = \frac{1}{4}n^2$, $n \in \mathbf{Z}$ and $c > 1$, $h = 0$. For example, at $h = 0$ we always have $L_{-1} \cdot 1 = 0$, but at the next level $L_{-2} \cdot 1 = T_{zz}$ is nonzero.

Let us make a few remarks about the nonunitary case. In the full matter CFT of string theory, the states $L_{-\{k\}}|h\rangle$ obtained from any primary state $|h\rangle$ *are* linearly independent when the momentum is nonzero. This can be seen by using the same light-cone decomposition used in the no-ghost proof of chapter 4. The term in L_{-n} of greatest N^{lc} is $k^-\alpha^+_{-n}$. These manifestly generate independent states; the upper triangular structure then guarantees that this independence holds also for the full Virasoro generators.

A representation of the Virasoro algebra with all of the $L_{-\{k\}}|h\rangle$ linearly independent is known as a *Verma module*. Verma modules exist at all values of c and h. Verma modules are particularly interesting when the dimension h takes a value $h_{r,s}$ such that the Kac determinant vanishes. The module then contains nonvanishing null states (states that are orthogonal to all states in the module). Acting on a null state with a Virasoro generator gives a null state again, since for any null $|v\rangle$ and for any state $|\psi\rangle$ in the module we have $\langle\psi|(L_n|v\rangle) = ((\langle\psi|L_n)|v\rangle) = 0$. The representation is thus reducible: the subspace of null states is left invariant by the Virasoro algebra.[1] The L_n for $n > 0$ must therefore annihilate the *lowest* null state, so this state is in fact primary, in addition to being a level rs descendant of the original primary state $|h_{r,s}\rangle$. That is, the $h_{r,s}$ Verma module contains an $h = h_{r,s} + rs$ Verma submodule. In some cases, including the special discrete values of c (15.1.15), there is an intricate pattern of nested submodules.

Clearly a Verma module can be unitary only at those values of c and h where nondegenerate unitary representations are allowed. At the (c, h) values with degenerate unitary representations, the unitary representation is obtained from the corresponding Verma module by modding out the null states.

As a final example consider the matter sector of string theory, $c = 26$. From the OCQ, we know that there are many null physical states at $h = 1$. This can be seen from the Kac formula as well. For $c = 26$, $\alpha_+ = 3i/6^{1/2}$,

---

[1] By contrast, a unitary representation is always irreducible. The reader can show that the lowest state in any invariant subspace would have to be orthogonal to itself, and therefore vanish.

$\alpha_- = 2i/6^{1/2}$, and so

$$h_{r,s} = \frac{25 - (3r + 2s)^2}{24} . \tag{15.1.16}$$

The corresponding null physical state is at

$$h = h_{r,s} + rs = \frac{25 - (3r - 2s)^2}{24} . \tag{15.1.17}$$

Any pair of positive integers with $|3r - 2s| = 1$ leads to a null physical state at $h = 1$. For example, the states $(r, s) = (1, 1)$ and $(1, 2)$ were constructed in exercise 4.2. With care, one can show that the number of null states implied by the Kac formula is exactly that required by the no-ghost theorem.

15.2 The conformal bootstrap

We now study the constraints imposed by conformal invariance on correlation functions on the sphere. In chapter 6 we saw that the Möbius subgroup, with three complex parameters, reduced the n-point function to a function of $n-3$ complex variables. The rest of the conformal symmetry gives further information: it determines all the correlation functions of descendant fields in terms of those of the primary fields.

To begin, consider the correlation function of the energy momentum tensor $T(z)$ with n primary fields $\mathcal{O}$. The singularities of the correlation function as a function of z are known from the $T\mathcal{O}$ OPE. In addition, it must fall as z^{-4} for $z \to \infty$, since in the coordinate patch $u = 1/z$, $T_{uu} = z^4 T_{zz}$ is holomorphic at $u = 0$. This determines the correlation function to be

$$\langle T(z)\mathcal{O}_1(z_1)\dots\mathcal{O}_n(z_n)\rangle_{S_2}$$
$$= \sum_{i=1}^{n}\left[\frac{h_i}{(z - z_i)^2} + \frac{1}{(z - z_i)}\frac{\partial}{\partial z_i}\right]\langle\mathcal{O}_1(z_1)\dots\mathcal{O}_n(z_n)\rangle_{S_2} . \tag{15.2.1}$$

A possible holomorphic addition is forbidden by the boundary condition at infinity. In addition, the asymptotics of order z^{-1}, z^{-2}, and z^{-3} must vanish; these are the same as the conditions from Möbius invariance, developed in section 6.7. The correlation function with several Ts is of the same form, with additional singularities from the TT OPE. Now make a Laurent expansion in $z - z_1$,

$$T(z)\mathcal{O}_1(z_1) = \sum_{k=-\infty}^{\infty} (z - z_1)^{k-2}L_{-k}\cdot\mathcal{O}_1(z_1) . \tag{15.2.2}$$

Then for $k \geq 1$, matching coefficients of $(z - z_1)^{k-2}$ on the right and left of the correlator (expectation value) (15.2.1) gives

$$\langle \, [L_{-k} \cdot \mathcal{O}_1(z_1)] \, \mathcal{O}_2(z_2) \dots \mathcal{O}_n(z_n) \, \rangle_{S_2} = \mathscr{L}_{-k} \langle \, \mathcal{O}_1(z_1) \dots \mathcal{O}_n(z_n) \, \rangle_{S_2} \, ,$$
(15.2.3)

where

$$\mathscr{L}_{-k} = \sum_{i=2}^{n} \left[\frac{h_i(k-1)}{(z_i - z_1)^k} - \frac{1}{(z_i - z_1)^{k-1}} \frac{\partial}{\partial z_i} \right] .$$
(15.2.4)

This extends to multiple generators, and to the antiholomorphic side,

$$\left\langle \, [L_{-k_1} \dots L_{-k_\ell} \tilde{L}_{-l_1} \dots \tilde{L}_{-l_m} \cdot \mathcal{O}_1(z_1)] \dots \mathcal{O}_n(z_n) \, \right\rangle_{S_2}$$
$$= \mathscr{L}_{-k_\ell} \dots \mathscr{L}_{-k_1} \tilde{\mathscr{L}}_{-l_m} \dots \tilde{\mathscr{L}}_{-l_1} \langle \, \mathcal{O}_1(z_1) \dots \mathcal{O}_n(z_n) \, \rangle_{S_2} \, .$$
(15.2.5)

The additional terms from the TT OPE do not contribute when all the k_i and l_i are positive. The correlator of one descendant and $n-1$ primaries is thus expressed in terms of that of n primaries. Clearly this can be extended to n descendants, though the result is more complicated because there are additional terms from the TT singularities.

Earlier we argued that the operator product coefficients were the basic data in CFT, determining all the other correlations via factorization. We see now that it is only the operator product coefficients of primaries that are necessary. It is worth developing this somewhat further for the four-point correlation. Start with the operator product of two primaries, with the sum over operators now broken up into a sum over conformal families i and a sum within each family,

$$\mathcal{O}_m(z, \bar{z}) \mathcal{O}_n(0,0) = \sum_{i,\{k,\tilde{k}\}} z^{-h_m - h_n + h_i + N} \bar{z}^{-\tilde{h}_m - \tilde{h}_n + \tilde{h}_i + \tilde{N}}$$
$$\times c^{i\{k,\tilde{k}\}}{}_{mn} L_{-\{k\}} \tilde{L}_{-\{\tilde{k}\}} \cdot \mathcal{O}_i(0,0) \, ,$$
(15.2.6)

where N is the total level of $\{k\}$. Writing the operator product coefficient $c^{i\{k,\tilde{k}\}}{}_{mn}$ as a three-point correlator and using the result (15.2.5) to relate this to the correlator of three primaries gives

$$c^{i\{k,\tilde{k}\}}{}_{mn} = \sum_{\{k',\tilde{k}'\}} \mathcal{M}^{-1}_{\{k\},\{k'\}} \mathcal{M}^{-1}_{\{\tilde{k}\},\{\tilde{k}'\}}$$
$$\times \mathscr{L}_{-\{k'\}} \tilde{\mathscr{L}}_{-\{\tilde{k}'\}} \langle \, \mathcal{O}_m(\infty, \infty) \mathcal{O}_n(1,1) \mathcal{O}_i(z_1, \bar{z}_1) \, \rangle_{S_2} \Big|_{z_1=0} \, .$$
(15.2.7)

To relate the operator product coefficient to a correlator we have to raise an index, so the inverse $\mathcal{M}^{-1}$ appears (with an appropriate adjustment for degenerate representations). The right-hand side is equal to the operator product of the primaries times a function of the coordinates and their derivatives, the latter being completely determined by the conformal invariance. Carrying out the differentiations in $\mathscr{L}_{-\{k'\}}$ and $\tilde{\mathscr{L}}_{-\{\tilde{k}'\}}$ and then summing leaves

$$c^{i\{k,\tilde{k}\}}{}_{mn} = \beta^{i\{k\}}_{mn} \tilde{\beta}^{i\{\tilde{k}\}}_{mn} c^i_{mn} . \tag{15.2.8}$$

The coefficient $\beta^{i\{k\}}_{mn}$ is a function of the weights h_m, h_n, and h_i and the central charge c, but is otherwise independent of the CFT.

Now use the OPE (15.2.6) to relate the four-point correlation to the product of three-point correlations,

$$\langle \mathcal{O}_j(\infty,\infty)\mathcal{O}_l(1,1)\mathcal{O}_m(z,\bar{z})\mathcal{O}_n(0,0)\rangle_{S_2} = \sum_i c^i_{jl}c_{imn}\mathscr{F}^{jl}_{mn}(i|z)\tilde{\mathscr{F}}^{jl}_{mn}(i|\bar{z}) , \tag{15.2.9}$$

where

$$\mathscr{F}^{jl}_{mn}(i|z) = \sum_{\{k\},\{k'\}} z^{-h_m-h_n+h_i+N} \beta^{i\{k\}}_{jl} \mathcal{M}_{\{k\},\{k'\}} \beta^{i\{k'\}}_{mn} . \tag{15.2.10}$$

This function is known as the conformal block, and is holomorphic except at $z = 0$, 1, and ∞. The steps leading to the decomposition (15.2.9) show that the conformal block is determined by the conformal invariance as a function of h_j, h_l, h_m, h_n, h_i, c, and z. One can calculate it order by order in z by working through the definition.

Recall that the single condition for a set of operator product coefficients to define a consistent CFT on the sphere is duality of the four-point function, the equality of the decompositions (15.2.9) in the $(jl)(mn)$, $(jm)(ln)$, and $(jn)(lm)$ channels. The program of solving this constraint is known as the *conformal bootstrap*. The general solution is not known. One limitation is that the conformal blocks are not known in closed form except for special values of c and h.

Beyond the sphere, there are the additional constraints of modular invariance of the zero-point and one-point functions on the torus. Here we will discuss only a few of the most general consequences. Separating the sum over states in the partition function into a sum over conformal families and a sum within each family yields

$$Z(\tau) = \sum_{i,\{k,\tilde{k}\}} q^{-c/24+h_i+N} \bar{q}^{-\tilde{c}/24+\tilde{h}_i+\tilde{N}}$$

$$= \sum_i \chi_{c,h_i}(q)\chi_{\tilde{c},\tilde{h}_i}(\bar{q}) . \tag{15.2.11}$$

Here

$$\chi_{c,h}(q) = q^{-c/24+h} \sum_{\{k\}} q^N \tag{15.2.12}$$

is the *character* of the (c, h) representation of the Virasoro algebra. For a Verma module the states generated by the L_{-k} are in one-to-one correspondence with the excitations of a free boson, generated by α_{-k}. Thus,

$$\chi_{c,h}(q) = q^{-c/24+h} \prod_{n=1}^{\infty} \frac{1}{1-q^n} \qquad (15.2.13)$$

for a nondegenerate representation. For degenerate representations it is necessary to correct this expression for overcounting. A generic degenerate representation would have only one null primary, say at level N; the representation obtained by modding out the resulting null Verma module would then have character $(1-q^N)q^{1/24}\eta(q)^{-1}$. For the unitary degenerate representations (15.1.15), with their nested submodules, the calculation of the character is more complicated.

In section 7.2 we found the asymptotic behavior of the partition function for a general CFT,

$$Z(i\ell) \overset{\ell \to 0}{\sim} \exp(\pi c/6\ell) \,, \qquad (15.2.14)$$

letting $c = \tilde{c}$. For a single conformal family, letting $q = \exp(-2\pi\ell)$,

$$\chi_{c,h}(q) \leq q^{h+(1-c)/24}\eta(i\ell)^{-1} \overset{\ell \to 0}{\sim} \ell^{1/2}\exp(\pi/12\ell) \,. \qquad (15.2.15)$$

Then for a general CFT

$$Z(i\ell) \leq \mathcal{N}\ell \exp(\pi/6\ell) \qquad (15.2.16)$$

as $\ell \to 0$, with $\mathcal{N}$ the total number of primary fields in the sum (15.2.11). Comparing this bound with the known asymptotic behavior (15.2.14), $\mathcal{N}$ can be finite only if $c < 1$. So, while we have been able to use conformal invariance to reduce sums over states to sums over primaries only, this remains an infinite sum whenever $c \geq 1$. The $c < 1$ theories, to be considered in the next section, stand out as particularly simple.

15.3 Minimal models

For fields in degenerate representations, conformal invariance imposes additional strong constraints on the correlation functions. Throughout this section we take $c \leq 1$, because only in this range do degenerate representations of positive h exist. We will not initially assume the CFT to be unitary, but the special unitary values of c will eventually appear.

Consider, as an example, a primary field $\mathcal{O}_{1,2}$ with weight

$$h = h_{1,2} = \frac{c-1}{24} + \frac{(\alpha_+ + 2\alpha_-)^2}{4} \,. \qquad (15.3.1)$$

For now we leave the right-moving weight $\tilde{h}$ unspecified. The vanishing descendant is

$$\mathcal{N}_{1,2} = \left[L_{-2} - \frac{3}{2(2h_{1,2}+1)} L_{-1}^2 \right] \cdot \mathcal{O}_{1,2} = 0 \,. \qquad (15.3.2)$$

Inserting this into a correlation with other primary fields and using the relation (15.2.5) expressing correlations of descendants in terms of those of primaries gives a partial differential equation for the correlations of the degenerate primary,

$$0 = \left\langle \mathcal{N}_{1,2}(z_1) \prod_{i=2}^{n} \mathcal{O}_i(z_i) \right\rangle_{S_2}$$

$$= \left[\mathcal{L}_{-2} - \frac{3}{2(2h_{1,2}+1)} \mathcal{L}_{-1}^2 \right] \mathcal{A}_n$$

$$= \left[\sum_{i=2}^{n} \frac{h_i}{(z_i - z_1)^2} - \sum_{i=2}^{n} \frac{1}{z_i - z_1} \frac{\partial}{\partial z_i} - \frac{3}{2(2h_{1,2}+1)} \frac{\partial^2}{\partial z_1^2} \right] \mathcal{A}_n \,, \qquad (15.3.3)$$

where

$$\mathcal{A}_n = \left\langle \mathcal{O}_{1,2}(z_1, \bar{z}_1) \prod_{i=2}^{n} \mathcal{O}_i(z_i, \bar{z}_i) \right\rangle_{S_2} \,. \qquad (15.3.4)$$

For $n = 4$, the correlation is known from conformal invariance up to a function of a single complex variable, and eq. (15.3.3) becomes an ordinary differential equation. In particular, setting to zero the z^{-1}, z^{-2}, and z^{-3} terms in the $T(z)$ expectation value (15.2.1) allows one to solve for $\partial/\partial z_{2,3,4}$ in terms of $\partial/\partial z_1$, with the result that eq. (15.3.3) becomes

$$\left[\sum_{i=2}^{4} \frac{h_i}{(z_i - z_1)^2} - \sum_{2 \leq i < j \leq 4} \frac{h_{1,2} - h_2 - h_3 - h_4 + 2(h_i + h_j)}{(z_i - z_1)(z_j - z_1)} \right.$$

$$\left. + \sum_{i=2}^{4} \frac{1}{z_i - z_1} \frac{\partial}{\partial z_1} - \frac{3}{2(2h_{1,2}+1)} \frac{\partial^2}{\partial z_1^2} \right] \mathcal{A}_4 = 0 \,. \qquad (15.3.5)$$

This differential equation is of hypergeometric form. The hypergeometric functions, however, are holomorphic (except for branch cuts at coincident points), while $\mathcal{O}_{1,2}$ has an unknown $\bar{z}_1$ dependence. Now insert the expansion (15.2.9), in which the four-point correlation is written as a sum of terms, each a holomorphic conformal block times a conjugated block. The conformal blocks satisfy the same differential equation (15.3.5) and so are hypergeometric functions. Being second order, the differential equation (15.3.5) has two independent solutions, and each conformal block is a linear combination of these.

This procedure generalizes to any degenerate primary field. The primary $\mathcal{O}_{r,s}$ will satisfy a generalization of the hypergeometric equation. This differential equation is of maximum order rs, coming from the term L_{-1}^{rs} in the null state $\mathcal{N}_{r,s}$. If the antiholomorphic weight is also degenerate, $\tilde{h} = h_{\tilde{r},\tilde{s}}$, the antiholomorphic conformal blocks satisfy a differential equation of order $\tilde{r}\tilde{s}$, so that

$$\left\langle \mathcal{O}_{r,s;\tilde{r},\tilde{s}}(z_1,\bar{z}_1) \prod_{i=2}^{4} \mathcal{O}_i(z_i,\bar{z}_i) \right\rangle_{S_2} = \sum_{i=1}^{rs} \sum_{j=1}^{\tilde{r}\tilde{s}} a_{ij} f_i(z) \tilde{f}_j(\bar{z}) , \qquad (15.3.6)$$

where $f_i(z)$ and $\tilde{f}_j(\bar{z})$ are the general solutions of the holomorphic and antiholomorphic equations. The constants a_{ij} are not determined by the differential equation. They are constrained by locality — the holomorphic and antiholomorphic functions each have branch cuts, but the product must be single-valued — and by associativity. We will describe below some theories in which it has been possible to solve these conditions.

Let us see how the differential equation constrains the operator products of $\mathcal{O}_{1,2}$. According to the theory of ordinary differential equations, the points $z_1 = z_i$ are regular singular points, so that the solutions are of the form $(z_1 - z_i)^{\kappa}$ times a holomorphic function. Inserting this form into the differential equation and examining the most singular term yields the characteristic equation

$$\frac{3}{2(2h_{1,2}+1)}\kappa(\kappa-1) + \kappa - h_i = 0 . \qquad (15.3.7)$$

This gives two solutions $(z_1 - z_i)^{\kappa_{\pm}}$ for the leading behavior as $z_1 \to z_i$; comparing this to the OPE gives

$$h_{\pm} = h_{1,2} + h_i + \kappa_{\pm} \qquad (15.3.8)$$

for the primary fields in the $\mathcal{O}_{1,2}\mathcal{O}_i$ product. Parameterizing the weight by

$$h_i = \frac{c-1}{24} + \frac{\gamma^2}{4} , \qquad (15.3.9)$$

the two solutions to the characteristic equation correspond to

$$h_{\pm} = \frac{c-1}{24} + \frac{(\gamma \pm \alpha_-)^2}{4} . \qquad (15.3.10)$$

These are the only weights that can appear in the operator product, so we have derived the *fusion rule*,

$$\mathcal{O}_{1,2}\mathcal{O}_{(\gamma)} = [\mathcal{O}_{(\gamma+\alpha_-)}] + [\mathcal{O}_{(\gamma-\alpha_-)}] ; \qquad (15.3.11)$$

we have labeled the primary fields other than $\mathcal{O}_{1,2}$ by the corresponding parameter γ. A fusion rule is an OPE without the coefficients, a list of the conformal families that are allowed to appear in a given operator product (though it is possible that some will in fact have vanishing coefficient).

For the operator $\mathcal{O}_{2,1}$, one obtains in the same way the fusion rule

$$\mathcal{O}_{2,1}\mathcal{O}_{(\gamma)} = [\mathcal{O}_{(\gamma+\alpha_+)}] + [\mathcal{O}_{(\gamma-\alpha_+)}] \ . \tag{15.3.12}$$

In particular, for the product of two degenerate primaries this becomes

$$\mathcal{O}_{1,2}\mathcal{O}_{r,s} = [\mathcal{O}_{r,s+1}] + [\mathcal{O}_{r,s-1}] \ , \tag{15.3.13a}$$

$$\mathcal{O}_{2,1}\mathcal{O}_{r,s} = [\mathcal{O}_{r+1,s}] + [\mathcal{O}_{r-1,s}] \ . \tag{15.3.13b}$$

For positive values of the indices, the families on the right-hand side are degenerate. In fact, only these actually appear. Consider the fusion rule for $\mathcal{O}_{1,2}\mathcal{O}_{2,1}$. By applying the rule (15.3.13a) we conclude that only $[\mathcal{O}_{2,2}]$ and $[\mathcal{O}_{2,0}]$ may appear in the product, while the rule (15.3.13b) allows only $[\mathcal{O}_{2,2}]$ and $[\mathcal{O}_{0,2}]$. Together, these imply that only $[\mathcal{O}_{2,2}]$ can actually appear in the product. This generalizes: only primaries $r \geq 1$ and $s \geq 1$ are generated. The algebra of degenerate conformal families thus closes, and iterated products of $\mathcal{O}_{1,2}$ and $\mathcal{O}_{2,1}$ generate all degenerate $\mathcal{O}_{r,s}$. This suggests that we focus on CFTs in which *all* conformal families are degenerate.

The values of r and s are still unbounded above, so that the operator algebra will generate an infinite set of conformal families. When $\alpha_-/\alpha_+ = -p/q$ is *rational*, the algebra closes on a finite set.[2] In particular, one then has

$$c = 1 - 6\frac{(p-q)^2}{pq} \ , \tag{15.3.14a}$$

$$h_{r,s} = \frac{(rq - sp)^2 - (p-q)^2}{4pq} \ . \tag{15.3.14b}$$

The point is that there is a reflection symmetry,

$$h_{p-r,q-s} = h_{r,s} \ , \tag{15.3.15}$$

so that each conformal family has at least *two* null vectors, at levels rs and $(p-r)(q-s)$, and its correlators satisfy *two* differential equations. The reflection of the conditions $r > 0$ and $s > 0$ is $r < p$ and $s < q$, so the operators are restricted to the range

$$1 \leq r \leq p - 1 \ , \quad 1 \leq s \leq q - 1 \ . \tag{15.3.16}$$

These theories, with a finite algebra of degenerate conformal families, are known as *minimal models*. They have been solved: the general solution of the locality, duality, and modular invariance conditions is known, and the operator product coefficients can be extracted though the details are too lengthy to present here.

---

[2] Note that $\alpha_+\alpha_- = -1$, and that $0 > \alpha_-/\alpha_+ > -1$.

Although the minimal models seem rather special, they have received
a great deal of attention, as examples of nontrivial CFTs, as prototypes
for more general solutions of the conformal bootstrap, as building blocks
for four-dimensional string theories, and because they describe the critical
behavior of many two-dimensional systems. We will return to several of
these points later.

Let us now consider the question of unitarity. A necessary condition
for unitarity is that all weights are nonnegative. One can show that this
is true of the weights (15.3.14) only for $q = p + 1$. These are precisely the
$c < 1$ representations (15.1.15) already singled out by unitarity:

$$p = m , \quad q = m + 1 . \tag{15.3.17}$$

Notice that these theories have been found and solved purely from symmetry, without ever giving a Lagrangian description. This is how they
were discovered, though various Lagrangian descriptions are now known;
we will mention several later. For $m = 3$, c is $\frac{1}{2}$ and there is an obvious
Lagrangian representation, the free fermion. The allowed primaries,

$$h_{1,1} = 0 , \quad h_{2,1} = \frac{1}{2} , \quad h_{1,2} = \frac{1}{16} , \tag{15.3.18}$$

are already familiar, being respectively the unit operator, the fermion ψ,
and the R sector ground state.

The full minimal model fusion rules can be derived using repeated
applications of the $\mathcal{O}_{2,1}$ and $\mathcal{O}_{1,2}$ rules and associativity. They are

$$\mathcal{O}_{r_1,s_1} \mathcal{O}_{r_2,s_2} = \sum [\mathcal{O}_{r,s}] , \tag{15.3.19a}$$

$$r = |r_1 - r_2| + 1, \ |r_1 - r_2| + 3, \dots ,$$

$$\min(r_1 + r_2 - 1, 2p - 1 - r_1 - r_2) , \tag{15.3.19b}$$

$$s = |s_1 - s_2| + 1, \ s_1 + s_2 + 3, \dots ,$$

$$\min(s_1 + s_2 - 1, 2q - 1 - s_1 - s_2) . \tag{15.3.19c}$$

For $\mathcal{O}_{p-1,1}$ only a single term appears in the fusion with any other field,
$\mathcal{O}_{p-1,1} \mathcal{O}_{r,s} = [\mathcal{O}_{p-r,s}]$. A primary with these properties is known as a *simple
current*. Simple currents have the useful property that they have definite
monodromy with respect to any other primary. Consider the operator
product of a simple current $J(z)$ of weight h with any primary,

$$J(z)\mathcal{O}_i(0) = z^{h_{i'} - h_i - h}[\mathcal{O}_{i'}(0) + \text{descendants}] , \tag{15.3.20}$$

where $J \cdot [\mathcal{O}_i] = [\mathcal{O}_{i'}]$. The terms with descendants bring in only integer
powers of z, so all terms on the right pick up a common phase

$$2\pi(h_{i'} - h_i - h) = 2\pi Q_i \tag{15.3.21}$$

when z encircles the origin. The charge Q_i, defined mod 1, is a discrete
symmetry of the OPE. Using the associativity of the OPE, the operator

product coefficient c^k_{ij} can be nonzero only if $Q_i + Q_j = Q_k$. Also, by taking repeated operator products of J with itself one must eventually reach the unit operator; suppose this occurs first for J^N. Then associativity implies that NQ_i must be an integer, so this is a $\mathbf{Z}_N$ symmetry. For the minimal models,

$$\mathcal{O}_{p-1,1}\mathcal{O}_{p-1,1} = [\mathcal{O}_{1,1}] \tag{15.3.22}$$

which is the identity, and so the discrete symmetry is $\mathbf{Z}_2$. Evaluating the weights (15.3.21) gives

$$Q_{r,s} = \frac{p(1-s) + q(1-r)}{2} \mod 1 . \tag{15.3.23}$$

For the unitary case (15.3.17), $\exp(2\pi i Q_{r,s})$ is $(-1)^{s-1}$ for m odd and $(-1)^{r-1}$ for m even.

Feigin–Fuchs representation

To close this section, we describe a clever use of CFT to generate integral representations of the solutions to the differential equations satisfied by the degenerate fields. Define

$$c = 1 - 24\alpha_0^2 \tag{15.3.24}$$

and consider the linear dilaton theory with the same value of the central charge,

$$T = -\frac{1}{2}\partial\phi\partial\phi + 2^{1/2}i\alpha_0\partial^2\phi . \tag{15.3.25}$$

The linear dilaton theory is not the same as a minimal model. In particular, the modes α_{-k} generate a Fock space of independent states, so the partition function is of order $\exp(\pi/6\ell)$ as $\ell \to 0$, larger than that of a minimal model. However, the correlators of the minimal model can be obtained from those of the linear dilaton theory. The vertex operator

$$V_\alpha = \exp(2^{1/2}i\alpha\phi) \tag{15.3.26}$$

has weight $\alpha^2 - 2\alpha\alpha_0$, so for

$$\alpha = \alpha_0 - \frac{\gamma}{2} \tag{15.3.27}$$

it is a primary of weight

$$\frac{c-1}{24} + \frac{\gamma^2}{4} . \tag{15.3.28}$$

For $\gamma = r\alpha_+ + s\alpha_-$ it is then degenerate, and its correlator satisfies the same differential equation as the corresponding minimal model primary.

There is a complication: the correlator

$$\langle V_{\alpha_1} V_{\alpha_2} V_{\alpha_3} V_{\alpha_4} \rangle \tag{15.3.29}$$

generally vanishes due to the conservation law

$$\sum_i \alpha_i = 2\alpha_0 \tag{15.3.30}$$

(derived in exercise 6.2). There is a trick which enables us to find a nonvanishing correlator that satisfies the same differential equation. The operators

$$J_{\pm} = \exp(2^{1/2} i \alpha_{\pm} \phi) \tag{15.3.31}$$

are of weight $(1,0)$, so the line integral

$$Q_{\pm} = \oint dz \, J_{\pm} \, , \tag{15.3.32}$$

known as a *screening charge*, is conformally invariant. Inserting $Q_+^{n_+} Q_-^{n_-}$ into the expectation value, the charge conservation condition is satisfied for

$$n_+ = \frac{1}{2} \sum_i r_i - 2 \, , \quad n_- = \frac{1}{2} \sum_i s_i - 2 \, . \tag{15.3.33}$$

Further, since the screening charges are conformally invariant, they do not introduce singularities into $T(z)$ and the derivation of the differential equation still holds. Thus, the minimal model conformal blocks are represented as contour integrals of the correlators of free-field exponentials, which are of course known. This is the *Feigin–Fuchs representation*. It is possible to replace $V_\alpha \to V_{2\alpha_0 - \alpha}$ in some of the vertex operators, since this has the same weight; one still obtains integer values of $n_\pm$, but this may reduce the number of screening charges needed. It may seem curious that the charges of the $(1,0)$ vertex operators are just such as to allow for integer $n_\pm$. In fact, one can work backwards, deriving the Kac determinant from the linear dilaton theory with screening charges.

The contours in the screening operators have not been specified — they may be any nontrivial closed contours (but must end on the same Riemann sheet where they began, because there are branch cuts in the integrand), or they may begin and end on vertex operators if the integrand vanishes sufficiently rapidly at those points. By various choices of contour one generates all solutions to the differential equations, as in the theory of hypergeometric functions. As noted before, one must impose associativity and locality to determine the actual correlation functions. The Feigin–Fuchs representation has been a useful tool in solving these conditions.

15.4 Current algebras

We now consider a Virasoro algebra L_k combined with a current algebra j_k^a. We saw in section 11.5 that the Virasoro generators are actually constructed from the currents. We will extend that discussion to make fuller use of the world-sheet symmetry.

Recall that a primary state $|r, i\rangle$ in representation r of g satisfies

$$L_m|r, i\rangle = j_m^a|r, i\rangle, \quad m > 0, \qquad (15.4.1a)$$

$$j_0^a|r, i\rangle = |r, j\rangle t_{r,ji}^a. \qquad (15.4.1b)$$

As in the case of the Virasoro algebra, we are interested in highest weight representations, obtained by acting on a primary state with the L_m and j_m^a for $m < 0$. As we have discussed, a CFT with a current algebra can always be factored into a Sugawara part and a part that commutes with the current algebra. We focus on the Sugawara part, where

$$T(z) = \frac{1}{(k + h(g))\psi^2} :jj(z): . \qquad (15.4.2)$$

Recall also that the central charge is

$$c^{g,k} = \frac{k \dim(g)}{k + h(g)} \qquad (15.4.3)$$

and that the weight of a primary state is

$$h_r = \frac{Q_r}{(k + h(g))\psi^2}. \qquad (15.4.4)$$

As in the Virasoro case, all correlations can be reduced to those of the primary fields. In parallel to the derivation of eq. (15.2.3), one finds

$$\langle (j_{-m}^a \cdot \mathcal{O}_1(z_1)) \mathcal{O}_2(z_2) \dots \mathcal{O}_n(z_n) \rangle_{S_2} = \mathcal{J}_{-m}^a \langle \mathcal{O}_1(z_1) \dots \mathcal{O}_n(z_n) \rangle_{S_2}, \quad (15.4.5)$$

where

$$\mathcal{J}_{-m}^a = -\sum_{i=2}^{n} \frac{t^{a(i)}}{(z_i - z_1)^m}, \qquad (15.4.6)$$

and so on for multiple raising operators. Here, $t^{a(i)}$ acts in the representation r_i on the primary $\mathcal{O}_i$; the representation indices on $t^{a(i)}$ and $\mathcal{O}_i$ are suppressed.

The Sugawara theory is solved in the same way as the minimal models. In particular, *all* representations are degenerate, and in fact contain null descendants of two distinct types. The first follows directly from the Sugawara form of T, which in modes reads

$$L_m = \frac{1}{(k + h(g))\psi^2} \sum_{n=-\infty}^{\infty} j_n^a j_{m-n}^a. \qquad (15.4.7)$$

For $m = -1$, this implies that any correlator of primaries is annihilated by

$$\mathcal{L}_{-1} - \frac{2}{(k+h(g))\psi^2} \sum_a t^{a(i)} \mathcal{J}^a_{-1} \ . \tag{15.4.8}$$

This is the *Knizhnik–Zamolodchikov (KZ) equation*,

$$\left[\frac{\partial}{\partial z_1} + \frac{2}{(k+h(g))\psi^2} \sum_a \sum_{i=2}^n \frac{t^{a(1)} t^{a(i)}}{z_1 - z_i} \right] \langle\, \mathcal{O}_1(z_1) \ldots \mathcal{O}_n(z_n) \,\rangle_{S_2} = 0 \ . \tag{15.4.9}$$

We have suppressed group indices on the primary fields, but by writing the correlator in terms of g-invariants, the KZ equation becomes a set of coupled first order differential equations — coupled because there is in general more than one g-invariant for given representations r_i. Exercise 15.5 develops one example. For the leading singularity $(z_1 - z_i)^\kappa$ as $z_1 \to z_i$, the KZ equation reproduces the known result (15.4.4) but does not give fusion rules. There is again a free-field representation of the current algebra (exercise 15.6), analogous to the Feigin–Fuchs representation of the Virasoro algebra.

The second type of null descendant involves the currents only, and does constrain the fusion rules. For convenience, let us focus on the case $g = SU(2)$. The results can then be extended to general g by examining the $SU(2)$ subalgebras associated with the various roots α. We saw in chapter 11 that the $SU(2)$ current algebra has at least two interesting $SU(2)$ Lie subalgebras, namely the global symmetry $j_0^\pm$, j_0^3 and the pseudospin

$$j_{-1}^+ \ , \quad j_0^3 - \frac{k}{2} \ , \quad j_1^- \ . \tag{15.4.10}$$

Now consider some primary field

$$|j, m\rangle \ , \tag{15.4.11}$$

which we have labeled by its quantum numbers under the global $SU(2)$. What are its pseudospin quantum numbers (j', m')? Since it is primary, it is annihilated by the pseudospin lowering operator, so $m' = -j'$. We also have $m' = m - k/2$, so $j' = k/2 - m$. Now, the pseudospin representation has dimension $2j' + 1$, so if we raise any state $2j' + 1$ times we get zero:

$$(j_{-1}^+)^{k-2m+1} |j, m\rangle = 0 \ . \tag{15.4.12}$$

This is the null descendant.

Now take the correlation of this descendant with some current algebra primaries and use the relation (15.4.5) between the correlators of descendants and primaries to obtain

$$0 = \left\langle (\mathscr{J}_{-1}^+)^{k-2m_1+1} \cdot \mathcal{O}_1(z_1) \dots \mathcal{O}_n(z_n) \right\rangle_{S_2}$$

$$= \left[-\sum_{i=2}^{n} \frac{t^{+(i)}}{z_i - z_1} \right]^{k-2m_1+1} \left\langle \mathcal{O}_1(z_1) \dots \mathcal{O}_n(z_n) \right\rangle_{S_2} . \qquad (15.4.13)$$

Notice that, unlike the earlier null equations, this one involves no derivatives and is purely algebraic. To see how this constrains the operator products, consider the three-point correlation. By considering the separate z_i dependences in eq. (15.4.13) one obtains

$$0 = \sum_{m_2, m_3} [(t^{+(2)})^{l_2}]_{m_2, n_2} [(t^{+(3)})^{l_3}]_{m_3, n_3} \left\langle \mathcal{O}_{j_1, m_1} \mathcal{O}_{j_2, m_2} \mathcal{O}_{j_3, m_3} \right\rangle_{S_2} , \qquad (15.4.14)$$

where we have now written out the group indices explicitly. This holds for all n_2 and n_3, and for

$$l_2 + l_3 \geq k - 2m_1 + 1 . \qquad (15.4.15)$$

The matrix elements of $(t^+)^l$ are nonvanishing for at least some $n_{2,3}$ if $m_2 \geq l_2 - j_2$ and $m_3 \geq l_3 - j_3$. Noting the restriction on $l_{2,3}$, we can conclude that the correlation vanishes when $m_2 + m_3 \geq k - 2m_1 + 1 - j_2 - j_3$. Using $m_1 + m_2 + m_3 = 0$ and taking $m_1 = j_1$ (the most stringent case) gives

$$\left\langle \mathcal{O}_{j_1, j_1} \mathcal{O}_{j_2, m_2} \mathcal{O}_{j_3, m_3} \right\rangle_{S_2} = 0 \quad \text{if} \quad j_1 + j_2 + j_3 > k . \qquad (15.4.16)$$

Although this was derived for $m_1 = j_1$, rotational invariance now guarantees that it applies for all m_1. Applying also the standard result for multiplication of $SU(2)$ representations, we have the fusion rule

$$[j_1] \times [j_2] = [|j_1 - j_2|] + [|j_1 - j_2| + 1] + \dots + [\min(j_1 + j_2, k - j_1 - j_2)] . \qquad (15.4.17)$$

Again there is a simple current, the maximum value $j = k/2$:

$$[j_1] \times [k/2] = [k/2 - j_1] . \qquad (15.4.18)$$

The corresponding $\mathbf{Z}_2$ symmetry is simply $(-1)^{2j}$.

Modular invariance

The spectrum of a $g \times g$ current algebra will contain some number $n_{r\tilde{r}}$ of each highest weight representation $|r, \tilde{r}\rangle$. The partition function is then

$$Z(\tau) = \sum_{r, \tilde{r}} n_{r\tilde{r}} \chi_r(q) \chi_{\tilde{r}}(q)^* , \qquad (15.4.19)$$

with the character defined by analogy to that for the conformal algebra, eq. (15.2.12). Invariance under $\tau \to \tau + 1$ amounts as usual to level matching, so $n_{r\tilde{r}}$ can be nonvanishing only when $h_r - h_{\tilde{r}}$ is an integer. Under $\tau \to -1/\tau$ the characters mix,

$$\chi_r(q') = \sum_{r'} S_{rr'} \chi_{r'}(q) , \qquad (15.4.20)$$

so the condition for modular invariance is the matrix equation

$$S^\dagger n S = n . \qquad (15.4.21)$$

The characters are obtained by considering all states generated by the raising operators, with appropriate allowance for degeneracy. Only the currents need be considered, since by the Sugawara relation the Virasoro generators do not generate any additional states. The calculation is then parallel to the calculation of the characters of finite Lie algebras, and the result is similar to the Weyl character formula. The details are too lengthy to repeat here, and we will only mention one simple classic result: the modular S matrix for $SU(2)$ at level k is

$$S_{jj'} = \left(\frac{2}{k+2} \right)^{1/2} \sin \frac{\pi(2j+1)(2j'+1)}{k+2} . \qquad (15.4.22)$$

The general solution to the modular invariance conditions is known. One solution, at any level, is the diagonal modular invariant for which each representation with $j = \tilde{j}$ appears once:

$$n_{j\tilde{j}} = \delta_{j\tilde{j}} . \qquad (15.4.23)$$

These are known as the A invariants. When the level k is even, there is another solution obtained by twisting with respect to $(-1)^{2j}$. One keeps the previous states with j integer only, and adds in a twisted sector where $\tilde{j} = k/2 - j$. For k a multiple of 4, j in the twisted sector runs over integers, while for $k + 2$ a multiple of 4, j in the twisted sector runs over half-integers:

$$n_{j\tilde{j}} = \delta_{j\tilde{j}} \Big|_{j\in\mathbf{Z}} + \delta_{k/2-j,\tilde{j}} \Big|_{j\in\mathbf{Z}+k/4} . \qquad (15.4.24)$$

These are known as the D invariants. For the special values $k = 10, 16, 28$ there are exceptional solutions, the E invariants. The A–D–E terminology refers to the simply-laced Lie algebras. The solutions are in one-to-one correspondence with these algebras, the Dynkin diagrams arising in the construction of the invariants.

Strings on group manifolds

Thus far the discussion has used only symmetry, without reference to a Lagrangian. There is an important Lagrangian example of a current algebra. Let us start with a simple case, a nonlinear sigma model with a three-dimensional target space,

$$S = \frac{1}{2\pi\alpha'} \int d^2z \, (G_{mn} + B_{mn}) \partial X^m \bar{\partial} X^n \,. \tag{15.4.25}$$

Let G_{mn} be the metric of a 3-sphere of radius r and let the antisymmetric tensor field strength be

$$H_{mnp} = \frac{q}{r^3} \epsilon_{mnp} \tag{15.4.26}$$

for some constant q; ϵ_{mnp} is a tensor normalized to $\epsilon_{mnp} \epsilon^{mnp} = 6$. The curvature is

$$R_{mn} = \frac{2}{r^2} G_{mn} \,. \tag{15.4.27}$$

To leading order in α', the nonvanishing beta functions (3.7.14) for this nonlinear sigma model are

$$\beta_{mn}^G = \alpha' G_{mn} \left(\frac{2}{r^2} - \frac{q^2}{2r^6} \right) \,, \tag{15.4.28a}$$

$$\beta^\Phi = \frac{1}{2} - \frac{\alpha' q^2}{4r^6} \,. \tag{15.4.28b}$$

The first term in β^Φ is the contribution of three free scalars. The theory is therefore conformally invariant to leading order in α' if

$$r^2 = \frac{|q|}{2} + O(\alpha') \,. \tag{15.4.29}$$

The central charge is

$$c = 6\beta^\Phi = 3 - \frac{6\alpha'}{r^2} + O\left(\frac{\alpha'^2}{r^4} \right) \,. \tag{15.4.30}$$

A 3-sphere has symmetry algebra $O(4) = SU(2) \times SU(2)$. In a CFT, we know that each current will be either holomorphic or antiholomorphic. Comparing with the $SU(2)$ Sugawara central charge

$$c = 3 - \frac{6}{k+2} \,, \tag{15.4.31}$$

the sigma model is evidently a Sugawara theory. One $SU(2)$ will be left-moving on the world-sheet and one right-moving.

The general analysis of current algebras showed that the level k is quantized. In the nonlinear sigma model it arises from the Dirac quantization condition. The argument is parallel to that in section 13.3. A nonzero total flux H is incompatible with $H = dB$ for a single-valued B. We can write the dependence of the string amplitude on this background as

$$\exp\left(\frac{i}{2\pi\alpha'} \int_M B \right) = \exp\left(\frac{i}{2\pi\alpha'} \int_N H \right) \,, \tag{15.4.32}$$

where M is the embedding of the world-sheet in the target space and N is any three-dimensional manifold in S_3 whose boundary is M. In order that this be independent of the choice of N we need

$$1 = \exp\left(\frac{i}{2\pi\alpha'}\int_{S_3} H\right) = \exp\left(\frac{\pi i q}{\alpha'}\right) . \tag{15.4.33}$$

Thus,

$$q = 2\alpha' n , \quad r^2 = \alpha' n \tag{15.4.34}$$

for integer n. More generally, $\int H$ over any closed 3-manifold in spacetime must be a multiple of $4\pi^2\alpha'$.

This is the desired quantization, and $|n|$ is just the level k of the current algebra. In particular, the one-loop central charge (15.4.30) becomes

$$c = 3 - \frac{6}{|n|} + O\left(\frac{1}{n^2}\right) , \tag{15.4.35}$$

agreeing with the current algebra result to this order.

The 3-sphere is the same as the $SU(2)$ group manifold, under the identification

$$g = x^4 + i x^i \sigma^i , \quad \sum_{i=1}^{4}(x^i)^2 = 1 . \tag{15.4.36}$$

The action (15.4.25) can be rewritten as the *Wess–Zumino–Novikov–Witten (WZNW) action*

$$S = \frac{|n|}{4\pi}\int_M d^2z \, \mathrm{Tr}(\partial g^{-1}\bar{\partial}g) + \frac{in}{12\pi}\int_N \mathrm{Tr}(\omega^3) , \tag{15.4.37}$$

where $\omega = g^{-1}dg$ is the *Maurer–Cartan 1-form*. Here M is the embedding of the world-sheet in the group manifold, and N is any 3-surface in the group manifold whose boundary is M. In this form, the action generalizes to any Lie group g. The second term is known as the *Wess–Zumino* term. The reader can check that

$$d(\omega^3) = 0 . \tag{15.4.38}$$

Therefore, locally on the group $\omega^3 = d\chi$ for some 2-form χ, and the Chern–Simons term can be written as a two-dimensional action

$$\frac{n}{12\pi}\int_M \mathrm{Tr}(\chi) . \tag{15.4.39}$$

As with the magnetic monopole, there is no such χ that is nonsingular on the whole space.

The variation of the WZNW action is[3]

$$\delta S = \frac{|n|}{2\pi} \int d^2z \, \mathrm{Tr} \, [\bar{\partial} g \, g^{-1} \partial (g^{-1} \delta g)]$$

$$= \frac{|n|}{2\pi} \int d^2z \, \mathrm{Tr} \, [g^{-1} \partial g \, \bar{\partial}(\delta g g^{-1})] \,. \qquad (15.4.40)$$

As guaranteed by conformal invariance, the global $g \times g$ symmetry

$$\delta g(z, \bar{z}) = i\epsilon_L g(z, \bar{z}) - ig(z, \bar{z})\epsilon_R \qquad (15.4.41)$$

is elevated to a current algebra,

$$\delta g(z, \bar{z}) = i\epsilon_L(z)g(z, \bar{z}) - ig(z, \bar{z})\epsilon_R(\bar{z}) \,. \qquad (15.4.42)$$

Left-multiplication is associated with a left-moving current algebra and right-multiplication with a right-moving current algebra. The currents are

$$|n|\mathrm{Tr}(\epsilon_R g^{-1} \partial g) \,, \quad |n|\mathrm{Tr}(\epsilon_L \bar{\partial} g g^{-1}) \,. \qquad (15.4.43)$$

Let us check that the Poisson bracket of two currents has the correct c-number piece. To get this, it is sufficient to expand

$$g = 1 + i(2|n|)^{-1/2}\phi_a \sigma^a + \dots \qquad (15.4.44)$$

and keep the leading terms in the Lagrangian density and currents,

$$\mathscr{L} = \frac{1}{4\pi}\partial\phi^a \bar{\partial}\phi^a + O(\phi^3) \,, \qquad (15.4.45a)$$

$$j_R^a = |n|^{1/2}\partial\phi^a + O(\phi^2) \,, \qquad (15.4.45b)$$

$$j_L^a = |n|^{1/2}\bar{\partial}\phi^a + O(\phi^2) \,. \qquad (15.4.45c)$$

The higher-order terms do not contribute to the c-number in the Poisson bracket. The kinetic term now has the canonical $\alpha' = 2$ normalization so the level $k = |n|$ follows from the normalization of the currents.

Which states appear in the spectrum? We can make an educated guess by thinking about large k, where the group manifold becomes more and more flat. The currents then approximate free boson modes so the primary states, annihilated by the raising operators, have no internal excitations — the vertex operators are just functions of g. The representation matrices form a complete set of such functions, so we identify

$$D_{ij}^r(g) = \mathcal{O}_i^r(z)\tilde{\mathcal{O}}_j^r(\bar{z}) \,. \qquad (15.4.46)$$

This transforms as the representation (r, r) under $g \times g$, so summing over all r gives the diagonal modular invariant. Recall that for each k the number of primaries is finite; $D_{ij}^r(g)$ for higher r evidently is not primary. This reasoning is correct for simply connected groups, but otherwise we must

---

[3] This is for $n > 0$; for $n < 0$ interchange z and $\bar{z}$.

exclude some representations and add in winding sectors. For example, $SU(2)/\mathbf{Z}(2) = O(3)$ leads to the D invariant. We can understand the restriction to even levels for the D invariant: $\int H$ on $SU(2)/\mathbf{Z}(2)$ is half of $\int H$ on $SU(2)$, so the coefficient must be even to give a well-defined path integral.

The group manifold example vividly shows how familiar notions of spacetime are altered in string theory. If we consider eight flat dimensions with both right- and left-moving momenta compactified on the E_8 root lattice, we obtain an $E_{8L} \times E_{8R}$ current algebra at level one. We get the same theory with *248 dimensions* forming the E_8 group manifold with unit H charge.

15.5 Coset models

A clever construction allows us to obtain from current algebras the minimal models and many new CFTs. Consider a current algebra G, which might be a sum of several factors (g_i, k_i). Let H be some subalgebra. Then as in the discussion of Sugawara theories we can separate the energy-momentum tensor into two pieces,

$$T^G = T^H + T^{G/H} . \tag{15.5.1}$$

The central charge of $T^{G/H}$ is

$$c^{G/H} = c^G - c^H . \tag{15.5.2}$$

For any subalgebra the Sugawara theory thus separates into the Sugawara theory of the subalgebra, and a new *coset* CFT. A notable example is

$$G = SU(2)_k \oplus SU(2)_1 , \quad c^G = 4 - \frac{6}{k+2} , \tag{15.5.3a}$$

$$H = SU(2)_{k+1} , \quad c^H = 3 - \frac{6}{k+3} , \tag{15.5.3b}$$

where the subscripts denote the levels. Here, the H currents are the sums of the currents of the two $SU(2)$ current algebras in G, $j^a = j^a_{(1)} + j^a_{(2)}$. Then the central charges

$$c^{G/H} = 1 - \frac{6}{(k+2)(k+3)} \tag{15.5.4}$$

are precisely those of the unitary minimal models with $m = k + 2$.

A representation of the G current algebra can be decomposed under the subalgebras,

$$\chi_r^G(q) = \sum_{r',r''} n_{r'r''}^r \chi_{r'}^H(q) \chi_{r''}^{G/H}(q), \tag{15.5.5}$$

where r is any representation of G, and r' and r'' respectively run over all H and G/H representations, with $n^r_{r'r''}$ nonnegative integers. For the minimal model coset (15.5.3), all unitary representations can be obtained in this way. The current algebra theories are rather well understood, so this is often a useful way to represent the coset theory. For example, while the Kac determinant gives necessary conditions for a minimal model representation to be unitary, the coset construction is regarded as having provided the existence proof, the unitary current algebra representations having been constructed directly. The minimal model fusion rules (15.3.19) can be derived from the $SU(2)$ current algebra rules (15.4.17), and the minimal model modular transformation

$$S_{rs,r's'} = \left[\frac{8}{(p+1)(q+1)} (-1)^{(r+s)(r'+s')} \right]^{1/2} \sin \frac{\pi rr'}{p} \sin \frac{\pi ss'}{q} \qquad (15.5.6)$$

can be obtained from the $SU(2)$ result (15.4.22). Further, the minimal model modular invariants are closely related to the $SU(2)$ A–D–E invariants.

Taking various G and H leads to a wealth of new theories. In this section and the next we will describe only some of the most important examples, and then in section 15.7 we discuss some generalizations. The coset construction can be regarded as gauging the subalgebra H. Conformal invariance forbids a kinetic term for the gauge field, and the equation of motion for this field then requires the H-charge to vanish, leaving the coset theory. This is the gauging of a continuous symmetry; equivalently, one is treating the H currents as constraints. Recall that gauging a discrete symmetry gave the orbifold (twisting) construction.

The *parafermionic theories* are:

$$\frac{SU(2)_k}{U(1)}, \quad c = 2 - \frac{6}{k+2}. \qquad (15.5.7)$$

Focusing on the $U(1)$ current algebra generated by j^3, by the OPE we can write this in terms of a left-moving boson H with standard normalization $H(z)H(0) \sim -\ln z$:

$$j^3 = i(k/2)^{1/2}\partial H, \quad T^H = -\frac{1}{2}\partial H \partial H. \qquad (15.5.8)$$

Operators can be separated into a free boson part and a parafermionic part. For the $SU(2)$ currents themselves we have

$$j^+ = \exp[iH(2/k)^{1/2}]\psi_1, \quad j^- = \exp[-iH(2/k)^{1/2}]\psi_1^\dagger, \qquad (15.5.9)$$

where ψ_1 is known as the parafermionic current. Subtracting the weight of the exponential, the current has weight $(k-1)/k$. One obtains further currents

$$:(j^+)^l := (j^+_{-1})^l \cdot 1 \equiv \exp(ilH(2/k)^{1/2})\psi_l \,, \tag{15.5.10}$$

with ψ_l having weight $l(k-l)/k$. The current algebra null vector (15.4.12) implies that ψ_l vanishes for $l > k$, which could also have been anticipated from its negative weight. The weight also implies that $\psi_0 = \psi_k = 1$, and from this one can also deduce that $\psi_l = \psi^\dagger_{k-l}$. The current algebra primaries similarly separate,

$$\mathcal{O}_{j,m} = \exp[imH(2/k)^{1/2}]\psi^j_m \,, \tag{15.5.11}$$

where ψ^j_m is a primary field of the parafermion algebra, and has weight $j(j+1)/(k+2) - m^2/k$.

Factoring out the OPE of the free boson, the operator products of the parafermionic currents become

$$\psi_l(z)\psi_{l'}(0) \approx z^{-2ll'/k}(\psi_{l+l'} + \ldots) \,. \tag{15.5.12}$$

This algebra is more complicated than those encountered previously, in that the currents have branch cuts with respect to each other. However, it is simple in one respect: each pair of currents has definite *monodromy*, meaning that all terms in the operator product change by the same phase, $\exp(-4\pi ill'/k)$, when one current circles the other. We will mention an application of the parafermion theories later.

For small k, the parafermion theories reduce to known examples. For $k = 1$, the parafermion central charge is zero and the parafermion theory trivial. In other words, at $k = 1$ the free boson is the whole $SU(2)$ current algebra: this is just the torus at its self-dual radius. For $k = 2$, the parafermion central charge is $\frac{1}{2}$, so the parafermion must be an ordinary free fermion. We recall from section 11.5 that $SU(2)$ at $k = 2$ can be represented in terms of three free fermions. The free boson H is obtained by bosonizing $\psi^{1,2}$, leaving ψ^3 as the parafermion. At $k = 3$ the parafermion central charge is $\frac{4}{5}$, identifying it as the $m = 5$ unitary minimal model.

Although constructed as $SU(2)$ cosets, the minimal models have no $SU(2)$ symmetry nor other any weight 1 primaries. In order for an operator from the G theory to be part of the coset theory, it must be nonsingular with respect to the H currents, and no linear combination of the currents $j^a_{(1)}$ and $j^a_{(2)}$ is nonsingular with respect to $j^a_{(1)} + j^a_{(2)}$. The situation becomes more interesting if we consider the bilinear invariants

$$:j^a_{(1)}j^a_{(1)}: \,, \quad :j^a_{(1)}j^a_{(2)}: \,, \quad :j^a_{(2)}j^a_{(2)}: \,. \tag{15.5.13}$$

In parallel with the calculations in exercise 11.7, the operator product of the H current with these bilinears is

$$\left[j^b_{(1)}(z) + j^b_{(2)}(z) \right] : j^a_{(i)} j^a_{(j)}(0) := \sum_{k=0}^{\infty} \frac{1}{z^{k+1}} \left[j^b_{k(1)} + j^b_{k(2)} \right] j^a_{-1(i)} j^a_{-1(j)} \cdot 1 . \quad (15.5.14)$$

The $k = 0$ term vanishes because the bilinear is G-invariant. For $k = 1$, commuting the lowering operator to the right gives a linear combination of $j^b_{-1(1)}$ and $j^b_{-1(2)}$. All higher poles vanish. Thus, there are three bilinear invariants and only two possible singularities, so one linear combination commutes with the H current and lies entirely within the coset theory. This is just the coset energy-momentum tensor $T^{G/H}$, which we already know.

For $SU(2)$ cosets that is the end of the story, but let us consider the generalization

$$G = SU(n)_{k_1} \oplus SU(n)_{k_2} , \quad c^G = (n^2 - 1)\left[\frac{k_1}{k_1 + n} + \frac{k_2}{k_2 + n} \right] ,$$
$$(15.5.15a)$$
$$H = SU(n)_{k_1+k_2} , \quad c^H = (n^2 - 1)\frac{k_1 + k_2}{k_1 + k_2 + n} .$$
$$(15.5.15b)$$

For $n \geq 3$ there is a symmetric cubic invariant

$$d^{abc} \propto \mathrm{Tr}(t^a\{t^b, t^c\}) , \quad (15.5.16)$$

which vanishes for $n = 2$. Similarly, for $n \geq 4$ there is an independent symmetric quartic invariant, and so forth. Using the cubic invariant, we can construct the four invariants $d^{abc} : j^a_{(i)} j^b_{(j)} j^c_{(k)} :$. The operator product with the H current has three possible singularities, $z^{-2} d^{abc} : j^b_{(j)} j^c_{(k)} :$, so there must be one linear combination $W(z)$ that lies in the coset theory. That is, the coset theory has a conserved spin-3 current. The states of the coset theory fall in representations of an *extended chiral algebra*, consisting of the Laurent modes of $T(z)$, $W(z)$, and any additional generators needed to close the algebra.

In general, the algebra contains higher spin currents as well. For example, the operator product $W(z)W(0)$ contains a spin-4 term involving the product of four currents. For the special case $n = 3$ and $k_2 = 1$, making use of the current algebra null vectors, the algebra of $T(z)$ and $W(z)$ actually closes without any new fields. It is the W_3 *algebra*, which in OPE form is

$$W(z)W(0) \sim \frac{c}{3z^6} + \frac{2}{z^4}T(0) + \frac{1}{z^3}\partial T(0) + \frac{3}{10z^2}\partial^2 T(0) + \frac{1}{15z}\partial^3 T(0)$$
$$+ \frac{16}{220 + 50c}\left(\frac{2}{z^2} + \frac{1}{z}\partial \right)[10 : T^2(0) : -3\partial^2 T(0)] . \quad (15.5.17)$$

In contrast to the various algebras we have encountered before, this one is *nonlinear*: the spin-4 term involves the square of $T(z)$. This is the only closed algebra containing only a spin-2 and spin-3 current and was first discovered by imposing closure directly. It has a representation theory parallel to that of the Virasoro algebra, and in particular has a series of unitary degenerate representations of central charge

$$c = 2 - \frac{24}{(k+3)(k+4)} . \tag{15.5.18}$$

The $(k_1, k_2, n) = (k, 1, 3)$ cosets produce these representations. As it happens, the first nontrivial case is $k = 1$, $c = \frac{4}{5}$, which as we have seen also has a parafermionic algebra. The number of extended chiral algebras is enormous, and they have not been fully classified.

15.6   Representations of the $N = 1$ superconformal algebra

All the ideas of this chapter generalize to the superconformal algebras. In this section we will describe only the basics: the Kac formula, the discrete series, and the coset construction.

A highest weight state, of either the R or NS algebra, is annihilated by L_n and G_n for $n > 0$. The representation is generated by L_n for $n < 0$ and G_n for $n \leq 0$. Each G_n acts at most once, since $G_n^2 = L_{2n}$. The Kac formula for the R and NS algebras can be written in a uniform way,

$$\det(\mathcal{M}^N)_{\mathrm{R,NS}} = (h - \epsilon\hat{c}/16)K_N \prod_{1 \leq rs \leq 2N} (h - h_{r,s})^{P_{\mathrm{R,NS}}(N - rs/2)} . \tag{15.6.1}$$

Here, ϵ is 1 in the R sector and 0 in the NS sector. The zeros are at

$$h_{r,s} = \frac{\hat{c} - 1 + \epsilon}{16} + \frac{1}{4}(r\hat{\alpha}_+ + s\hat{\alpha}_-)^2 , \tag{15.6.2}$$

where $r - s$ must be even in the R sector and odd in the NS sector. We have defined $\hat{c} = 2c/3$ and

$$\hat{\alpha}_\pm = \frac{1}{4}\left[(1 - \hat{c})^{1/2} \pm (9 - \hat{c})^{1/2}\right] . \tag{15.6.3}$$

The multiplicity of each zero is again the number of ways a given level can be reached by the raising operators of the theory,

$$\prod_{n=1}^{\infty} \frac{1 + q^{n-1}}{1 - q^n} = \sum_{k=0}^{\infty} P_{\mathrm{R}}(k)q^k , \tag{15.6.4a}$$

$$\prod_{n=1}^{\infty} \frac{1 + q^{n-1/2}}{1 - q^n} = \sum_{k=0}^{\infty} P_{\mathrm{NS}}(k)q^k . \tag{15.6.4b}$$

Unitary representations are allowed at

$$\hat{c} \geq 1 , \quad h \geq \epsilon\frac{\hat{c}}{16} , \tag{15.6.5}$$

and at the discrete series

$$c = \frac{3}{2} - \frac{12}{m(m+2)} , \quad m = 2, 3, \dots ,$$

$$= 0, \frac{7}{10}, 1, \frac{81}{70}, \dots , \tag{15.6.6a}$$

$$h = h_{r,s} \equiv \frac{[r(m+2) - sm]^2 - 4}{8m(m+2)} + \frac{\epsilon}{16} , \tag{15.6.6b}$$

where $1 \leq r \leq m - 1$ and $1 \leq s \leq m + 1$.

A coset representation for the $N = 1$ unitary discrete series is

$$G = SU(2)_k \oplus SU(2)_2 , \quad H = SU(2)_{k+2} . \tag{15.6.7}$$

The central charge is correct for $m = k + 2$. The reader can verify that the coset theory has $N = 1$ world-sheet supersymmetry: using the free fermion representation of the $k = 2$ factor, one linear combination of the $(\frac{3}{2}, 0)$ fields $j^a_{(1)}\psi^a$ and $i\epsilon^{abc}\psi^a\psi^b\psi^c$ is nonsingular with respect to the H current and is the supercurrent of the coset theory.

For small m, some of these theories are familiar. At $m = 2$, c vanishes and we have the trivial theory. At $m = 3$, $c = \frac{7}{10}$, which is the $m = 4$ member of the Virasoro unitary series. At $m = 4$, $c = 1$; this is the free boson representation discussed in section 10.7.

15.7 Rational CFT

We have seen that holomorphicity on the world-sheet is a powerful property. It would be useful if a general local operator of weight $(h, \tilde{h})$ could be divided in some way into a holomorphic $(h, 0)$ field times an antiholomorphic $(0, \tilde{h})$ field, or a sum of such terms. The conformal block expression (15.2.9) shows the sense in which this is possible: by organizing intermediate states into conformal families, the correlation function is written as a sum of terms, each holomorphic times antiholomorphic. While this was carried out for the four-point function on the sphere, it is clear that the derivation can be extended to n-point functions on arbitrary Riemann surfaces. For example, the conformal blocks of the zero-point function on the torus are just the characters,

$$Z(\tau) = \sum_{i,\bar{j}} n_{i\bar{j}} \chi_i(q) \chi_{\bar{j}}(q)^* , \tag{15.7.1}$$

where $n_{i\bar{j}}$ counts the number of times a given representation of the left and right algebras appears in the spectrum.

When the sum is infinite this factorization does not seem particularly helpful, but when the sum is finite it is. In fact, in all the examples discussed in this section, and in virtually all known exact CFTs, the sum is finite. What is happening is that the spectrum, though it must contain an infinite number of Virasoro representations for $c \geq 1$, consists of a finite number of representations of some larger extended chiral algebra. This is the definition of a *rational conformal field theory (RCFT)*.

It has been conjectured that all rational theories can be represented as cosets, and that *any* CFT can be arbitrarily well approximated by a rational theory (see exercise 15.9 for an example). If so, then we are close to constructing the general CFT, but the second conjecture in particular seems very optimistic.

We will describe here a few of the general ideas and results. The basic objects in RCFT are the conformal blocks and the *fusion rules*, nonnegative integers N_{ij}^k which count the number of ways the representations i and j can be combined to give the representation k. For the Virasoro algebra, we know that two representations can be combined to give a third in a unique way: the expectation value of the primaries determines those of all descendants. For other algebras, N_{ij}^k may be greater than 1. For example, even for ordinary Lie algebras there are two ways to combine two adjoint **8**s of $SU(3)$ to make another adjoint, namely d^{abc} and f^{abc}. As a result, the same holds for the corresponding current algebra representations: $N_{\mathbf{88}}^{\mathbf{8}} = 2$.

Repeating the derivation of the conformal blocks, for a general algebra the number of independent blocks $\mathscr{F}_{ij}^{kl}(r|z)$ is

$$N_{ijkl} = N_{ij}^r N_{rkl} \,, \tag{15.7.2}$$

where the repeated index is summed. Indices are lowered with $N_{ij}^0 = N_{ij}$, zero denoting the identity representation. One can show that for each i, N_{ij} is nonvanishing only for a single j. This defines the conjugate representation, $N_{i\bar{i}} = 1$. In the minimal models and $SU(2)$ current algebra, all representations are self-conjugate, but for $SU(n)$, $n > 2$ for example, they are not. By associativity, the s-channel conformal blocks $\mathscr{F}_{ij}^{kl}(r|z)$ are linearly related to the t-channel blocks $\mathscr{F}_{il}^{jk}(r|1-z)$. The number of independent functions must be the same in each channel, so the fusion rules themselves satisfy an associativity relation,

$$N_{ij}^r N_{rkl} = N_{ik}^r N_{rjl} = N_{il}^r N_{rjk} \,. \tag{15.7.3}$$

We will now derive two of the simpler results in this subject, namely that the weights and the central charge must in fact be rational numbers in an RCFT. First note that the conformal blocks are not single-valued on the original Riemann surface — they have branch cuts — but they are single-valued on the covering space, where a new sheet is defined whenever one vertex operator circles another. Any series of moves that brings the vertex operators back to their original positions and sheets must leave the conformal blocks invariant. For example,

$$\tau_1\tau_2\tau_3\tau_4 = \tau_{12}\tau_{13}\tau_{23} \,, \tag{15.7.4}$$

where $\tau_{i\cdots j}$ denotes a *Dehn twist*, cutting open the surface on a circle containing the indicated vertex operators, rotating by 2π and gluing. To see this, examine for example vertex operator 1. On the right-hand side, the combined effect of τ_{12} and τ_{13} is for this operator to circle operators 2 and 3 and to rotate by 4π. On the left, this is the same as the combined effect of τ_4 (which on the sphere is the same as τ_{123}) and τ_1. Eq. (15.7.4) is an N_{ijkl}-dimensional matrix equation on the conformal blocks. For example,

$$\tau_1 : \quad \mathscr{F}_{ij}^{kl}(r|z) \to \exp(2\pi i h_i)\mathscr{F}_{ij}^{kl}(r|z) \,, \tag{15.7.5a}$$

$$\tau_{12} : \quad \mathscr{F}_{ij}^{kl}(r|z) \to \exp(2\pi i h_r)\mathscr{F}_{ij}^{kl}(r|z) \,. \tag{15.7.5b}$$

On the other hand, τ_{13} is not diagonal in this basis, but rather in the dual basis $\mathscr{F}_{il}^{jk}(r|1-z)$.

In order to get a basis-independent statement, take the determinant of eq. (15.7.4) and use (15.7.5) to get

$$N_{ijkl}(h_i + h_j + h_k + h_l) - \sum_r (N_{ij}^r N_{rkl} + N_{ik}^r N_{rjl} + N_{il}^r N_{rjk})h_r \in \mathbf{Z} \,. \tag{15.7.6}$$

This step is possible only when the number $\mathscr{N}$ of primaries is finite.

There are many more equations than weights. Focusing on the special case $i = j = k = l$ gives

$$\sum_r N_{ii}^r N_{rii}(4h_i - 3h_r) \in \mathbf{Z} \,. \tag{15.7.7}$$

This is $\mathscr{N} - 1$ equations for $\mathscr{N} - 1$ weights, where $\mathscr{N}$ is the number of primaries; the weight h_0 is always 0, and the $i = 0$ equation is trivial. Let us consider the example of $SU(2)$ current algebra at level 3, where there are four primaries, $j = 0, \frac{1}{2}, 1, \frac{3}{2}$. From the general result (15.4.17), the nonzero fusion rules of the form N_{ii}^r are

$$N_{00}^0 = N_{1/2,1/2}^1 = N_{1/2,1/2}^1 = N_{11}^0 = N_{11}^1 = N_{3/2,3/2}^0 = 1 \,. \tag{15.7.8}$$

Thus we find that

$$8h_{1/2} - 3h_1 \,, \quad 5h_1 \,, \quad 4h_{3/2} \tag{15.7.9}$$

are all integers, which implies that the weights are all rational. These results are consistent with the known weights $j(j+1)/(k+2)$. The reader can show that eqs. (15.7.7) are always nondegenerate and therefore require the weights to be rational.[4]

---

[4] We are assuming that all the N_{iiii} are nonzero. More generally, one can derive a similar relation with $N_{i\bar{\imath}\bar{\imath}}$, which is always positive.

For the central charge, consider the zero-point function on the torus. The covering space here is just Teichmüller space, on which one may check that

$$S^4 = (ST)^3 = 1 . \tag{15.7.10}$$

The determinant of this implies that

$$1 = [(\det S)^4]^{-3}[(\det S \det T)^3]^4 = (\det T)^{12} . \tag{15.7.11}$$

The transformation T acts on the characters as

$$T : \quad \chi_i(q) \to \exp[2\pi i(h_i - c/24)]\chi_i(q) . \tag{15.7.12}$$

Thus,

$$\frac{\mathcal{N}c}{2} - 12\sum_i h_i \in \mathbf{Z} , \tag{15.7.13}$$

and the rationality of c follows from that of the weights.

The consistency conditions for RCFT have been developed in a systematic way. Let us just mention some of the most central results. The first is the *Verlinde formula*, which determines the fusion rules in terms of the modular transformation S:

$$N^i_{jk} = \sum_r \frac{S^r_j S^r_k S^{\dagger i}_r}{S^r_0}. \tag{15.7.14}$$

Indices are lowered with N^0_{ij}. The second is *naturalness*: any operator product coefficient that is allowed by the full chiral algebra is actually nonzero.[5] The third result describes all possible modular invariants (15.7.1): either $n_{i\bar{\jmath}} = \delta_{ij}$ (the diagonal invariant), or $n_{i\bar{\jmath}} = \delta_{i\omega(\bar{\jmath})}$, where $\omega(\bar{\jmath})$ is some permutation symmetry of the fusion rules. The latter two results are not quite as useful as they sound, because they only hold with respect to the full chiral algebra of the theory. As we have seen in the W algebra coset example, this may be larger than one realizes.

Finally, let us mention a rather different generalization of the coset idea. Suppose we have a current algebra G, and we consider all $(2,0)$ operators formed from bilinears in the currents,

$$T' = L_{ab} :j^a j^b : . \tag{15.7.15}$$

The condition that the TT OPE has the correct form for an energy-momentum tensor, and therefore that the modes of T form a Virasoro algebra, is readily found. It is the *Virasoro master equation*,

$$L_{ab} = 2L_{ac}k^{cd}L_{db} - L_{cd}L_{ef}f^{ce}_a f^{df}_b - L_{cd}f^{ce}_f f^{df}_{(a} L_{b)e} , \tag{15.7.16}$$

---

[5] This precise statement holds only when the N^i_{jk} are restricted to the values 0 and 1; otherwise, it requires some refinement.

where k^{ab} is the coefficient $1/z^2$ in the current–current OPE. The central charge is

$$c = 2k^{ab}L_{ab} \ . \tag{15.7.17}$$

We already know some solutions to this: the Sugawara tensor for G, or for any subalgebra H of G. Remarkably, the set of solutions is very much larger: for $G = SU(3)_k$, the number has been estimated as $\frac{1}{4}$ *billion* for each k. For each solution the G theory separates into two decoupled theories, with energy-momentum tensors T' and $T^G - T'$. Some of these may be equivalent to known theories, but others are new and many have irrational central charge.

15.8 Renormalization group flows

Consistent string propagation requires a conformally invariant world-sheet theory, but there are several reasons to consider the relation of CFTs to the larger set of all two-dimensional field theories. First, CFT also has application to the description of critical phenomena, where the parameters can be varied away from their critical values. Second, there is a rich mathematical and physical interplay between conformal theories and nearby nonconformal ones, each illuminating the other. Third, conformally invariant theories correspond to string backgrounds that satisfy the classical equations of motion. One might then guess that the proper setting for quantum string theory would be a path integral over all background field configurations — that is, over all two-dimensional quantum field theories. This last is more speculative; it is related to other formulations of string field theory, a subject discussed briefly in chapter 9.

In this section we will develop some general results relating conformal and nonconformal theories. In the next we will discuss some examples and applications. Once again, this is an enormous subject and we can only sketch a few of the central ideas and results.

Scale invariance and the renormalization group

Consider the scale transformation

$$\delta_s z = \epsilon z \tag{15.8.1}$$

on a world-sheet with flat metric $g_{ab} = \delta_{ab}$. Alternatively we could keep the coordinates fixed and scale up the metric,

$$\delta_s g_{ab} = 2\epsilon g_{ab} \ . \tag{15.8.2}$$

In either form the net change (3.4.6) in the action and measure is

$$-\frac{\epsilon}{2\pi} \int d^2\sigma \, T^a_{\ a}(\sigma) \, . \tag{15.8.3}$$

A flat world-sheet theory will therefore be scale-invariant provided that

$$T^a_{\ a} = \partial_a \mathcal{K}^a, \tag{15.8.4}$$

for some local operator $\mathcal{K}^a$.

Scale invariance plays an important role in many parts of physics. One expects that the extreme low energy limit of any quantum field theory will approach a scale-invariant theory. This has not been proven in general, but seems to be true in all examples. The scale-invariant theory may be trivial: if all states are massive then at low enough energy nothing is left. Consider for example a statistical mechanical system. The Boltzmann sum is the same as the Euclidean path integral in quantum field theory. This may have an energy gap for generic values of the parameters and so be trivial at long distance, but when the parameters are tuned to send the gap to zero (a second order phase transition) it is described by a nontrivial scale-invariant theory.

The term *nontrivial* in this context is used in two different ways. The broad usage (which is applied in the previous paragraph) means any field theory without an energy gap, so that there are states of arbitrarily small nonzero energy. A narrower usage reserves the term for scale-invariant theories with interactions that remain important at all distances, as opposed to those whose low energy limit is equivalent to that of a free field theory.

Scale and conformal invariances are closely related. The scale transformation rescales world-sheet distances by a constant factor, leaving angles and ratios of lengths invariant. A conformal transformation rescales world-sheet distances by a position-dependent factor; on a very small patch of the world-sheet it looks like a scale transformation. In particular, conformal transformations leave angles of intersection between curves invariant. Comparing the condition (15.8.4) with the condition $T^a_{\ a} = 0$ for conformal invariance, one sees that it is possible in principle for a theory to be scale-invariant without being conformally-invariant. However, it is difficult to find examples. Later in the section we will prove that for compact unitary CFTs in two dimensions scale invariance does imply conformal invariance. Exercise 15.12 gives a nonunitary counterexample.

This is of some importance in dimensions greater than two. In the previous chapter we encountered two nontrivial (in the narrow sense) scale-invariant theories. The first was the $d = 4$, $N = 4$ gauge theory. The second was the $d = 6$ $(2,0)$ tensionless string theory, which arose on coincident IIA or M-theory 5-branes. Both are believed to be conformally invariant.

In quantum field theory, the behavior of matrix elements under a rigid scale transformation is governed by a differential equation, the *renormalization group equation*. Let us derive such an equation. Consider a general quantum field theory in d-dimensional spacetime; spacetime here corresponds to the string world-sheet, which is the case $d = 2$. The scale transformation of a general expectation value is

$$\epsilon^{-1}\delta_s\left\langle \prod_m \mathscr{A}_{i_m}(\sigma_m) \right\rangle = -\frac{1}{2\pi} \int d^2\sigma \left\langle T^a{}_a(\sigma) \prod_m \mathscr{A}_{i_m}(\sigma_m) \right\rangle$$

$$- \sum_n \Delta_{i_n}{}^j \left\langle \mathscr{A}_j(\sigma_n) \prod_{m\neq n} \mathscr{A}_{i_m}(\sigma_m) \right\rangle , \qquad (15.8.5)$$

where $\mathscr{A}_i$ is a complete set of local operators. The second term is from the action of the scale transformation on the operators,

$$\delta_s\mathscr{A}_i(\sigma) = -\Delta_i{}^j \mathscr{A}_j(\sigma) . \qquad (15.8.6)$$

The integrated trace of the energy-momentum tensor can be expanded in terms of the complete set,

$$\int d^d\sigma\, T^a{}_a = -2\pi \sum_i{}' \int d^d\sigma\, \beta^i(g)\mathscr{A}_i . \qquad (15.8.7)$$

The prime on the sum indicates that it runs only over operators with dimension less than or equal to d, because this is the dimension of the energy-momentum tensor. We can similarly write a general renormalizable action as a sum over all such terms

$$S = \sum_i{}' g^i \int d^dx\, \mathscr{A}_i(\sigma) . \qquad (15.8.8)$$

Here g^i is a general notation that includes the interactions as well as the masses and the kinetic term normalizations. The expansions (15.8.7) and (15.8.8) can be used to rewrite the scale transformation (15.8.5) as the *renormalization group equation*,

$$\epsilon^{-1}\delta_s\left\langle \prod_m \mathscr{A}_{i_m}(\sigma_m) \right\rangle = -\sum_i{}' \beta^i(g)\frac{\partial}{\partial g^i}\left\langle \prod_m \mathscr{A}_{i_m}(\sigma_m) \right\rangle$$

$$- \sum_n \Delta_{i_n}{}^j \left\langle \mathscr{A}_j(\sigma_n) \prod_{m\neq n} \mathscr{A}_{i_m}(\sigma_m) \right\rangle . \qquad (15.8.9)$$

There may also be contact terms between $T^a{}_a$ and the other operators, and terms from the g_i-derivative acting on the local operators. These are dependent on definitions (the choice of renormalization scheme) and can all be absorbed into the definition of $\Delta_i{}^j$. Eq. (15.8.9) states that a scale transformation is equivalent to a change in the coupling plus a mixing of operators. As one looks at longer distances the couplings and operators flow.

The Zamolodchikov c-theorem.

Without conformal invariance, T_{zz} is not holomorphic, its modes do not generate a Virasoro algebra, and the central charge c is not defined. Nevertheless, c has a useful extension to the space of all two-dimensional field theories.

Define

$$F(r^2) = z^4 \langle T_{zz}(z, \bar{z}) T_{zz}(0, 0) \rangle , \qquad (15.8.10a)$$

$$G(r^2) = 4z^3 \bar{z} \langle T_{zz}(z, \bar{z}) T_{z\bar{z}}(0, 0) \rangle , \qquad (15.8.10b)$$

$$H(r^2) = 16z^2 \bar{z}^2 \langle T_{z\bar{z}}(z, \bar{z}) T_{z\bar{z}}(0, 0) \rangle . \qquad (15.8.10c)$$

Rotational invariance implies that these depend only on $r^2 = z\bar{z}$, as indicated. From conservation, $\bar{\partial} T_{zz} + \partial T_{z\bar{z}} = 0$, one finds that

$$4\dot{F} + \dot{G} - 3G = 0 , \quad 4\dot{G} - 4G + \dot{H} - 2H = 0 , \qquad (15.8.11)$$

where a dot denotes differentiation with respect to $\ln r^2$. The *Zamolodchikov C function* is the combination

$$C = 2F - G - \frac{3}{8}H . \qquad (15.8.12)$$

This has the property

$$\dot{C} = -\frac{3}{4}H. \qquad (15.8.13)$$

In a unitary theory H can be written as a sum of absolute squares by inserting a complete set of states, and so is nonnegative. The result (15.8.13) shows that the physics changes in a monotonic way as we look at longer and longer distances. Also, C is stationary if and only if the two-point function of $T_{z\bar{z}}$ with itself is zero, implying (by a general result in unitary quantum field theory) that $T_{z\bar{z}}$ itself vanishes identically. The theory is then conformally invariant and C becomes precisely c.

The monotonicity property also implies that the theory at long distance will approach a stationary point of C and therefore a CFT. Again, this is intuitively plausible: at long distances the theory should forget about underlying distance scales. In general this is likely to happen in the trivial sense that all fields are massive and only the empty $c = 0$ theory remains. However, if massless degrees of freedom are present due to some combination of symmetry and the tuning of parameters, the c-theorem implies that their interactions will be conformally invariant. We should emphasize that the unitarity and compactness are playing a role; in the more general case there do exist counterexamples (exercise 15.12).

Like c, the C function seems to represent some generalized measure of the density of states. The monotonicity is then very plausible: a massive field would contribute to the number of degrees of freedom measured at short distance, but drop out at distances long compared to its Compton wavelength. In spite of this intuitive interpretation, there seems to be no simple generalization of the C function to $d > 2$. However, the principle that the long distance limit of any quantum field theory is conformally invariant still seems to hold under broad conditions.

Conformal perturbation theory

Now let us consider adding small conformally-noninvariant terms to the action of a CFT,

$$S = S_0 + \lambda^i \int d^2z\, \mathcal{O}_i \,, \tag{15.8.14}$$

where S_0 is the action of the CFT. For convenience we focus on the case that the perturbations are primary fields, but the results are easily generalized. The λ^i are the earlier couplings g^i minus the value at the conformal point.

The main question is how the physics in the perturbed theory depends on scale. Consider the following operator product, which arises in first order perturbation theory for correlations of the energy-momentum tensor:

$$- T_{zz}(z,\bar{z})\, \lambda^i \int d^2w\, \mathcal{O}_i(w,\bar{w}) \,. \tag{15.8.15}$$

We have

$$\partial_{\bar{z}} T_{zz}(z)\mathcal{O}_i(w,\bar{w})$$
$$= \partial_{\bar{z}}\left[(z-w)^{-2}h_i + (z-w)^{-1}\partial_w\right]\mathcal{O}_i(w,\bar{w})$$
$$= -2\pi h_i \partial_z \delta^2(z-w)\mathcal{O}_i(w,\bar{w}) + 2\pi\delta^2(z-w)\partial_w\mathcal{O}_i(w,\bar{w}) \,. \tag{15.8.16}$$

Integrating this, the first order perturbation (15.8.15) implies that perturbation leads to

$$\partial_{\bar{z}} T_{zz}(z,\bar{z}) = 2\pi\lambda_i(h_i - 1)\partial_z \mathcal{O}_i(z,\bar{z}) \,. \tag{15.8.17}$$

As expected, the energy-momentum tensor is no longer holomorphic, unless the perturbation is of weight $h_i = 1$. The energy-momentum tensor must still be conserved,

$$\partial_{\bar{z}} T_{zz} + \partial_z T_{\bar{z}z} = 0 \,. \tag{15.8.18}$$

Inspection of the divergence (15.8.17) thus identifies

$$T_{\bar{z}z} = 2\pi\lambda^i(1 - h_i)\mathcal{O}_i(z,\bar{z}) . \qquad (15.8.19)$$

We assume that the perturbations are rotationally invariant, $h_i = \tilde{h}_i$, so that T_{ab} remains symmetric.

Referring back to the renormalization group, we have

$$\beta^i = 2(h_i - 1)\lambda^i , \qquad (15.8.20)$$

so that a rescaling of lengths by ϵ is equivalent to a rescaling of the couplings,

$$\delta\lambda^i = 2\epsilon(1 - h_i)\lambda^i . \qquad (15.8.21)$$

A perturbation with $h_i > 1$ is thus termed *irrelevant,* because its effect drops away at long distance and we return to the conformal theory. A perturbation with $h_i < 1$ is termed *relevant.* It grows more important at low energies, and we move further from the original conformal theory. A perturbation with $h_i = 1$ is termed *marginal.*

Now let us go to the next order in g. Consider first the case that the perturbations $\mathcal{O}_i$ are all of weight $(1,1)$, marginal operators. Second order perturbation theory will then involve the operator product

$$\frac{1}{2}\int d^2z\, \mathcal{O}_i(z,\bar{z})\int d^2w\, \mathcal{O}_j(w,\bar{w}) , \qquad (15.8.22)$$

the factor of $\frac{1}{2}$ coming from the expansion of $\exp(-S)$. The part of the OPE that involves only marginal operators is

$$\mathcal{O}_i(z,\bar{z})\mathcal{O}_j(w,\bar{w}) \sim \frac{1}{|z - w|^2}c^k{}_{ij}\mathcal{O}_k(w,\bar{w}) , \qquad (15.8.23)$$

so the second order term (15.8.22) will have a logarithmic divergence when $z \to w$,

$$2\pi\int\frac{dr}{r}c^k{}_{ij}\int d^2w\, \mathcal{O}_k(w,\bar{w}) . \qquad (15.8.24)$$

The divergence must be cut off at the lower end, introducing a scale into the problem and breaking conformal invariance. At the upper end, the scale is set by the distance at which we are probing the system. We can read off immediately the scale dependence: if we increase the scale of measurement by a factor $1 + \epsilon$, the log increases by ϵ. This is equivalent to shifting the couplings by

$$\delta\lambda^k = -2\pi\epsilon c^k{}_{ij}\lambda^i\lambda^j . \qquad (15.8.25)$$

In other words,

$$\beta^k = 2\pi c^k{}_{ij}\lambda^i\lambda^j . \qquad (15.8.26)$$

As an application, suppose that we are interested in perturbations that preserve conformal invariance. We have the familiar necessary condition that the perturbation be a (1,1) tensor, but now we see that there are further conditions: conformal invariance will be violated to second order in λ unless

$$c^k{}_{ij}\lambda^i\lambda^j = 0 \tag{15.8.27}$$

for all (1,1) operators k.

Now we wish to go to second order in λ for perturbations that are not marginal. At weak coupling, the order λ^2 term is important only if the first order term is small — that is, if the coupling is nearly marginal. To leading order in $h_i - 1$, we can just carry over our result for $O(\lambda^2)$ in the marginal case. Combining the contributions (15.8.20) and (15.8.26), we then have

$$\beta^i = 2(h_i - 1)\lambda^i + 2\pi c^k{}_{ij}\lambda^i\lambda^j , \tag{15.8.28}$$

with corrections being higher order in $h_i - 1$ or λ^i. Let us also work out the C function. With $T_{\bar{z}\bar{z}} = -\pi\beta^i\mathcal{O}_i$, the result (15.8.13) for the C function becomes to leading order

$$\dot{C} = -12\pi^2\beta^i\beta^j G_{ij} , \tag{15.8.29}$$

where

$$G_{ij} = z^2\bar{z}^2 \langle \mathcal{O}_i(z,\bar{z})\mathcal{O}_j(0,0) \rangle \tag{15.8.30}$$

is evaluated at $\lambda^i = 0$. Observe that

$$\beta^i = \frac{\partial}{\partial\lambda_i}U(g) , \tag{15.8.31a}$$

$$U(g) = (h_i - 1)\lambda^i\lambda_i + \frac{2\pi}{3}c_{ijk}\lambda^i\lambda^j\lambda^k , \tag{15.8.31b}$$

indices being lowered with G_{ij}. Using this and $\beta^i = -2\lambda^i$ gives

$$\dot{C} = 24\pi^2\beta_j\lambda^j = 24\pi^2\dot{U} . \tag{15.8.32}$$

This integrates to

$$C = c + 24\pi^2 U \tag{15.8.33}$$

with c being the central charge at the conformal point $\lambda^i = 0$.

Now let us apply this to the case of a single slightly relevant operator,

$$\dot{\lambda} = (1 - h)\lambda - \pi c_{111}\lambda^2, \tag{15.8.34}$$

normalized so that $G_{11} = 1$. If λ starts out positive it grows, but not indefinitely: the negative second order term cuts off the growth. At long distance we arrive at a *new* conformal theory, with coupling

$$\lambda' = \frac{1-h}{\pi c_{111}} \,. \tag{15.8.35}$$

From the string spacetime point of view, we can interpret $U(\lambda)$ as a potential energy for the light field corresponding to the world-sheet coupling λ, and the two conformal theories correspond to the two stationary points of the cubic potential. Note that $\lambda = 0$ is a local maximum: relevant operators on the world-sheet correspond to tachyons in spacetime. The central charge of the new fixed point is

$$c' = c - 8\frac{(1-h)^3}{c_{111}^2} \,. \tag{15.8.36}$$

15.9   Statistical mechanics

The partition function in classical statistical mechanics is

$$Z = \int [dq] \, \exp(-\beta H) \,, \tag{15.9.1}$$

where the integral runs over configuration space, β is the inverse temperature, and the Hamiltonian H is the integral of a local density. This has a strong formal similarity to the path integral for Euclidean quantum theory,

$$Z = \int [d\phi] \, \exp(-S/\hbar) \,. \tag{15.9.2}$$

In the statistical mechanical case, the configuration is a function of the spatial dimensions only, so that statistical mechanics in d *spatial* dimensions resembles quantum field theory in d *spacetime dimensions*. An obvious difference between the two situations is that in the statistical mechanical case there is generally an underlying discrete structure, while in relativistic field theory and on the string world-sheet we are generally interested in a continuous manifold.

There is a context in statistical mechanics in which one essentially takes the continuum limit. This is in *critical phenomena,* in which some degrees of freedom have correlation lengths very long compared to the atomic scale, and the discrete structure is no longer seen. In this case, the statistical ensemble is essentially identical to a relativistic field theory. Let us discuss the classic example, the Ising model. Here one has an array of spins on a square lattice in two dimensions, each spin σ_i taking the values ± 1. The energy is

$$H = -\sum_{\text{links}} \sigma_i \sigma_{i'} \,. \tag{15.9.3}$$

The sum runs over all nearest-neighbor pairs (links). The energy favors adjacent pairs being aligned. When β is small, so that the temperature is large, the correlations between spins are weak and short-range,

$$\langle \sigma_i \sigma_j \rangle \sim \exp[-|i-j|/\xi(\beta)] \tag{15.9.4}$$

as the distance $|i-j|$ goes to infinity. For sufficiently large β the $\mathbf{Z}_2$ symmetry $\sigma_i \to -\sigma_i$ is broken and there is long-range order,

$$\langle \sigma_i \sigma_j \rangle \sim v^2(\beta) + \exp[-|i-j|/\xi'(\beta)] . \tag{15.9.5}$$

For both small and large β the fluctuations are short-range. However, the transition between these behaviors is second order, both $\xi(\beta)$ and $\xi'(\beta)$ going to infinity at the critical value β_c. At the critical point the falloff is power law rather than exponential,

$$\langle \sigma_i \sigma_j \rangle \sim |i-j|^{-\eta} , \quad \beta = \beta_c . \tag{15.9.6}$$

The long-wavelength fluctuations at this point should be described by a continuum path integral. The value of the *critical exponent* η is known from the exact solution of the Ising model to be $\frac{1}{4}$. This cannot be deduced from any classical reasoning, but depends in an essential way on the nonlinear interactions between the fluctuations.

To deduce the CFT describing the critical theory, note the global symmetry of the Ising model, the $\mathbf{Z}_2$ symmetry $\sigma_i \to -\sigma_i$. We have a whole family of CFTs with this symmetry, the minimal models. For reasons to be explained below, the correct minimal model is the first nontrivial one, $m = 3$ with $c = \frac{1}{2}$. The nontrivial primary fields of this theory, taking into account the identification (15.3.15), are

$$\mathcal{O}_{1,1} : h = 0 , \quad \mathcal{O}_{1,2} : h = \frac{1}{16} , \quad \mathcal{O}_{1,3} : h = \frac{1}{2} . \tag{15.9.7}$$

Under the $\mathbf{Z}_2$ (15.3.23), $\mathcal{O}_{1,2}$ is odd and the other two are even. In particular, the Ising spins, being odd under $\mathbf{Z}_2$, should evidently be identified as

$$\sigma_i \to \sigma(z,\bar{z}) = \mathcal{O}_{1,2}(z)\tilde{\mathcal{O}}_{1,2}(\bar{z}) . \tag{15.9.8}$$

The left- and right-moving factors must be the same to give a rotationally invariant operator. There are separate $\mathbf{Z}_2$s acting on the left- and right-moving theories, but all operators have equal left and right charges so we can take either one. The expectation value

$$\langle \sigma(z,\bar{z})\sigma(0,0) \rangle \propto (z\bar{z})^{-2h} = (z\bar{z})^{-1/8} \tag{15.9.9}$$

agrees with the exact solution for the critical exponent η.

The $m = 3$ minimal model is equivalent to the free massless Majorana fermion. Indeed, Onsager solved the Ising model by showing that it could be rewritten in terms of a free fermion on a lattice, which in general is massive but which becomes massless at β_c. Note that $\mathcal{O}_{1,3}$ has the correct dimension to be identified with the fermion field, and $\mathcal{O}_{1,2}$ has the correct dimension to be the R sector ground state vertex operator for a single Majorana fermion.

Incidentally, the solubility of the Ising model for general β can be understood directly from the CFT. Changing the temperature is equivalent to adding

$$\mathcal{O}_{1,3}(z)\tilde{\mathcal{O}}_{1,3}(\bar{z}) \tag{15.9.10}$$

to the action. This is the only relevant perturbation that is invariant under the $\mathbf{Z}_2$ symmetry. This perturbation breaks the conformal invariance, but it can be shown from the OPEs of the CFT that a spin-4 current constructed from T_{zz}^2 is still conserved. The existence of a symmetry of spin greater than 2 in a *massive* theory is sufficient to allow a complete solution. Of course, in the present case the perturbation (15.9.10) is just a mass for the free fermion, but for other CFTs without such a simple Lagrangian description this more abstract approach is needed.

The requirement that operators have integer spin means that we can only pair the same conformal family on the right and left. For the theory quantized on the circle, this corresponds to the A modular invariant discussed earlier,

$$[\mathcal{O}_{1,1}\tilde{\mathcal{O}}_{1,1}] + [\mathcal{O}_{1,2}\tilde{\mathcal{O}}_{1,2}] + [\mathcal{O}_{1,3}\tilde{\mathcal{O}}_{1,3}] . \tag{15.9.11}$$

In terms of the free fermion theory this is the diagonal GSO projection.

For two-dimensional critical theories with few enough degrees of freedom that the central charge is less than one, the classification of unitary representations of the Virasoro algebra completely determines the possible critical exponents: they must be given by one of the minimal models.[6] For this reason this same set of CFTs arises from many different short-distance theories. Let us mention one such context, which illustrates the relation among all the unitary minimal models through the $\mathbf{Z}_2$ symmetry they share. We noted that the $m = 3$ theory has only one relevant perturbation that is invariant under $\mathbf{Z}_2$. We therefore identified this with a variation of the temperature away from the critical point. The operator $\mathcal{O}_{1,1}\tilde{\mathcal{O}}_{1,1}$ is just the identity and adding it to the action has a trivial effect. The operator $\mathcal{O}_{1,2}\tilde{\mathcal{O}}_{1,2}$ is odd under $\mathbf{Z}_2$ and corresponds to turning on a magnetic field that breaks the $\sigma_i \to -\sigma_i$ symmetry. For the minimal model at general m there are $m - 2$ nontrivial relevant $\mathbf{Z}_2$-invariant operators. This corresponds to *multicritical behavior*. To reach such a model one must tune $m - 2$ parameters precisely.

---

[6] There is a caveat: the CFTs that arise in statistical physics need not be unitary. Unitarity in that context is related to a property known as reflection positivity, which holds in most but not all systems of interest.

For example, take the Ising model with thermally equilibrated (annealed) vacancies, so that each spin σ can take values ± 1 or 0, the last corresponding to an empty site. When the density ρ of vacancies is small, the behavior is much like the Ising model, with the same critical behavior at some point $\beta_c(\rho)$. However, when the vacancy density reaches a critical value ρ_c, then at $\beta_c(\rho_c)$ there are independent long-range fluctuations of the spin and *density*. This is known as the *tricritical Ising model,* tricritical referring to the need to adjust two parameters to reach the critical point. Since there are more long-range degrees of freedom than in the Ising model, we might expect the critical theory to have a greater central charge. The tricritical Ising model has been identified with the next minimal model, $m = 4$ with $c = \frac{7}{10}$. This generalizes: with spins (also called 'heights') taking $m-1$ values, there is a multicritical point obtained by adjusting $m-2$ parameters which is described by the corresponding minimal model. In fact, every CFT we have described in this chapter can be obtained as the critical limit of a lattice theory, and indeed of a solvable lattice theory. It is quite likely that every rational theory can be obtained from a solvable lattice theory.

A different generalization of the Ising model is the $\mathbf{Z}_k$ Ising model (the clock model). Here the spins take k values $\sigma_i = \exp(2\pi i n/k)$ for $n = 0, 1, \ldots, k-1$, and there is a $\mathbf{Z}_k$ symmetry $\sigma_i \to \exp(2\pi i/k)\sigma_i$. The energy is

$$H = -\sum_{\text{links}} \text{Re}(\sigma_i \sigma_{i'}^*) . \tag{15.9.12}$$

Again there is a critical point at a value β_c. The critical behavior is described by the $\mathbf{Z}_k$ parafermion theory. The $\mathbf{Z}_k$ parafermions describe a generic critical system in which the fluctuations transform under a $\mathbf{Z}_k$ symmetry.

Several of the low-lying minimal models can be realized in different ways. The $m = 5$ theory is obtained as a four-height $\mathbf{Z}_2$ model or a $\mathbf{Z}_3$ Ising model. It is also known as the three-state Potts model, referring to a different generalization of the Ising model (spins taking k values with a permutation symmetry S_k) which happens to be the same as the $\mathbf{Z}_k$ generalization when $k = 3$. The $m = 6$ model can be obtained as a five-height $\mathbf{Z}_2$ model or as a tricritical point of the $\mathbf{Z}_3$ Potts/Ising model with vacancies. In fact the $m = 3, 4, 5, 6$ theories have all been realized experimentally, usually in systems of atoms adsorbed on surfaces. Since the $m = 4$ model is also the $m = 3$ minimal model of the $N = 1$ supersymmetric series, this is in a sense the first experimental realization of supersymmetry. (Some atomic and nuclear systems have an approximate Fermi/Bose symmetry, but this is a nonrelativistic algebra whose closure does not involve the translations.)

Landau–Ginzburg models

To complete this section, we will give a slightly different Lagrangian description of the minimal models. To study the long-wavelength behavior of the Ising model, we can integrate out the individual spins and work with a field $\phi(z, \bar{z})$ representing the average spin over a region of many sites. This field takes essentially continuous values, rather than the original discrete ones. The first few terms in the Lagrangian density for ϕ would be

$$\mathcal{L} = \partial\phi\bar{\partial}\phi + \lambda_1\phi^2 + \lambda_2\phi^4 \; . \tag{15.9.13}$$

At $\lambda_1 = 0$ the tree-level mass of the field ϕ is zero. We thus identify λ_1 as being proportional to $\beta_c - \beta$, with $\lambda_1 = 0$ being the critical theory, the $m = 3$ minimal model.

This is the *Landau–Ginzburg* description. The original idea was that the classical potential for ϕ represented the free energy of the system. Now one thinks of this as the effective Lagrangian density for a full quantum (or thermal) path integral. The quantum or thermal fluctuations cannot be neglected. In some systems, though not here, they change the transition from continuous to discontinuous, so that there is no critical behavior. In general they significantly modify the scaling properties (critical exponents).

Now add a $\lambda_3\phi^6$ term and tune λ_1 and λ_2 to zero. We might expect a different critical behavior — the potential is flatter than before, so will have more states below a given energy, but it is still positive so there will be fewer states than for a free scalar. In other words, we guess that c is more than $\frac{1}{2}$ and less than 1. It is natural to identify this with the next minimal model, the $m = 4$ tricritical Ising model, since the number of relevant $\mathbf{Z}_2$-invariant perturbations is two. Similarly, we guess that the Landau–Ginzburg model whose leading potential is ϕ^{2m-2} represents the mth minimal model.

Representing the minimal models by a strongly interacting quantum field theory seems to have little quantitative value, but it gives an intuitive picture of the operator content. To start we guess that ϕ corresponds to the operator of lowest dimension, namely $\mathcal{O}_{2,2}$. Also, we guess that we have the diagonal theory, so the left-moving representation is the same as the right-moving one, and we indicate only the latter. Now, to find ϕ^2, use the fusion rule

$$\mathcal{O}_{2,2}\mathcal{O}_{2,2} = [\mathcal{O}_{1,1}] + [\mathcal{O}_{3,1}] + [\mathcal{O}_{3,3}] + [\mathcal{O}_{1,3}] \; . \tag{15.9.14}$$

The first term is the identity; we guess that ϕ^2 is the remaining operator of lowest dimension, namely $\mathcal{O}_{3,3}$. Taking further products with $\mathcal{O}_{2,2}$, we identify

$$\phi^n = \mathcal{O}_{n+1,n+1} \; , \quad 0 \leq n \leq m - 2 \; . \tag{15.9.15}$$

This terminates due to the upper bound (15.3.16), $r \leq m - 1$. The lowest term in $\phi \cdot \phi^{m-2}$ is $\mathcal{O}_{m,m-2}$ which by reflection is $\mathcal{O}_{1,2}$. We then continue

$$\phi^{m-1+n} = \mathcal{O}_{n+1,n+2} , \quad 0 \leq n \leq m - 3 . \tag{15.9.16}$$

All this guesswork can be checked in various ways. One check is that the $\mathbf{Z}_2$ symmetry assignment (15.3.23), namely $(-1)^s$ for m odd and $(-1)^r$ for m even, matches that of ϕ^n. As another check, where is the next monomial ϕ^{2m-3}? The product $\phi \cdot \phi^{2m-4}$ leads to no new primaries. This is just right: the equation of motion is

$$m\lambda_m \phi^{2m-3} = \partial\bar{\partial}\phi = L_{-1}\tilde{L}_{-1} \cdot \phi , \tag{15.9.17}$$

so this operator is a descendant. The powers (15.9.15) and (15.9.16) are all the relevant primary operators.

What happens if we add a relevant perturbation to the Lagrangian for the mth minimal model? The Landau–Ginzburg picture indicates that adding ϕ^{2k-2} causes the theory to flow to the kth minimal model. Let us consider in particular ϕ^{2m-4} for m large. This is

$$\mathcal{O}_{m-1,m-2} = \mathcal{O}_{1,3} , \quad h = 1 - \frac{2}{m+1} , \tag{15.9.18}$$

which is nearly marginal. Thus we can apply the formalism of the previous section. From the fusion rule

$$\mathcal{O}_{1,3}\mathcal{O}_{1,3} = [\mathcal{O}_{1,1}] + [\mathcal{O}_{1,3}] + [\mathcal{O}_{1,5}] , \tag{15.9.19}$$

the only nearly marginal operator in $\mathcal{O}_{1,3}\mathcal{O}_{1,3}$ is $\mathcal{O}_{1,3}$ itself, so we are in precisely the single-operator situation worked out in the last paragraph of the previous section. Thus, we can construct a new conformal theory by a small $\mathcal{O}_{1,3}$ perturbation of the minimal model. The Landau–Ginzburg picture indicates that this is the next minimal model down. We can compute the central charge from the c-theorem. Taking from the literature the value $c_{111} = 4/3^{1/2}$ for the large-m minimal model yields

$$c' = c - \frac{12}{m^3} . \tag{15.9.20}$$

For large m this is indeed the difference between the central charges of successive minimal models.

Exercises

15.1 Evaluate $\det(\mathscr{M}^3)$ and compare with the Kac formula.

15.2 Derive eqs. (15.2.3) and (15.2.5) for the expectation value of a descendant.

15.3 Work out the steps outlined in the derivation of eq. (15.2.9) to find explicitly the $N = 0$ and $N = 1$ terms in $\mathscr{F}^{jl}_{mn}(i|z)$.

15.4 Verify that the discrete symmetries associated with the simple currents are as asserted below eqs. (15.3.23) and (15.4.18) for the unitary minimal models and the $SU(2)$ WZNW models.

15.5 (a) For the $SU(n)$ current algebra at level k, consider the four-point function with two insertions in the representation $(\mathbf{n}, \mathbf{n})$ and two in the representation $(\bar{\mathbf{n}}, \bar{\mathbf{n}})$. Find the KZ equation for the $SU(n)$ invariants.
(b) Find the general solution for $k = 1$ and determine the coefficients using associativity and locality. Compare this with the free-boson representation.
(c) Do the same for general k; the solution involves hypergeometric functions.

15.6 The *Wakimoto representation* is a free-field representation for the $SU(2)$ current algebra, analogous to the Feigin–Fuchs representation of the minimal models. Show that the following currents form an $SU(2)$ current algebra of level $k = q^2 - 2$:

$$J^+ = iw/2^{1/2} , \quad J^3 = iq\partial\phi/2^{1/2} - w\chi ,$$
$$J^- = i[w\chi^2 + (2 - q^2)\partial\chi]/2^{1/2} + q\chi\partial\phi .$$

Here w, χ are a commuting $\beta\gamma$ system and ϕ is a free scalar. Show that the Sugawara energy momentum tensor corresponds to the $\beta\gamma$ theory with $h_w = 1$ and $h_\chi = 0$, and with ϕ being a linear dilaton theory of appropriate central charge.

15.7 For the coset construction of the minimal models, combine primary fields from the two factors in G to form irreducible representations of $SU(2)$. Subtract the weight of the corresponding primary of H and show that the resulting weight is one of the allowed weights for the minimal model. Not all minimal model primaries are obtained in this way; some are excited states in the current algebras.

15.8 Repeat the previous exercise for the coset construction of the minimal $N = 1$ superconformal theories.

15.9 For the periodic scalar at any radius, the analysis in section 15.2 shows that the spectrum contains an infinite number of conformal families. Show, however, that if R^2/α' is rational, the partition function is a sum of a finite number of factors, each one holomorphic times antiholomorphic in τ. Show that at these radii there is an enlarged chiral algebra.

15.10 Apply the result (15.7.7) to the $SU(2)$ current algebra at $k = 4$. Show that the resulting relations are consistent with the actual weights of the $SU(2)$ primaries.

15.11 Verify the Verlinde formula (15.7.14) for the $SU(2)$ modular transformation (15.4.22). In this case indices are raised with the identity matrix.

15.12 For the general massless closed string vertex operator, we found the condition for Weyl invariance in section 3.6. Find the weaker condition for invariance under rigid Weyl transformations, and find solutions that have only this smaller invariance.

16
Orbifolds

In the final four chapters we would like to see how compactification of string theory connects with previous ideas for unifying the Standard Model. Our primary focus is the weakly coupled $E_8 \times E_8$ heterotic string, whose compactification leads most directly to physics resembling the Standard Model. At various points we consider other string theories and the effects of strong coupling. In addition, compactified string theories have interesting nonperturbative dynamics, beyond that which we have seen in ten dimensions. In the final chapter we discuss some of the most interesting phenomena.

The two main issues are specific constructions of four-dimensional string theories and general results derived from world-sheet and spacetime symmetries. Our approach to the constructions will generally be to present only the simplest examples of each type, in order to illustrate the characteristic physics of compactified string theories. On the other hand, we have collected as many of the general results as possible.

String compactifications fall into two general categories. The first are based on free world-sheet CFTs, or on CFTs like the minimal models that are solvable though not free. For these one can generally determine the exact tree-level spectrum and interactions. The second category is compactification in the geometric sense, taking the string to propagate on a smooth spacetime manifold some of whose dimensions are compact. In general one is limited to an expansion in powers of α'/R_c^2, with R_c being the characteristic radius of compactification. This is in addition to the usual expansion in the string coupling g. Commonly in a moduli space of smooth compactifications there will be special points (or subspaces) described by free CFTs. Thus the two approaches are complementary, one giving a very detailed picture at special points and the other giving a less detailed but global picture. Some of the solvable compactifications have no such geometric interpretation.

In this chapter we discuss free CFTs and in the next geometric compactification. Again, the literature in each case is quite large and a full account is far beyond the scope of this book.

16.1 Orbifolds of the heterotic string

In section 8.5 we discussed orbifolds, manifolds obtained from flat spacetime by identifying points under a discrete group H of symmetries. Although these manifolds generally have singularities, the resulting string theories are well behaved. The effect of the identification is to add twisted closed strings to the Hilbert space and to project onto invariant states.

We start with the ten-dimensional $E_8 \times E_8$ string, with H a subgroup of the Poincaré × gauge group. An element of H will act on the coordinates as a rotation θ and translation v,

$$X^m \to \theta^{mn} X^n + v^m , \qquad (16.1.1)$$

where $m, n = 4, \dots, 9$. For a four-dimensional theory H will act trivially on X^μ for $\mu = 0, \dots, 3$. In order to preserve world-sheet supersymmetry the twist must commute with the supercurrent, and so its action on the right-moving fermions is

$$\tilde\psi^m \to \theta^{mn} \tilde\psi^n . \qquad (16.1.2)$$

In addition it acts on the current algebra fermions as a gauge rotation γ,

$$\lambda^A \to \gamma^{AB} \lambda^B . \qquad (16.1.3)$$

Here we are considering gauge rotations γ^{AB} which are in the manifest $SO(16) \times SO(16)$ subgroup of $E_8 \times E_8$. The full element is denoted $(\theta, v; \gamma)$. Just as the fixed points can be thought of as points of singular spacetime curvature, a nontrivial γ can be thought of as singular gauge curvature at the fixed points.

Ignoring the gauge rotation, the set of all elements (θ, v) forms the *space group* S. In the twisted theory the strings are propagating on the space $M^4 \times K$, where

$$K = R^6/S . \qquad (16.1.4)$$

Because the elements of S in general have fixed points, this space is an orbifold.

Ignoring the translation as well as the gauge rotation leaves the *point group* P, the set of all rotations θ appearing in the elements of the twist group. An orbifold is called Abelian or non-Abelian according to whether the point group is Abelian or non-Abelian.

The subgroup of S consisting of pure translations $(1, v)$ is an Abelian group Λ. An alternative description of the orbifold is to twist first by Λ

to form a particular 6-torus,

$$T^6 = R^6/\Lambda .$$ (16.1.5)

The space group multiplication law

$$(\theta, w) \cdot (1, v) \cdot (\theta, w)^{-1} = (1, \theta v) ,$$ (16.1.6)

implies that the group

$$\bar{P} \equiv S/\Lambda$$ (16.1.7)

is a symmetry of the 6-torus. This is the same as the point group P except that some elements include translations. One can now twist the torus by $\bar{P}$ to form the orbifold

$$K = T^6/\bar{P} .$$ (16.1.8)

We can assume that the identity element in spacetime appears only with the identity in the gauge group, as $e = (1, 0; 1)$. This is no loss of generality, because if there were additional elements of the form $(1, 0; \gamma)$, one could first twist on the subgroup consisting of these pure gauge twists to obtain a different ten-dimensional theory, or perhaps a different description of the same theory, and then twist this theory under the remaining group which has no pure gauge twists. By closure it follows that each element (θ, v) of the space group appears with a unique gauge element $\gamma(\theta, v)$, and that these have the multiplication law

$$\gamma(\theta_1, v_1)\gamma(\theta_2, v_2) = \gamma((\theta_1, v_1) \cdot (\theta_2, v_2)) .$$ (16.1.9)

That is, there is a homomorphism from the space group to the gauge group.

Modular invariance

Modular invariance requires that the projection onto H-invariant states be accompanied by the addition of twisted states for each $h \in H$:

$$\varphi(\sigma^1 + 2\pi) = h \cdot \varphi(\sigma^1) ,$$ (16.1.10)

where φ stands for a generic world-sheet field. The resulting sum over path integral sectors is naively modular-invariant. However, we know from the example of the superstring in chapter 10 that modular invariance can be spoiled by phases in the path integral. In particular, the phase under $\tau \rightarrow \tau + 1$ is determined by the level mismatch, the difference $L_0 - \tilde{L}_0$ mod 1. In fact, for Abelian orbifolds it has been shown that this is the only potential obstruction to modular invariance.

To see how this works, consider the spectrum in the sector with twist h. Let N be the smallest integer such that $h^N = 1$; we then call this a

Z$_N$ *twist*. We can always choose the axes so that the rotation is of the form

$$\theta = \exp[2\pi i(\phi_2 J_{45} + \phi_3 J_{67} + \phi_4 J_{89})] . \tag{16.1.11}$$

Define the complex linear combinations

$$Z^i = 2^{-1/2}(X^{2i} + iX^{2i+1}) , \quad i = 2, 3, 4 , \tag{16.1.12}$$

with

$$Z^{\bar{i}} \equiv \overline{Z^i} = 2^{-1/2}(X^{2i} - iX^{2i+1}) . \tag{16.1.13}$$

The periodicity is then

$$Z^i(\sigma + 2\pi) = \exp(2\pi i\phi_i)Z^i(\sigma) . \tag{16.1.14}$$

Taking the same complex basis for the $\tilde{\psi}^m$ gives

$$\tilde{\psi}^i(\sigma + 2\pi) = \exp[2\pi i(\phi_i + v)]\tilde{\psi}^i(\sigma) \tag{16.1.15}$$

with $v = 0$ in the R sector and $v = \frac{1}{2}$ in the NS sector. The supercurrent is then periodic or antiperiodic in the usual way depending on v. The oscillators have the following mode numbers:

$$\alpha^i : n + \phi_i , \quad \alpha^{\bar{i}} : n - \phi_i , \tag{16.1.16a}$$

$$\tilde{\alpha}^i : n - \phi_i , \quad \tilde{\alpha}^{\bar{i}} : n + \phi_i , \tag{16.1.16b}$$

$$\tilde{\psi}^i : n - \phi_i \text{ (R) }, \ n - \phi_i + \tfrac{1}{2} \text{ (NS) }, \tag{16.1.16c}$$

$$\tilde{\psi}^{\bar{i}} : n + \phi_i \text{ (R) }, \ n + \phi_i + \tfrac{1}{2} \text{ (NS) }. \tag{16.1.16d}$$

For a single element, the gauge twist can always be taken in the block-diagonal $U(1)^{16}$ subgroup,

$$\gamma = \text{diag}[\exp(2\pi i\beta_1), \ldots, \exp(2\pi i\beta_{16})] . \tag{16.1.17}$$

This acts on the complex linear combinations $\lambda^{K\pm} = 2^{-1/2}(\lambda^{2K-1} \pm i\lambda^{2K})$ as

$$\lambda^{K\pm} \rightarrow \exp(\pm 2\pi i\beta_K)\lambda^{K\pm} . \tag{16.1.18}$$

The oscillators $\lambda^{K\pm}$ thus have mode numbers $n \mp \beta_K$ in the R sector of the current algebra, and $n \mp \beta_K + \frac{1}{2}$ in the NS sector.

Because $h^N = 1$ we can write

$$\phi_i = \frac{r_i}{N} , \quad \beta_K = \frac{s_K}{N} , \tag{16.1.19}$$

for integers r_i and s_K. Actually, we can say a bit more, because the various R sectors are in spinor representations and so contain eigenvalues

$$\frac{1}{2}\sum_{i=2}^{4} \phi_i , \quad \frac{1}{2}\sum_{K=1}^{8} \beta_K , \quad \frac{1}{2}\sum_{K=9}^{16} \beta_K . \tag{16.1.20}$$

Thus we have the mod 2 conditions

$$\sum_{i=2}^{4} r_i = \sum_{K=1}^{8} s_K = \sum_{K=9}^{16} s_K = 0 \mod 2 . \tag{16.1.21}$$

To be precise, if these are not satisfied then h^N is a nontrivial twist of the ten-dimensional theory, and so just changes the starting point.

Consider first the sector (R,R,R), labeled by the periodicities of the two sets of current algebra fermions and the supercurrent. Recall the general result that a complex boson with mode numbers $n + \theta$ has zero-point energy

$$\frac{1}{24} - \frac{1}{8}(2\theta - 1)^2 , \tag{16.1.22}$$

and a complex fermion has the negative of this. The above discussion of modes then gives the level mismatch as

$$L_0 - \tilde{L}_0 = -\sum_{i=2}^{4}(N^i + \tilde{N}^i + \tilde{N}^i_\psi)\phi_i - \sum_{K=1}^{16} N^K \beta_K$$

$$-\frac{1}{2}\sum_{i=2}^{4} \phi_i(1 - \phi_i) + \frac{1}{2}\sum_{K=1}^{16} \beta_K(1 - \beta_K) \mod 1 . \tag{16.1.23}$$

Here N^i counts the number of α^i excitations minus the number of $\alpha^{\bar{i}}$ excitations, and so on.

The oscillator part of $L_0 - \tilde{L}_0$ is a multiple of $1/N$, and the zero-point part a multiple of $1/2N^2$, so that in general there are *no* states for which $L_0 - \tilde{L}_0$ is an integer. Suppose, however, that the zero-point contribution is actually a multiple of $1/N$,

$$-\frac{1}{2}\sum_{i=2}^{4} \phi_i(1 - \phi_i) + \frac{1}{2}\sum_{K=1}^{16} \beta_K(1 - \beta_K) = \frac{m}{N} \tag{16.1.24}$$

for integer m. Then imposing on the excitation numbers the condition

$$\sum_{i=2}^{4}(N^i + \tilde{N}^i + \tilde{N}^i_\psi)\phi_i + \sum_{K=1}^{16} N^K \beta_K = \frac{m}{N} \mod 1 \tag{16.1.25}$$

leaves only states with integer $L_0 - \tilde{L}_0$. The left-hand side is just the transformation of the oscillators under h, so this condition is the projection onto h-invariant states. In particular, the phase of h in the twisted sector is determined by the zero-point energy (16.1.24).

Now consider the sector (R,R,NS). The $\tilde{\psi}$ modes are shifted by one-half, so the level mismatch is equal to the earlier value (16.1.23) plus

$$\delta = \frac{1}{2}\left(-\tilde{N}_\psi - 1 + \sum_{i=2}^{4} \phi_i\right), \qquad (16.1.26)$$

where the first term is from the excitations and the last two are from the change in the zero-point energy. Level matching again requires that

$$\delta = k/N \qquad (16.1.27)$$

for some integer k. The first two terms add to an integer due to the GSO projection, and then (16.1.27) follows from the mod 2 conditions (16.1.21). Level matching in all other sectors follows in the same way from conditions (16.1.21) and (16.1.24). The latter can also be rephrased

$$\sum_{i=2}^{4} r_i^2 - \sum_{K=1}^{16} s_K^2 = 0 \mod 2N. \qquad (16.1.28)$$

For Abelian orbifolds, as long as there are any states for which the level mismatch (16.1.23) is an integer, then by imposing the projection (16.1.25) one obtains a consistent theory. For non-Abelian orbifolds there are additional conditions.

Other free CFTs

The orbifolds above can be thought of as arising from the ten-dimensional theory in one step, twisting by the full space group, or in two, twisting first by the translations to make a toroidal theory and then twisting by the point group. The second construction can be made more general as follows. Represent the current algebra in bosonic form, so the toroidal theory has a momentum lattice of signature (22,6). Many lattices have symmetries that rotate the left and the right momenta independently, as opposed to the above construction in which $\theta_L = \theta_R$ on the (6,6) spacetime momenta. These more general theories are known as *asymmetric orbifolds*. Though there is no longer a geometric interpretation in terms of propagation on a singular space, the construct is consistent in CFT and in string theory.

Another construction is to fermionize all the internal coordinates, giving 44 left-movers and 18 right-movers. Since the Lorentz invariance is broken one can take arbitrary combinations of independent R and NS boundary conditions on the 62 fermions, subject to the constraints of modular invariance, locality of the OPE, and so on. Alternatively, join the real fermions into $22 + 9$ complex fermions and take sectors with independent aperiodicities $\exp(2\pi i v)$ for each fermion. In spite of appearances this is not strictly more general, because in the first case one can have combinations of boundary conditions such that the fermions cannot be put into pairs having the same boundary conditions in all sectors. The ten-dimensional E_8 theory from section 11.3 is an example with such essentially real fermions. One can also take some fermions of each type. The general consistent theory is known.

The supercurrent $\tilde{T}_F$ is now written purely in terms of fermions. For example, a single $X\tilde{\psi}$ CFT becomes a theory of three fermions with $\tilde{T}_F = i\tilde{\chi}_1\tilde{\chi}_2\tilde{\chi}_3$. The boundary conditions must be correlated so that all terms in the supercurrent are simultaneously R or NS. It is interesting to ask what is the most general $\tilde{T}_F$ that can be constructed from free fermions alone. A general $(0, \frac{3}{2})$ tensor would be

$$\tilde{T}_F = i \sum_{I,J,K=1}^{18} \tilde{\chi}_I \tilde{\chi}_J \tilde{\chi}_K c_{IJK} \, . \tag{16.1.29}$$

The conditions for the $\tilde{T}_F \tilde{T}_F$ OPE to generate a superconformal algebra are easily solved. The requirement that there be no four-fermi term in the OPE is

$$c_{IJM}c_{KLM} + c_{JKM}c_{ILM} + c_{KIM}c_{JLM} = 0 \, . \tag{16.1.30}$$

This is the Jacobi identity, requiring c_{IJK} to be the structure constants of a Lie algebra. The condition that the $\bar{z}^{-1}$ term in the OPE be precisely $2\tilde{T}_B$ is then

$$18c_{IKL}c_{JKL} = \delta_{IJ}. \tag{16.1.31}$$

This fixes the normalization of c_{IJK}, and requires the algebra to be semisimple (no Abelian factors). The dimension of the group is the number of fermions, 18. There are three semisimple groups of dimension 18, namely $SU(2)^6$, $SU(3) \times SO(5)$, and $SU(4) \times SU(2)$.

Another construction is to bosonize all fermions including the $\tilde{\psi}^\mu$ to form a lattice of signature (22,9), and then to make a Narain-like construction. Again $\tilde{T}_F$ can be generalized, to a sum of terms of the form

$$e^{ik \cdot X_R} \, , \ k^2 = 6/\alpha' \; ; \quad e^{il \cdot X_R}\bar{\partial}X_R \, , \ l^2 = 2/\alpha' \, . \tag{16.1.32}$$

Obviously there are overlaps among these constructions, though often one or the other description is more convenient. The fermionic construction in particular has been employed by a number of groups. We will be able to see a great deal of interesting spacetime physics even in the simplest orbifold models, so we will not develop these generalizations further.

16.2 Spacetime supersymmetry

We have seen that in consistent string theories there is a symmetry that relates fermions to bosons. The important question is whether in the real world this symmetry is spontaneously broken at very high energy, or whether part of it survives down to the weak interaction scale, the energy that can be reached by particle accelerators. In fact there is a strong argument, independent of string theory, for expecting that exactly one $d = 4$ supersymmetry survives and is spontaneously broken near the weak scale.

The argument has to do with the self-energies of elementary particles. The energy in the field of a charged point particle diverges at short distance. If we suppose that this is cut off physically at some distance l then naively the self-energy is

$$\delta m \approx \frac{\alpha}{l} , \tag{16.2.1}$$

with $\alpha = e^2/4\pi$ the fine structure constant. The electron is known to be pointlike down to at least 10^{-16} cm, implying that the energy (16.2.1) is more than 10^3 times the actual electron mass. However, it has been known since the 1930s that relativistic quantum effects reduce the simple classical estimate (16.2.1) to

$$\delta m \approx \alpha m \ln \frac{1}{ml} . \tag{16.2.2}$$

Taking l to be near the Planck scale, the logarithm is of order 50 and the self-energy, taking into account numerical factors, is roughly 20% of the actual mass of the electron. For quarks the effect is larger due to the larger $SU(3)$ coupling, so that the self-energy is of order the mass itself. In simple grand unified theories the bottom quark and tau lepton are in the same multiplet and have equal 'bare' masses, but the inclusion of the self-energies accounts to good accuracy for the observed ratio

$$\frac{m_b}{m_\tau} \approx 3 . \tag{16.2.3}$$

This is a successful test of grand unification, though less impressive than the unification of the gauge couplings because it is more model-dependent and because the ratio is not known with the same precision.

This leaves one problem in the Standard Model, the Higgs boson. This is the only scalar, and the only particle for which the estimate (16.2.1) is not reduced by relativistic quantum effects. If the Higgs boson remains pointlike up to energies near the Planck scale as in ordinary grand unified theories, then the self-energy is roughly 15 orders of magnitude larger than the actual mass. We have to suppose that the bare mass cancels this correction to an accuracy of roughly one part in 10^{30}, because it is actually the mass-squared that adds. This seems quite unsatisfactory, especially in light of the very physical way we are able to think about the other self-energies.

One possible resolution of this *naturalness problem* is that the Higgs scalar is not pointlike but actually composite on a scale not far from the weak scale. This is the idea of technicolor theories; it has not been ruled out but has not led to convincing models. A second is that there is some other effect that cancels the self-energy. Indeed, this is the case in supersymmetric theories. The Higgs mass-squared comes from the superpotential, and as discussed in section B.2 this is not renormalized: the self-energy is canceled by a fermionic loop amplitude, at least down to the scale of supersymmetry breaking.

For this reason theories with supersymmetry broken near the weak scale have received a great deal of attention, both in particle phenomenology and in string theory. The $d = 4$ supersymmetry algebra must be $N = 1$ because the gauge-couplings in the Standard Model are chiral. As discussed in section B.2, the $N = 2$ and larger algebras do not allow this.

Supersymmetric string theories are also attractive because as we will see later supersymmetry in spacetime implies a much-enlarged symmetry on the world-sheet, and so the construction and solution of these CFTs has gone much farther than for the nonsupersymmetric theories. Also, non-supersymmetric string theories usually, though not always, have tachyons in their spectra. Finally, the order-by-order supersymmetric cancellation of the vacuum energy means that there are no tadpole divergences and the perturbation theory is finite at each order.

It is still a logical possibility that all the supersymmetry of string theory is broken at the string scale, and even that the low energy limit of string theory is a technicolor theory. Low energy supersymmetry and string theory are independent ideas: either might be right and the other wrong. However, the discovery of low energy supersymmetry would be an encouraging sign that these ideas are in the right direction. Also, the measurement of the many new masses and couplings of the superpartners would give new windows onto higher energy physics. Given the important role that supersymmetry plays at short distance, and the phenomenological reasons for expecting supersymmetry near the weak scale, it is reasonable to hope that of all the new phenomena that accompany string theory supersymmetry will be directly visible.

What then are the conditions for an orbifold compactification to have an unbroken $N = 1$ supersymmetry? Let us consider first the case that the point group is $\mathbf{Z}_N$ so that it is generated by a single element of the form (16.1.11). This acts on the supersymmetries as

$$Q_\alpha \to D(\phi)_{\alpha\beta} Q_\beta , \tag{16.2.4}$$

where $D(\phi)$ is the spinor representation of the rotation. In the usual s-basis this is

$$Q_s \to \exp(2\pi i s \cdot \phi)Q_s \,. \tag{16.2.5}$$

The (s_2, s_3, s_4) run over all combinations of $\pm\frac{1}{2}$, each combination appearing twice. Thus if

$$\phi_2 + \phi_3 + \phi_4 = 0 \tag{16.2.6}$$

with the ϕs otherwise generic, there will be four unbroken supersymmetries, namely those with $s_2 = s_3 = s_4$. Three-quarters of the original 16 supersymmetries of the heterotic string are broken. Other possibilities such as $\phi_2 + \phi_3 - \phi_4 = 0$ give equivalent physics.

Note that this discussion is quite similar to the discussion of the supersymmetry of rotated D-branes in section 13.4. As there, we can express the result in a more general way. Since the rotation takes the Z^i into linear combinations of themselves, it lies in a $U(3)$ subgroup of the $SO(6)$ rotational symmetry of the six orbifold dimensions. The condition (16.2.6) states that the rotation actually lies in $SU(3)$. Under

$$SO(9,1) \to SO(3,1) \times SO(6) \to SO(3,1) \times SU(3) \,, \tag{16.2.7}$$

the **16** decomposes as derived in section B.1,

$$\mathbf{16} \to (\mathbf{2},\mathbf{4}) + (\bar{\mathbf{2}},\bar{\mathbf{4}}) \to (\mathbf{2},\mathbf{3}) + (\mathbf{2},\mathbf{1}) + (\bar{\mathbf{2}},\bar{\mathbf{3}}) + (\bar{\mathbf{2}},\mathbf{1}) \,. \tag{16.2.8}$$

If $P \subset SU(3) \subset SO(6)$, the generators $(\mathbf{2},\mathbf{1})$ and $(\bar{\mathbf{2}},\mathbf{1})$ will survive the orbifold projection and there will be unbroken $N = 1$ supersymmetry. Similarly the stricter condition

$$\phi_2 + \phi_3 = \phi_4 = 0 \tag{16.2.9}$$

implies that

$$P \subset SU(2) \subset SU(3) \subset SO(6) \,. \tag{16.2.10}$$

In this case there will be unbroken $N = 2$ supersymmetry.

16.3 Examples

The main example we will consider is based on a $\mathbf{Z}_3$ orbifold of the torus. The lattice Λ for the $\mathbf{Z}_3$ orbifold is generated by the six translations

$$t_i : \quad Z^i \to Z^i + R_i \,, \tag{16.3.1a}$$

$$u_i : \quad Z^i \to Z^i + \alpha R_i \,, \quad \alpha = \exp(2\pi i/3) \tag{16.3.1b}$$

The lattice in one complex plane is shown in figure 16.1, with R_i the lattice spacing. For $R_i = \alpha'^{1/2}$ this is the root lattice of $SU(3)$, so up to rescaling

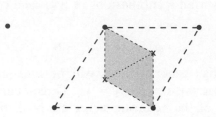

Fig. 16.1. A two-dimensional lattice invariant under rotations by $\pi/3$. A unit cell is indicated. The two points indicated by $\times$ are invariant under the combination of a $2\pi/3$ rotation and a lattice translation, as are the corner points of the unit cell. A fundamental region for the orbifold identification is shaded. One can think of the orbifold space as formed by folding the shaded region on the dotted line and identifying the edges.

of the Z^i, Λ is the root lattice of $SU(3) \times SU(3) \times SU(3)$. This is invariant under independent six-fold rotations of each $SU(3)$ lattice.

For the $\mathbf{Z}_3$ orbifold, the point group consists of a simultaneous three-fold rotation of all three lattices, the $\mathbf{Z}_3$ group $\{1, r, r^2\}$ generated by

$$r: \quad Z^2 \to \alpha Z^2 , \quad Z^3 \to \alpha Z^3 , \quad Z^4 \to \alpha^{-2} Z^4 . \qquad (16.3.2)$$

In the notation (16.1.11) this is

$$\phi_i = (\tfrac{1}{3}, \tfrac{1}{3}, -\tfrac{2}{3}) , \qquad (16.3.3)$$

which satisfies the mod 2 condition and leaves $N = 1$ supersymmetry unbroken.

Initially we will consider the simple case that there are no Wilson lines. That is, the translations Λ are not accompanied by gauge twists:[1] they are of the form $g = (1, v; 1)$. The gauge twist must satisfy the mod 2 and level-matching conditions. An easy way to do this is to have the gauge rotation act on the gauge fermions in exactly the same way as the spacetime rotation (16.1.11) acts on the $\tilde{\psi}$,

$$\beta_K = (\phi_2, \phi_3, \phi_4, 0^5; 0^8) = (\tfrac{1}{3}, \tfrac{1}{3}, -\tfrac{2}{3}, 0^5; 0^8) . \qquad (16.3.4)$$

This is called *embedding the spin connection in the gauge connection*. The two terms in the level-matching condition (16.1.24) then cancel automati-

---

[1] The gauge twists accompanying transformations with fixed points are not referred to as Wilson lines, because in a sense they *do* produce a local field strength, a delta function at the fixed point.

cally. One way to think about this is to note that the nontrivial part of the world-sheet theory is parity-invariant, which allows a coordinate-invariant Pauli–Villars regulator. We will analyze the spectrum of this model, and discuss more general gauge twists later.

Examining the untwisted sector first, we have to impose the $\mathbf{Z}_3$ projection on the states of the toroidal compactification. We assume that none of the R_i are $\alpha'^{1/2}$, to avoid extra massless states from the $SU(3)$ roots. The eigenvalues of $h = (r, 1; \gamma)$ are powers of α. We first classify the massless left- and right-moving states by their eigenvalues.

On the left-moving side are

$$\alpha^0 : \quad \alpha^\mu_{-1}|0\rangle , \quad |a\rangle \in (\mathbf{8}, \mathbf{1}, \mathbf{1}) + (\mathbf{1}, \mathbf{78}, \mathbf{1}) + (\mathbf{1}, \mathbf{1}, \mathbf{248}) , \quad (16.3.5a)$$

$$\alpha^1 : \quad \alpha^i_{-1}|0\rangle , \quad |a\rangle \in (\mathbf{3}, \mathbf{27}, \mathbf{1}) , \quad (16.3.5b)$$

$$\alpha^2 : \quad \alpha^{\bar{i}}_{-1}|0\rangle , \quad |a\rangle \in (\bar{\mathbf{3}}, \overline{\mathbf{27}}, \mathbf{1}) . \quad (16.3.5c)$$

For the states with an α_{-1} oscillator excited the eigenvalue comes from the rotation r. The states from the current algebra have been denoted by their group index $|a\rangle$, without reference to a specific (fermionic or bosonic) representation. These states have been decomposed according to their transformation under

$$SU(3) \times E_6 \times E_8 \subset E_8 \times E_8 . \quad (16.3.6)$$

This decomposition was given in section 11.4 and a derivation outlined in exercise 11.5. The $SU(3)$ acts on the first three complex gauge fermions $\lambda^{1+,2+,3+}$. The gauge rotation (16.3.4) acts on any state as

$$\exp[2\pi i(q_1 + q_2 - 2q_3)/3] , \quad (16.3.7)$$

where the q_K are the eigenvalues of the state under $U(1)^{16}$. This is an element of $SU(3)$, in fact of the center of $SU(3)$, acting as α on any element of the $\mathbf{3}$ and α^2 on any element of the $\bar{\mathbf{3}}$.

On the right-moving side h acts only through the rotation r, giving

$$\alpha^0 : \quad \tilde{\psi}^\mu_{-1/2}|0\rangle_{\mathrm{NS}} , \quad |\tfrac{1}{2}, \mathbf{1}\rangle_{\mathrm{R}} , \quad |-\tfrac{1}{2}, \bar{\mathbf{1}}\rangle_{\mathrm{R}} , \quad (16.3.8a)$$

$$\alpha^1 : \quad \tilde{\psi}^i_{-1/2}|0\rangle_{\mathrm{NS}} , \quad |\tfrac{1}{2}, \mathbf{3}\rangle_{\mathrm{R}} , \quad (16.3.8b)$$

$$\alpha^2 : \quad \tilde{\psi}^{\bar{i}}_{-1/2}|0\rangle_{\mathrm{NS}} , \quad |-\tfrac{1}{2}, \bar{\mathbf{3}}\rangle_{\mathrm{R}} . \quad (16.3.8c)$$

Here μ runs over the noncompact transverse dimensions $2, 3$. We have labeled the fermionic states by their four-dimensional helicity s_1 and by their $SU(3) \subset SO(6)$ transformation. In terms of the spins (s_2, s_3, s_4) the $\mathbf{1}$, $\bar{\mathbf{3}}$, $\mathbf{3}$, and $\bar{\mathbf{1}}$ consist of states with zero, one, two, or three $-\tfrac{1}{2}$s respectively.

Now pair up left- and right-moving states, looking first at the bosons. In the sector $\alpha^0 \cdot \alpha^0$ are

$$\alpha^\mu_{-1}\tilde{\psi}^\nu_{-1/2}|0\rangle_{\mathrm{NS}} , \quad (16.3.9)$$

which are the four-dimensional graviton, dilaton, and axion, as well as

$$\tilde{\psi}^{\mu}_{-1/2}|a\rangle_{\text{NS}} \ , \quad a \in (\mathbf{8,1,1}) + (\mathbf{1,78,1}) + (\mathbf{1,1,248}) \ . \tag{16.3.10}$$

These are $SU(3) \times E_6 \times E_8$ gauge bosons. The gauge group is just the subgroup left invariant by the twist. In the sector $\alpha^1 \cdot \alpha^2$ there are neutral scalars of the form

$$\alpha^i_{-1} \tilde{\psi}^{\bar{\jmath}}_{-1/2} |0,0\rangle_{\text{NS}} \tag{16.3.11}$$

and scalars

$$\tilde{\psi}^{\bar{\jmath}}_{-1/2}|a\rangle_{\text{NS}} \ , \quad a \in (\mathbf{3,27,1}) \ . \tag{16.3.12}$$

The sector $\alpha^2 \cdot \alpha^1$ contributes a conjugate set of states. The neutral scalars are from the internal modes of the graviton and antisymmetric tensor. In particular, the symmetric combinations are the moduli for a flat internal metric of the form

$$G_{i\bar{\jmath}} dZ^i dZ^{\bar{\jmath}}. \tag{16.3.13}$$

A metric of this form, with no $dZ^i dZ^j$ or $dZ^{\bar{\imath}} dZ^{\bar{\jmath}}$ components, is known as *Hermitean*.

The fermions are the superpartners of these. The bosonic states in each line of the right-moving spectrum (16.3.8) are replaced by the fermionic states in the same line. In the sector $\alpha^0 \cdot \alpha^0$ are the states

$$\alpha^{2\pm i3}_{-1}|s_1,\mathbf{1}\rangle_{\text{R}} \ , \tag{16.3.14}$$

which are the gravitinos with helicity $\pm\frac{3}{2}$ and the dilatinos with helicity $\pm\frac{1}{2}$. The other components of the ten-dimensional gravitino are in the sectors $\alpha^0 \cdot \alpha^{1,2}$ and are removed by the projection, consistent with the earlier deduction that the theory has $N = 1$ supersymmetry. The other spinors with helicity $\frac{1}{2}$ are from the sector $\alpha^2 \cdot \alpha$:

$$|a,\tfrac{1}{2},\mathbf{3}\rangle_{\text{R}} \ , \quad a \in (\bar{\mathbf{3}},\overline{\mathbf{27}},\mathbf{1}) \ , \tag{16.3.15a}$$

$$\alpha^{\bar{\imath}}_{-1}|\tfrac{1}{2},\mathbf{3}\rangle_{\text{R}} \ . \tag{16.3.15b}$$

Now consider the twisted sectors. There are 27 equivalence classes with rotation r, corresponding to the elements

$$h = r t^{n_2}_2 t^{n_3}_3 t^{n_4}_4 \ , \quad n_i \in \{0,1,2\}. \tag{16.3.16}$$

The inverses of these give 27 classes with rotation r^2. The classes are in one-to-one correspondence with the fixed points,[2] which are at

$$Z^i = \frac{\exp(i\pi/6)}{3^{1/2}}(n_2 R_2, n_3 R_3, n_4 R_4) \ . \tag{16.3.17}$$

---

[2] This one-to-one correspondence does not hold for more complicated space groups.

These are all related by translation, and so give 27 copies of the same spectrum. Thus we need only analyze the class $h = (r, 1; \gamma)$.

We analyze strings twisted by the rotation r. Starting on the right-moving side, in the R sector the μ oscillators are all integer moded, while the i oscillators have mode numbers $n + \frac{2}{3}$. The zero-point energy vanishes by the usual cancellation between Bose and Fermi contributions in the R sector. The only fermionic zero modes are from the spacetime fermions, $\tilde{\psi}_0^{2 \pm i3}$, so there are two ground states $| \pm \frac{1}{2} \rangle_{h,\mathrm{R}}$. To figure out which survives the GSO projection we look at the bosonized vertex operators. As in eq. (10.3.25), states of a spinor field in a sector with periodicity $\varphi(\sigma_1 + 2\pi) = \exp(2\pi i \zeta)\varphi(\sigma_1)$ have vertex operators

$$e^{isH} , \quad s = \tfrac{1}{2} - \zeta \bmod 1 , \quad e^{i\tilde{s}\tilde{H}} , \quad \tilde{s} = -\tfrac{1}{2} + \zeta \bmod 1 , \qquad (16.3.18)$$

for left- or right-movers respectively. This follows from the OPE of the bosonized spinor with the vertex operator. The vertex operators for the R sector twisted states then have

$$\exp(i\tilde{s}_a \tilde{H}_a) , \quad \tilde{\mathbf{s}} = (\pm\tfrac{1}{2}, -\tfrac{1}{6}, -\tfrac{1}{6}, -\tfrac{1}{6}) . \qquad (16.3.19)$$

The GSO projection as defined in chapter 10 is

$$\exp[\pi i(\tilde{s}_1 + \tilde{s}_2 + \tilde{s}_3 + \tilde{s}_4)] = 1 . \qquad (16.3.20)$$

Thus it is the state

$$| + \tfrac{1}{2} \rangle_{h,\mathrm{R}} \qquad (16.3.21)$$

that remains.

In the NS sector, the fermionic modes are shifted by $\frac{1}{2}$ to $n + \frac{1}{6}$. The zero-point energy is $\frac{1}{36}$ for a complex boson of shift $\frac{1}{3}$ or $\frac{2}{3}$, and $-\frac{1}{72}$ for a complex boson of shift $\frac{1}{6}$ or $\frac{5}{6}$, and the negative in either case for a fermion, giving

$$-\frac{2}{24} - \frac{2}{48} + \frac{3}{36} + \frac{3}{72} = 0 . \qquad (16.3.22)$$

The only massless state is then the ground state

$$|0\rangle_{h,\mathrm{NS}} . \qquad (16.3.23)$$

On the left-moving side, we will figure out the spectrum in the fermionic formulation. In the (R,NS) sector the three twisted complex fermions have mode numbers $n + \frac{1}{3}$, and the zero-point energy is

$$-\frac{2}{24} + \frac{3}{36} - \frac{3}{36} + \frac{10}{24} - \frac{16}{48} = 0 . \qquad (16.3.24)$$

There are ten fermionic zero modes, $\lambda_0^{I\pm}$ for $I = 7, \ldots, 16$, so there are 32 ground states forming a **16** and **$\overline{\mathbf{16}}$** of $SO(10)$. Again examining the vertex operator as in eq. (16.3.19), the current algebra GSO projection is

$$\sum_{K=1}^{8} q_K \in 2\mathbf{Z} \qquad (16.3.25)$$

in terms of the $U(1)^8$ charges. The vertex operators (16.3.18) imply that $q_1 = q_2 = q_3 = \frac{1}{6}$, and the projection then picks out the $\overline{\mathbf{16}}$.

The other current algebra sector with massless states is (NS,NS), with zero-point energy

$$-\frac{2}{24} + \frac{3}{36} + \frac{3}{72} - \frac{10}{48} - \frac{16}{48} = -\frac{1}{2} . \qquad (16.3.26)$$

There are several massless states,

$$\lambda_{-1/6}^{1+} \lambda_{-1/6}^{2+} \lambda_{-1/6}^{3+} |0\rangle_{\text{NS,NS}} , \quad \lambda_{-1/2}^{I} |0\rangle_{\text{NS,NS}} , \ 7 \leq I \leq 16 , \quad (16.3.27a)$$

$$\lambda_{-1/6}^{K+} \alpha_{-1/3}^{\bar{J}} |0\rangle_{\text{NS,NS}} , \ K = 1,2,3 . \qquad (16.3.27b)$$

The states in the first line are a singlet and a $\mathbf{10}$ of $SO(10)$, combining with the $\overline{\mathbf{16}}$ from the (R,NS) sector to form a $\overline{\mathbf{27}}$ of E_6. The nine states in the second line transform as three $\mathbf{3}$s of the gauge $SU(3)$, distinguished from one another by the index $\bar{J}$. In all, the left-moving spectrum contains the massless states

$$(\mathbf{1}, \overline{\mathbf{27}}, \mathbf{1}) + (\mathbf{3}, \mathbf{1}, \mathbf{1})^3 . \qquad (16.3.28)$$

As we have seen from the discussion of modular invariance, the h-projection is equivalent to level matching, so we can match either right-moving state (16.3.21) or (16.3.23) with any left-moving state (16.3.28). The classes (16.3.16) with rotation r give 27 copies of this spectrum, while the twisted sectors with rotation r^2 give the antiparticles.

Connection with grand unification

One of the factors in the low energy gauge group is E_6. As discussed in section 11.4, this is a possible grand unified group for the Standard Model. In E_6 unification, a generation of quarks and leptons is in the $\mathbf{27}$ or $\overline{\mathbf{27}}$ of E_6. Which representation we call the $\mathbf{27}$ and which the $\overline{\mathbf{27}}$ is a matter of convention. These are precisely the representations appearing in the $\mathbf{Z}_3$ orbifold: the helicity $\frac{1}{2}$ states that are charged under E_6 are all in the $\overline{\mathbf{27}}$ of E_6. The untwisted states (16.3.15a) comprise nine generations, forming a triplet of the gauge $SU(3)$ and a triplet of $SU(3) \in SO(6)$, and each twisted sector (16.3.28) with rotation r contributes one $\overline{\mathbf{27}}$, for 36 in all. Notice in particular that the matter is chiral, the helicity $+\frac{1}{2}$ and $-\frac{1}{2}$ states carrying different representations of the gauge group. The GSO projection correlates the spacetime helicity with the internal components of the spin, while the twist contains both a spacetime and a gauge rotation and so correlates the internal spin with the gauge quantum numbers.

Of course this model has too many generations to be realistic, but just the same it is interesting to look at how the gauge symmetry would be reduced to the Standard Model $SU(3) \times SU(2) \times U(1)$. As we will discuss later, the $SU(3)$ symmetry can be broken by the twisted sector states $(\mathbf{3}, \mathbf{1}, \mathbf{1})$. There are no light states that carry both the E_6 and E_8 gauge quantum numbers, so if the Standard Model is embedded in the former the latter is *hidden,* detectable only through gravitational strength interactions. We will see later that this can have important effects, but for now we can ignore it. This leaves the E_6 factor. From experience with grand unified theories, one might expect that this could be broken to the Standard Model gauge group by the Higgs mechanism, the expectation value of a scalar field. However, that is not possible here. All scalars with E_6 charge are in the $\mathbf{27}$ representation or its conjugate, and it is not possible to break E_6 to the Standard Model gauge group with this representation. Consulting the decomposition (11.4.25), there are two components of the $\mathbf{27}$ that are neutral under $SU(3) \times SU(2) \times U(1)$, but even if both have expectation values the gauge symmetry is broken only to $SU(5)$. To break $SU(5)$ to $SU(3) \times SU(2) \times U(1)$, the smallest possible representation is the adjoint $\mathbf{24}$, but this is not contained in the $\mathbf{27}$ of E_6. We will see in chapter 18 that this is a general property of level one current algebras. The current algebras here are at level one just as in ten dimensions, because the orbifold projection does not change their OPEs.

There are still several ways to break to the Standard Model gauge group. One is to include Wilson lines on the original torus. The full twist group of the orbifold is generated by the four elements

$$h_1 = (r, 0; \gamma) \,, \quad h_2 = (1, t_2; \gamma_2) \,, \quad h_3 = (1, t_3; \gamma_3) \,, \quad h_4 = (1, t_4; \gamma_4) \,,$$
(16.3.29)

where the translations are now accompanied by gauge rotations. The gauge twists are highly constrained. For example, $t_2 t_3 = t_3 t_2$ implies that $\gamma_2 \gamma_3 = \gamma_3 \gamma_2$ by the homomorphism property (no pure gauge twists). Also, $r^3 = 1$ implies that $\gamma^3 = 1$, while $(rt_i)^3 = 1$ implies that $(\gamma \gamma_i)^3 = 1$, and so on. Further, all the elements

$$h_1 h_2^{n_2} h_3^{n_3} h_4^{n_4}$$
(16.3.30)

must satisfy the mod 2 and level-matching conditions. Unlike the simple toroidal compactification, the general solution is not known; the number of inequivalent solutions has been estimated to be at least 10^6. Various examples resembling the Standard Model have been found. We will give one below.

We should note that if the low energy $SU(3) \times SU(2) \times U(1)$ is embedded

Table 16.1. *Allowed gauge twists for the $\mathbf{Z}_3$ orbifold, and the resulting gauge groups.*

	β_K	gauge group
(i)	$(\frac{1}{3}, \frac{1}{3}, -\frac{2}{3}, 0^5; 0^8)$	$E_6 \times SU(3) \times E_8$
(ii)	$(0^8; 0^8)$	$E_8 \times E_8$
(iii)	$(\frac{1}{3}, \frac{1}{3}, 0^6; -\frac{2}{3}, 0^7)$	$E_7 \times U(1) \times SO(14) \times U(1)$
(iv)	$(\frac{1}{3}, \frac{1}{3}, -\frac{2}{3}, 0^5; \frac{1}{3}, \frac{1}{3}, -\frac{2}{3}, 0^5)$	$E_6 \times SU(3) \times E_6 \times SU(3)$
(v)	$(\frac{1}{3}, \frac{1}{3}, \frac{1}{3}, \frac{1}{3}, \frac{2}{3}, 0^3; \frac{2}{3}, 0^7)$	$SU(9) \times SO(14) \times U(1)$

in the standard way in E_6, then the usual grand unified prediction $\sin^2 \theta_w = \frac{3}{8}$ still holds with Wilson line breaking even though there is no scale at which the theory looks like a four-dimensional unified theory. The reason is the inheritance principle that orbifold projections do not change the couplings of untwisted states such as the gauge bosons.

A different route to symmetry breaking is to use higher level current algebras. One way to construct an orbifold model of this type is to start with an orbifold that has two copies of the same group. For example, embedding the spin connection in each E_8 (twist (iv) in table 16.1), leaves an unbroken $SU(3) \times E_6 \times SU(3) \times E_6$. Add a twist that has the effect of interchanging the two E_6s so that only the diagonal E_6

$$j^a = j_{(1)}^a + j_{(2)}^a \tag{16.3.31}$$

survives. The z^{-2} term in the OPE is additive, so the level is now $k = 2$. The resulting model has larger representations which can break the unified group down to the Standard Model. Realistic models of this type have been constructed. The higher level and Wilson line breakings have an important difference in terms of the scale of symmetry breaking, as we discuss further in chapter 18.

Generalizations

Staying with the $\mathbf{Z}_3$ orbifold but considering more general gauge twists, there are five inequivalent solutions to the mod 2 and level-matching conditions. These are shown in table 16.1, the first twist being the solution (16.3.4) with gauge rotation equal to spacetime rotation. One realistic model with Wilson lines uses the twist (v) in the table, with

$$\gamma_2 = (0^7, \tfrac{2}{3}; 0, \tfrac{1}{3}, \tfrac{1}{3}, 0^6) \,, \tag{16.3.32a}$$

$$\gamma_3 = 0 \,, \tag{16.3.32b}$$

$$\gamma_4 = (\tfrac{1}{3}, \tfrac{1}{3}, \tfrac{1}{3}, \tfrac{2}{3}, \tfrac{1}{3}, 0, \tfrac{1}{3}, \tfrac{1}{3}; \tfrac{1}{3}, \tfrac{1}{3}, 0^6) \,. \tag{16.3.32c}$$

With Wilson lines the gauge twist (16.3.30) is in general different at each fixed point, so the spectra in the different twisted sectors are no longer the same. Because $\gamma_3 = 0$ in this example, the 27 fixed points fall into nine sets of three, and in fact the Wilson lines have been chosen so as to reduce the number of generations to three. The gauge group is $SU(3) \times SU(2) \times U(1)^5$, with a hidden $SO(10) \times U(1)^3$. The chiral matter comprises precisely three generations, accompanied by a number of nonchiral $SU(2)$ doublets (Higgs fields) and $3 + \bar{3}$s of $SU(3)$. There are also some massless fields coupling to the hidden gauge group, and some singlet fields. The obvious problems with this model are the extra $U(1)$ gauge symmetries, and the extra color triplets which can mediate baryon decay. Some of the singlets are moduli, and in certain of the flat directions the extra $U(1)$s are broken and the triplets heavy. Of course, given the enormous number of consistent CFTs, as well as the large number of free parameters (moduli) in each, string theory will not have real predictive power until the dynamics that selects the vacuum is understood. We will say more about this later.

Another orbifold is a square lattice in each plane with the $\mathbf{Z}_4$ rotation

$$r' : \quad Z^2 \to iZ^2 , \quad Z^3 \to iZ^3 , \quad Z^4 \to i^{-2}Z^4 . \tag{16.3.33}$$

Let us again embed the spin connection in the gauge connection. This will share certain features with the $\mathbf{Z}_3$ orbifold. In particular, the gauge twist is again in $SU(3)$ so the unbroken group will include an $E_6 \times E_8$ factor, and the spin-$\frac{1}{2}$ states will again be in the **27** and the $\overline{\mathbf{27}}$. This will hold for any model with rotation in $SU(3)$ and with spin connection embedded in gauge connection. The extra gauge factor depends on the model; here it is $SU(2) \times U(1)$ rather than $SU(3)$. Another difference is that the modulus

$$dZ^4 d\bar{Z}^4 \tag{16.3.34}$$

now survives the twist, in addition to the mixed components

$$dZ^i d\bar{Z}^{\bar{\jmath}} . \tag{16.3.35}$$

This corresponds to a change in the complex structure of the compactified dimensions: when this modulus is turned on the metric is no longer Hermitean, though it becomes Hermitean again by redefining the Z^i. The moduli (16.3.34) are thus known as complex structure moduli. The moduli (16.3.35) are known as Kähler moduli, for reasons to be explained in the next chapter.

A third difference is that one finds that the helicity-$\frac{1}{2}$ states include *both* **27**s and the $\overline{\mathbf{27}}$s. We will see in later chapters that this is correlated with the appearance of the two kinds of moduli. These are generations and antigenerations, the latter having the opposite chirality. In the Standard Model there are no antigenerations, but these can obtain mass by pairing with some of the generations when some scalar fields are given expectation values. The net number of generations is the difference, in which by definition the generations are whichever of the **27**s and $\overline{\bf 27}$s are more numerous.

World-sheet supersymmetries

There is an important general pattern which will apply beyond the orbifold example. The supercurrent for the compact CFT can be separated into two pieces that separately commute with the twist,

$$i\sum_{i=2}^{4} \bar{\partial} Z^i \tilde{\psi}^{\bar{\imath}} \,, \quad i\sum_{i=2}^{4} \bar{\partial} Z^{\bar{\imath}} \tilde{\psi}^i \,. \tag{16.3.36}$$

These, together with the energy-momentum tensor and the current

$$\sum_{i=2}^{4} \tilde{\psi}^i \tilde{\psi}^{\bar{\imath}} \,, \tag{16.3.37}$$

form a right-moving $N = 2$ superconformal algebra. This is a global symmetry of the internal CFT. We will see in chapter 19 that there is a close connection between (0,2) supersymmetry on the world-sheet and $N = 1$ supersymmetry in spacetime.

When in addition the gauge twist is equal to the spacetime twist, we can do the same thing with the $\lambda^{K\pm}$ for $K = 1, 2, 3$, forming the left-moving supercurrents

$$i\sum_{i=2}^{4} \partial Z^i \lambda^{(i-1)-} \,, \quad i\sum_{i=2}^{4} \partial Z^{\bar{\imath}} \lambda^{(i-1)+} \,. \tag{16.3.38}$$

In this case the compact part of the world-sheet theory separates into 26 free current algebra fermions and a $(c, \tilde{c}) = (9, 9)$ CFT which has (2,2) world-sheet supersymmetry. String theories of this type are highly constrained, as we will see in chapter 19.

16.4   Low energy field theory

It is interesting to look in more detail at the low energy field theory resulting from the $\mathbf{Z}_3$ orbifold.

Untwisted states

For the untwisted fields of the orbifold compactification, we can determine the low energy effective action without a stringy calculation. The action for these fields follows directly from the ten-dimensional action via the inheritance principle. The ten-dimensional bosonic low energy action

$$S_{\text{het}} = \frac{1}{2\kappa_{10}^2} \int d^{10}x \, (-G)^{1/2} e^{-2\Phi} \left[R + 4\partial_\mu \Phi \partial^\mu \Phi - \frac{1}{2}|\tilde{H}_3|^2 - \frac{\alpha'}{4}\text{Tr}_v(|F_2|^2) \right]$$

(16.4.1)

is determined entirely by supersymmetry, with

$$\tilde{H}_3 = dB_2 - \frac{\alpha'}{4}\text{Tr}_v(A_1 \wedge dA_1 - 2iA_1 \wedge A_1 \wedge A_1/3) \, .$$

(16.4.2)

The trace is normalized to the vector representation of $SO(16)$.

It is very instructive to carry out this exercise. Insert into the action those fields that survive the $\mathbf{Z}_3$ projection,

$$G_{\mu\nu} \, , \ B_{\mu\nu} \, , \ \Phi \, , \ G_{i\bar{j}} \, , \ B_{i\bar{j}} \, , \ A_\mu^a \, , \ A_{ij\bar{x}} \, , \ A_{\bar{i}jx} \, .$$

(16.4.3)

We have subdivided the gauge generators into a in the adjoint of $SU(3) \times E_6 \times E_8$, jx in the $(\mathbf{3}, \mathbf{27}, \mathbf{1})$, and $\bar{j}\bar{x}$ in the $(\bar{\mathbf{3}}, \overline{\mathbf{27}}, \mathbf{1})$. Now dimensionally reduce by requiring the fields to be slowly varying functions of the x^μ and to be independent of the x^m.

Let us first ignore the ten-dimensional gauge field. The reduction is then a special case of that for the bosonic string in chapter 8,

$$S = \frac{1}{2\kappa_4^2} \int d^4x \, (-G)^{1/2} \left[R - 2\partial_\mu \Phi_4 \partial^\mu \Phi_4 - \frac{1}{2}e^{-4\Phi_4}|H_3|^2 \right.$$
$$\left. - \frac{1}{2}G^{i\bar{j}}G^{k\bar{l}}(\partial_\mu G_{i\bar{l}}\partial^\mu G_{\bar{j}k} + \partial_\mu B_{i\bar{l}}\partial^\mu B_{\bar{j}k}) \right] \, .$$

(16.4.4)

We have defined the four-dimensional dilaton

$$\Phi_4 = \Phi - \frac{1}{4}\det G_{mn} \, .$$

(16.4.5)

We have also made a Weyl transformation to the four-dimensional Einstein metric

$$G_{\mu\nu\,\text{Einstein}} = e^{-2\Phi_4}G_{\mu\nu} \, ;$$

(16.4.6)

henceforth in this chapter this metric is used implicitly. This action differs from the bosonic reduction (8.4.2) in that the projection has removed the Kaluza–Klein and antisymmetric tensor gauge bosons and the ij and $\bar{i}\bar{j}$ components of the internal metric and antisymmetric tensor. Note that

$$G_{i\bar{j}} = G_{\bar{j}i} = G_{\bar{j}i}^* = G_{i\bar{j}}^* \, ,$$

(16.4.7a)

$$B_{i\bar{j}} = -B_{\bar{j}i} = -B_{\bar{j}i}^* = B_{i\bar{j}}^* \, .$$

(16.4.7b)

The action must be of the general form (B.2.28) required by $N = 1$ supersymmetry. To make the comparison we must first convert the antisymmetric tensor to a scalar as in section B.4,[3]

$$-\frac{1}{2}\int d^4x\,(-G)^{1/2}e^{-4\Phi_4}|H_3|^2 + \int a\,dH_3$$

$$\rightarrow -\frac{1}{2}\int d^4x\,(-G)^{1/2}e^{4\Phi_4}\partial_\mu a\partial^\mu a \ . \qquad (16.4.8)$$

The action then takes the form

$$\frac{1}{2\kappa_4^2}\int d^4x\,(-G)^{1/2}\left[R - \frac{2\partial_\mu S^*\partial^\mu S}{(S+S^*)^2} - \frac{1}{2}G^{i\bar{j}}G^{k\bar{l}}\partial_\mu T_{i\bar{l}}\partial^\mu T_{\bar{j}k}\right] , \qquad (16.4.9)$$

where

$$S = e^{-2\Phi_4} + ia \ , \qquad T_{i\bar{j}} = G_{i\bar{j}} + B_{i\bar{j}} \ . \qquad (16.4.10)$$

This is of the supergravity form (B.2.28) with the Kähler potential

$$\kappa_4^2 K = -\ln(S + S^*) - \ln\det(T_{i\bar{j}} + T^*_{i\bar{j}}) \ . \qquad (16.4.11)$$

The index i in eq. (B.2.28) is the same as the pair $i\bar{j}$ in eq. (16.4.9).

Now add the four-dimensional gauge field. In addition to its kinetic term, this appears in the Bianchi identity for the field strength $\tilde{H}$, so that the left-hand side of eq. (16.4.8) becomes

$$-\frac{1}{2}\int d^4x\,(-G)^{1/2}e^{-4\Phi_4}|\tilde{H}_3|^2 + \int a\left[d\tilde{H}_3 + \frac{\alpha'}{4}\mathrm{Tr}_{\mathrm{v}}(F_2 \wedge F_2)\right] \ . \qquad (16.4.12)$$

After Poincaré duality the additional terms in the action are

$$-\frac{1}{4g_4^2}\int e^{-2\Phi_4}\mathrm{Tr}_{\mathrm{v}}(|F_2|^2) + \frac{1}{2g_4^2}\int a\,\mathrm{Tr}_{\mathrm{v}}(F_2 \wedge F_2) \qquad (16.4.13)$$

with $g_4^2 = 4\kappa_4^2/\alpha'$. This is of the supergravity form (B.2.28) with the gauge kinetic term

$$f_{ab} = \frac{\delta_{ab}}{g_4^2}S \ . \qquad (16.4.14)$$

Finally add the scalars coming from the ten-dimensional gauge field. The calculations are a bit longer and are left to the references. The final result is that the Kähler potential is modified to

$$\kappa_4^2 K = -\ln(S + S^*) - \ln\det\left[T_{i\bar{j}} + T^*_{i\bar{j}} - \alpha'\mathrm{Tr}_{\mathrm{v}}(A_i A^*_j)\right] , \qquad (16.4.15)$$

there is a superpotential

$$W = \epsilon^{ijk}\mathrm{Tr}_{\mathrm{v}}(A_i\,[A_j, A_k]) \ , \qquad (16.4.16)$$

---

[3] One could instead use Poincaré duality to write the supergravity action using an antisymmetric tensor. This is known as the *linear multiplet* formalism and appears often in the string literature.

and the gauge kinetic term is unchanged. The superpotential accounts for the potential energy from the reduction of $F_{mn}F^{mn}$. We have kept the scalars in matrix notation. In components this becomes

$$W = \epsilon^{ijk}\epsilon^{\bar{l}\bar{m}\bar{n}}d^{\bar{x}\bar{y}\bar{z}}A_{i\bar{l}\bar{x}}A_{j\bar{m}\bar{y}}A_{k\bar{n}\bar{z}} . \tag{16.4.17}$$

Here $d^{\bar{x}\bar{y}\bar{z}}$ is the $\mathbf{27}^3$ invariant of E_6, which has just the right form to give rise to the quark and lepton masses.

We will see in chapter 18 that several features found in this example actually apply to the tree-level effective action of every four-dimensional heterotic string theory.

T-duality

The original toroidal compactification had T-duality $O(22,6,\mathbf{Z})$. The subgroup of this that commutes with the $\mathbf{Z}_3$ twist will survive as a T-duality of the orbifold theory. In this case it is an $SU(3,3,\mathbf{Z})$ subgroup. It is interesting to look at the special case that $T_{i\bar{j}}$ is diagonal,

$$T_{i\bar{j}} = T_i\delta_{ij} , \quad \text{no sum on } i , \tag{16.4.18}$$

and work only to second order in the A_i. The Kähler potential becomes

$$\kappa_4^2 K = -\ln(S+S^*) - \sum_i \ln(T_i+T_i^*) + \alpha' \sum_i \frac{\text{Tr}_v(A_iA_i^*)}{T_i+T_i^*} . \tag{16.4.19}$$

For this form of $T_{i\bar{j}}$ the lattice is a product of three two-dimensional lattices. We analyzed the T-duality of a two-dimensional toroidal lattice in section 8.4, finding it to be essentially $PSL(2,\mathbf{Z}) \times PSL(2,\mathbf{Z})$. The first factor acts on τ, which characterizes the shape (complex structure) of the torus, while the second acts on ρ, which characterizes the size of the torus and the $B_{i\bar{i}}$ background. In the $\mathbf{Z}_3$ orbifold the twist fixes the shape, so $\tau = \exp(\pi i/3)$, while $\rho = iT_i$ in each plane. Thus there is a $PSL(2,\mathbf{Z})^3$ T-duality subgroup that acts as

$$T_i \to \frac{a_iT_i - ib_i}{ic_iT_i + d_i} , \quad a_id_i - b_ic_i = 1 . \tag{16.4.20}$$

This takes

$$T_i + T_i^* \to \frac{T_i + T_i^*}{|ic_iT_i + d_i|^2} . \tag{16.4.21}$$

The second term in the Kähler potential (16.4.19) is not invariant under this, changing by

$$\kappa_4^2 K \to \kappa_4^2 K + \text{Re}\left[\sum_i \ln(ic_iT_i + d_i)\right] . \tag{16.4.22}$$

This does not affect the kinetic terms because it is the real part of a holomorphic function; in other words, this is a Kähler transformation (B.2.32). The final term in the Kähler potential is invariant provided that

$$A_i \to \frac{A_i}{ic_i T_i + d_i} \, . \tag{16.4.23}$$

The superpotential (16.4.16) then transforms as

$$W \to \frac{W}{\prod_{i=2}^{4}(ic_i T_i + d_i)} \, . \tag{16.4.24}$$

This is consistent with the general Kähler transformation (B.2.33).

The space of untwisted moduli is the subspace of the toroidal moduli space that is left invariant by $\mathbf{Z}_3$. For the moduli $T_{i\bar{j}}$ this is

$$\frac{SU(3,3)}{SU(3) \times SU(3) \times SU(3,3,\mathbf{Z})} \, . \tag{16.4.25}$$

There are also flat directions for the matter fields A_i, giving a larger coset in all. The full moduli space for the untwisted fields is the product of this space with the dilaton–axion moduli space

$$\frac{SU(1,1)}{U(1) \times PSL(2,\mathbf{Z})} \, . \tag{16.4.26}$$

For orbifolds having complex structure moduli (16.3.34), the T-duality group would contain an additional $PSL(2,\mathbf{Z})$ acting on the complex structure moduli U. Various subsequent expressions are appropriately generalized. In particular the moduli space is a product of three cosets: one for the dilaton, one for the Kähler moduli, and one for the complex structure moduli.

Twisted states

For the untwisted states we were able to learn a remarkable amount from general arguments, without detailed calculations. To find the effective action for the twisted states it is necessary to do some explicit calculations with twisted state vertex operators. These methods are well developed but are too detailed for the scope of this book, so we will simply cite a few of the most interesting results.

The main one has to do with the E_6 singlet states in each twisted sector,

$$\lambda^{K+}_{-1/6} \alpha^{\bar{j}}_{-1/3} |0\rangle_{\text{NS,NS}} \, , \quad K = 1, 2, 3 \, , \tag{16.4.27}$$

transforming as three triplets of the gauge $SU(3)$. The result is that these do not appear in the superpotential, and as a consequence the potential has a flat direction with an interesting geometric interpretation. The potential for these modes comes only from the $SU(3)$ D-term. Defining the field $M_{K\bar{J}}$ associated with these states, the D-term (B.2.20) is

$$D^a \propto M^*_{K\bar{J}} t^a_{KL} M_{L\bar{J}} = \text{Tr}(M^\dagger t^a M) , \qquad (16.4.28)$$

where t^a are the fundamental $SU(3)$ matrices. Since the t^a run over a complete set of traceless matrices, D^a can vanish for all a only if

$$MM^\dagger = \rho^2 I \quad \Rightarrow \quad M = \rho U , \qquad (16.4.29)$$

with I the identity, ρ a real constant, and U unitary. The matrix U can be taken to the identity by an $SU(3)$ gauge rotation. Thus there is a one-parameter family of vacua, along which the $SU(3)$ symmetry is completely broken.

These vacua can be understood as compactification on manifolds in which the orbifold singularity has been smoothed out (*blown up*); ρ is the radius of curvature. Thus the orbifold is a limit of the smooth spaces that we will discuss in the next chapter. Indeed, it is known that for some values of the moduli these spaces have orbifold singularities. The orbifold construction shows that the physics remain well-behaved even when the geometry appears to be singular.

The existence of the flat direction ρ can be understood as a general consequence of (2,2) world-sheet supersymmetry, the subject of chapter 19. For compactifications with less world-sheet supersymmetry similar results often hold but they are more model-dependent. We noted above that the number of consistent solutions for orbifolds with Wilson lines is a large number, of order 10^6. This is typical for free CFT constructions. However, when one takes into account that these are embedded in a larger space of smooth compactifications, many of them lie within the same moduli space and the number of distinct moduli spaces is much smaller. As we will discuss in chapter 19, with the inclusion of nonperturbative effects the number of disconnected vacua becomes smaller still.

The moduli spaces for the smooth geometries are in general more complicated and less explicitly known than the cosets that parameterize the orbifolds. The CFT corresponding to a general background of the twisted moduli is not free, because the twisted vertex operators are rather complicated. Expanding in powers of the twisted fields, the first few terms can be determined by considering string scattering amplitudes. For example, denoting a general twisted field (modulus or generation) by C_α, the leading correction to the Kähler potential takes the form

$$C_\alpha C^*_\alpha \prod_{i=2}^{4} (T_j + T^*_j)^{n^i_\alpha} . \qquad (16.4.30)$$

The constants n^i_α, known as *modular weights*, can be determined from the scattering amplitudes, and for general orbifold theories are given in the references. The invariance of the Kähler metric implies the T-duality transformation

$$C_\alpha \to C_\alpha \prod_{i=2}^{4} (ic_i T_i + d_i)^{n_\alpha^i} . \tag{16.4.31}$$

Threshold corrections

The effective action obtained above receives corrections from string loops. The most important of these is the one-loop correction to the gauge coupling, the *threshold corrections* from loops of heavy particles. The Standard Model gauge couplings are known to sufficient accuracy that predictions from unification are sensitive to this correction. Also, the dependence of f_{ab} on fields other than S comes only from one loop, and we will see later that this has an important connection with supersymmetry breaking.

To one-loop accuracy the physical gauge coupling at a scale μ can be written

$$\frac{1}{g_a^2(\mu)} = \frac{Sk_a}{g_4^2} + \frac{b_a}{16\pi^2} \ln \frac{m_{\rm SU}^2}{\mu^2} + \frac{1}{16\pi^2} \tilde{\Delta}_a , \tag{16.4.32}$$

where SU stands for *string unification*. The subscript on g_a denotes a specific factor in the gauge group, whereas that on g_4 denotes the dimension. The first term on the right is the tree-level coupling; in the present case the current algebra level is $k_a = 1$, but for future reference we give the more general form, to be discussed in chapter 18. The second is due to the running of the coupling below the string scale, with the coefficient b_a being related to the renormalization group beta function by

$$\beta_a = \frac{b_a g_a^3}{16\pi^2} . \tag{16.4.33}$$

The final term $\tilde{\Delta}_a$ is the threshold correction. It depends on the masses of all the string states, and therefore on the moduli.

A great deal is known about $\tilde{\Delta}_a$. To calculate it directly one considers the torus amplitude in a constant background field $F_{\mu\nu}^a$, which appears in the world-sheet action in the form $F_{\mu\nu}^a j^a X^\mu \bar{\partial} X^\nu$. This can be simplified by the same sort of manipulations as we used in section 12.6 to obtain explicit loop amplitudes, though the details are longer and we just sketch the results. It is useful to separate the threshold correction as follows:

$$\tilde{\Delta}_a = \Delta_a + 16\pi^2 k_a Y . \tag{16.4.34}$$

The second term has the same dependence k_a on the gauge group as does the tree-level term. It therefore does not affect the predictions for ratios of couplings or for the unification scale, though it is important for some purposes as we will mention below.

The term Δ_a is given by an integral over moduli space,

$$\Delta_a = \int_\Gamma \frac{d^2\tau}{\tau_2} [\mathscr{B}_a(\tau, \bar{\tau}) - b_a] \,. \tag{16.4.35}$$

Here the function $\mathscr{B}_a(\tau, \bar{\tau})$ is related to a trace over the string spectrum weighted by Q_a^2, with Q_a the gauge charge. The limit of $\mathscr{B}_a(\tau, \bar{\tau})$ as $\tau \to i\infty$ is just b_a, so this integral converges; the term b_a was subtracted out by the matching onto the low energy field theory behavior. Also,

$$m_{\mathrm{SU}} = \frac{2 \exp[(1-\gamma)/2]}{3^{3/4}(2\pi\alpha')^{1/2}} \,, \tag{16.4.36}$$

where $\gamma \approx 0.577$ is Euler's constant. We will discuss the physical meaning of this scale in chapter 18. The correction Y is also given by an integral over moduli space; the calculation and final expression are somewhat more complicated than for Δ_a, due in part to the need to separate IR divergences.

For orbifold compactifications, Δ_a can be evaluated in closed form. Let us point out one important general feature. The path integral on the torus includes a sum over the twists h_1 and h_2 in the two directions. If these are generic, so that they lie in $SU(3)$ but not in any proper subgroup (in other words, if they leave only $N = 1$ supersymmetry unbroken), then they effectively force the fields in the path integral to lie near some fixed point. The path integral is therefore insensitive to the shape of the *spacetime* torus and so is independent of the untwisted moduli. If on the other hand h_1 and h_2 lie in $SU(2) \subset SU(3)$, leaving $N = 2$ unbroken, then the amplitude can depend on the moduli. An example is the $\mathbf{Z}_4$ orbifold (16.3.33), in a sector in which $h_1 = 1$ and $h_2 = r'^2$. In particular h_2 acts as

$$r'^2: \quad Z^2 \to -Z^2 \,, \quad Z^3 \to -Z^3 \,, \quad Z^4 \to +Z^4 \,. \tag{16.4.37}$$

The field Z^4 is completely untwisted and so can wander over the whole spacetime torus. The threshold correction correspondingly depends on both the Kähler and complex structure moduli, T_4 and U_4. Finally, if $h_1 = h_2 = 1$ so that they leave $N = 4$ unbroken, then the threshold correction vanishes due to the $N = 4$ supersymmetry.

The actual form of the threshold correction is

$$\Delta_a = c_a - \sum_i \frac{b_a^i |P^i|}{|P|} \left\{ \ln\left[(T_i + T_i^*)|\eta(T_i)|^4\right] + \ln\left[(U_i + U_i^*)|\eta(U_i)|^4\right] \right\} \,, \tag{16.4.38}$$

with c_a independent of the moduli and η the Dedekind eta function. The sum runs over all pairs (h_1^i, h_2^i) that leave $N = 2$ unbroken. Here P is the orbifold point group, P^i is the discrete group generated by h_1^i and h_2^i, and $|P^i|$ and $|P|$ are the orders of these groups. Also, b_a^i is the beta function coefficient for the $N = 2$ theory on the T^6/P^i orbifold, and T_i and U_i are the moduli for the fixed plane. For the $\mathbf{Z}_3$ orbifold there are no $N = 2$ sectors and the result is a constant whose value is quite small, of order 5%.

At tree level, g_a^{-2} is the real part of the holomorphic function f in the gauge kinetic term. Noting that

$$\ln\left[(T_i + T_i^*)|\eta(T_i)|^4\right] = \ln(T_i + T_i^*) + 4\,\mathrm{Re}[\ln \eta(T_i)] \,, \qquad (16.4.39)$$

the same is not true for the one-loop coupling. The second term can arise from a holomorphic one-loop contribution $4 \ln \eta(T_i)$ to the function f in the effective local action obtained by integrating out massive string states (the *Wilsonian* action). The term $\ln(T_i + T_i^*)$ is due to explicit massless states. This is a general feature in supersymmetric quantum theory: it is the Wilsonian action, not the physical couplings, that has holomorphicity properties and satisfies nonrenormalization theorems. On the other hand, the physical couplings (16.4.38) are T-duality-invariant as one would expect.

Note on the other hand that the Wilsonian f is *not* T-duality-invariant, because it omits the term $\ln(T_i + T_i^*)$. This can be understood as follows. The various massless fields (including their fermionic components) transform nontrivially under T-duality due to their modular weights. This leads to an anomaly in the T-duality transformation, which is canceled by the explicit transformation of f. In fact, for orbifolds the moduli dependence of the full threshold correction $\tilde{\Delta}_a$ can be determined from holomorphicity and the cancellation of the T-duality anomaly. It has the same functional form as Δ_a but with coefficients given by sums over the modular weights.

Exercises

16.1 Find the massless spectrum of the $SO(32)$ heterotic string on the $\mathbf{Z}_3$ orbifold.

16.2 Find the massless spectrum of the $E_8 \times E_8$ heterotic string on the $\mathbf{Z}_4$ orbifold (16.3.33).

16.3 Find the massless spectrum of the six-dimensional $E_8 \times E_8$ heterotic string on the orbifold $T^4/\mathbf{Z}_2$,

$$X^m \to -X^m \,, \quad m = 6, 7, 8, 9 \,.$$

Determine the unbroken $d = 6$ supersymmetry and the supersymmetry multiplets of the massless states.

16.4 Repeat the previous exercise for the orbifold $T^4/\mathbf{Z}_3$,

$$Z^i \to \exp(2\pi i/3)Z^i \,, \quad i = 3, 4 \,.$$

If you do both this and the previous exercise, compare the spectra. These two orbifolds are special cases of the same *K3 surface,* to be discussed further in chapter 19.

16.5–16.7 Repeat the previous three exercises for the type IIA string.

16.8–16.10 Repeat the same three exercises for the type IIB string.

17

Calabi–Yau compactification

The study of compactification on smooth manifolds requires new, geometric, tools. A full introduction to this subject and its application to string theory would be a long book in itself. What we wish to do in this chapter is to present just the most important results, with almost all calculations and derivations omitted.

17.1 Conditions for $N = 1$ supersymmetry

We will assume four-dimensional Poincaré invariance. The metric is then of the form

$$G_{MN} = \begin{bmatrix} f(y)\eta_{\mu\nu} & 0 \\ 0 & G_{mn}(y) \end{bmatrix} . \tag{17.1.1}$$

We denote the noncompact coordinates by x^μ with $\mu, \nu = 0, \ldots, 3$ and the compact coordinates by y^m with $m, n = 4, \ldots, 9$. The indices M, N run over all coordinates, $0, \ldots, 9$. The other potentially nonvanishing fields are $\Phi(y)$, $\widetilde{H}_{mnp}(y)$, and $F_{mn}(y)$.

It is convenient to focus from the start on backgrounds that leave some supersymmetry unbroken. The condition for this is that the variations of the Fermi fields are zero. This is discussed further in appendix B, in connection with eq. (B.2.25). For the $d = 10$, $N = 1$ supergravity of the heterotic string these variations are

$$\delta\psi_\mu = \nabla_\mu \varepsilon , \tag{17.1.2a}$$

$$\delta\psi_m = \left(\partial_m + \frac{1}{4}\Omega^-_{mnp}\Gamma^{np} \right)\varepsilon , \tag{17.1.2b}$$

$$\delta\chi = \left(\Gamma^m \partial_m \Phi - \frac{1}{12}\Gamma^{mnp}\widetilde{H}_{mnp} \right)\varepsilon , \tag{17.1.2c}$$

$$\delta\lambda = F_{mn}\Gamma^{mn}\varepsilon . \tag{17.1.2d}$$

These are the variations of the gravitino, dilatino, and gaugino respectively. As in the corresponding nonlinear sigma model (12.3.30), the spin connection constructed from the metric appears in combination with the 3-form field strength,

$$\Omega^{\pm}_{MNP} = \omega_{MNP} \pm \frac{1}{2}H_{MNP} \,. \tag{17.1.3}$$

Under the decomposition $SO(9,1) \to SO(3,1) \times SO(6)$, the **16** decomposes as

$$\mathbf{16} \to (\mathbf{2},\mathbf{4}) + (\bar{\mathbf{2}},\bar{\mathbf{4}}) \,. \tag{17.1.4}$$

Thus a Majorana–Weyl **16** supersymmetry parameter can be written

$$\varepsilon(y) \to \varepsilon_{\alpha\beta}(y) + \varepsilon^{*}_{\alpha\beta}(y) \,, \tag{17.1.5}$$

where the indices on $\varepsilon_{\alpha\beta}$ transform respectively as $(\mathbf{2},\mathbf{4})$. If there is any unbroken supersymmetry, then by $SO(3,1)$ rotations we can generate further supersymmetries and so reach the form

$$\varepsilon_{\alpha\beta} = u_{\alpha}\zeta_{\beta}(y) \tag{17.1.6}$$

for an arbitrary Weyl spinor u. Each internal spinor $\zeta_{\beta}(y)$ for which $\delta(\text{fermions})= 0$ thus gives one copy of the minimum $d = 4$ supersymmetry algebra.

The conditions that the variations (17.1.2) vanish for some spinor $\zeta_{\beta}(y)$ can be solved to obtain conditions on the background fields. Again, we quote the results without going through the calculations. Until the last section of this chapter we will make the additional assumption that the antisymmetric tensor field strength (often called the *torsion* in the literature) vanishes,

$$\tilde{H}_{mnp} = 0 \,. \tag{17.1.7}$$

From the vanishing of $\delta\chi$ one can then deduce that if there is any unbroken supersymmetry then the dilaton is constant,

$$\partial_m\Phi = 0 \,. \tag{17.1.8}$$

The vanishing of $\delta\psi_{\mu}$ next implies that

$$G_{\mu\nu} = \eta_{\mu\nu} \,, \tag{17.1.9}$$

forbidding a y-dependent scale factor. The vanishing of $\delta\psi_m$ then implies that

$$\nabla_m\zeta = 0 \,, \tag{17.1.10}$$

so that ζ is covariantly constant on the internal space. This is a strong condition. It implies, for example, that

$$[\nabla_m, \nabla_n]\zeta = \frac{1}{4}R_{mnpq}\Gamma^{pq}\zeta = 0 \,. \tag{17.1.11}$$

This means that the components Γ^{pq} that appear are not general $SO(6)$ rotations but must lie in a subgroup leaving one component of the spinor invariant. The subgroup with this property is $SU(3)$. In eq. (B.1.49) we show that under $SO(6) \to SU(3)$, the spinor decomposes $\mathbf{4} \to \mathbf{3} + \mathbf{1}$, so that if $R_{mnpq}\Gamma^{pq}$ is in this $SU(3)$ then there will be an invariant spinor.

The existence of a covariantly constant spinor ζ is thus the condition that the manifold have $SU(3)$ *holonomy*. In other words, under parallel transport around a closed loop, a spinor (or any other covariant quantity) comes back to itself not with an arbitrary rotation but with a rotation in $SU(3) \subset SO(6)$. This is the same as the condition for $N = 1$ supersymmetry in orbifolds. To see this, transport a spinor from any point to its image under the orbifold rotation: this is a closed loop on the orbifold. The orbifold is locally flat, but to compare the spinor to its original value we must rotate back. Thus the orbifold point group is the holonomy, and as we found in chapter 16, a point group in $SU(3)$ gives unbroken $d = 4$, $N = 1$ supersymmetry. Similarly, $SU(2)$ holonomy leaves a second spinor invariant and so gives an unbroken $d = 4$, $N = 2$ supersymmetry.

The final supersymmetric variation $\delta\lambda^a$ vanishes if $F^a_{mn}\Gamma^{mn}$ is also an $SU(3)$ rotation. Writing the indices on F_{mn} in terms of the complex indices transforming under $SU(3)$, this means that

$$F_{ij} = F_{\bar{i}\bar{j}} = 0 \,, \qquad G^{ij}F_{i\bar{j}} = 0 \,. \tag{17.1.12}$$

In addition we must impose the Bianchi identities on the various field strengths. In particular, for the torsion this is

$$d\tilde{H}_3 = \frac{\alpha'}{4}\left[\mathrm{tr}(R_2 \wedge R_2) - \mathrm{Tr}_v(F_2 \wedge F_2)\right] \,. \tag{17.1.13}$$

For vanishing $\tilde{H}$, this condition is quite strong, and the only solution seems to be to set R_2 and F_2 essentially equal. That is, consider $SO(6) \subset SO(16) \subset E_8$, and require the gauge connection to be equal to the spin connection ω_μ of the Lorentz $SO(6)$. This is referred to as embedding the spin connection in the gauge connection, generalizing the same idea in the orbifold. Recall that the corrections (17.1.13) were deduced in section 12.3 from anomalies on the world-sheet. When the spin connection is embedded in the gauge connection, six of the current algebra λ^A couple in the same way as the $\tilde{\psi}^m$. The relevant part of the world-sheet theory is then parity-invariant, accounting for the cancellation of anomalies.

With the spin connection embedded in the gauge connection, the conditions (17.1.12) for the vanishing of the gaugino variation follow from $SU(3)$ holonomy. The Bianchi identity for the field strength also follows from that for the curvature. It remains to consider the equations of motion. We might have begun with these, but it is easiest to save them for the end because at this point they are automatically satisfied. With vanishing torsion and a constant dilaton the field equations reduce to

$$R_{mn} = 0 , \quad \nabla^m F_{mn} = 0 . \qquad (17.1.14)$$

These can be shown to follow respectively from $SU(3)$ holonomy and the conditions (17.1.12).

We should remember that the field equations and supersymmetry variations in this section are only the leading terms in an expansion in derivatives, $\alpha'^{1/2}\partial_m$ being the dimensionless parameter. The conditions we have found are therefore correct when the length scale R_c of the compactified manifold is large compared to the string scale. However, we will see in section 17.5 that many of the conclusions have a much wider range of validity.

17.2 Calabi–Yau manifolds

To summarize, we found in the last section that under the assumption of vanishing torsion, the compactified dimensions must form a space of $SU(3)$ holonomy. In this section we present some of the relevant mathematics. Again, we give only definitions and results, without derivations. All manifolds in this section are assumed to be compact.

Real manifolds

We need to introduce the ideas of *cohomology* and *homology*. The exterior derivative d introduced in section B.4 is nilpotent, $d^2 = 0$. As with the BRST operator, this allows us to define a cohomology. A p-form ω_p is *closed* if $d\omega_p = 0$ and *exact* if $\omega_p = d\alpha_{p-1}$ for some $(p-1)$-form. A closed p-form can always be written locally in the form $d\alpha_{p-1}$, but not necessarily globally. Thus we define the pth *de Rham cohomology* of a manifold K,

$$H^p(K) = \frac{\text{closed } p\text{-forms on } K}{\text{exact } p\text{-forms on } K} . \qquad (17.2.1)$$

The dimension of $H^p(K)$ is the *Betti number* b_p. The Betti numbers depend only on the topology of the space. In particular, the *Euler number* is

$$\chi(K) = \sum_{p=0}^{d} (-1)^p b_p . \qquad (17.2.2)$$

The operator

$$\Delta_d = *d * d + d * d * = (d + *d*)^2 \qquad (17.2.3)$$

is a second order differential on p-forms which reduces to the Laplacian in flat space. The Poincaré $*$ is defined in section B.4. A p-form is said to be *harmonic* if $\Delta_d \omega = 0$. It can be shown that the harmonic p-forms are in one-to-one correspondence with the group $H^p(K)$: each equivalence class contains exactly one harmonic form. Using the Poincaré dual one can turn a harmonic p-form into a harmonic $(d-p)$-form. This is the Hodge $*$ map between $H^p(K)$ and $H^{d-p}(K)$, and implies

$$b_p = b_{d-p} \,. \tag{17.2.4}$$

For submanifolds of K one can define the boundary operator δ, which is also nilpotent. Rather than on a submanifold N itself it is useful to focus on the corresponding integral

$$\int_N \tag{17.2.5}$$

since these form a vector space: we can consider arbitrary real linear combinations, called *chains*.[1] We can then define *closed* and *exact* with respect to δ; a closed chain is a *cycle*. The *simplicial homology* for p-dimensional submanifolds (p-chains) is

$$H_p(K) = \frac{\text{closed } p\text{-chains in } K}{\text{exact } p\text{-chains in } K} \,. \tag{17.2.6}$$

That is, it consists of closed submanifolds that are not themselves boundaries.

There is a one-to-one correspondence between $H^p(K)$ and $H_{d-p}(K)$. For any p-form ω_p there is a $(d-p)$-cycle $N(\omega)$ with the property that

$$\int_K \omega_p \wedge \alpha_{d-p} = \int_{N(\omega)} \alpha_{d-p} \tag{17.2.7}$$

for all closed $(d-p)$-forms.

Complex manifolds

A complex manifold is an even-dimensional manifold, $d = 2n$, such that we can form n complex coordinates z^i and the transition functions

$$z'^i(z^j) \tag{17.2.8}$$

are holomorphic between all pairs of patches. Specifically, this is a *complex n-fold*. We have encountered this idea for $n = 1$ on the string world-sheet. Two complex manifolds are equivalent if there is a one-to-one holomorphic map between them. As we have seen in the case of Riemann surfaces, a manifold of given topology can have more than one inequivalent complex

---

[1] This will define *real* homology; by analogy one can define integer homology, complex homology, and so on.

structure. A *Hermitean metric* on a complex manifold is one for which

$$G_{ij} = G_{\bar{i}\bar{j}} = 0 \,. \tag{17.2.9}$$

On a complex manifold we can define (p, q)-forms as having p anti-symmetric holomorphic indices and q antisymmetric antiholomorphic indices,

$$\omega_{i_1 \cdots i_p \bar{j}_1 \cdots \bar{j}_q} \,. \tag{17.2.10}$$

The relative order of the different types of index is not important and can always be taken as shown. We can similarly separate the exterior derivative, $d = \partial + \bar{\partial}$, where

$$\partial = dz^i \partial_i \,, \quad \bar{\partial} = d\bar{z}^{\bar{i}} \partial_{\bar{i}} \,. \tag{17.2.11}$$

Then ∂ and $\bar{\partial}$ take (p, q)-forms into $(p+1, q)$-forms and $(p, q+1)$-forms respectively. Each is nilpotent,

$$\partial^2 = \bar{\partial}^2 = 0 \,. \tag{17.2.12}$$

Thus we can define the *Dolbeault cohomology*

$$H_{\bar{\partial}}^{p,q}(K) = \frac{\bar{\partial}\text{-closed } (p, q)\text{-forms in } K}{\bar{\partial}\text{-exact } (p, q)\text{-forms in } K} \,. \tag{17.2.13}$$

The dimension of $H_{\bar{\partial}}^{p,q}(K)$ is the *Hodge number* $h^{p,q}$.

Using the inner product

$$\int d^n z \, d^n \bar{z} \, (G)^{1/2} G^{\bar{i}i'} \cdots G^{j\bar{j}'} \cdots (\omega_{i \cdots \bar{j} \cdots})^* \omega_{i' \cdots \bar{j}' \cdots} \,, \tag{17.2.14}$$

one defines the adjoints $\partial^\dagger$ and $\bar{\partial}^\dagger$ and the Laplacians

$$\Delta_\partial = \partial \partial^\dagger + \partial^\dagger \partial \,, \quad \Delta_{\bar{\partial}} = \bar{\partial} \bar{\partial}^\dagger + \bar{\partial}^\dagger \bar{\partial} \,. \tag{17.2.15}$$

Then the $\Delta_{\bar{\partial}}$-harmonic (p, q)-forms are in one-to-one correspondence with $H_{\bar{\partial}}^{p,q}(K)$.

Kähler manifolds

Kähler manifolds are complex manifolds with a Hermitean metric of a special form. The additional restriction can be stated in several ways. Define the *Kähler form*

$$J_{1,1} = iG_{i\bar{j}} dz^i d\bar{z}^{\bar{j}} \,. \tag{17.2.16}$$

One way to define a Kähler manifold is that the Kähler form is closed,

$$dJ_{1,1} = 0 \,. \tag{17.2.17}$$

A second is that parallel transport takes holomorphic indices only into holomorphic indices. In other words, the holonomy is in $U(n) \subset SO(2n)$.

A final equivalent statement is that the metric is locally of the form

$$G_{i\bar{\jmath}} = \frac{\partial}{\partial z^i}\frac{\partial}{\partial \bar{z}^j}K(z,\bar{z}) \ . \tag{17.2.18}$$

The *Kähler potential* $K(z,\bar{z})$ need not be globally defined. The potential

$$K'(z,\bar{z}) = K(z,\bar{z}) + f(z) + f(z)^* \tag{17.2.19}$$

gives the same metric, and it may be necessary to take different potentials in different patches. We are now focusing on the spacetime geometry, but we have seen this same idea in *field space* in eq. (B.2.32).

For Kähler metrics the various Laplacians become identical,

$$\Delta_d = 2\Delta_{\bar{\partial}} = 2\Delta_{\partial} \ . \tag{17.2.20}$$

Then the cohomologies

$$H^{p,q}_{\bar{\partial}}(K) = H^{p,q}_{\partial}(K) \equiv H^{p,q}(K) \tag{17.2.21}$$

are the same. The Hodge and Betti numbers are therefore also related,

$$b_k = \sum_{p=0}^{k} h^{p,k-p} \ . \tag{17.2.22}$$

Complex conjugation gives

$$h^{p,q} = h^{q,p} \tag{17.2.23}$$

and the Hodge $*$ gives

$$h^{n-p,n-q} = h^{p,q} \ . \tag{17.2.24}$$

Since the Kähler form is closed it is in $H^{1,1}(K)$. Its equivalence class is known as the *Kähler class* and is always nontrivial. Taking a basis ω_A for $H^{1,1}(K)$, we can expand

$$J_{1,1} = \sum_A v^A \omega_{1,1\,A} \ , \tag{17.2.25}$$

and the real parameters v^A label the Kähler class.

Manifolds of $SU(3)$ holonomy

A manifold has $SU(3)$ holonomy if and only if it is Ricci-flat and Kähler. While there are many examples of Kähler manifolds, there are few explicit examples of Ricci-flat Kähler metrics. There is, however, an important existence theorem. For a Kähler manifold, only the mixed components $R_{i\bar{\jmath}}$ of the Ricci tensor are nonzero. Further, the *Ricci form*

$$\mathscr{R}_{1,1} = R_{i\bar{\jmath}}dz^i d\bar{z}^j \tag{17.2.26}$$

is closed, $d\mathcal{R}_{1,1} = 0$. It therefore defines an equivalence class in $H^{1,1}(K)$. With normalization $\mathcal{R}_{1,1}/2\pi$, this is known as the *first Chern class* c_1. Obviously this class is trivial for a Ricci-flat manifold. The hard theorem, conjectured by Calabi and proved by Yau, is that for any Kähler manifold with $c_1 = 0$ there exists a unique Ricci-flat metric with a given complex structure and Kähler class. A vanishing first Chern class, $c_1 = 0$, means that $\mathcal{R}_{1,1}$ is exact. A Kähler manifold with $c_1 = 0$ is known as a *Calabi–Yau manifold*.

Another theorem states that a Kähler manifold has $c_1 = 0$ if and only if there is a nowhere vanishing holomorphic $(3,0)$-form $\Omega_{3,0}$. The $(3,0)$-form is covariantly constant in the Ricci-flat metric. It further can be shown that

$$h^{p,0} = h^{3-p,0} \,. \tag{17.2.27}$$

For any complex manifold $h^{0,0} = 1$, corresponding to the constant function. Finally, for a Calabi–Yau manifold of exactly $SU(3)$ holonomy and not a subgroup, it can be shown that

$$b_1 = h^{1,0} = h^{0,1} = 0 \,. \tag{17.2.28}$$

Using the various properties above, all the Hodge numbers of a Calabi–Yau 3-fold are fixed by just two independent numbers, $h^{1,1}$ and $h^{2,1}$. The full set of Hodge numbers is conventionally displayed as *Hodge diamond*,

$$
\begin{array}{ccccccc}
& & & h^{3,3} & & & \\
& & h^{3,2} & & h^{2,3} & & \\
& h^{3,1} & & h^{2,2} & & h^{1,3} & \\
h^{3,0} & & h^{2,1} & & h^{1,2} & & h^{0,3} \\
& h^{2,0} & & h^{1,1} & & h^{0,2} & \\
& & h^{1,0} & & h^{0,1} & & \\
& & & h^{0,0} & & &
\end{array}
\;=\;
\begin{array}{ccccccc}
& & & 1 & & & \\
& & 0 & & 0 & & \\
& 0 & & h^{1,1} & & 0 & \\
1 & & h^{2,1} & & h^{2,1} & & 1 \\
& 0 & & h^{1,1} & & 0 & \\
& & 0 & & 0 & & \\
& & & 1 & & &
\end{array}
\,. \tag{17.2.29}
$$

In particular, the Euler number (17.2.2) is

$$\chi = 2(h^{1,1} - h^{2,1}) \,. \tag{17.2.30}$$

Examples

An even-dimensional torus is a Calabi–Yau manifold but an uninteresting one: the holonomy is trivial. To break to $N = 1$ supersymmetry we need nontrivial $SU(3)$ holonomy. The $\mathbf{Z}_3$ orbifold of T^6 has this property but is not a manifold, having orbifold singularities. A smooth Calabi–Yau space can be produced by blowing up all the singularities, as follows. The *Eguchi–Hanson space* EH_3 has three complex coordinates w_i with metric

$$G_{i\bar{j}} = \left(1 + \frac{\rho^6}{r^6}\right)^{1/3}\left[\delta_{i\bar{j}} - \frac{\rho^6 w_i \bar{w}_{\bar{j}}}{r^2(\rho^6 + r^6)}\right], \tag{17.2.31}$$

where $r^2 = w_i \bar{w}_{\bar{i}}$ and ρ is a constant that sets the scale of the geometry. After the identification

$$w_i \cong \exp(2\pi i/3) w_i \,, \tag{17.2.32}$$

this becomes an everywhere smooth space which is asymptotically R^6/Z_3. This is the same as the geometry around the T^6/Z_3 orbifold fixed points. Each orbifold fixed point can be replaced by a small EH_3 to give a smooth Calabi–Yau space. The $(1, 1)$-forms are the nine $dz_i d\bar{z}_{\bar{j}}$ and the 27 blow-up modes $\partial G_{i\bar{j}}/\partial \rho$ from varying the sizes of the EH_3s. There is only one complex structure, so

$$h^{1,1} = 36 \,, \quad h^{2,1} = 0 \,, \quad \chi = 72 \,. \tag{17.2.33}$$

A second construction starts with *complex projective space CP^n*, formed by taking $n + 1$ complex coordinates and identifying

$$(z_1, z_2, \ldots, z_{n+1}) \cong (\lambda z_1, \lambda z_2, \ldots, \lambda z_{n+1}) \tag{17.2.34}$$

for any complex λ. The identification is important because it makes the space compact. The space CP^n is Kähler but not Calabi–Yau; many Calabi–Yau manifolds can be obtained from it as submanifolds. In particular, let G be a homogeneous polynomial in the z^i,

$$G(\lambda z_1, \ldots, \lambda z_{n+1}) = \lambda^k G(z_1, \ldots, z_{n+1}) \tag{17.2.35}$$

for some k. The submanifold of CP^n defined by

$$G(z_1, \ldots, z_{n+1}) = 0 \tag{17.2.36}$$

is a Kähler manifold of complex dimension $n - 1$. It can be shown that this submanifold has vanishing c_1 for $k = n + 1$, so that a quintic polynomial in CP^4, which is $(n, k) = (4, 5)$, gives a good manifold for string compactification. This manifold can be shown to have

$$h^{1,1} = 1 \,, \quad h^{2,1} = 101 \,, \quad \chi = -200 \,. \tag{17.2.37}$$

The unique Kähler modulus is the overall scale of the manifold. The complex structure moduli correspond to the parameters in the polynomial G, which after taking into account linear coordinate redefinitions number $9!/(5! \cdot 4!) - 25 = 101$.

Obvious generalizations include starting with a product of CP^n spaces, requiring several polynomials to vanish, and using weighted projective spaces where coordinates scale by different powers of λ. One can also divide by a discrete symmetry. For example, a particular case of the quintic polynomial in CP^4,

$$z_1^5 + z_2^5 + z_3^5 + z_4^5 + z_5^5 = 0 \,, \tag{17.2.38}$$

has a $Z_5 \times Z_5$ symmetry which is freely acting, meaning that it has no fixed points. Since the Euler number χ can be written as an integral over the curvature, identifying by this $\mathbf{Z}_5 \times \mathbf{Z}_5$ reduces χ by a factor of 25, to $\chi = -8$. Identifying by a symmetry with fixed points produces a space with orbifold singularities. These can be blown up, but the Euler number is then not simply obtained by dividing by the order of the group, because of the curvature at the blow-ups.

Another example is the *Tian–Yau space*. This is formed from two copies of CP^3 with coordinates z_i and w_i by imposing three polynomial equations: one cubic in z, one cubic in w, and one linear in z and linear in w. This has $\chi = -18$, and there is a freely-acting $\mathbf{Z}_3$ symmetry which can reduce this to $\chi = -6$.

World-sheet supersymmetry

With the spin connection embedded in the gauge connection, the interacting part of the world-sheet theory is invariant under parity, which interchanges ψ^i with $\lambda^{(i-1)+}$ for $i = 2, 3, 4$. Since the heterotic theory has a $(0, 1)$ superconformal symmetry, the parity symmetry implies that it is enlarged to $(1, 1)$.

For any metric $G_{mn}(y)$, the superfield formalism of section 12.3 allows us to write a nonlinear sigma model having (1,1) supersymmetry. If in addition the metric is Kähler, then there is actually (2,2) supersymmetry. One way to see this is to observe that this is the condition for the mixed components $\omega_{ai}^{\bar{j}}$ and $\omega_{a\bar{i}}^{j}$ of the spin connection to vanish, and therefore for the world-sheet action (12.3.30) to be invariant under a $U(1)$ rotation of the complex fermions,

$$\tilde{\psi}^i \to \exp(i\theta)\tilde{\psi}^i \,. \tag{17.2.39}$$

The right-moving supercurrent then separates into two terms

$$iG_{\bar{i}j}\bar{\partial}X^{\bar{i}}\tilde{\psi}^j + iG_{i\bar{j}}\bar{\partial}X^i\tilde{\psi}^{\bar{j}} \,, \tag{17.2.40}$$

which have opposite charges under the $U(1)$ symmetry and so must be separately conserved. The left-moving supercurrent also separates. Another way to see the enlarged supersymmetry is by dimensional reduction of $d = 4$, $N = 1$ supersymmetry. As discussed in section B.2, this supersymmetry requires that the field space be Kähler; dimensional reduction takes the four generators of $d = 4$, $N = 1$ into $d = 2$ (2,2).

When the metric satisfies the stronger condition of $SU(n)$ holonomy, the sigma model is conformally invariant, the supersymmetries are extended to superconformal symmetries, and the $U(1)$ global symmetry is extended to left- and right-moving $U(1)$ current algebras.

For Calabi–Yau compactification, the interacting part of the world-sheet theory is a $(c, \tilde{c}) = (9, 9)$ CFT. Since the spin connection is embedded in the gauge connection, the six interacting λ^A couple in the same way as supersymmetric fermions ψ^m. Thus, as with the orbifold example, the world-sheet theory has (2,2) superconformal symmetry. In chapters 18 and 19 we will study this world-sheet symmetry systematically. We will see that a minimum of (0,2) supersymmetry on the world-sheet is necessary in order to have *spacetime* supersymmetry. We will also see that the extra left-moving supersymmetry of Calabi–Yau compactification is responsible for a great deal of special structure.

17.3 Massless spectrum

We now look at the spectrum of fluctuations around the background. We will use lower case a, g, b, and ϕ to distinguish the fluctuations from the background fields. The various wave operators separate into noncompact and internal pieces, for example

$$\nabla_M\nabla^M = \partial_\mu\partial^\mu + \nabla_m\nabla^m \,, \tag{17.3.1a}$$

$$\Gamma_M\nabla^M = \Gamma_\mu\partial^\mu + \Gamma_m\nabla^m \,. \tag{17.3.1b}$$

The solutions similarly separate into a sum over functions of x^μ times a complete set of functions of y^m. Massless fields in four dimensions arise from those modes of the ten-dimensional massless fields that are annihilated by the internal part of the wave operator.

We start with the ten-dimensional gauge field. The ten-dimensional index separates $M \to \mu, i, \bar{\imath}$. Similarly the adjoint decomposes under

$$E_8 \times E_8 \;\to\; SU(3) \times E_6 \times E_8 \tag{17.3.2}$$

into

$$a: \; (\mathbf{1}, \mathbf{78}, \mathbf{1}) + (\mathbf{1}, \mathbf{1}, \mathbf{248}) \,, \tag{17.3.3a}$$

$$ix: \; (\mathbf{3}, \mathbf{27}, \mathbf{1}) \,, \quad \bar{\imath}\bar{x}: \; (\bar{\mathbf{3}}, \overline{\mathbf{27}}, \mathbf{1}) \,, \quad i\bar{\jmath}: \; (\mathbf{8}, \mathbf{1}, \mathbf{1}) \,. \tag{17.3.3b}$$

That is, a denotes the adjoint of $E_6 \times E_8$, x the $\mathbf{27}$ of E_6 and i, j the $\mathbf{3}$ of $SU(3)$. We use the same index for the $\mathbf{3}$ of the gauge $SU(3)$ and the spacetime $SU(3)$ because their connections are the same.

We denote the various components of the gauge fluctuation as $a_{M,X}$ with X any of the gauge components (17.3.3). The massless modes of the form $a_{\mu,X}$ are the unbroken gauge fields in four dimensions. These arise from gauge symmetries that commute with the background fields. Since the latter are in $SU(3)$, the four-dimensional gauge symmetry is $E_6 \times E_8$, meaning $X = a$. In terms of the wave operator, the internal part acting on $a_{\mu,X}$ is the scalar Laplacian $\nabla^m\nabla_m$ with gauge-covariant derivative. It has zero modes only for fields that are neutral under the background gauge fields. Comparing with the $\mathbf{Z}_3$ orbifold, the low energy symmetry $SU(3)$

is absent. This is consistent with the analysis (16.4.29) of the blowing-up modes, which were seen to break the $SU(3)$ symmetry.

The field $a_{i,a}$ can be regarded as a $(1,0)$-form, with only the index i coupling to the background connection, and the relevant wave operator is in fact Δ_d. The number of zero modes is then $h^{1,0}$, which vanishes for $SU(3)$ holonomy.

The field $a_{i,jx}$ is not a $(2,0)$-form because the tangent space and gauge indices are not antisymmetrized. However, by using the metric and the antisymmetric three-form we can produce

$$a_{\bar{i}\bar{l}\bar{m}x} = a_{i,jx}G^{j\bar{k}}\Omega_{\bar{k}\bar{l}\bar{m}} , \qquad (17.3.4)$$

which is a $(1,2)$-form on the indices $\bar{i}\bar{l}\bar{m}$. The relevant wave operator is again Δ_d so the number of zero modes is $h^{2,1}$. These fields are scalars in the $\mathbf{27}$ of E_6.

The field $a_{i,j\bar{x}}$ is a $(1,1)$-form and the relevant wave operator is again Δ_d. The number of zero modes is $h^{1,1}$. These fields are scalars in the $\overline{\mathbf{27}}$ of E_6.

The field $a_{i,j\bar{k}}$ cannot be written as a (p,q)-form and the number of massless modes is not given by a Hodge number. This field can be regarded as a 1-form (the index i) transforming as a generator of the Lorentz group (the indices $j\bar{k}$); the corresponding cohomology is denoted $H^1(\text{End}\,T)$. Because these are neutral under E_6 they are less directly relevant to the low energy physics than the charged fields. We will discuss some of their physics in section 17.6.

The zero modes of $a_{\bar{i},X}$ are the conjugates of those of $a_{i,X}$.

The massless modes of the gaugino must be the same as those of $a_{M,X}$ by supersymmetry. This is related directly to the $SU(3)$ holonomy. Under $SO(9,1) \to SO(3,1) \times SU(3)$,

$$\mathbf{16} \;\to\; (\mathbf{2},\mathbf{1})+(\mathbf{2},\mathbf{3})+(\bar{\mathbf{2}},\bar{\mathbf{1}})+(\bar{\mathbf{2}},\bar{\mathbf{3}}) . \qquad (17.3.5)$$

The $(\mathbf{2},\mathbf{1})$ is neutral under the tangent space group and so couples in the same way as a^μ, providing the four-dimensional gauginos. The $(\mathbf{2},\mathbf{3})$ couples in the same way as a_i and so provides the fermionic partners of those scalars. Thus there are $h^{2,1}$ $\mathbf{27}$s and $h^{1,1}$ $\overline{\mathbf{27}}$s in the $\mathbf{2}$ of $SO(3,1)$. The spectrum is chiral and the net number of generations minus antigenerations is

$$|h^{2,1} - h^{1,1}| = \frac{|\chi|}{2} . \qquad (17.3.6)$$

This is 36 for the blown-up orbifold, just as for the singular orbifold, and 100 for the quintic in CP^4. However, dividing by $\mathbf{Z}_5 \times \mathbf{Z}_5$ reduces the latter number to a more reasonable 4, while the Tian–Yau space has a net of 3 generations. The relation (17.3.6) can be understood from an index theorem for the Dirac equation.

Dividing by a discrete group is also useful for breaking the E_6 symmetry. We saw for the orbifold that this could be done by Wilson lines, gauge backgrounds that are locally trivial but give a net rotation around closed curves. The Calabi–Yau spaces produced by polynomial equations in projective spaces are simply connected, but dividing by a freely-acting group produces nontrivial closed curves running from a point to its image. Adding a Wilson line means that the string theory is twisted by the product of the spacetime symmetry and a gauge rotation W. The nontrivial curves produced in dividing by $\mathbf{Z}_n$ have the property that if traversed n times they become closed paths on the original (covering) space, which are all topologically trivial. Thus the Wilson line must also satisfy $W^n = 1$.

For a freely-acting group, adding Wilson lines does not change the net number of generations. However, the different quark and lepton multiplets of a given generation in general come from different **27**s of the untwisted theory. Thus, while the inheritance principle requires the Standard Model gauge couplings to satisfy E_6 relations, the Yukawa couplings of the quarks and leptons in general do not. This is good because the E_6 relations for the gauge couplings (which are the same as the $SU(5)$ relations) work rather well, while those for the Yukawa couplings are more mixed, with only the heaviest generation ratio m_b/m_τ working well. This may also help to account for the stability of the proton, as the Higgs couplings that give mass to the quarks and leptons are no longer related to couplings of color triplet scalars that might mediate baryon decay.

Now we consider the bosonic supergravity fields, g_{MN}, b_{MN}, and ϕ. The components with all indices noncompact, $g_{\mu\nu}$, $b_{\mu\nu}$, and ϕ, each have a single zero mode (the constant function) giving the corresponding field in four dimensions.

The components $g_{\mu i}$ and $b_{\mu i}$ are $(1,0)$-forms on the internal space and so have no zero modes because $h^{1,0} = 0$. In particular, massless modes of $g_{\mu i}$ would be Kaluza–Klein gauge bosons, which are in one-to-one correspondence with the continuous symmetries of the internal space. It can be shown that a Calabi–Yau manifold has no continuous symmetries.

The components g_{ij} correspond to changes in the complex structure, since a coordinate change would be needed to bring the metric back to Hermitean form. This field is symmetric and so not a (p,q)-form, but by the same trick as for $a_{i\bar{l}\bar{m}x}$ we can form

$$g_{i\bar{l}\bar{m}} = g_{ij}G^{j\bar{k}}\Omega_{\bar{k}\bar{l}\bar{m}} \,. \tag{17.3.7}$$

The wave operator is Δ_d and so the number of complex structure moduli is $h^{2,1}$. These are complex fields, with $g_{\bar{\imath}\bar{\jmath}}$ being the conjugate. The field b_{ij} is a $(2,0)$-form and has $h^{2,0} = 0$ zero modes.

The fluctuation $g_{i\bar{\jmath}}$ is a $(1,1)$-form, and the wave operator is Δ_d. Thus it gives rise to $h^{1,1}$ real moduli. The field $b_{i\bar{\jmath}}$ also is a $(1,1)$-form and gives $h^{1,1}$ real moduli. These combine to form $h^{1,1}$ complex fields.

The full massless spectrum is

- $d = 4$, $N = 1$ supergravity: $G_{\mu\nu}$ and the gravitino.
- The dilaton–axion chiral superfield S.
- Gauge bosons and gauginos in the adjoint of $E_6 \times E_8$.
- $h^{2,1}$ chiral superfields in the **27** of E_6.
- $h^{1,1}$ chiral superfields in the **$\overline{27}$** of E_6.
- $h^{2,1}$ chiral superfields for the complex structure moduli.
- $h^{1,1}$ chiral superfields for the Kähler moduli.
- Some number of E_6 singlets from $H^1(\mathrm{End}\,T)$.

17.4 Low energy field theory

We would now like to deduce the effective four-dimensional action for the massless fields. We emphasize again that the actual calculations are omitted, but we will outline the method and the results. The general $d = 4$, $N = 1$ supersymmetric action depends on two holomorphic functions, the gauge kinetic term and the superpotential, and one general function, the Kähler potential. We will show in the next chapter that the gauge kinetic term is the same in all heterotic string compactifications, and so we need determine only the other two functions. In this section and the next we will ignore the E_6 singlet fields from $H^1(\mathrm{End}\,T)$, setting their values to zero.

We consider the low energy effective field theory at string tree level. For now we assume the compactification radius to be large compared to the string length, so that we can restrict attention to the massless fields of the ten-dimensional theory and also ignore higher dimension terms in the effective action. This is the *field-theory approximation*. In the next section we consider corrections to this approximation.

Expand each ten-dimensional field in a complete set of eigenfunctions $f_m(y)$ of the appropriate wave operator on the internal space, schematically

$$\varphi(x, y) = \sum_m \phi_m(x) f_m(y) \, . \tag{17.4.1}$$

Insert this into the ten-dimensional action and integrate over the internal

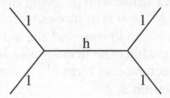

Fig. 17.1.   Quartic interaction among light fields induced by integrating out a heavy field.

coordinates to obtain the four-dimensional Lagrangian density,

$$\mathcal{L}_4(\phi) = \int d^6y \, \mathcal{L}_{10}(\varphi) \,. \qquad (17.4.2)$$

This still depends on all the functions $\phi_m(x)$, the infinite number of massive fields as well as the finite number of massless ones. Split $\varphi(x, y)$ into 'light' and 'heavy' parts,

$$\varphi = \varphi_l + \varphi_h, \qquad (17.4.3)$$

according to whether f_m has a zero or nonzero eigenvalue under the internal wave operator. We want to integrate φ_h out so as to obtain an effective action for the finite number of four-dimensional fields in φ_l. The simplest approach would be to set $\varphi_h = 0$ in $\mathcal{L}_4$, but this is not quite right. Since we are at string tree level we can treat the problem classically: what we must do is extremize the action with respect to φ_h with φ_l fixed. The result is the effective action for φ_l. As a schematic example, consider the following terms

$$m\varphi_h^2 + g\varphi_h\varphi_l^2 \,. \qquad (17.4.4)$$

Setting φ_h to its extremum $-g\varphi_l^2/2m$ leaves the effective interaction

$$-\frac{g^2}{4m}\varphi_l^4 \qquad (17.4.5)$$

for the light fields. Figure 17.1 shows the corresponding Feynman graph. This is known as a Kaluza–Klein correction to the low energy action. It is easy to see that these always involve at least four light fields. With an interaction $\varphi_h\varphi_l$ we could induce a quadratic or cubic term, but this is absent by definition. It is an off-diagonal mass term mixing the light and heavy fields, but the latter are defined to be eigenstates of zero mass. The terms that we will be interested in contain two or three light fields and so we can ignore the Kaluza–Klein corrections.

Let us first consider the fields associated with $(1, 1)$-forms, beginning with the superpotential for the $\overline{27}$s of E_6. We will focus on the renormalizable terms, which are at most cubic in the fields. The quadratic terms vanish because the $\overline{27}$s are massless, and the linear terms vanish because their presence would imply that the background is not supersymmetric by eq. (B.2.25); actually both terms are forbidden by E_6 as well. Thus we are interested in terms that are precisely cubic. These are related to four-dimensional Yukawa couplings. The relevant expansions for φ_1 are

$$a_{i,\bar{j}\bar{x}}(x,y) = \sum_A \phi^A{}_{\bar{x}}(x)\omega_{Ai\bar{j}}(y) , \qquad (17.4.6a)$$

$$\lambda_{i,\bar{j}\bar{x}}(x,y) = \sum_A \lambda^A{}_{\bar{x}}(x)\omega_{Ai\bar{j}}(y) , \qquad (17.4.6b)$$

where A runs over a complete set of nontrivial (1,1)-forms; henceforth summation convention is used for this index. A four-dimensional Weyl spinor index on λ is suppressed. Inserting these expansions into the action, the ten-dimensional term

$$\int d^6 y \, \text{Tr}_v(\bar{\lambda}\Gamma^m[A_m,\lambda]) \qquad (17.4.7)$$

becomes

$$d^{\bar{x}\bar{y}\bar{z}} \bar{\lambda}^A{}_{\bar{x}} \lambda^B{}_{\bar{y}} \phi^C{}_{\bar{z}} \int_K \omega_{1,1A} \wedge \omega_{1,1B} \wedge \omega_{1,1C} . \qquad (17.4.8)$$

Here $d^{\bar{x}\bar{y}\bar{z}}$ is the E_6 invariant for $\overline{27} \cdot \overline{27} \cdot \overline{27}$. The superpotential is then

$$W(\phi) = d^{\bar{x}\bar{y}\bar{z}} \phi^A{}_{\bar{x}} \phi^B{}_{\bar{y}} \phi^C{}_{\bar{z}} \int_K \omega_{1,1A} \wedge \omega_{1,1B} \wedge \omega_{1,1C} . \qquad (17.4.9)$$

The wedge product of the internal wavefunctions is a (3,3)-form and so can be integrated over the internal space without using the metric.

This part of the superpotential is independent of all moduli, and so is topological. To make this explicit, we use the correspondence (17.2.7) between 2-forms and 4-cycles. Take a basic N_A of nontrivial 4-cycles, and let ω_A be the corresponding basis of 2-forms. Three 4-cycles will generically intersect in isolated points. We can therefore define the intersection number, the total number of intersections weighted by orientation; this is a topological invariant.[2] A standard result from topology relates the intersection number of N_A, N_B, and N_C to the integral of the wedge product:

$$\#(N_A, N_B, N_C) = \int_K \omega_{1,1A} \wedge \omega_{1,1B} \wedge \omega_{1,1C} . \qquad (17.4.10)$$

Thus the superpotential is determined by these integers. Typically many of the intersection numbers vanish for topological reasons unrelated to

---

[2] If the cycles do not intersect only in isolated points, as is obviously the case if for example two are the same, one can make them do so by deforming them within the same homology class. This then defines the intersection number.

symmetry; this may be useful in understanding the stability of the proton and the rich texture of the Yukawa couplings in the Standard Model. For later reference we mention that there is a dual basis N^A of 2-cycles such that

$$\#(N^A, N_B) = \int_{N^A} \omega_{1,1\,B} = \delta^A{}_B \,. \tag{17.4.11}$$

For the (1,1) Kähler moduli the superpotential is zero. The static Calabi–Yau space solves the field equations for any value of the moduli, so the potential and therefore the superpotential for these vanishes.

Now we consider the Kähler potential, starting with the (1,1) moduli. These have the expansion

$$(g_{i\bar{j}} + b_{i\bar{j}})(x, y) = \sum_A T^A(x)\omega_{Ai\bar{j}}(y) \,. \tag{17.4.12}$$

The four-dimensional kinetic term is obtained from the ten-dimensional kinetic term by inserting this expansion. The result is

$$G_{A\bar{B}} = \frac{1}{V} \int d^6 y \, (\det G)^{1/2} G^{i\bar{k}} G^{l\bar{j}} \omega_{Ai\bar{j}} \omega^*_{B k \bar{l}} \,. \tag{17.4.13}$$

The integral can be related to the one appearing in the superpotential by using the Kähler form $J_{1,1}$ defined in (17.2.16). Parameterize the Kähler moduli space by $h^{1,1}$ complex numbers T^A,

$$J_{1,1} + iB_{1,1} = T^A \omega_{1,1\,A} \,, \quad T^A = v^A + ib^A \,. \tag{17.4.14}$$

Then after some calculation,

$$G_{A\bar{B}} = -\frac{\partial^2}{\partial T^A \partial T^{B*}} \ln W(v) \,, \tag{17.4.15}$$

where $2v^A = T^A + T^{A*}$ and

$$W(v) = \#(N_A, N_B, N_C)v^A v^B v^C = \int_K J_{1,1} \wedge J_{1,1} \wedge J_{1,1} \,. \tag{17.4.16}$$

This is just the superpotential, evaluated at $\phi = v$; it is also equal to the volume of the Calabi–Yau space. The $N = 1$ spacetime supersymmetry requires that this metric be Kähler.[3] The expression (17.4.15) gives the metric on Kähler moduli space directly in Kähler form, with

$$K_1(T, T^*) = -\ln W(v) \,. \tag{17.4.17}$$

Thus the Kähler potential for the moduli is determined in terms of the superpotential W. This is a very special property, which we will see later to be a consequence of the (2,2) world-sheet supersymmetry. The Kähler

---

[3] Kähler, Kähler, everywhere. Note that in some places it is the geometry of the compactification that is referred to, while here it is the geometry of the low energy scalar field space.

potential depends on the Kähler moduli and so is not a topological invariant, but it is quasitopological in the sense that its dependence on these moduli is determined by topological data. Note that it is independent of the complex structure moduli, another consequence of the (2,2) world-sheet supersymmetry.

The metric for the $\overline{27}$ kinetic terms is closely related,

$$G'_{A\bar{B}} = \exp[\kappa_4^2(K_2 - K_1)/3]G_{A\bar{B}} \qquad (17.4.18)$$

with K_2 to be defined below. The Kähler potential for these fields is then $G'_{A\bar{B}}\phi^A\phi^{B*}$.

Similar results hold for the (2, 1)-forms, though the precise statements and the derivations (which are again omitted) are somewhat more intricate. Expand

$$a_{i,jx}(x,y) = \frac{1}{2}\sum_a \chi^a_x(x)\omega_{aik\bar{l}}(y)\Omega^{\bar{k}\bar{l}}_j(y) , \qquad (17.4.19a)$$

$$\lambda_{i,jx}(x,y) = \frac{1}{2}\sum_a \lambda^a_x(x)\omega_{aik\bar{l}}(y)\Omega^{\bar{k}\bar{l}}_j(y) , \qquad (17.4.19b)$$

where a runs over the (1,2)-forms. In this case it is the kinetic term for the moduli that has a simple expression in terms of forms,

$$
\begin{aligned}
G_{a\bar{b}} &= -\frac{\displaystyle\int_K \omega_{1,2a} \wedge \omega^*_{1,2a}}{\displaystyle\int_K \Omega_{3,0} \wedge \Omega^*_{3,0}} \\
&= -\frac{\partial}{\partial X^a}\frac{\partial}{\partial X^{\bar{a}}}K_2(X, X^*) ,
\end{aligned} \qquad (17.4.20)
$$

with

$$K_2(X, X^*) = \ln\left(i\int_K \Omega_{3,0} \wedge \Omega^*_{3,0}\right) . \qquad (17.4.21)$$

Here X^a are coordinates for the moduli space of complex structures, $a = 1, \ldots, h^{2,1}$, and $X^{\bar{a}} = X^{a*}$.

To relate the superpotential to the Kähler potential it is useful to take *special coordinates* on moduli space. The Betti number b_3 is $2h^{2,1} + 2$. One can always find a basis of 3-cycles

$$\{A^I, B_J\} , \quad I, J = 0, \ldots h^{2,1} \qquad (17.4.22)$$

such that the intersection numbers are

$$\#(A^I, B_J) = \delta^I_J , \quad \#(A^I, A^J) = \#(B_I, B_J) = 0 . \qquad (17.4.23)$$

The corresponding (1,2)-forms are $\alpha_{1,2I}$ and $\beta^J_{1,2}$. Thus we can define

$$Z^I = \int_{A^I} \Omega_{3,0} . \qquad (17.4.24)$$

These $h^{2,1}+1$ complex numbers are one too many to serve as coordinates on the complex structure moduli space. However, there is no natural normalization for Ω, so we must identify

$$(Z^0, Z^1, \ldots, Z^n) \cong (\lambda Z^0, \lambda Z^1, \ldots, \lambda Z^n) \tag{17.4.25}$$

with $n = h^{2,1}$, and this projective space has the correct number of coordinates. The integrals

$$\mathscr{G}_I(Z) = \int_{B_I} \Omega_{3,0} \tag{17.4.26}$$

then cannot be independent variables; for given topology the $\mathscr{G}_I$ are known functions of the Z^J. These can be determined in terms of a single function $\mathscr{G}(Z)$,

$$\mathscr{G}_I = \frac{\partial \mathscr{G}}{\partial Z^I} \,, \qquad \mathscr{G}(\lambda Z) = \lambda^2 \mathscr{G}(Z) \,. \tag{17.4.27}$$

The nonprojective coordinates are then $X^a = Z^a/Z^0$ for $a = 1, \ldots, h^{2,1}$.

The Kähler potential (17.4.21) for the complex structure moduli can be expressed in terms of $\mathscr{G}$,

$$K_2(Z, Z^*) = \ln \mathrm{Im}(Z^{I*} \partial_I \mathscr{G}(Z)) \,. \tag{17.4.28}$$

So also can the superpotential,

$$W(Z, \chi) = \frac{\chi^a \chi^b \chi^c}{3!} \frac{\partial^3 \mathscr{G}(Z)}{\partial Z^a \partial Z^b \partial Z^c} \,. \tag{17.4.29}$$

The matter metric is again slightly different from that for the moduli,

$$G'_{a\bar{b}} = \exp[\kappa_4^2(K_1 - K_2)/3] G_{a\bar{b}} \,. \tag{17.4.30}$$

The intersection numbers (17.4.23) are invariant under a symplectic change of basis,

$$\begin{bmatrix} A'^I \\ B'_J \end{bmatrix} = S \begin{bmatrix} A^I \\ B_J \end{bmatrix} \tag{17.4.31}$$

for $S \in Sp(h^{2,1} + 1, \mathbf{Z})$. The new coordinates

$$\begin{bmatrix} Z'^I \\ \mathscr{G}'_J \end{bmatrix} = S \begin{bmatrix} Z^I \\ \mathscr{G}_J \end{bmatrix} \tag{17.4.32}$$

are then another set of special coordinates for the same moduli space.

To summarize, the low energy effective action is determined in terms of two holomorphic functions, $W(T)$ and $\mathscr{G}(Z)$. Each of these is determined in turn by the topology of the Calabi–Yau manifold and can be calculated by well-developed methods from analytic geometry. Notice that the actual Ricci-flat metric is never used — a good thing, as the explicit form is not known in any nontrivial example.

17.5 Higher corrections

Thus far we have considered only the leading term in an expansion in
α'/R_c^2. We now consider the corrections, remaining in this chapter at the
string tree level. The ten-dimensional action derived from string theory
has an infinite series of higher derivative corrections, each derivative accompanied by $\alpha'^{1/2}$. These terms can be deduced from the momentum
expansion of the tree-level scattering amplitudes. Alternatively, they can
be obtained from the higher loop corrections to the world-sheet beta functions, where again the expansion parameter is α'/R_c^2. The supersymmetry
transformations are given by a similar series.

The most immediate questions would seem to be whether the Calabi–
Yau manifolds solve the full field equations, and whether they remain
supersymmetric. Actually they do not. They continue to solve the field
equations when the terms quadratic and cubic in the curvatures and field
strengths are included in the action, but with the inclusion of the quartic
terms (corresponding to four loops in the world-sheet sigma model) they
in general do not. However, this is not really the right question. Rather,
as in any perturbation theory, we need to know whether the solution
can be corrected order-by-order so as to solve the field equations at
each order. It is not trivial that this is possible — as in other forms of
perturbation theory there is a danger of vanishing denominators — but
it has been shown to be possible from an analysis of the detailed form of
the corrections to the beta functions.

Remarkably, this same result obtained by a rather technical world-sheet
argument can be obtained much more easily and usefully from an analysis
of the spacetime effective action — a common theme in supersymmetric
theories. First note that regardless of whether the Calabi–Yau space can
be corrected to give an exact solution, we can still study the physics for
nearby configurations by the method of the previous section. Expand the
fields as background plus fluctuation, separate the fluctuations into light
and heavy, and integrate out the heavy fields to obtain an effective action
for the light fields. Corrections in the α'/R_c^2 expansion give additional
terms in the low energy action. Now, an important point is that any mass
scale appearing in these terms will be the compactification energy R_c^{-1}
times a power of the small parameter α'/R_c^2. Thus there is still a clean
separation between the light and heavy fields, and it makes sense to discuss
the effective action for the former.

The final key point is that this low energy effective action must be
supersymmetric. Because the full theory is supersymmetric, any breaking
must be spontaneous rather than explicit. To see this another way, note
that as $\alpha'/R_c^2 \to 0$ with R_c fixed, supersymmetry is restored and so the

gravitino becomes massless; near this limit it remains one of the light fields. However, the only consistent theory of light spin-$\frac{3}{2}$ particles is spontaneously broken supergravity.

We will see in the next chapter that the gauge kinetic term receives no corrections at string tree level to any order of α'/R_c^2. All corrections to the low energy effective action must then appear in the Kähler potential or the superpotential. We can now state the criterion for the corrections to spoil the solution: they must produce a correction to superpotential that depends only on the moduli, $\delta W(T, Z)$. In this case there will be a potential for the moduli, which at general points will not be stationary so that most or all of the previous static supersymmetric solutions are gone.

Now let us argue that this is impossible. Consider how a string amplitude depends on the moduli b^A for the $B_{i\bar{j}}$ background, eq. (17.4.14). This background enters into the string amplitude as

$$\frac{1}{2\pi\alpha'} \int_M B_{1,1} = \frac{n_A b^A}{2\pi\alpha'} . \tag{17.5.1}$$

Since the background $B_{1,1}$ form is closed, the integral depends only on the topology of the embedding of the world-sheet M in spacetime. The embedding is equivalent to a sum $n_A N^A$ of generators of $H_2(K)$, and so the integral follows as in eq. (17.4.11). Now, world-sheet perturbation theory is an expansion around the configuration $X^\mu(\sigma) = $ constant, which is topologically trivial. To all orders of perturbation theory, $n_A = 0$ and the amplitudes are independent of b^A. There are thus $h_{1,1}$ symmetries

$$T^A \to T^A + i\epsilon^A . \tag{17.5.2}$$

Since the superpotential must be holomorphic in T^A, this implies that it is actually independent of T^A.

To obtain a nonrenormalization theorem, let us write

$$T^A = c^A T \tag{17.5.3}$$

with the c^A fixed complex numbers, and focus on the dependence on T. Varying T at fixed c^A rescales $G_{i\bar{j}}$ and so scales the size of K while holding its shape fixed. Thus the world-sheet perturbation expansion parameter is T^{-1}. Since the superpotential is holomorphic in T, it can receive no corrections in world-sheet perturbation theory.

Thus the terms that might destabilize the vacuum and break supersymmetry cannot be generated, and so the Calabi–Yau solution can be perturbatively corrected to all orders to give a static supersymmetric background. A potential could also be generated by a Fayet–Iliopoulos term in the more general case that the gauge group includes a $U(1)$ factor. A separate argument excludes this; we postpone the discussion to section 18.7. It also follows that the superpotential for the matter fields, though calculated above by means specific to the $\alpha'/R_c^2 \to 0$ limit, is exact to all orders in α'/R_c^2.

Ordinarily the Kähler potential does not have similar nonrenormalization properties because it is not holomorphic, and so could have an arbitrary dependence on $T + T^*$. However, the presence of (2,2) superconformal symmetry on the world-sheet puts strong additional constraints on the theory. The action obtained in the field-theory approximation of the previous section had two notable properties. First, there was no superpotential for the moduli. Second, the full low energy action was determined by two holomorphic functions, one depending only on the Kähler moduli and the other only on the complex structure moduli. We will see in chapter 19 that these properties actually follow from (2,2) world-sheet supersymmetry and so are exact properties of the string tree-level action in Calabi–Yau compactification. The nonrenormalization of the superpotential then implies the same for the Kähler potential, and the full effective action found in the field-theory approximation is exact to all orders in α'/R_c^2, except for one term to be discussed in chapter 19.

For future reference, note that the Kähler potential (17.4.16) for the overall scale T is

$$ -3\ln(T + T^*) , \tag{17.5.4} $$

up to instanton corrections that are exponentially small in T.

Instanton corrections

The discussion above does not exclude the possibility of corrections that are nonperturbative on the string world-sheet. Indeed, these do break the shift symmetries (17.5.2). Consider a *world-sheet instanton,* meaning a topologically nontrivial embedding of the world-sheet in spacetime: the string world-sheet wraps around some noncontractible surface in spacetime. The n_A defined in eq. (17.5.1) are then nonzero and the amplitude depends on b^A, breaking the $T^A \to T^A + i\epsilon^A$ symmetry.

To see whether these can affect the superpotential, compare the Polyakov action

$$ \frac{1}{2\pi\alpha'} \int d^2z \, G_{i\bar{j}}(\partial_z Z^i \partial_{\bar{z}} Z^{\bar{j}} + \partial_z Z^{\bar{i}} \partial_{\bar{z}} Z^j) \tag{17.5.5} $$

to

$$ \frac{1}{2\pi\alpha'} \int_M J_{1,1} = \frac{1}{2\pi\alpha'} \int d^2z \, G_{i\bar{j}}(\partial_z Z^i \partial_{\bar{z}} Z^{\bar{j}} - \partial_z Z^{\bar{i}} \partial_{\bar{z}} Z^j) $$

$$ = \frac{n_A v^A}{2\pi\alpha'} . \tag{17.5.6} $$

The two terms in the action (17.5.5) are nonnegative, so the action is bounded below by $|n_A v^A|/2\pi\alpha'$. When this bound is attained, then either $\partial_{\bar{z}} Z^i$ or $\partial_{\bar{z}} Z^{\bar{\imath}}$ vanishes, and the embedding of the world-sheet in spacetime is a *holomorphic instanton*. In this case, the Polyakov action and the coupling to B combine to give the path integral factor

$$\exp(-n_A T^A/2\pi\alpha') \tag{17.5.7}$$

or its conjugate. This is holomorphic in T^A and so can appear in the super-potential: holomorphic instantons can, and do, correct the superpotential. In particular they correct the cubic terms in $\overline{\mathbf{27}}^3$. By (2,2) supersymmetry they also then correct the metric for the (1,1)-forms. However, study of the detailed form of the instanton amplitudes, in particular the fermion zero modes, shows that they cannot generate a superpotential for the moduli fields alone and so do not destabilize the solution. Again, this will be understood later as a consequence of (2,2) symmetry: no superpotential for the moduli can be generated.

Instantons cannot correct the metric for the (1,2)-forms. Instanton corrections depend as in eq. (17.5.7) on the (1,1) modulus T, and the metric for the (1,2)-forms cannot depend on (1,1) moduli (another consequence of (2,2) supersymmetry). They cannot then correct the $\overline{\mathbf{27}}^3$ superpotential either. The low energy action for the (1,2)-forms, though obtained in the field theory limit, is exact at string tree level. The low energy action for the (1,1)-forms receives instanton corrections.

17.6 Generalizations

Let us now consider the E_6 singlets from $H^1(\text{End } T)$. In particular, are there flat directions for these fields? Also, are there flat directions for the charged fields in the **27**s and $\overline{\mathbf{27}}$s? In each case the massless fields originate from the compact components of the ten-dimensional gauge field, so flat directions would correspond to varying the gauge field away from the 'spin connection = gauge connection' form assumed so far. The Bianchi identity (17.1.13) then implies that generically the torsion $\tilde{H}$ must be nonvanishing, so these flat directions take us outside the vanishing-torsion ansatz with which we began. Also, with the spin connection unrelated to the gauge connection there are in general no longer any left-moving supersymmetries on the world-sheet, and the world-sheet supersymmetry is reduced to (0, 2). We will thus refer to the fields parameterizing these potential flat directions as (0,2) moduli.

The analysis of the general solution is somewhat more intricate than for vanishing torsion, but again there are existence theorems to the effect that under appropriate topological conditions solutions exist in the field theory limit. The nonrenormalization theorem above still applies, so that these remain solutions to all orders of world-sheet perturbation theory. Nonperturbatively, instantons in (0, 2) backgrounds have fewer fermion zero modes and there is no general argument forbidding a superpotential for the (0,2) moduli. Initially it was believed that a superpotential would generically appear and destabilize most (0,2) vacua. However, it is now known that many of the (0,2) directions are exactly flat, so that the typical (2,2) moduli space is embedded in a larger moduli space of (0,2) theories. There are likely also moduli spaces of (0,2) theories that are not connected to any (2,2) theories.

The understanding of (0,2) theories is much less complete than for (2,2) theories, and the analysis of them is more intricate. We will therefore not discuss them in any detail, though some of the methods to be developed in chapter 19 for (2,2) theories are also useful in the (0,2) case.

We would like to mention briefly some phenomenological features of the (0,2) vacua. We have emphasized that the (2,2) theories look much like a grand unified Standard Model, with an E_6 gauge group and matter in the **27**. Under the $SU(3) \times SU(2) \times U(1)$ subgroup of E_6, the **27** contains 15 states with chiral gauge couplings, having the precise quantum numbers of a generation of quarks and leptons, and 12 with parity-symmetric couplings. The latter can have $(SU(3) \times SU(2) \times U(1))$-invariant mass terms and so can be much more massive than the weak scale. Indeed, these extra states in the **27** can mediate baryon decay, so they must be much heavier than the weak scale. Other arguments based on the running of the gauge couplings and the lightness of the Standard Model neutrinos also suggest that the extra states are quite massive. In addition to the extra states within each **27**, typical (2,2) theories have both **27**s and $\overline{\bf 27}$s, which from the low energy point of view correspond to generations with left- and right-handed weak interactions. Although it is possible that some right-handed 'mirror generations' exist near the weak scale, this seems unlikely for a number of reasons. Fortunately the gauge symmetry allows a **27** and a $\overline{\bf 27}$ to pair up and become massive.

Although these various masses are allowed by the low energy gauge symmetry, we need a specific mechanism for generating them. As long as we stay within the (2,2) theories, even adding Wilson lines to break the E_6 symmetry, the general properties of these theories guarantee that the quantum numbers of the low energy fields add up to complete multiplets of E_6. However, along the (0,2) directions the extra states can become massive. In addition to the $\bf 27^3$ and $\overline{\bf 27}^3$ terms already discussed, the lowest order superpotential contains terms $\bf 1^3$ and $\bf 1 \cdot 27 \cdot \overline{27}$, and together these have the potential to generate all the needed masses.

The (0,2) moduli from the **27**s and $\overline{\bf 27}$s may be useful for another related reason. In most examples, such as those discussed in section 17.2, the freely-acting discrete symmetry is Abelian, for example $\mathbf{Z}_3$. The Wilson lines must have the same algebra as the space group, so they commute and can be taken to lie in a $U(1)^6$ subalgebra of E_6. Since the low energy group is that part of E_6 that commutes with the Wilson lines, it contains at least this $U(1)^6$ and so has rank 6, as compared to the rank 4 of the Standard Model: the closest we can come in this way to the Standard Model is $SU(3) \times SU(2) \times U(1)^3$. The additional $U(1)$s might be broken somewhat above the weak scale, but again there are problems; in particular the extra 12 states in the **27** are chiral under the additional $U(1)$s so this prevents them from becoming very massive. One way to break these symmetries is to twist by a *non-Abelian* discrete group, but another is to give expectation values to (0,2) moduli from the **27** and **$\overline{27}$**. Clearly this breaks some of the E_6 symmetry, and in fact it necessarily reduces the rank of the gauge group. The group theory in section 11.4 shows that the **27** contains two singlets of $SU(3) \times SU(2) \times U(1)$. If both of these have expectation values they break E_6 to the minimal grand unified group $SU(5)$, and combined with Wilson line breaking this can give the Standard Model gauge group.

Thus it is an attractive possibility that our vacuum is given by turning on some of the (0,2) moduli of a Calabi–Yau compactification. We should emphasize, however, that if one considers the large set of (0,2) theories that can be constructed by asymmetric orbifolds or free fermions, only a small subset of these have a close resemblance to the Standard Model.

No exercises

The nature of this chapter, all results and no derivations, does not lend itself to exercises. The reader who wishes to learn more should consult the references.

18

Physics in four dimensions

We have now studied two kinds of four-dimensional string theory, based on orbifolds and on Calabi–Yau manifolds. We saw that the low energy physics of the weakly coupled heterotic string resembles a unified version of the Standard Model rather well. In this chapter we present general results, valid for any compactification. In most of this chapter we are concerned with weakly coupled heterotic string theories, but at various points we will discuss how the results are affected by the new understanding of strongly coupled strings.

18.1   Continuous and discrete symmetries

An important result holding in all string theories is that *there are no continuous global symmetries; any continuous symmetries must be gauged.* We start with the bosonic string. Associated with any symmetry will be a world-sheet charge

$$Q = \frac{1}{2\pi i} \oint (dz\, j_z - d\bar{z}\, j_{\bar{z}}) . \tag{18.1.1}$$

This is to be a symmetry of the physical spectrum and so it must be conformally invariant. Thus j_z transforms as a $(1,0)$ tensor and $j_{\bar{z}}$ as a $(0,1)$ tensor. We can then form the two vertex operators

$$j_z \bar{\partial} X^\mu e^{ik\cdot X} , \quad \partial X^\mu j_{\bar{z}} e^{ik\cdot X} . \tag{18.1.2}$$

These create massless vectors coupling to the left- and right-moving parts of the charge Q. Thus the left- and right-moving parts of Q each give rise to a spacetime gauge symmetry. If Q is carried only by fields moving in one direction, then only one of the currents and only one of the vertex operators is nonvanishing. Turning the construction around, any local symmetry in spacetime gives rise to a global symmetry on the world-sheet.

327

For type I or II strings the same argument holds immediately if we use superspace, writing

$$Q = \frac{1}{2\pi i} \oint (dz \, d\theta \, J - d\bar{z} \, d\bar{\theta} \, \tilde{J}) \,. \tag{18.1.3}$$

Superconformal invariance requires that J be a $(\frac{1}{2}, 0)$ tensor superfield and $\tilde{J}$ a $(0, \frac{1}{2})$ tensor superfield. Combined with $\tilde{\psi}^\mu$ or ψ^μ respectively, these give gauge boson vertex operators, so again this is a gauge symmetry in spacetime. The same is true for the heterotic string, using the bosonic argument on one side and the supersymmetric argument on the other.

The absence of continuous global symmetries has often been imposed as an aesthetic criterion by model builders in field theory, and we see that it is realized in string theory. There is a slight loophole in the argument, which we will discuss later in the section.

We have seen in the examples from earlier chapters that string theories generally have discrete symmetries at special points in moduli space. It is harder to generalize about whether these are local or global symmetries because the difference is subtle for a discrete symmetry: there is no associated gauge boson in the local case. The meaning of a discrete local symmetry was discussed in section 8.5 in the context of the field theory on the world-sheet. The simplest way to verify that a discrete symmetry is local is to find a point in moduli space where it is enlarged to a continuous gauge symmetry. For example, this is the case for the T-duality of the bosonic and heterotic strings. To see what this would mean, consider a spacetime with x^8 and x^9 periodic, with the radius R_8 a function of x^9. Then $R_8(x^9)$ need not be strictly periodic; rather, it could also be that

$$R_8(2\pi R_9) = \alpha'/R_8(0) \,. \tag{18.1.4}$$

This is the essence of a discrete gauge symmetry: that on nontrivial loops fields need be periodic only up to a gauge transformation. Since T-duality is embedded in the larger U-duality of the type II theory, the latter must be a gauge symmetry as well. Thus we could have a similar aperiodicity in the IIB string coupling, for example:

$$\Phi(2\pi R_9) = -\Phi(0) \,, \quad g(2\pi R_9) = 1/g(0) \,. \tag{18.1.5}$$

It is not clear that this is true of all discrete symmetries in string theory, but it seems quite likely.

P, C, T, and all that

We would like to discuss briefly the breaking of the discrete spacetime symmetries P, C, and T in string theory.

Parity symmetry P is invariance under reflection of any one coordinate, say $X^3 \to -X^3$. It is not a good symmetry of the Standard Model, being violated by the gauge interactions. Classifying particles moving in the 1-direction by the helicity $\Sigma^{23} = s_1$, the helicity $+\frac{1}{2}$ states form some gauge representation r_+, and the helicity $-\frac{1}{2}$ states some representation r_-. Parity takes the helicity $s_1 \to -s_1$, and so is a good symmetry only if $r_+ = r_-$. In the Standard Model it appears (barring the discovery of new massive states with the opposite gauge couplings) that $r_+ \neq r_-$: the gauge couplings are *chiral*.

Let us consider the situation in string theory, starting with the ten-dimensional heterotic string. In ten dimensions states are labeled by their $SO(8)$ representation. Parity again reverses the spinor representations **8** and **8′**, and is a good symmetry only if the corresponding gauge representations are the same, $r = r'$. For the heterotic string, r is the adjoint representation while r' is empty, so the gauge couplings are chiral and there is no parity symmetry. To see how this arises, note that the heterotic string action and world-sheet supercurrent (or BRST charge) are invariant if we combine the reflection $X^3 \to -X^3$ with $\psi^3 \to -\psi^3$. However, this also flips the sign of $\exp(\pi i \tilde{F})$ in the R sector, and so it is not a symmetry of the theory because the GSO projection restricts the spectrum to $\exp(\pi i \tilde{F}) = +1$.

Although the ten-dimensional spectrum is chiral, compactification to four dimensions can produce a nonchiral spectrum. This is true of toroidal compactification, for example, as one sees from the discussion in section 11.6. The point is that the theory is invariant under simultaneous reflection of one spacetime and one internal coordinate, say X^3 and X^9, as well as their partners ψ^3 and ψ^9. This is a symmetry of the action, supercurrent, and GSO projection, and so of the full theory. From the ten-dimensional point of view, it is a rotation by π in the (3,9) plane, but from the four-dimensional point of view it is a reflection of the 3-axis, combined with an internal action which gives negative intrinsic parity to the 9-oscillators. This symmetry reverses the momenta $k^9_{R,L}$, which are the charges under the corresponding Kaluza–Klein gauge symmetries, while leaving the other internal momenta invariant. Strictly speaking, it is therefore not a pure parity operation (which by the usual definition leaves gauge charges invariant) or a CP transformation (which inverts all charges), but something in between.

In the $\mathbf{Z}_3$ orbifold example, the spectrum was found to be chiral. The orbifold twist removes all parity symmetries. Notice that simultaneous reflection of $X^{3,5,7,9}$, which takes $Z^i \leftrightarrow Z^{\bar{\imath}}$, satisfies $Pr = r^2P$ and so commutes with the twist projection. However, to extend this action to the various spinor fields requires that P reflect $\psi^{3,5,7,9}$ and $\lambda^{2,4,6}$ as well. This acts on an odd number of the λ fermions and so does not commute with the current algebra GSO projection. The combined effect of the orbifold twist and the ψ and λ GSO projections removes all parity symmetry and leaves a chiral spectrum. Chiral gauge couplings arise in many other kinds of string compactification.

There is one interesting general remark. The chirality of the spectrum can be expressed in terms of a mathematical object known as an *index*. Separate $\exp(\pi i \tilde{F})$ into a spacetime part and an internal part, $\tilde{F} = \tilde{F}_4 + \tilde{F}_K$. For massless fermions moving in the 1-direction, $2s_1 = -i\exp(\pi i \tilde{F}_4)$, which in turn is equal to $i\exp(\pi i \tilde{F}_K)$ due to the GSO projection. For massless R sector states the internal part is annihilated by G_0, so the net chirality (number of helicity $+\frac{1}{2}$ states minus helicity $-\frac{1}{2}$ states) in a given irreducible representation r is

$$N_{+\frac{1}{2},r} - N_{-\frac{1}{2},r} = \text{Tr}_{r,\ker(G_0)}[i\exp(\pi i \tilde{F}_K)] , \qquad (18.1.6)$$

the trace running over all states in the internal CFT which are in the representation r and are annihilated by G_0. One can now drop the last restriction on the trace,

$$N_{+\frac{1}{2},r} - N_{-\frac{1}{2},r} = \text{Tr}_r[i\exp(\pi i \tilde{F}_K)] . \qquad (18.1.7)$$

The point is that any state $|\psi\rangle$ with a positive eigenvalue v under G_0^2 is always paired with a state $G_0|\psi\rangle$ of opposite $\exp(\pi i \tilde{F}_K)$, so these states make no net contribution to the trace. The state $G_0|\psi\rangle$ cannot vanish because $G_0 G_0 |\psi\rangle = v|\psi\rangle$.

Such a trace is known as an index: this can be defined whenever one has a Hermitean operator G_0 anticommuting with a unitary operator $\exp(\pi i \tilde{F}_K)$. The index has the important property that it is invariant under continuous changes of the CFT. Under such a change, the eigenvalues v of G_0^2 change continuously, but the trace of $\exp(\pi i \tilde{F}_K)$ at $v = 0$ remains invariant because states can only move away from $v = 0$ in pairs with opposite $\exp(\pi i \tilde{F}_K)$. This invariance can also be understood from the spacetime point of view: a continuous change in the background fields can give mass to some previously massive states, but to make a massive representation one must combine states of opposite helicity.[1] Using this invariance, the index may often be calculated by deforming to a convenient limit. There is one subtlety that comes up in some examples: the index may change in certain limits due to states running off to infinity in field space.

Charge conjugation C leaves spacetime invariant but conjugates the gauge generators. In the Standard Model this is again broken by the

---

[1] This is one of those statements that, surprisingly, need no longer hold at strong coupling. We will discuss this further in sections 19.7 and 19.8.

gauge couplings of the fermions. For C invariance to hold, the fermion representations must satisfy $r_+ = \bar{r}_+$ and $r_- = \bar{r}_-$. CPT invariance, to be discussed below, implies that $r_+ = \bar{r}_-$ so that chiral gauge couplings violate C as well as P. Thus the orbifold example also violates C.

The combination CP takes $r_+ \to \bar{r}_-$ and so is automatically a symmetry of the gauge couplings as a consequence of CPT. In the Standard Model Lagrangian, CP is broken by phases in the fermion–Higgs Yukawa couplings. In the $\mathbf{Z}_3$ orbifold example, the transformation that reverses $X^{3,5,7,9}$, $\psi^{3,5,7,9}$, and all of the λ^I for I odd is a symmetry of the action, the BRST charge, and all projections. From the point of view of the four-dimensional theory this is CP, because the action on the λ^I changes the sign of all the diagonal generators, which is charge conjugation. The $\mathbf{Z}_3$ orbifold is thus CP-invariant. However, recall that there were many moduli. These included the flat metric background $G_{i\bar{j}}dZ^i dZ^{\bar{j}}$. The operation CP takes $G_{i\bar{j}} \to G_{ij}$. Reality of the metric requires $G_{i\bar{j}}$ to be Hermitean, while CP requires it to be real. The generic Hermitean $G_{i\bar{j}}$ is not real, so CP is broken almost everywhere in moduli space. One must also consider other possible CP operations, such as adding discrete rotations of some of the Z^i, or permutations of the Z^i, to the transformation. These will be symmetries at special points in moduli space, but are again broken generically. This is also true for most other string compactifications: there will be CP-invariant vacua, but some of the many moduli will be CP-odd so that CP-invariance is spontaneously broken at generic points.

It is interesting to note that CP, like the discrete symmetries discussed earlier, is a gauge symmetry. The operation described above can be thought of as rotations by π in the (3,5) and (7,9) planes, combined with a gauge rotation. These are all part of the local symmetry of the ten-dimensional theory, though this is partly spontaneously broken by the compactification.

In local, Lorentz-invariant, quantum field theory the combination CPT is always an exact symmetry. It is easy to show that CPT is a symmetry of string perturbation theory, using essentially the same argument as is used to prove the CPT theorem in field theory. Consider the operation θ that reverses $X^{0,3}$ and $\psi^{0,3}$. If we continue to Euclidean time this is just a rotation by π in the (iX^0, X^3) plane and so is obviously a symmetry. The analytic continuation is well behaved because $X^{0,3}$ and $\psi^{0,3}$ are free fields. Clearly θ includes parity and time-reversal. To see that it also implies charge conjugation, recall that a vertex operator $\mathcal{V}$ with $k^0 < 0$ creates a string in the initial state, while a vertex operator with $k^0 > 0$ destroys a string in the final state. If $\mathcal{V}$ carries some charge q it creates a string of charge q. The operation θ does not act on the charges, so $\theta \cdot \mathcal{V}$ also has charge q and so destroys a string of charge $-q$. Thus, θ takes a string in the in-state to the C-, P-, and T-reversed string in the out-state.

To make this slightly more formal, recall from section 9.1 that the

S-matrix is given schematically by

$$\langle \alpha, \text{out} | \beta, \text{in} \rangle = \left\langle \overline{\mathcal{V}}_\alpha \mathcal{V}_\beta \right\rangle , \qquad (18.1.8)$$

where to be concise we have only indicated one vertex operator in each of the initial and final states. Then by acting with θ this becomes

$$\langle \alpha, \text{out} | \beta, \text{in} \rangle = \left\langle \theta \cdot \overline{\mathcal{V}}_\alpha \, \theta \cdot \mathcal{V}_\beta \right\rangle = \langle \theta \bar{\beta}, \text{out} | \theta \bar{\alpha}, \text{in} \rangle . \qquad (18.1.9)$$

The CPT operation is antiunitary,

$$\langle CPT \cdot \beta, \text{out} | CPT \cdot \alpha, \text{in} \rangle = \langle \alpha, \text{out} | \beta, \text{in} \rangle , \qquad (18.1.10)$$

so we see that CPT is θ combined with the conjugation of the vertex operator.

This argument is formulated in string perturbation theory. Elsewhere we have encountered results that hold to all orders of perturbation theory but are spoiled by nonperturbative effects. Without a nonperturbative formulation of string theory we cannot directly extend the CPT theorem, but we can 'prove' it by the strategy that we have used elsewhere: assert that the low energy physics of string theory is governed by quantum field theory, and then cite the CPT theorem from the latter. Still, there may be surprises; we can hope that when string theory is better understood it will make some distinctive non-field-theoretic prediction for observable physics.

The spin-statistics theorem is often discussed alongside the CPT theorem. The discussion in section 10.6 for free boson theories is easily generalized. Consider a basis of Hermitean (1,1) operators $\mathscr{A}_i$ with definite Σ^{01} eigenvalue s_0 and $\beta\gamma$ ghost number q. Now consider the OPE of such an operator with itself. In any unitary CFT, a simple positivity argument shows that the leading term in the OPE of a Hermitean operator with itself is the unit operator. Then

$$\mathscr{A}_i(z, \bar{z}) \mathscr{A}_i(0, 0) \sim (z\bar{z})^{-2} \bar{z}^{2(q+q^2-s_0^2)} \exp(2q\tilde{\phi} + 2is_0\tilde{H}_0) , \qquad (18.1.11)$$

where the z- and $\bar{z}$-dependence follows from the weight $\tilde{h} = 2(q + q^2 - s_0^2)$ of the exponential. For NS states, with integer spacetime spin, s_0 and q are integers, while for R states, with half-integer spacetime spin, they are half-integer. It follows that the operator product (18.1.11) is symmetric in the NS sector and antisymmetric in the R sector. The spacetime spin is thus correlated with world-sheet statistics, and the spacetime spin-statistics theorem then follows as in section 10.6. Again this is a rather narrow and technical way to establish this result.

The strong CP problem

In the Standard Model action *CP* violation can occur in two places, the fermion–Higgs Yukawa couplings and the theta terms

$$S_\theta = \frac{\theta}{8\pi^2} \int \mathrm{Tr}(F_2 \wedge F_2) \,. \tag{18.1.12}$$

This is θ times the instanton number, the trace normalized to the **n** of $SU(n)$. For the weak $SU(2)$ and $U(1)$ gauge interactions the fluctuations of the gauge field are small and the effect of S_θ is negligible, but for the strongly coupled $SU(3)$ gauge field the nontrivial topological sectors make significant contributions. The result is *CP* violation proportional to θ in the strong interactions. The limits on the neutron electric dipole moment imply that

$$|\theta| < 10^{-9} \,. \tag{18.1.13}$$

The *CP*-violating phases in the fermion–Higgs couplings are known from kaon physics not to be much less than unity. Understanding the small value of θ is the *strong CP problem*.

One proposed solution, *Peccei–Quinn (PQ) symmetry*, is automatically incorporated in string theory. In eq. (16.4.13) we found the coupling

$$\frac{1}{2g_4^2} \int aF_2^a \wedge F_2^a \,. \tag{18.1.14}$$

Aside from this term, the action is invariant under

$$a \to a + \epsilon \,, \tag{18.1.15}$$

known as PQ symmetry. The field a, which would be massless if the symmetry (18.1.15) were exact, is the axion. The axion and the θ-parameter appear only in the combination $\theta + 8\pi^2 a/g_4^2$, so θ has no physical effect: it can be absorbed in a redefinition of a. The effective physical value θ_{eff} is $\theta + 8\pi^2 \langle a \rangle / g_4^2$. The strong interaction produces a potential for a, which is minimized precisely at $\theta_{\mathrm{eff}} = 0$ because at this point the various contributions to the path integral add coherently. The weak interactions induce a nonzero value, but this is acceptably small.

The axion a is known as the *model-independent axion* because the coupling (18.1.14) is present in every four-dimensional string theory: the amplitude with one $B_{\mu\nu}$ vertex operator and two gauge vertex operators does not depend on the compactification. Unfortunately, the model-independent axion may not solve the strong *CP* problem. There are likely to be several non-Abelian gauge groups below the string scale. Low energy string theories typically have hidden gauge groups larger than $SU(3)$, and the corresponding strong interaction scales are $\Lambda_{\mathrm{hidden}} > \Lambda_{\mathrm{QCD}}$. We will see later in the chapter that this is a likely source of supersymmetry breaking.

The model-independent axion couples to all gauge fields. The gauge group with the largest scale Λ gives the largest contribution, so that the axion sets the θ-parameter for *that* gauge group approximately to zero. In a CP-violating theory, the θ-parameters for the different gauge groups will in general differ, so that θ_{QCD} remains large. Nonperturbative effects at the string scale may also contribute to the axion potential.

Another difficulty is cosmological. The axion a, being closely related to the graviton and dilaton, couples with gravitational strength κ. In other words, the *axion decay constant* is close to the Planck length. A decay constant this small leads to an energy density in the axion field today that is too large; it takes a rather nonstandard cosmology to evade this bound.

Both problems might be evaded if there were additional axions with appropriate decay constants. In Calabi–Yau compactifications there are shift symmetries (17.5.2) of the $B_{i\bar{j}}$ background, $T^A \rightarrow T^A + i\epsilon^A$. Further, the threshold corrections discussed in section 16.4 induce the coupling (18.1.14) to the gauge fields. However, these are only approximate PQ symmetries, because world-sheet instantons generate interactions proportional to

$$\exp(-n_A T^A/2\pi\alpha') = \exp[-n_A(v^A + ib^A)/2\pi\alpha'] \ . \qquad (18.1.16)$$

These spoil the PQ symmetries and generate masses for the axions b^A. There is some suppression if $v^A/2\pi\alpha'$ is large, and possibly additional suppression from light fermion masses, which appear in relating the instanton amplitudes to the actual axion mass. However, the suppression must be very large, so that the axion mass from this source is well below the QCD scale, if this is to solve the strong CP problem.

In the type I and II theories the scalars from the R–R sector are also potential axions. As discussed in section 12.1, their amplitudes vanish at zero momentum, implying a symmetry $C \rightarrow C + \epsilon$ for each such scalar. In addition they can have the necessary couplings to gauge fields. They receive mass from D-instanton effects.

In summary we have potentially three kinds of axion — model-independent, $B_{i\bar{j}}$, and R–R — which receive mass from three kinds of instanton — field theory, world-sheet, and Dirichlet. Not surprisingly, one can show that these are related by various string dualities. It may be that in some regions of parameter space the axions are light enough to solve the strong CP problem. There may also be additional approximate PQ symmetries from light fermions coupling to some of the strong groups. Or it may be that the solution to the strong CP problem lies in another direction, depending on details of the origin of CP violation.

Incidentally, these PQ symmetries are continuous global symmetries, seemingly violating the result obtained earlier. The loophole is that the world-sheet charge Q *vanishes* in each case — strings do not carry any of the PQ charges. We know this for the R–R charges; for the others it follows because the axion vertex operator at zero momentum is a total derivative. However, since in each case these are not really symmetries, being violated by the various instanton effects, the general conclusion about continuous global symmetries evidently still holds.

The arguments thus far are based on our understanding of perturbative string theory, but it is likely that the conclusion also holds at strong coupling. If a symmetry is exact at large g, it remains a symmetry as g is taken into the perturbative regime, since this is just a particular point in field space. At weak coupling it can then take one of two forms. It could be visible in string perturbation theory, meaning that it holds at each order of perturbation theory; it is then covered by the above discussion. Or, it could hold only in the full theory; the duality symmetries are of this type, but these are all discrete symmetries.

18.2 Gauge symmetries

Gauge and gravitational couplings

In sections 12.3 and 12.4 we obtained the relation between the gauge and gravitational couplings of the heterotic string in ten dimensions:

$$g_{10}^2 = \frac{4\kappa_{10}^2}{\alpha'} . \tag{18.2.1}$$

If we compactify, then by the usual dimensional reduction

$$g_4^2 = g_{10}^2/V , \quad \kappa_4^2 = \kappa_{10}^2/V , \tag{18.2.2}$$

with V the compactification volume. The relation between the parameters in the four-dimensional action is then the same,

$$g_4^2 = \frac{4\kappa_4^2}{\alpha'} . \tag{18.2.3}$$

Also, the actual physical values of the couplings depend on the dilaton as[2] e^{Φ_4}, but this enters in the same way on each side so that

$$g_{YM}^2 = \frac{4\kappa^2}{\alpha'} . \tag{18.2.4}$$

This derivation is valid only in the field-theory limit, but with one generalization it holds for any four-dimensional string theory. For gauge bosons

---

[2] When $\langle\Phi_4\rangle \neq 0$, the rescaling (16.4.6) changes the background value of the metric. To study the physics in a given background, as we are doing in this chapter, one should instead rescale

$$G'_{\mu\nu\,\text{Einstein}} = \exp[-2(\Phi_4 - \langle\Phi_4\rangle)]G_{\mu\nu} ,$$

and the coefficient of the gravitational action is then the physical coupling $\kappa = \exp(\langle\Phi_4\rangle)\kappa_4$.

with polarizations and momenta in the four noncompact directions, the explicit calculation (12.4.13) of the three-gauge-boson amplitude involves only the four-dimensional and current algebra fields and so is independent of the rest of the theory. The only free parameter is the parameter $\hat{k}$ from the current algebra, which appeared in the three-gauge-boson amplitude as $\hat{k}^{-1/2}$. Thus the general result is

$$g_{YM}^2 = \frac{4\kappa^2}{\hat{k}\alpha'} . \tag{18.2.5}$$

For completeness[3] let us recall that $\hat{k}$ is the coefficient of $z^{-2}\delta^{ab}$ in the $j^a j^b$ OPE, and that the gauge field Lagrangian density is defined to be

$$-\frac{1}{4g_{YM}^2} F_{\mu\nu}^a F^{a\mu\nu} . \tag{18.2.6}$$

The parameter $\hat{k}$ differs from the quantized level of the current algebra through the convention for the normalization of the gauge generators, which can be parameterized in terms of the length-squared of a long root, $\psi^2 = 2\hat{k}/k$. The common current algebra convention is $\psi^2 = 2$ so that $\hat{k} = k$. The common particle physics convention is that the inner product for $SO(n)$ groups is the trace in the vector representation, and the inner product for $SU(n)$ groups is twice the trace in the fundamental representation. Both of these give $\psi^2 = 1$ so that $\hat{k} = \frac{1}{2}k$. We should emphasize that it is the quantized level k that matters physically — for example, it determines the allowed gauge representations — but that when we deal with expressions that require a normalization of the generators (like the gauge action) it is generally the parameter $\hat{k}$ that appears.

It is interesting to consider the corresponding relation in open string theory. The ten-dimensional coupling was obtained in eq. (13.3.31),

$$\frac{g_{YM}^2}{\kappa} = 2(2\pi)^{7/2}\alpha' \quad \text{(type I, } d = 10) . \tag{18.2.7}$$

Under compactification this becomes

$$\frac{g_{YM}^2}{\kappa} = \frac{2(2\pi)^{7/2}\alpha'}{V^{1/2}} \quad \text{(type I, } d < 10) . \tag{18.2.8}$$

Unlike the closed string relation, this depends on the compactification volume.

---

[3] We feel compelled to be precise about the factors of 2, but most readers will want to skip such digressions as this paragraph.

Gauge quantum numbers

For a gauge group based on a current algebra of level k, only certain representations can be carried by the massless states. The total left-moving weight h of the matter part of any vertex operator is unity. Since the energy-momentum tensor is additive,

$$T_B = T_B^s + T_B' , \qquad (18.2.9)$$

the contribution of the current algebra to h is at most unity. This leaves two possibilities. Either the current algebra state is a primary field with $h \leq 1$, or it is a descendant of the form

$$j_{-1}^a \cdot 1 = j^a . \qquad (18.2.10)$$

Let us consider the latter case first. The current j^a has $h = 1$, so for bosons the remainder of the matter vertex operator has weight $(0, \frac{1}{2})$. One possibility is ψ^μ, which just gives the gauge boson states. There could also be $(0, \frac{1}{2})$ fields from the internal CFT, but we will see later in the section that this is inconsistent with having any chiral gauge interactions. For fermions the remainder of the matter vertex operator would have weight $(0, \frac{5}{8})$. This combines with the $\beta\gamma$ ghost vertex operator $e^{-\tilde{\phi}/2}$ to give a $(0, 1)$ current. This is a spacetime spinor, and so is the world-sheet current associated with a spacetime supersymmetry. Thus there are massless fermions of this type only if the theory is supersymmetric, in which case they are the gauginos.

For massless states based on current algebra primaries, the restriction (11.5.43) limits the representations that may appear. For $SU(2)$ at $k = 1$ only the **1** and **2** are allowed, while for $SU(3)$ at $k = 1$ only the **1**, **3**, and $\bar{\mathbf{3}}$ are allowed.

In the Standard Model, there are several notable patterns in the gauge quantum numbers of the quarks and leptons: replication of generations, chirality, quantization of the electric charge, and absence of large ('exotic') representations of $SU(2)$ and $SU(3)$. We have seen in the orbifold and Calabi–Yau examples that multiple generations arise frequently in four-dimensional string theories. This is an attractive feature of higher-dimensional theories in general. The generations arise from massless excitations that differ in the compact dimensions but have the same spacetime quantum numbers. Chirality was discussed in section 18.1, and quantization of electric charge will be discussed in section 18.4. Finally, the absence of exotics, the fact that only the **1** and **2** of $SU(2)$ and the **1**, **3**, and $\bar{\mathbf{3}}$ of $SU(3)$ are found, is 'explained' by string theory if we assume that these gauge symmetries arise from $k = 1$ current algebras. Also, the only scalar in the Standard Model is the $SU(2)$ doublet Higgs scalar, and from tests of this model it is known that no more than

$O(1\%)$ of the $SU(2) \times U(1)$ breaking can come from larger representations.

Unfortunately, this is not a firm prediction of string theory. While the simplest four-dimensional string theories have $k = 1$, there is still an enormous number of tree-level string vacua with higher level current algebras. Also, as discussed in section 16.3, $k = 1$ is impossible if a grand unified group remains below the string scale. For $SU(5)$ only the representations **1**, **5**, **5̄**, **10**, and **10̄** are allowed, for $SO(10)$ only **1**, **16**, **16̄**, and **10**, and for E_6 only **1**, **27**, and **27̄**. In each case this includes the representations carried by the quarks, leptons, and the Higgs scalar that breaks the electroweak symmetry, but not the representations needed to break the unified group to $SU(3) \times SU(2) \times U(1)$. The latter are allowed for levels $k \geq 2$. We will return to this point in the next section.

Right-moving gauge symmetries

Thus far we have considered gauge symmetries carried by the left-moving degrees of freedom of the heterotic string. For these the conformal invariance leads to a current algebra. For gauge symmetries carried by the right-movers, the superconformal algebra plus gauge symmetry give rise to a *superconformal current algebra (SCCA)*. The matter part of the gauge boson vertex operator in the -1 picture is

$$\partial X^\mu \tilde{\psi}^a e^{ik \cdot X} \tag{18.2.11}$$

with $\tilde{\psi}^a$ a weight $(0, \frac{1}{2})$ superconformal tensor field. Then

$$\tilde{G}_{-1/2} \cdot \tilde{\psi}^a = \tilde{\jmath}^a \tag{18.2.12}$$

is a $(0, 1)$ field. It is nontrivial because

$$\tilde{G}_{1/2} \cdot \tilde{\jmath}^a = 2\tilde{L}_0 \cdot \tilde{\psi}^a = \tilde{\psi}^a . \tag{18.2.13}$$

Also, $\tilde{\jmath}^a$ is a conformal tensor, annihilated by $\tilde{L}_n$ for $n > 0$, though not a superconformal tensor. The $\tilde{\jmath}^a$ thus form a right-moving current algebra.

We take the current algebra to be based on a simple group g at level k, and for simplicity use the current algebra normalization (which is no problem, because we are about to see that these gauge symmetries will never appear in particle physics!). Using the Jacobi identity we can fill in the rest of the operator products,

$$\tilde{\psi}^a(\bar{z})\tilde{\psi}^b(0) \sim \frac{k\delta^{ab}}{\bar{z}} , \qquad (18.2.14a)$$

$$\tilde{\jmath}^a(\bar{z})\tilde{\psi}^b(0) \sim \frac{if^{abc}}{\bar{z}}\tilde{\psi}^c(0) , \qquad (18.2.14b)$$

$$\tilde{T}_F(\bar{z})\tilde{\psi}^a(0) \sim \frac{1}{\bar{z}}\tilde{\jmath}^a(0) , \qquad (18.2.14c)$$

$$\tilde{\jmath}^a(\bar{z})\tilde{\jmath}^b(0) \sim \frac{k\delta^{ab}}{\bar{z}^2} + \frac{if^{abc}}{\bar{z}}\tilde{\jmath}^c(0) , \qquad (18.2.14d)$$

$$\tilde{T}_F(\bar{z})\tilde{\jmath}^a(0) \sim \frac{1}{\bar{z}^2}\tilde{\psi}^a(0) + \frac{1}{\bar{z}}\bar{\partial}\tilde{\psi}^a(0) . \qquad (18.2.14e)$$

In particular, the $\tilde{\psi}^a$ are free right-moving fields with a nonstandard normalization.

We can now carry out a generalization of the Sugawara construction. The $\tilde{\jmath}\tilde{\psi}$ product implies that if we define

$$\tilde{\jmath}^a = \tilde{\jmath}^a_\psi + \tilde{\jmath}'^a , \qquad (18.2.15)$$

where

$$\tilde{\jmath}^a_\psi = -\frac{i}{2k}f^{abc}\tilde{\psi}^b\tilde{\psi}^c , \qquad (18.2.16)$$

then $\tilde{\jmath}'^a$ is nonsingular with respect to the $\tilde{\psi}^a$. It follows that there are actually two current algebras. One is built out of the $\tilde{\psi}^a$ and has current $\tilde{\jmath}^a_\psi$ and level $k_\psi = h(g)$. The other commutes with the $\tilde{\psi}^a$ and has current $\tilde{\jmath}'^a$ and level $k' = k - k_\psi$. We see that $k \geq h(g)$, with equality if and only if $\tilde{\jmath}'^a$ is trivial.

As in the Sugawara construction we can separate $\tilde{T}_F$,

$$\tilde{T}_F = \tilde{T}^s_F + \tilde{T}''_F , \qquad (18.2.17)$$

where

$$\tilde{T}^s_F = -\frac{i}{6k^2}f^{abc}\tilde{\psi}^a\tilde{\psi}^b\tilde{\psi}^c + \frac{1}{k}\tilde{\psi}^a\tilde{\jmath}'^a \qquad (18.2.18)$$

and $\tilde{T}''_F$ is nonsingular with respect to $\tilde{\psi}^a$ and $\tilde{\jmath}'^a$. Further,

$$\tilde{T}_B = \tilde{T}^\psi_B + \tilde{T}'_B + \tilde{T}''_B , \qquad (18.2.19)$$

with

$$\tilde{T}^\psi_B = -\frac{1}{2k}\tilde{\psi}^a\bar{\partial}\tilde{\psi}^a , \qquad (18.2.20a)$$

$$\tilde{T}'_B = \frac{1}{2(k' + h(g))} :\tilde{\jmath}'\tilde{\jmath}': . \qquad (18.2.20b)$$

The remainders $\tilde{T}_F''$ and $\tilde{T}_B''$ are nonsingular with respect to both $\tilde{\psi}^a$ and $\tilde{\jmath}'^a$. The CFT thus separates into three pieces, with central charges

$$\tilde{c}^\psi = \frac{\dim(g)}{2}, \quad \tilde{c}' = \frac{k' \dim(g)}{k' + h(g)}, \quad \tilde{c}'' = \tilde{c} - \tilde{c}^\psi - \tilde{c}'. \tag{18.2.21}$$

The SCFT separates into only two pieces, because $\tilde{\psi}^a$ and $\tilde{\jmath}'^a$ are coupled in the supercurrent. In particular, the central charge for the $\tilde{\psi}\tilde{\jmath}'$ SCFT is

$$\tilde{c}^\psi + \tilde{c}' = \frac{(3k' + h(g)) \dim(g)}{2(k' + h(g))}. \tag{18.2.22}$$

This lies in the range

$$\frac{\dim(g)}{2} \leq \tilde{c}^\psi + \tilde{c}' \leq \frac{3 \dim(g)}{2}. \tag{18.2.23}$$

The lower bound is reached only when $\tilde{\jmath}'^a$ vanishes, and the upper only for an Abelian algebra.

For an Abelian SCCA, the non-Abelian terms in the OPE (18.2.14) vanish. In particular, $\tilde{\jmath}_\psi$ vanishes and $k = k'$, so a nontrivial theory requires that $k' \neq 0$. We can then normalize the currents to set $k = k' = 1$. Writing the current as the derivative of a free boson, $\tilde{\jmath} = i\bar{\partial}H$, gives

$$\tilde{T}_F^s = i\tilde{\psi}\bar{\partial}H, \quad \tilde{T}_B^\psi + \tilde{T}_B' = -\frac{1}{2}\tilde{\psi}\bar{\partial}\tilde{\psi} - \frac{1}{2}\bar{\partial}H\bar{\partial}H. \tag{18.2.24}$$

If there is a right-moving gauge symmetry below the string scale the gauge boson vertex operator must be periodic, and so the fermionic currents $\tilde{\psi}^a$ must always have the same periodicity as the supercurrent $\tilde{T}_F$. This defines an *untwisted* SCCA.

One can derive strong results restricting the relevance of right-moving gauge symmetries to physics. In the $(1, 0)$ heterotic string,

1. If there are any massless fermions, then there are no non-Abelian SCCAs.

2. All massless fermions are neutral under any Abelian SCCA gauge symmetries.

3. If any fermions have chiral gauge couplings, then there are no SCCAs.

The first two results are sufficient to imply that the Standard Model $SU(3) \times SU(2) \times U(1)$ gauge symmetries must come from the left-moving gauge symmetries in heterotic string theory. If, as it appears, the $SU(3) \times SU(2) \times U(1)$ gauge couplings are chiral, then there are no right-moving gauge symmetries at all.

To show these, consider the vertex operator for any massless spin-$\frac{1}{2}$ state, whose matter part is

$$S_\alpha \mathcal{V}_K e^{ik_\mu X^\mu} . \tag{18.2.25}$$

Here S_α is a spin field for the four noncompact dimensions, leaving a weight $(1, \frac{3}{8})$ operator $\mathcal{V}_K$ from the internal theory. The Ramond generator $\tilde{G}_0$ is Hermitean, implying that

$$\tilde{G}_0^2 = \tilde{L}_0 - \frac{\tilde{c}}{24} \geq 0 \tag{18.2.26}$$

in any unitary SCFT. The internal theory here has central charge 9, and so the internal part $\mathcal{V}_K$ of any massless spin-$\frac{1}{2}$ state saturates the inequality. Incidentally, this also implies that there can never be fermionic tachyons. Further, if the internal theory decomposes into a sum of SCFTs, $G_0 = \sum_i G_0^i$, then the same argument requires that

$$\tilde{L}_0^i = \frac{\tilde{c}^i}{24} \tag{18.2.27}$$

within *each* SCFT.

Now suppose that one of these SCFTs is a non-Abelian SCCA. In the R sector the $\tilde{\psi}^a$ and $\tilde{\jmath}^a$ are periodic. Then $\tilde{L}_0^\psi + \tilde{L}_0'$ is bounded below by the zero-point energy $\frac{1}{16} \dim(g)$ of the $\tilde{\psi}^a$, and

$$\tilde{L}_0^\psi + \tilde{L}_0' - \frac{\tilde{c}^\psi + \tilde{c}'}{24} \geq \frac{h(g)\dim(g)}{24(k' + h(g))} > 0 . \tag{18.2.28}$$

This is strictly positive for all states, so massless fermions are impossible and the first result is established. For an Abelian SCCA, the same form holds with $k' = 1$ and $h(g) = 0$, so equality is possible. However, the term $\frac{1}{2}\jmath_0\jmath_0$ in $\tilde{L}_0'$ makes an additional positive contribution unless the charge $\jmath_0$ is zero for the state, establishing the second result.

The equivalence (18.2.24) means that a $U(1)$ SCCA algebra has the same world-sheet action as a flat dimension. Further, as noted above, for an SCCA associated with a gauge interaction the periodicity of the fermionic current ψ is the same as that of the ψ^μ. Then if there is a $U(1)$ SCCA the massless R sector ground states will be the same as those of a *five*-dimensional theory. The $SO(4,1)$ spinor representation **4** decomposes into one four-dimensional representation of each chirality, **2** + $\bar{\mathbf{2}}$, so the massless states come in pairs of opposite chirality. In other words, the $SO(4,1)$ spin $\psi_0^\mu \psi_0$ commutes with the GSO projection and (in the massless sector) with the superconformal generators, and so takes massless physical states into massless physical states of the opposite four-dimensional chirality. This establishes the third result, and shows that heterotic string vacua with right-moving gauge symmetries are not relevant to the Standard Model.

Gauge symmetries of type II strings

Now let us consider the possibility of getting the Standard Model from the type II string. Here, both sides are supersymmetric, so the vertex operators of gauge bosons are of one of the two forms

$$\psi^a \tilde{\psi}^\mu e^{ik \cdot X} \, , \quad \psi^\mu \tilde{\psi}^a e^{ik \cdot X} \, , \tag{18.2.29}$$

where ψ^a is associated with a left-moving SCCA and $\tilde{\psi}^a$ with a right-moving SCCA. For example, one could take the internal theory to consist of 18 right-moving and 18 left-moving fermions with trilinear supercurrents (16.1.29). This leads to gauge algebra $g_R \times g_L$ with g_R and g_L each of dimension 18. This can then be broken to the Standard Model by twists. This seems much more economical than the heterotic string, where the dimension of the gauge group can be much larger. However, we will see that the Standard Model does not quite fit into the type II string theory.

The same analysis as used in the heterotic string shows that only one of the two types of gauge boson (18.2.29) may exist. If there are chiral fermions in the R–NS sector there can be no left-moving SCCA, and if there are chiral fermions in the NS–R sector there can be no right-moving SCCA. In order to have both chiral fermions and gauge symmetries, the fermions must all come from one sector, say R–NS, and the gauge symmetries all from right-moving SCCAs.

Now let us see that this does not leave room for the Standard Model. To be precise, it is impossible to have an $SU(3) \times SU(2) \times U(1)$ gauge symmetry with massless $SU(3)$ triplet and $SU(2)$ doublet fermions. The internal part of any massless state has weight $\tilde{h} = \frac{1}{2}$. This restricts the current algebra part to be either a primary state of the SCCA, annihilated by all the $\tilde{\psi}^a_r$ and $\tilde{J}'^a_n$ for $r, n > 0$, or of the form $\tilde{\psi}^a_{-1/2}|1\rangle$. The latter is a gaugino, in the adjoint representation, so the triplets and doublets must be primary states instead. By the same argument as in the conformal case, the allowed representations for the primary states are restricted according to the level k' of the current $\tilde{J}'^a$ of the SCCA, so that $k' \geq 1$ in both the $SU(2)$ and $SU(3)$ factors in order to have doublets and triplets respectively. Noting that the central charge (18.2.22) increases with k', the total central charge of the SCCAs is

$$\tilde{c} \geq \frac{8}{2} + \tilde{c}^{SU(3),1} + \frac{3}{2} + \tilde{c}^{SU(2),1} + \tilde{c}^{U(1)} = 4 + 2 + \frac{3}{2} + 1 + \frac{3}{2}$$
$$= 10 \, . \tag{18.2.30}$$

This exceeds the total $\tilde{c} = 9$ of the internal theory, so there is a contradiction.

This is an elegant argument, using only the world-sheet symmetries. However, progress in string duality has made its limitations clearer. Since all string theories are connected by dualities, we would expect that non-perturbatively a spectrum that can be obtained in one string theory can be obtained in any other. The most obvious limitation of the argument is that it applies only to vacua without D-branes, because the latter would have additional open string states. One might also wonder whether some or all of the Standard Model states can originate not as strings but as D-branes. As long as string perturbation theory is valid then all D-branes and other nonperturbative states should have masses that diverge as $g \to 0$, so that string perturbation theory gives a complete account of the physics at any fixed energy. However, we will see in the next chapter that D-branes can become massless at some points in moduli space, and that this is associated with a breakdown of string perturbation theory.

18.3 Mass scales

There are a number of important mass scales in string theory:

1. The *gravitational* scale $m_{\mathrm{grav}} = \kappa^{-1} = 2.4 \times 10^{18}$ GeV, at which quantum gravitational effects become important; this is somewhat more useful than the Planck mass, which is a factor of $(8\pi)^{1/2}$ greater.

2. The *electroweak* scale m_{ew}, the scale of $SU(2) \times U(1)$ breaking, $O(10^2)$ GeV.

3. The *string* scale $m_{\mathrm{s}} = \alpha'^{-1/2}$, the mass scale of excited string states.

4. The *compactification* scale $m_{\mathrm{c}} = R_{\mathrm{c}}^{-1}$, the characteristic mass of states with momentum in the compact directions.

5. The *grand unification* scale m_{GUT}, at which the $SU(3) \times SU(2) \times U(1)$ interactions are united in a simple group.

6. The *superpartner* scale m_{sp}, the mass scale of the superpartners of the Standard Model particles.

In this section we consider relations among these scales. Of course, there may be additional scales. The unification of the gauge group may take place in several steps, and there may be other intermediate scales at which new degrees of freedom appear. Also, these scales may not all be relevant. For example, when the internal CFT is a sigma model on a manifold large compared to the string scale, the idea of compactification applies. There are states with masses-squared of order $m_{\mathrm{c}}^2 \ll m_{\mathrm{s}}^2$, states which would be massless in the noncompact theory and which have internal momenta of order m_{c}. However, as m_{c} increases to m_{s} these states become indistinguishable from the various 'stringy' states, and compactification is not so meaningful. The internal CFT may have several equivalent descriptions as a quantum field theory, with 'internal excitations' and 'stringy states' interchanging roles. Similar remarks apply to the grand unification and supersymmetry scales.

For most of the discussion we will assume explicitly that the string theory is weakly coupled, and that the Standard Model gauge couplings remain perturbative up to the string scale. In this case it is possible to make some fairly strong statements. As we know from chapter 14, strong coupling opens up many new dynamical possibilities. The consequences for physics in four dimensions have not been fully explored; we will make a few comments at the end of the section.

The relation between the string and gravitational scales follows from the relation (18.2.5) between the couplings,

$$\frac{m_{\rm s}}{m_{\rm grav}} = g_{\rm YM}(k/2)^{1/2} \,. \tag{18.3.1}$$

The quantities on the right are not too far from unity, so the string and gravitational scales are comparable. In the minimal supersymmetric model to be discussed below, the coupling $g_{\rm YM}$ at high energy is of order 0.7; for $k = 1$ this gives $m_{\rm s} \approx 1.2 \times 10^{18}$ GeV . This result is shown graphically in figure 18.1: plotted as a function of energy E are the four-dimensional gauge coupling $\alpha_{\rm YM} = g_{\rm YM}^2/4\pi$ and the corresponding dimensionless gravitational coupling $\kappa^2 E^2$. The scale where these meet is the expected scale of unification of the gravitational and gauge interactions, the string scale.

Now consider the compactification scale. Suppose that there are k dimensions compactified at some scale $m_{\rm c} \ll m_{\rm s}$. Between the scales $m_{\rm c}$ and $m_{\rm s}$, physics is described by a $(4 + k)$-dimensional field theory, in which a gauge coupling α_{4+k} has dimension m^{-k} and the gravitational coupling G_{4+k} has dimension m^{-k-2}. The behaviors of the dimensionless couplings $\alpha_{4+k}E^k$ and $G_{4+k}E^{k+2}$ are indicated in figure 18.1 by dashed lines. The gauge coupling rises rapidly from its four-dimensional value $\alpha_{\rm YM}$. Our assumption that the coupling remains weak up to the string scale then implies that the latter is not far above the compactification scale (in this section 'scale' always refers to energy, rather than the reciprocal length). Also, it presumably does not make sense for the compactification scale to be greater than the string scale, as illustrated by T-duality for toroidal compactification. Thus the string, gravitational, and compactification scales are reasonably close to one another. In open string theory, the quantitative relation (18.2.8) between the scales is different, but the reader can show that with the weak-coupling assumption these three scales are again close to one another.

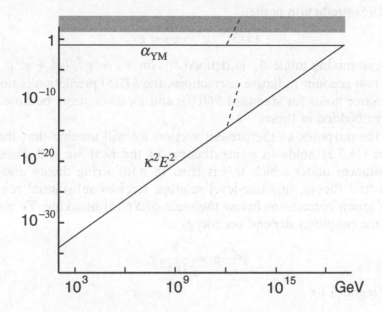

Fig. 18.1. The dimensionless gauge and gravitational couplings as a function of energy. On the scale of this graph we neglect the differences between gauge couplings and the running of these couplings. The dashed curves illustrate the effect of a compactification scale below the Planck scale, at 10^{12} GeV in this example (the slopes correspond to all six compact dimensions being at this same scale, and are reduced if there are fewer). The shaded region indicates the breakdown of perturbation theory.

Next consider the unification scale. First let us review $SU(5)$ unification of the Standard Model. The Standard Model gauge group $SU(3) \times SU(2) \times U(1)$ can be embedded in the **5** representation of $SU(5)$, with $SU(3)$ being the upper 3×3 block, $SU(2)$ the lower 2×2 block, and $U(1)$ hypercharge the diagonal element

$$\frac{Y}{2} = \text{diag}\left(-\frac{1}{3}, -\frac{1}{3}, -\frac{1}{3}, \frac{1}{2}, \frac{1}{2}\right). \tag{18.3.2}$$

The $SU(n)$ generators for the fundamental representation **n** are conventionally normalized $\text{Tr}(t^a t^b) = \frac{1}{2}\delta^{ab}$. This is also true for $U(1)$ if we define $t^{U(1)} = (\frac{3}{5})^{1/2}\frac{1}{2}Y$, in which case $SU(5)$ symmetry implies

$$g_3 = g_2 = g_1 = g_{SU(5)} \tag{18.3.3}$$

for the $SU(3) \times SU(2) \times U(1)$ couplings. The hypercharge coupling g' is defined by

$$\tfrac{1}{2}g'Y = g_{U(1)}t^{U(1)} \quad \Rightarrow \quad g' = (3/5)^{1/2}g_1 . \tag{18.3.4}$$

The $SU(5)$ prediction is then

$$(5/3)^{1/2}g' = g_2 = g_3 . \tag{18.3.5}$$

The weak mixing angle θ_w is defined by $\sin^2 \theta_w = g'^2/(g_2^2 + g'^2)$. Before taking into account radiative corrections, the $SU(5)$ prediction is $\sin^2 \theta_w = \frac{3}{8}$. The same holds for standard $SO(10)$ and E_6 unification, because $SU(5)$ is just embedded in these.

For the purposes of the present section we will assume that the same relation (18.3.5) holds in string theory; in the next we will discuss the circumstances under which this is true. In both string theory and grand unified field theory, this tree-level relation receives substantial renormalization group corrections below the scale of $SU(5)$ breaking. To one-loop order, the couplings depend on energy as

$$\mu \frac{\partial}{\partial \mu} g_i = \frac{b_i}{16\pi^2} g_i^3 . \tag{18.3.6}$$

This integrates to

$$\alpha_i^{-1}(\mu) = \alpha_i^{-1}(m_{\text{GUT}}) + \frac{b_i}{4\pi} \ln(m_{\text{GUT}}^2/\mu^2) , \tag{18.3.7}$$

where $\alpha_i = g_i^2/4\pi$. For a non-Abelian group the constant b_i is

$$b_i = -\frac{11}{3} T_g + \frac{1}{3} \underbrace{\sum T_r}_{\text{complex scalars}} + \frac{2}{3} \underbrace{\sum T_r}_{\substack{\text{Weyl} \\ \text{fermions}}} , \tag{18.3.8}$$

where $\text{Tr}(t_r^a t_r^b) = T_r \delta^{ab}$ and $T_g = T_{r=\text{adjoint}}$. For a $U(1)$ group the result is the same with $T_g = 0$ and T_r replaced by q^2.

The couplings at the weak interaction scale M_Z are $\alpha_1^{-1} \approx 59$, $\alpha_2^{-1} \approx 30$, and $\alpha_3^{-1} \approx 9$. Extrapolating the couplings $\alpha_i(\mu)$ as in eq. (18.3.7), $SU(5)$ unification makes the prediction (18.3.3) that at some scale m_{GUT} they become equal. This is often expressed as a prediction for $\sin^2 \theta_w(m_Z)$: use $\alpha_1^{-1}(m_Z)$ and $\alpha_3^{-1}(m_Z)$ to solve for m_{GUT} and α_{GUT}, and then extrapolate downwards to obtain a prediction for $\alpha_2^{-1}(m_Z)$. The prediction depends on the spectrum of the theory through the beta function (18.3.8).[4] For the minimal $SU(5)$ unification of the Standard Model,

$$\sin^2 \theta_w(m_Z) = 0.212 \pm 0.003 . \tag{18.3.9}$$

For the minimal supersymmetric Standard Model, which consists of the Standard Model plus a second Higgs doublet plus the supersymmetric

---

[4] The experiment and theory are sufficiently precise that one must take into account the two-loop beta function, threshold effects at the weak and unified scales, and other radiative corrections to the weak interaction.

partners of these,

$$\sin^2 \theta_w(m_Z) = 0.234 \pm 0.003 \ . \qquad (18.3.10)$$

The experimental value is

$$\sin^2 \theta_w(m_Z) = 0.2313 \pm 0.0003 \ . \qquad (18.3.11)$$

The minimal nonsupersymmetric model is clearly ruled out. On the other hand, the agreement between the minimal supersymmetric $SU(5)$ prediction and the actual value is striking, considering that *a priori* $\sin^2 \theta_w(m_Z)$ could have been anywhere between 0 and 1. The agreement between the supersymmetric prediction and the actual value means that the three gauge couplings meet, with

$$m_{\text{GUT}} = 10^{16.1 \pm 0.3} \text{ GeV} \ , \quad \alpha_{\text{GUT}}^{-1} \approx 25 \ . \qquad (18.3.12)$$

In the nonsupersymmetric case, the disagreement with $\sin^2 \theta_w(m_Z)$ implies that the three couplings do not meet at a single energy, but meet pairwise at three energies ranging from 10^{13} GeV to 10^{17} GeV.

To a first approximation, the unification scale (18.3.12) is fairly close to the string scale and so to the compactification and gravitational scales. This is also necessary for the stability of the proton. The running of the couplings is shown pictorially in figure 18.2. We should note that a direct comparison of the string and unification scales is not appropriate at the level of accuracy of the extrapolation (18.3.12). Rather, we should compare the measured couplings to a full one-loop string calculation: this is just the calculation (16.4.32). Ignoring for now the threshold correction, this relation is of the form (18.3.7) with the string unification scale (16.4.36)

$$m_{\text{SU}} = k^{1/2} g_{\text{YM}} \times 5.27 \times 10^{17} \text{ GeV} \rightarrow 3.8 \times 10^{17} \text{ GeV} \ . \qquad (18.3.13)$$

We have inserted the relation (18.3.1) between the gauge and gravitational scales and then carried out the numerical evaluation using the unified coupling (18.3.12) and assuming $k = 1$. The resulting discrepancy between the string unification scale and the value in minimal SUSY unification is a factor of 30. This is larger than the experimental uncertainty, but small compared to the fifteen orders of magnitude difference between the electroweak scale and the string scale. This suggests that the unification and string scales are actually one and the same, so that not just the three gauge couplings but also the gravitational coupling meet at a single point; the apparent difference between the unification and string scales would then be due to some small additional correction.

Before discussing what such a correction might be, let us consider the consequences if the two scales actually are separated. This means that there is a range $m_{\text{GUT}} < E < m_s$ in which physics is described by a grand unified *field* theory, with $SU(3) \times SU(2) \times U(1)$ contained in $SU(5)$ or another

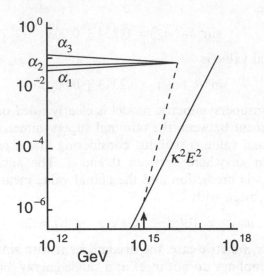

Fig. 18.2. The unification of the gauge couplings in the minimal supersymmetric unified model, and the near-miss of the gravitational coupling. The dashed line shows the potential effect of an extra dimension of the form $S_1/\mathbf{Z}_2$ at the scale indicated by the arrow.

simple group. This theory is presumably four-dimensional, because even a factor of 30 difference between the string and compactification scales is difficult to accommodate. The unified group must then be broken to $SU(3) \times SU(2) \times U(1)$ by the usual Higgs mechanism. As we have discussed in the previous section, this is not possible if the underlying current algebra is level one, because a Higgs scalar in the necessary representation cannot be lighter than the string scale. There do exist higher level string models in which such a separation of scales is possible.

An intermediate possibility is partial unification, embedding $SU(3) \times SU(2) \times U(1)$ in one of

$$SU(5)' \times U(1) \subset SO(10) , \qquad (18.3.14a)$$

$$SU(4) \times SU(2)_L \times SU(2)_R \subset SO(10) , \qquad (18.3.14b)$$

$$SU(3)_C \times SU(3)_L \times SU(3)_R \subset E_6 . \qquad (18.3.14c)$$

The group $SU(5)' \times U(1)$ is known as *flipped* $SU(5)$. Color $SU(3)$ and weak $SU(2)$ are embedded in $SU(5)$ in the usual way, but hypercharge is a linear combination of a generator from $SU(5)$ and the $U(1)$ generator. String models based on flipped $SU(5)$ have been studied in some detail. The group $SU(4) \times SU(2)_L \times SU(2)_R$ is known as *Pati–Salam unification*. Color $SU(3)$ is in the $SU(4)$ factor, weak $SU(2)$ is $SU(2)_L$, and hypercharge is a linear combination of a generator from $SU(4)$ and a generator from $SU(2)_R$. In the $SU(3)^3$ group, sometimes called *trinification*, color is $SU(3)_C$, weak $SU(2)$ is in $SU(3)_L$, and hypercharge is a linear combination of generators from $SU(3)_L$ and $SU(3)_R$. When G is one of these partially unified groups and is embedded in a simple group as indicated in eq. (18.3.14), then the Standard Model group within G has the same embedding as in simple unification. The tree-level prediction for $\sin^2\theta_w(m_Z)$ is therefore again $\frac{3}{8}$, but the running of the couplings will of course be different between m_{GUT} and m_s. These partially unified groups can all be broken to the Standard Model by Higgs fields that are allowed at level one.

Now let us consider the corrections that might eliminate the difference between m_{GUT} and m_{SU}. The quoted uncertainties in the grand unified predictions come primarily from the uncertainty in the measured value of α_3, and in the supersymmetric case from the unknown masses of the superpartners. There is a far greater uncertainty implicit in the assumption that the spectrum below the unification scale is minimal. Adding a few extra light fields, either at the electroweak scale or at an intermediate scale, can change the running by an amount sufficient to bring the unification scale up to the string scale.

There is also a threshold correction due to loops of string-mass fields. This is a function of the moduli, as in the orbifold example (16.4.38),

$$\Delta_a = c_a - \sum_i \frac{b_a^i |G^i|}{|G|} \ln \left[(T_i + T_i^*)|\eta(T_i)|^4 (U_i + U_i^*)|\eta(U_i)|^4 \right] . \quad (18.3.15)$$

Although this correction reflects a sum over the infinite set of string states, its numerical value is rather small for values of the moduli of order 1. It can become large if the moduli become large. For example,

$$\Delta_a \approx \sum_i \frac{b_a^i |G^i|}{|G|} \frac{\pi(T_i + T_i^*)}{6} \quad (18.3.16)$$

for large T_i, from the asymptotics of the eta function. For large enough T_i, in those models where the correction has the correct sign, this can account for the apparent difference between the string and unification scales.

Finally, in more complicated string models the tree-level predictions may be different and so also the predicted unification scale. We will discuss this somewhat in the next section.

All of these modifications have the drawback that a change large enough to raise the unification scale to the string scale will generically change the prediction for $\sin^2\theta_w$ by an amount greater than the experimental and theoretical uncertainty, so that the excellent agreement is partly accidental. Since the gauge couplings already meet, it would be simple and economical to leave them unchanged and instead change the energy dependence of the gravitational coupling so that it meets the other three. However, this seems impossible, since the 'running' of the gravitational coupling $\kappa^2 E^2$ is just dimensional analysis: the gravitational interaction is essentially classical below the string scale and quantum effects do not affect its energy dependence.

This is one point where the new dynamical ideas arising from strongly coupled string theory can make a difference. One way to change the dimensional analysis is to change the dimension! It does not help to have a low compactification scale of the ordinary sort: as shown in figure 18.1, all the couplings increase more rapidly but they do not meet any sooner. Consider, however, the *strongly coupled* $E_8 \times E_8$ heterotic string compactified on a Calabi–Yau space K. From the discussion in chapter 14, this is the eleven-dimensional M-theory compactified on a product space

$$K \times \frac{S_1}{\mathbf{Z}_2} \,. \tag{18.3.17}$$

The scales of the two factors are independent; let us suppose that the space $S_1/\mathbf{Z}_2$ is larger, so that its mass scale R_{10}^{-1} lies below the unification scale. The point is that the gauge and matter fields live on the boundary of this space, which remains four-dimensional, while the gravitational field lives in the five-dimensional bulk. The effect is as shown in figure 18.2: the gauge couplings evolve as in four dimensions, while the gravitational coupling has a kink. For an appropriate value of R_{10}, all four couplings meet at a point.

With the only data points being the low energy values of the gauge couplings, there is no way to distinguish between these various alternatives. If in fact supersymmetry is found at particle accelerators, then measurement of the superpartner masses will allow similar renormalization group extrapolations and may enable us to unravel the 'fine structure' at the string scale.

This brings us to the next scale, which is m_{sp}. The lower limits on the various charged and strongly interacting superpartners are of order 10^2 GeV. If supersymmetry is the solution to the hierarchy problem, the cancellation of the quantum corrections to the Higgs mass requires that the splitting between the Standard Model particles and their superpartners be not much larger than this,

$$10^2 \text{ GeV} \lesssim m_{\mathrm{sp}} \lesssim 10^3 \text{ GeV} \,. \tag{18.3.18}$$

Of all the new phenomena associated with string theory, supersymmetry is the one that is likely to be directly accessible to particle accelerators.

Finally, we should ask why the supersymmetry and electroweak scales lie so far below the others; we will discuss this briefly in section 18.8.

18.4 More on unification

In this section we collect a number of additional results on the relation between string theory and grand unification.

The first issue is the condition under which the grand unified relation $g_1 = g_2 = g_3$ holds in string theory at tree level. This is obviously the case in theories where a unified group remains unbroken below the string scale. It is also true if, as in the orbifold and Calabi–Yau cases, a unified group is broken at the string or compactification scale by twists. Although there is no scale at which the world looks like a four-dimensional grand unified theory, the inheritance principle guarantees that the equality of the tree-level couplings persists after the twist.

More generally one can make some statements just from current algebra arguments. The current algebra relation (18.2.5) between the gravitational coupling and any single gauge coupling implies that for the $SU(2)$ and $SU(3)$ gauge couplings

$$\frac{\alpha_2}{\alpha_3} = \frac{\hat{k}_3}{\hat{k}_2} = \frac{k_3}{k_2} \, . \tag{18.4.1}$$

Thus the grand unified prediction $\alpha_2 = \alpha_3$ holds whenever the levels of the $SU(3)$ and $SU(2)$ current algebras are equal. In any case one expects that the levels are small integers, models with large levels having complicated spectra, so that if the levels are not equal their ratio differs substantially from unity. Since the unification scale can be determined from any pair of couplings, this implies a large change in the unification scale, spoiling the near-equality between the unification and string scales. Thus it is likely that, whatever the levels of the $SU(2)$ and $SU(3)$ current algebras, they are equal.

For the $U(1)$ coupling there is no similar statement, because there is no level to give an absolute normalization to the current. One general result concerns the common situation that there is a continuous moduli space of vacua, all with an unbroken $U(1)$ symmetry: if there are chiral fermions, then at tree level the coupling g_1, and so also $\sin^2 \theta_w$, is the same for all the connected vacua. To see this, write the $U(1)$ current algebra in terms of a left-moving boson $H(z)$. Let us consider how H might appear in the vertex operator for the modulus that interpolates between the vacua. The $U(1)$ is assumed to be unbroken for all vacua, so the vertex operator must be invariant under $H \rightarrow H + \epsilon$ — it can only contain derivatives of H. Dimensionally, the only operator that can then appear in a massless vertex operator is ∂H, and the whole matter vertex operator must be

$$\partial H \tilde{\psi} e^{ik \cdot X} \tag{18.4.2}$$

for some $(0, \frac{1}{2})$ superconformal tensor $\tilde{\psi}$. However, we know from section 18.2 that such tensors are inconsistent with chirality, so H cannot appear in the vertex operator at all. Expectation values of the $U(1)$ current are then independent of the modulus, and therefore so is the gauge coupling.

A related issue is the quantization of electric charge. An isolated fractional multiple of the electron charge has never been seen in nature. The Standard Model has fractionally charged quarks, of course, but these are confined in hadrons of integer charge. It is therefore useful to work with

$$Q' = Q_{\mathrm{EM}} + \frac{T}{3}, \qquad (18.4.3)$$

where the triality T, defined mod 3, is $+1$ for an $SU(3)$ $\mathbf{3}$ and -1 for a $\bar{\mathbf{3}}$. One can take T to be the $SU(3)$ generator which is $\mathrm{diag}(1, 1, -2)$ in the $\mathbf{3}$ representation. Quarks are confined in states with $T = 0 \bmod 3$, so for all isolated states $Q' = Q_{\mathrm{EM}} \bmod 1$. The charge Q' has been defined so as to be an integer for all Standard Model fields, so it follows that Q_{EM} is an integer for all isolated states.

Now consider this issue in string theory, starting with some special cases. If there is an $SU(5)$ gauge group below the string scale, there can be no isolated fractional charges. In the $SU(5)$ $\mathbf{5}$, the charge

$$Q' = Q_{\mathrm{EM}} + \frac{1}{3}T = \frac{1}{2}Y + I_3 + \frac{1}{3}T \qquad (18.4.4)$$

is

$$\mathrm{diag}\left(-\frac{1}{3}, -\frac{1}{3}, -\frac{1}{3}, \frac{1}{2}, \frac{1}{2}\right) + \mathrm{diag}\left(0, 0, 0, \frac{1}{2}, -\frac{1}{2}\right) + \mathrm{diag}\left(\frac{1}{3}, \frac{1}{3}, -\frac{2}{3}, 0, 0\right)$$

$$= \mathrm{diag}(0, 0, -1, 1, 0). \qquad (18.4.5)$$

Since Q' is an integer for all states in the $\mathbf{5}$ and all representations can be obtained as tensor products of $\mathbf{5}$s, Q' is an integer for all states and so Q_{EM} is an integer for all isolated states.

Now consider the case in which there is a level one $SU(5)$ current algebra at the string scale, broken by twists to $SU(3) \times SU(2) \times U(1)$. Let us represent this current algebra by free fermions $\lambda^{K\pm}$ for $K = 4, \ldots, 8$, with $SU(3)$ acting on $K = 4, 5, 6$ and $SU(2)$ acting on $K = 7, 8$ (the numbering is kept consistent with the orbifold and Calabi–Yau chapters). The current corresponding to Q' is thus

$$j' = \lambda^{6-}\lambda^{6+} - \lambda^{7-}\lambda^{7+} = i\partial(H_7 - H_6). \qquad (18.4.6)$$

In a sector with boundary conditions

$$\lambda^{K+}(\sigma_1 + 2\pi) = \exp(2\pi i \nu_K)\lambda^{K+}(\sigma_1), \qquad (18.4.7)$$

the bosonized vertex operator

$$\exp\left[i\sum_{K}(1/2 - v_K)H_K\right] \tag{18.4.8}$$

has charge

$$Q' = v_6 - v_7 . \tag{18.4.9}$$

Thus there will be isolated fractional charges if there are twisted sectors with $v_6 \neq v_7$. In fact there *must* be such sectors. Consider the gauge boson associated with the current $\lambda^{6+}\lambda^{7-}$. This carries the $SU(3) \times SU(2)$ representation $(\mathbf{3},\mathbf{2})$ and is one of the $SU(5)$ bosons that is removed by the twists that break the $SU(5)$ symmetry. One of the twists must therefore have $\exp[2\pi i(v_6 - v_7)] \neq 1$, and the corresponding twisted sector has fractional Q'.

The lightest fractionally charged particle must be stable due to charge conservation. The number of fractional charges in ordinary matter is known to be less than 10^{-20} per nucleon. If fractionally charged particles of mass m were in thermal equilibrium in the early universe at temperatures $T > m$, it is estimated that annihilation would only reduce their present abundance to approximately 10^{-9} per nucleon. Whether this is a problem depends critically on the masses of the fractionally charged states, whether all are near the string scale or whether some are near the weak scale. If all the fractional charges are superheavy then the situation is very similar to that with magnetic monopoles in grand unified theories. Diluting the density of relic monopoles was one of the original motivations for inflationary cosmology; this would also sufficiently dilute the fractional charges. It may also be the case that the universe was never hot enough to produce string-scale states thermally. Fractionally charged particles with masses near the weak scale are a potentially severe problem, unless they are charged under a new strongly coupled gauge symmetry and so confined.

In Calabi–Yau compactification the fractionally charged states are superheavy. The twist that breaks $SU(5)$ is accompanied by a freely-acting spacetime symmetry, so that any string in the twisted sector of the gauge group will be stretched in spacetime. In orbifold compactifications there can be massless fractionally charged states from the twisted sectors, but the Calabi–Yau result suggests that superheavy masses are more generic.

Let us mention a generalization of the previous result. If the $SU(3)$ and $SU(2)$ gauge symmetries are at level one, and the tree-level value of $\sin^2\theta_{\mathrm{w}}$ is the $SU(5)$ value $\frac{3}{8}$, and $SU(5)$ is broken to $SU(3) \times SU(2) \times U(1)$, then there are states of fractional Q'. To see this, write the $SU(3) \times SU(2) \times U(1)$

current algebra in terms of free bosons, the diagonal currents being[5]

$$j^3_{SU(3)} = \frac{i}{2}\partial(H^4 - H^5)\,, \tag{18.4.10a}$$

$$j^8_{SU(3)} = \frac{i}{2 \times 3^{1/2}}\partial(H^4 + H^5 - 2H^6)\,, \tag{18.4.10b}$$

$$j^3_{SU(2)} = \frac{i}{2}\partial(H^7 - H^8)\,, \tag{18.4.10c}$$

$$j_{Y/2} = \frac{i}{6}\partial[-2(H^4 + H^5 + H^6) + 3(H^7 + H^8)]\,. \tag{18.4.10d}$$

The current $j_{Y/2}$ is normalized so that the z^{-2} term in the $j_{Y/2}j_{Y/2}$ operator product is $\frac{5}{3}$ times that of the non-Abelian currents, giving the tree-level value $\sin^2\theta_{\rm w} = \frac{3}{8}$. Then

$$j' = j_{Y/2} + j^3_{SU(2)} + \frac{2}{3^{1/2}}j^8_{SU(3)} = i\partial(H^7 - H^6) \tag{18.4.11}$$

just as above, and $Q' = k^7 - k^6$. If Q' were an integer for all states, then the $(1,0)$ operator

$$\exp[i(H^6 - H^7)] \tag{18.4.12}$$

would have single-valued OPEs with respect to all vertex operators. However, this would mean that the current algebra is larger than the assumed $SU(3) \times SU(2) \times U(1)$; in fact, closure of the OPE gives a full $SU(5)$ algebra and gauge group. So under the assumptions given there must be fractional charges. This is more general than the earlier result, the assumption of a twisted $SU(5)$ current algebra having been replaced by a weaker assumption about the weak mixing angle.

There are various further generalizations. By an extension of the above argument it can be shown that if the current algebras are level one, and there are no states of fractional Q', and $SU(5)$ is broken, then the tree-level $\sin^2\theta_{\rm w}$ must take one of the values $\frac{3}{20}, \frac{3}{32}, \frac{3}{44}, \ldots$. To make these values consistent with experiment takes a very nonstandard running of the couplings, suggesting that either the current algebras are higher level or that supermassive fractional charges should be expected to exist. One can also obtain constraints on higher level models, but they are less restrictive. We mention in passing that at higher levels we cannot use the same free-boson representation of the current algebras. Rather, simple currents, defined below eq. (15.3.19), play the role that exponentials of free fields play in the level one case.

---

[5] Only four free bosons are needed to represent the current algebra — the linear combination $H^4+H^5+H^6+H^7+H^8$ does not appear. The notation is chosen to correspond to the bosonization of the earlier free Fermi representation.

If unconfined fractional charges do exist, electric charge is quantized in a unit e/n smaller than the electron charge. The Dirac quantization condition implies that any magnetic monopole must have a magnetic charge which is an integer multiple of $2\pi n/e$. Various classical monopole solutions exist in string theories, and one expects that the minimum value allowed by the Dirac quantization is attained. Discovery of a monopole with charge $2\pi/e$ would imply the nonexistence of fractional charges, and so have implications for string theory through the above theorems.

The final issues are proton decay and neutrino masses. The details here are rather model-dependent, but we will outline some of the general issues. Two of the successes of the Standard Model are that it explains the stability of the proton and the lightness of the neutrinos. The most general renormalizable action with the fields and gauge symmetries of the Standard Model has no terms that violate baryon number B. This is termed an *accidental symmetry*, meaning that the long life of the proton is indirectly implied by the gauge symmetries. The allowed $\Delta B \neq 0$ terms of lowest dimension are some four-fermion interactions. These will be induced in grand unified theories by exchange of heavy gauge (X) bosons. The operators have dimension 6, so the amplitude goes as M_X^{-2}, and an estimate of the resulting proton lifetime is

$$\tau_P \approx \left(\frac{M_X}{10^{15} \text{ GeV}}\right)^4 \times 10^{31\pm1} \text{ years.} \qquad (18.4.13)$$

The experimental bound is of order 10^{32} years, so this is an interesting rate although very sensitive to the unification scale. Similarly, a mass for the Weyl neutrinos would violate lepton number, and L is another accidental symmetry of the Standard Model.

In supersymmetric theories there are gauge-invariant dimension 3, 4, and 5 operators that violate B and/or L. These are the superpotential terms

$$\mu_1 H_1 L$$
$$+ \eta_1 U^c D^c D^c + \eta_2 Q L D^c + \eta_3 L L E^c$$
$$+ \frac{\lambda_1}{M} Q Q Q L + \frac{\lambda_2}{M} U^c U^c D^c E^c + \frac{\lambda_3}{M} L L H_2 H_2 . \qquad (18.4.14)$$

Here Q, U^c, D^c, L, and E^c are chiral superfields, containing respectively the left-handed quark doublet, anti-up quark, anti-down quark, lepton doublet, and the positron; H_1 and H_2 are chiral superfields containing the two Higgs scalars needed in the supersymmetric Standard Model. Gauge and generation indices are omitted. The dimension 3 term in the first line would generate a neutrino mass and so it must be that $\mu_1 \leq 10^{-3}$ GeV, which is small compared to the weak scale and minuscule compared to the unification scale. The terms in the second line are of dimension 4, unsuppressed by heavy mass scales, and their dimensionless coefficients must be very small. For example, the first two terms together can induce proton decay, so $\eta_1\eta_2 \leq 10^{-24}$. The terms in the third line are of dimension 5, suppressed by one power of mass; the proton decay limit $\lambda_{1,2}/M \leq 10^{-25}$ GeV^{-1} requires a combination of heavy scales and small coefficients, while the lightness of the neutrino implies that $\lambda_3/M \leq 10^{-13}$ GeV^{-1}. Thus any supersymmetric theory needs discrete symmetries to eliminate almost completely the dimension 3 and 4 terms and at least to suppress the dimension 5 terms unless they are not proportional to small Yukawa couplings. Several groups have argued that the necessary symmetries exist in various classes of string vacua. In many examples these seem to be associated with an additional $U(1)$ gauge interaction broken in the TeV energy range.

There is at least one respect in which string theories, or at least higher-dimensional theories, may have an advantage over other supersymmetric unified theories. The $SU(2)$ doublet Higgs scalar that breaks the weak interaction must have a mass of order the electroweak scale, while its color triplet GUT partners can mediate proton decay and so must have masses near the unification scale. It is possible to arrange the necessary mass matrix for these states without fine tuning, but the models in general seem rather contrived. String theory provides another solution. When an $SU(5)$ current algebra symmetry is broken by twists, the low energy states do not in general fit into complete multiplets of the unified symmetry: some of the states are simply projected away. This is true somewhat more generally for any higher-dimensional gauge theory compactified to $d = 4$ with the gauge symmetry broken at the compactification scale by Wilson lines. In these cases one keeps certain attractive features, such as the unification of the gauge interactions and the prediction of mixing angle, but the undesired Higgs triplet need not be present.

18.5 Conditions for spacetime supersymmetry

Consider any four-dimensional string theory with $N = 1$ spacetime supersymmetry. We will show that there must be a right-moving $N = 2$ world-sheet superconformal symmetry, generalizing the results found in the orbifold and Calabi–Yau examples.

The current for spacetime supersymmetry is

$$\mathfrak{J}_\alpha = e^{-\tilde{\phi}/2}\tilde{S}_\alpha\tilde{\Sigma}\,, \quad \mathfrak{J}_{\dot\alpha} = e^{-\tilde{\phi}/2}\tilde{S}_{\dot\alpha}\tilde{\bar\Sigma}\,. \tag{18.5.1}$$

We have separated the four-dimensional spin field into its **2** and $\bar{\mathbf{2}}$ components, denoted respectively by undotted and dotted indices. The four-dimensional spin fields have opposite values of $\exp(\pi i\tilde{F})$, so the internal parts $\tilde{\Sigma}$ and $\tilde{\bar{\Sigma}}$ must also have opposite values by the GSO projection. These are the vertex operators for the ground states of the compact CFT. They must each be of weight $(0, \frac{3}{8})$ in order that the total currents have weight $(0, 1)$. As shown in section 18.2, this is the minimum weight for a field in this sector, and so $\tilde{G}_0$ annihilates both $\tilde{\Sigma}$ and $\tilde{\bar{\Sigma}}$.

The single-valuedness of the OPEs of $\mathfrak{J}_\alpha$ and $\mathfrak{J}_{\dot{\alpha}}$ implies that

$$\tilde{\Sigma}(\bar{z})\tilde{\bar{\Sigma}}(0) = \bar{z}^{-3/4} \cdot \text{single-valued} , \qquad (18.5.2a)$$

$$\tilde{\Sigma}(\bar{z})\tilde{\Sigma}(0) = \bar{z}^{3/4} \cdot \text{single-valued} , \qquad (18.5.2b)$$

in order to cancel the branch cuts from the other factors. By unitarity, the coefficient of the unit operator in the OPE

$$\tilde{\Sigma}(\bar{z})\tilde{\bar{\Sigma}}(0) = \bar{z}^{-3/4} \left(1 + \frac{\bar{z}}{2}\mathfrak{J} + \ldots \right) \qquad (18.5.3)$$

cannot vanish, and so can be normalized to 1 as shown. The point of the following argument will be to show that the second term is also nonvanishing, so that there is an additional conserved current $\mathfrak{J}$.

The OPE of supersymmetry currents is

$$\mathfrak{J}_\alpha(\bar{z})\mathfrak{J}_{\dot{\beta}}(0) \sim \frac{1}{2^{1/2}\bar{z}}(C\Gamma^\mu)_{\alpha\dot{\beta}}e^{-\tilde{\phi}}\tilde{\psi}_\mu(0) . \qquad (18.5.4)$$

As required by the supersymmetry algebra, the residue on the right-hand side is the spacetime momentum current; this is in the -1 picture $e^{-\tilde{\phi}}\tilde{\psi}^\mu$ just as in the ten-dimensional equation (12.4.18). It also follows from the supersymmetry algebra that the OPE $\mathfrak{J}_\alpha\mathfrak{J}_\beta$ of two undotted currents is nonsingular, implying that

$$\tilde{\Sigma}(\bar{z})\tilde{\Sigma}(0) = O(\bar{z}^{3/4}) . \qquad (18.5.5)$$

The four-point function is then

$$\left\langle \tilde{\Sigma}(\bar{z}_1)\tilde{\bar{\Sigma}}(\bar{z}_2)\tilde{\Sigma}(\bar{z}_3)\tilde{\bar{\Sigma}}(\bar{z}_4) \right\rangle = \left(\frac{\bar{z}_{13}\bar{z}_{24}}{\bar{z}_{12}\bar{z}_{14}\bar{z}_{23}\bar{z}_{34}} \right)^{3/4} f(\bar{z}_1, \bar{z}_2, \bar{z}_3, \bar{z}_4) , \qquad (18.5.6)$$

where the OPEs as various points become coincident imply that f is a holomorphic function of its arguments. The $\bar{z}^{-3/4}$ behavior as any of the $(0, \frac{3}{8})$ fields is taken to infinity then implies that f is bounded at infinity and so a constant. Taking the limit of the four-point function as $\bar{z}_{12} \to 0$, the term of order $\bar{z}_{12}^{-3/4}$ implies that $f = 1$. The term of order $\bar{z}_{12}^{1/4}$ then implies

$$\left\langle \mathfrak{J}(\bar{z}_2)\tilde{\Sigma}(\bar{z}_3)\tilde{\bar{\Sigma}}(\bar{z}_4) \right\rangle = \frac{3\bar{z}_{34}^{1/4}}{2\bar{z}_{23}\bar{z}_{24}} , \qquad (18.5.7)$$

so that in particular $\mathfrak{J}$ is nonzero. The further limits $\bar{z}_{23} \to 0$, $\bar{z}_{24} \to 0$, and

$\bar{z}_{34} \to 0$ then reveal that

$$\tilde{\jmath}(\bar{z})\tilde{\Sigma}(0) \sim \frac{3}{2\bar{z}}\tilde{\Sigma}(0) \,, \tag{18.5.8a}$$

$$\tilde{\jmath}(\bar{z})\tilde{\bar{\Sigma}}(0) \sim -\frac{3}{2\bar{z}}\tilde{\bar{\Sigma}}(0) \,, \tag{18.5.8b}$$

$$\tilde{\jmath}(\bar{z})\tilde{\jmath}(0) \sim \frac{3}{\bar{z}^2} \,. \tag{18.5.8c}$$

As in the discussion of bosonization, the $\tilde{\jmath}\tilde{\jmath}$ OPE implies that the expectation values of the current can be written in terms of those of a right-moving boson $\tilde{H}$,

$$\tilde{\jmath}(\bar{z}) = 3^{1/2}i\bar{\partial}\tilde{H}(\bar{z}) \,. \tag{18.5.9}$$

The energy-momentum tensor separates into one piece constructed from the current and another commuting with it,

$$\tilde{T}_B = -\frac{1}{2}\bar{\partial}\tilde{H}\bar{\partial}\tilde{H} + \tilde{T}'_B \,. \tag{18.5.10}$$

The $\tilde{\jmath}\tilde{\Sigma}$ OPE implies that

$$\tilde{\Sigma} = \exp(3^{1/2}i\tilde{H}/2)\tilde{\Sigma}' \,, \tag{18.5.11}$$

with $\tilde{\Sigma}'$ commuting with the current. The weight of the exponential is $(0, \frac{3}{8})$, the same as that of $\tilde{\Sigma}$ itself, so $\tilde{\Sigma}'$ is of weight $(0,0)$ and must be the identity. Thus the R ground state operators are functions only of the free field,

$$\tilde{\Sigma} = \exp(3^{1/2}i\tilde{H}/2) \,, \quad \tilde{\bar{\Sigma}} = \exp(-3^{1/2}i\tilde{H}/2) \,. \tag{18.5.12}$$

Now consider the supercurrent T_F of the compact CFT. Since $\tilde{\Sigma}$ and $\tilde{\bar{\Sigma}}$ are primary fields in the R sector and are annihilated by $\tilde{G}_0$, we have

$$\tilde{T}_F(\bar{z})\tilde{\Sigma}(0) = O(\bar{z}^{-1/2}) \,, \quad \tilde{T}_F(\bar{z})\tilde{\bar{\Sigma}}(0) = O(\bar{z}^{-1/2}) \,. \tag{18.5.13}$$

Using the explicit form (18.5.12), this implies

$$\tilde{T}_F = \tilde{T}_F^+ + \tilde{T}_F^- \,, \tag{18.5.14a}$$

$$\tilde{T}_F^+ \propto \exp(i\tilde{H}/3^{1/2}) \,, \quad \tilde{T}_F^- \propto \exp(-i\tilde{H}/3^{1/2}) \,. \tag{18.5.14b}$$

In other words,

$$\tilde{\jmath}(\bar{z})\tilde{T}_F^+(0) \sim \frac{1}{\bar{z}}\tilde{T}_F^+(0) \,, \quad \tilde{\jmath}(\bar{z})\tilde{T}_F^-(0) \sim -\frac{1}{\bar{z}}\tilde{T}_F^-(0) \,. \tag{18.5.15}$$

Applying the Jacobi identity, one obtains the full $(0, 2)$ superconformal OPE (11.1.4).

To summarize, the existence of $N = 1$ supersymmetry in *spacetime* implies the existence of $N = 2$ right-moving superconformal symmetry on the *world-sheet*. That is, there is at least $(0,2)$ superconformal symmetry.

The various components of the spacetime supersymmetry current are now known explicitly in terms of free scalar fields; for example

$$\jmath_{\frac{1}{2}\frac{1}{2}} = \exp\left[\tfrac{1}{2}(-\tilde{\phi} + i\tilde{H}_0 + i\tilde{H}_1 + 3^{1/2}i\tilde{H})\right] . \qquad (18.5.16)$$

Single-valuedness of this current with any vertex operator thus implies that all states have integer charge under

$$\jmath_{\text{GSO}} = \frac{1}{2}\bar{\partial}(-\tilde{\phi} + i\tilde{H}_0 + i\tilde{H}_1 + 3^{1/2}i\tilde{H}) . \qquad (18.5.17)$$

This integer charge condition is the generalization of the GSO projection.

The converse holds as well: if the $(0,1)$ world-sheet supersymmetry of the heterotic string is actually embedded in a $(0,2)$ or larger algebra, and if all states carry integer charge under the current $\tilde{J}$, then the theory has spacetime supersymmetry. The argument is simple: if there is an $N = 2$ right-moving supersymmetry, then by bosonizing the current $\tilde{J}$ we can construct the operator (18.5.16). This is a $(0,1)$ field, a world-sheet current. By the integer charge assumption it is local with respect to all the vertex operators, and so has a well-defined action on the physical states. It is a spacetime spinor and so corresponds to a spacetime supersymmetry. Lorentz and CPT invariance generate the remaining components of the supersymmetry current (18.5.1). Combining these currents with ∂X^μ gives the gravitino vertex operators, so the supersymmetry is local.

The same argument can be applied to extended spacetime supersymmetry. The analysis is a bit longer and is left to the references, but we summarize the results. If there is $N = 2$ spacetime supersymmetry in the heterotic string, then the right-moving internal CFT separates into two pieces. The first, with $\tilde{c} = 3$, is a specific $(0,2)$ superconformal theory: two free scalars and two free fermions forming the standard $(0,2)$ superfield discussed in section 11.1. The second, with $\tilde{c} = 6$, must have $(0,4)$ supersymmetry but is otherwise arbitrary. If there is $N = 4$ spacetime supersymmetry, then the right-moving internal CFT consists precisely of six free scalars and six free fermions — in other words, it is a toroidal theory.

18.6 Low energy actions

In section 16.4 we obtained the low energy effective action for the $\mathbf{Z}_3$ orbifold. Several important features of that action actually hold at string tree level for all four-dimensional string theories with $N = 1$ supersymmetry:

1. The Kähler potential is $-\kappa^{-2}\ln(S + S^*)$ plus terms independent of S.

2. The superpotential is independent of S.

3. The nonminimal gauge kinetic term is

$$f_{ab} = \frac{2\hat{k}_a \delta_{ab}}{g_4^2} S \ . \tag{18.6.1}$$

Such general results are not surprising from a world-sheet point of view. The vertex operators for Φ_4 and a involve only the noncompact free fields X^μ and $\tilde{\psi}^\mu$, which are independent of the compactification. The gauge boson vertex operators involve only these fields and the $(1,0)$ gauge currents, which again are universal up to the coefficient $\hat{k}$.

Rather than a detailed world-sheet derivation, it is very instructive to give a derivation based on the spacetime effective action. The introduction (16.4.12) of the axion field depends only on the four-dimensional fields and so is always valid. Under a shift $a \to a + \epsilon$ the action changes only by a term proportional to

$$\int F_2 \wedge F_2 \ . \tag{18.6.2}$$

This is a topological invariant and vanishes in perturbation theory. In perturbation theory there is then a PQ symmetry

$$S \to S + i\epsilon \ . \tag{18.6.3}$$

Second, there is a scale invariance: under

$$S \to tS \ , \quad G_{\mu\nu 4E} \to tG_{\mu\nu 4E} \ , \tag{18.6.4}$$

with the other bosonic fields invariant, the action changes by

$$S \to tS \ . \tag{18.6.5}$$

This is just the statement that a constant dilaton only appears in the world-sheet action multiplying the world-sheet Euler number. The scaling (18.6.4) of the metric arises because the Einstein metric differs from the string metric by a function of the dilaton.

The PQ symmetry requires that the Kähler potential depend only on $S + S^*$. In the kinetic term for S, the metric contributes a scaling t and so this term must be homogeneous in S; this determines the form given above for the Kähler potential.[6] In the gauge kinetic term, the metric contributes no net t-dependence so f_{ab} must scale as t; by holomorphicity it must be proportional to S. The PQ symmetry then requires that it depend on no other fields, in order that the variation ϵ multiply the topological term (18.6.2). The dependence on $\hat{k}_a$ was obtained in section 18.2. It is

---

[6] Scale invariance seems to allow an additional term $(C + C^*)\ln(S + S^*)$, where C is any other superfield. To rule this out we appeal to the world-sheet argument that an off-diagonal metric $G_{C\bar{S}}$ is impossible because the CFT factorizes.

often conventional to choose the additive normalization of the dilaton and the multiplicative normalization of the axion to eliminate g_4,

$$f_{ab} = \delta_{ab} \frac{S}{8\pi^2} \, . \tag{18.6.6}$$

The physical value of the coupling is then

$$\frac{g_{YM}^2}{8\pi^2} = \frac{1}{\text{Re}\langle S \rangle} \, . \tag{18.6.7}$$

PQ invariance and the holomorphicity of the superpotential together require that the superpotential be independent of S. This is precisely consistent with the scaling of the action. To see this consider the term

$$\int d^4x \, (-G_{4E})^{1/2} \exp(\kappa^2 K) \left(K^{\bar{i}j} W_{;i}^* W_{;j} - 3\kappa^2 W^* W \right) \tag{18.6.8}$$

in the potential (B.2.29). There is a scale-dependence t^2 from the metric and t^{-1} from $\exp(\kappa^2 K)$, and so the action has the correct scaling if the superpotential is scale-invariant.

One of the great strengths of this kind of argument is that it gives information to all orders of perturbation theory, and even nonperturbatively. An L-loop term in the effective action will scale as

$$S_L \to t^{1-L} S_L \, . \tag{18.6.9}$$

It follows from consideration of the potential again that an L-loop term in the superpotential scales as t^{-L}. PQ invariance requires $(S + S^*)^{-L}$ while holomorphicity requires S^{-L}, so only tree level is allowed, $L = 0$. This is an easy demonstration of one of the most important nonrenormalization theorems. The original proof in field theory involved detailed graphical manipulations; a parallel argument can be constructed in string perturbation theory using contour arguments. This nonrenormalization theorem has many important consequences. For example, particle masses or Yukawa couplings that vanish at tree level also vanish to all orders in perturbation theory (except in certain cases where D-terms are renormalized, as discussed in the next section).

For the gauge kinetic term f an L-loop contribution will scale as t^{1-L}. Again it must be holomorphic and PQ-invariant, allowing only $L = 1$, or $L = 0$ with the precise field dependence S. Thus, aside from this tree-level term f receives only one-loop corrections.[7] With $N = 1$ supersymmetry there are no such constraints on the Kähler potential because it need not

---

[7] Such statements are often rather subtle in that one must be precise about what is not being renormalized. The discussion in section 16.4 of the physical coupling versus the Wilsonian action illustrates some of the issues.

be holomorphic. An L-loop term $(S + S^*)^{-L}$ times any function of the other fields is allowed.

The PQ symmetry is broken by nonperturbative effects because the integral of $F_2 \wedge F_2$ is nonzero for a topologically nontrivial instanton field. The superpotential and gauge kinetic terms can then receive corrections, which can often be determined exactly. We will see an example of a nonperturbative superpotential below.

One final point: there is a useful general result about the metric for the space of scalar fields. Suppose we have a compactification with some moduli ϕ^i, which we take to be real. The world-sheet Lagrangian density $\mathscr{L}_{ws}$ is a function of the ϕ^i. One result of the analysis of string perturbation theory in chapter 9 was that the Zamolodchikov metric $\langle\!\langle \ | \ \rangle\!\rangle$, which is the two-point function on the sphere, determines the normalization of the vertex operators. In other words, the inner product of the string states created by ϕ^i and ϕ^j is

$$\mathscr{G}_{ij} = \left\langle\!\!\!\left\langle \frac{\partial \mathscr{L}_{ws}}{\partial \phi^i} \middle| \frac{\partial \mathscr{L}_{ws}}{\partial \phi^j} \right\rangle\!\!\!\right\rangle . \qquad (18.6.10)$$

This implies that the kinetic term for these fields is

$$\frac{1}{2}\mathscr{G}_{ij}\partial_\mu\phi^i\partial^\mu\phi^j . \qquad (18.6.11)$$

Thus the Zamolodchikov metric *is* the metric on moduli space. This result does not depend on having world-sheet supersymmetry, although in this case we have the additional information that the manifold is complex and Kähler.

18.7 Supersymmetry breaking in perturbation theory

Supersymmetry breaking at tree level

Now we would like to consider the spontaneous breaking of supersymmetry, with particular attention to the fact that the supersymmetry breaking scale is far below the string scale. The first question is whether it is possible to find examples having this property at string tree level. In fact it seems to be essentially impossible to do so.

Here is an example which illustrates the main issue. Consider the heterotic string on a simple cubic torus, $X^m \cong X^m + 2\pi R_m$ for $m = 4, \ldots, 9$, except that the translation in the 7-direction is accompanied by a $\pi/2$ rotation in the (8,9) plane. In other words, the (7,8,9)-directions form a cube with opposite faces identified, with a $\pi/2$ twist between one pair of opposite faces. This fits in the general category of orbifold models. However, the space is nonsingular because the combined rotation and translation has no fixed points. The rotation

$$(\phi_2, \phi_3, \phi_4) = (0, 0, \tfrac{1}{2}\pi) \qquad (18.7.1)$$

is not in $SU(3)$ and so all the supersymmetries are broken. However, there is a limit, $R_7 \to \infty$, where the identification in the 7-direction becomes irrelevant and supersymmetry is restored. More explicitly, the effect of the twist is that $p_7 R_7$ for any state is shifted from integer values by an amount proportional to the spin s_4, thus splitting the boson and fermion masses. This is the *Scherk–Schwarz mechanism*. The mass-squared splittings are of order R_7^{-2} and so go to zero as the 7-direction decompactifies. The obvious problem with this is that the supersymmetry breaking scale is tied to the compactification scale, which is inconsistent with the discussion in section 18.3. This linking of the supersymmetry breaking and compactification scales appears to be a generic problem with tree-level supersymmetry breaking. We could avoid it in the above example by taking instead the angle $\phi_4 \to 0$; however, crystallographic considerations limit ϕ_4 to a finite set of discrete values. Note that a twist acting on $\tilde{\psi}^m$ without acting on X^m would be a symmetry of the CFT for any values of ϕ_4, but would not commute with $\tilde{T}_F$ and so would render the theory inconsistent.

There is a theorem that greatly restricts the possibilities for a large ratio of scales at tree level. The simplest way to obtain such a ratio would be to start with a supersymmetric vacuum and turn on a modulus that breaks the supersymmetry. Vacua in the neighborhood of the supersymmetric point would then have arbitrarily small breaking. However, this situation is not possible. If there is a continuous family of string vacua with vanishing cosmological constant, then either all members of the family are spacetime supersymmetric, or none is. We will give both a world-sheet and a spacetime demonstration of this.

On the world-sheet, we know that the supersymmetric point has (0,2) supersymmetry with a quantized $U(1)$ charge. As we move away from this point either the supersymmetry must be broken to (0,1), which in particular implies that the $U(1)$ in the (0,2) algebra is broken, or we must shift the quantization of the charge. To obtain either effect the vertex operator for the modulus must depend on the boson $\tilde{H}$. It can be shown that this is impossible; the argument makes rather detailed use of the (0,2) world-sheet algebra so we defer it to the next chapter.

For the spacetime argument, let us denote the modulus as t, with $t = 0$ the supersymmetric point. The condition that the potential (B.2.29) be flat is

$$(\partial_t \partial_{\bar{t}} K)^{-1} |\partial_t W + \kappa^2 \partial_t K W|^2 = 3\kappa^2 |W|^2 . \qquad (18.7.2)$$

We assume that the modulus is neutral so that the D-term potential vanishes, but the argument can be extended to the case that it is not. Physically, the metric $\partial_t \partial_{\bar{t}} K$ must be nonvanishing and nonsingular. As a differential equation for W, the condition (18.7.2) then implies that if W vanishes for any t then it vanishes for all t, as claimed. This shows that a continuous family of string vacua with zero cosmological constant cannot include both supersymmetric and nonsupersymmetric states in any theory with $N = 1$ supergravity, independent of string theory.

The Scherk–Schwarz mechanism gives arbitrarily small supersymmetry breaking, but the supersymmetric point $R_7 = \infty$ is at infinite distance. This evades the theorem but it is also what makes this example uninteresting. One could try to evade the theorem with a small discrete rather than continuous parameter. For example, the Sugawara $SU(2)$ theories have $c = 3 - 6/(k + 2)$ with k an integer, and so cluster arbitrarily closely to $c = 3$ as $k \to \infty$. However, all attempts based on free, solvable, or smooth compactifications have run into the decompactification problem.

Supersymmetry breaking in the loop expansion

The conditions for unbroken supersymmetry are

$$W(\phi) = \partial_i W(\phi) = D^a(\phi, \phi^*) = 0 \,. \tag{18.7.3}$$

Now let us suppose that these conditions are satisfied at tree level and ask whether loop corrections can lead to them being violated. We know that the superpotential does not receive loop corrections, so the first two conditions will continue to hold to all orders. For non-Abelian D-terms, the vanishing of the D^a is implied by the gauge symmetry, so the key issue is the $U(1)$ D-terms.

The D-term potential is

$$V = \mathrm{Re}[(S/8\pi^2) + f_1(T)]\frac{D^2}{2} \tag{18.7.4a}$$

$$D = \frac{1}{\mathrm{Re}[(S/8\pi^2) + f_1(T)]}\left(2\xi - i\kappa^2 K_{,i}\frac{\delta\phi^i}{\delta\lambda}\right) \,. \tag{18.7.4b}$$

Here $\delta\phi^i/\delta\lambda$ is the $U(1)$ variation of the given scalar ϕ^i. We have used what we know about the gauge kinetic term — the threshold correction f_1 is included for completeness, but it is subleading and makes no difference in the following discussion. The scaling property (18.6.9) (which includes the scaling of the $(-G)^{1/2}$ in the action) implies that an L-loop contribution to the potential scales as t^{-L-1} and therefore as

$$S^{-L-1} \,. \tag{18.7.5}$$

Consider first the possibility of a nonzero Fayet–Iliopoulos term ξ being generated in perturbation theory. Expanding in powers of $1/S$, the leading term in the potential is of order $\xi^2/\mathrm{Re}(S)$. This is a *tree-level* effect, and so by assumption is absent.

Now consider the effect of gauging the PQ symmetry associated with S,

$$\delta S = iq\delta\lambda . \tag{18.7.6}$$

With the known form of the Kähler potential for S, the leading potential is

$$V \propto \frac{q^2}{(S+S^*)^3} . \tag{18.7.7}$$

This is a two-loop effect, so D itself is a one-loop effect. To see the significance of the variation (18.7.6), consider the effect on the PQ coupling

$$\delta \frac{1}{4\pi^2} \int \mathrm{Im}(S)F_2^a \wedge F_2^a = \frac{q\delta\lambda}{4\pi^2} \int F_2^a \wedge F_2^a . \tag{18.7.8}$$

This is not gauge-invariant but has just the right form to cancel against a one-loop anomaly in the gauge transformation, if the low energy fermion spectrum produces one. In fact, many compactifications do have anomalous spectra, and the anomaly is canceled by the variation (18.7.8) in a four-dimensional version of the Green–Schwarz mechanism. This is accompanied by cancellation of a gravitational anomaly. The induced D-term is proportional to $\mathrm{Tr}(Q)$, the total $U(1)$ charge of all massless left-handed fermions.

Thus $D \neq 0$ precisely if $\mathrm{Tr}(Q) \neq 0$, and then the supersymmetry of the original configuration is broken by a one-loop effect. The important question is whether the system can relax to a nearby supersymmetric configuration. The full D-term, including the other charged fields, is

$$D = \frac{q}{(S+S^*)} + \sum_{\phi^i \neq S} q_i \phi^{i*} \phi^i \tag{18.7.9}$$

and the potential is proportional to the square of this. If we can give the various ϕ^i small expectation values, of order $(S+S^*)^{-1/2}$, such that the D-term is set to zero while preserving $W = \partial_i W = 0$, then there is a supersymmetric minimum near the original configuration. In fact, in the known examples this is the case. Notice that while supersymmetry is restored, the new vacuum is qualitatively different from the original one. In particular, the $U(1)$ gauge symmetry is now broken by the expectation value of e^S, and the gauge boson is massive. Being a one-loop effect, the gauge boson mass-squared is of order $g^2/8\pi^2$ times the string scale. Thus the one-loop D-term produces a modest hierarchy of scales; this might be useful, for example, in accounting for the pattern of quark and lepton masses. Other massless particles may also become massive due to the shift in the ϕ^i. These are effects that cannot occur with only F-terms in the potential.

It is also interesting to consider the case that the PQ-like symmetry associated with the (1,1) moduli T^A is gauged,

$$\delta T^A = iq^A \delta \lambda \,. \tag{18.7.10}$$

To leading order in S the potential is then

$$V = \frac{(q^A \partial_A K)^2}{(S + S^*)} \,. \tag{18.7.11}$$

This is a tree-level effect. We are assuming that we have a supersymmetric tree-level solution, which is still possible on the submanifold of moduli space where $q^A \partial_A K = 0$. The would-be moduli orthogonal to this submanifold are all massive. There is a natural origin for the gauge transformation (18.7.10). The imaginary part of T^A is the integral of B_2 over the 2-cycle N^A. In the heterotic string the gauge variation of B_2 is proportional to $\mathrm{Tr}(\delta \lambda F_2)$, so if the $U(1)$ field strength has an expectation value there is a transformation

$$\delta T^A \propto i \delta \lambda \int_{N^A} F_2 \,. \tag{18.7.12}$$

This is automatically absent for Calabi–Yau compactification, because the integral of the flux measures the first Chern class. This is also another example of the difficulty of breaking supersymmetry by a small amount at tree level. It might seem that we could break the supersymmetry of the $q^A \partial_A K \neq 0$ vacua slightly by making F_2 small, but the integral of F_2 over any 2-cycle must satisfy a Dirac quantization condition. By a generalization of the monopole argument, $q_i \int F_2$ must be a multiple of 2π, where F_2 is proportional to any $U(1)$ generator of $E_8 \times E_8$, and q_i runs over the $U(1)$ charges of all heterotic string states.

18.8   Supersymmetry beyond perturbation theory

An example

In the previous section, we saw that a vacuum that is supersymmetric at tree level usually remains supersymmetric to all orders of perturbation theory. Remarkably, it is known that in most tree-level $N = 1$ vacua the supersymmetry is broken spontaneously by nonperturbative effects. Our understanding of nonperturbative string theory is still limited, but below the string scale we can work in the effective quantum field theory. In fact, there is a reasonably coherent understanding of nonperturbative breaking of supersymmetry in field theory, and the low energy theories emerging from the string theory are typically of the type in which this breaking occurs. This subject is quite involved; there are several symmetry-breaking mechanisms (gaugino condensation, instantons, composite goldstinos), and a variety of techniques are needed to unravel the physics. Fortunately, we can get a good idea of the issues by focusing on the simplest mechanism, gluino condensation, in the simplest $N = 1$ vacua.

Consider any (2,2) compactification, with the visible E_6 possibly broken by Wilson lines. The hidden E_8 generally has a large negative beta function

$$\beta_8 = \frac{b_8}{16\pi^2} g_{E_8}^3 \,, \quad -b_8 \gg 1 \,. \tag{18.8.1}$$

The running coupling is

$$g_{E_8}^2(\mu) = \frac{8\pi^2}{\text{Re}(S) + b_8 \ln(m_s/\mu)} \tag{18.8.2}$$

(for the present discussion we are not concerned about the small numerical difference between m_s and m_{SU}), and so becomes strong at a scale

$$\Lambda_8 = m_s \exp[-\text{Re}(S)/|b_8|] \,. \tag{18.8.3}$$

This is below the string scale but above the scale where any of the visible sector groups become strong. Just as with quarks in QCD, the strong attraction causes the gauginos to condense,

$$|\langle (\bar{\lambda}\lambda)_{\text{hidden}} \rangle| \approx \Lambda_8^3 \,. \tag{18.8.4}$$

Here and below '$\approx$' means up to numerical coefficients. As in QCD this condensate breaks a chiral symmetry, but in the pure supersymmetric gauge theory (gauge fields and gauginos only) it is known *not* to break supersymmetry.

In string theory at tree level the fields of the hidden E_8 couple to precisely one other light superfield, namely S. We have discussed the coupling of the dilaton and the axion to the field strength, but in addition supersymmetry requires a coupling between the auxiliary field and the gauginos

$$\kappa F_S (\bar{\lambda}\lambda)_{\text{hidden}} \,. \tag{18.8.5}$$

At scales below Λ_8 this looks like an effective interaction

$$\kappa F_S \langle (\bar{\lambda}\lambda)_{\text{hidden}} \rangle \approx F_S \kappa m_{SU}^3 \exp(-3S/|b_8|) \,. \tag{18.8.6}$$

From the general $N = 1$ action (B.2.16) this implies an effective superpotential[8]

$$W \approx \kappa m_{SU}^3 \exp(-3S/|b_8|) \,. \tag{18.8.7}$$

This superpotential is nonperturbative, vanishing at large S faster than any power of $1/S$. This is an example of the violation of a perturbative

---

[8] This must be holomorphic in S, whereas the scale Λ_8 depends on $\text{Re}(S)$. The point is that the phase of the condensate depends on the axion in just such a way as to account for the difference.

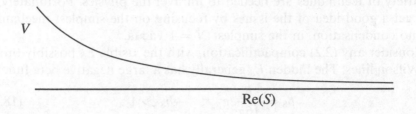

Fig. 18.3. The potential in a simple model of gluino condensation, as a function of the dilaton with other moduli held fixed.

nonrenormalization theorem by nonperturbative effects. This superpotential is not PQ-invariant, which is consistent with the earlier discussion.

What is more, this superpotential breaks supersymmetry. At tree level and to all orders of perturbation theory, the vacuum is supersymmetric for any value of S. Nonperturbatively,

$$F_S = \frac{\partial W}{\partial S} \approx \kappa m_{\mathrm{SU}}^3 \exp(-3S/|b_8|) \qquad (18.8.8)$$

is nonzero, which is the criterion (B.2.25) for the breaking of supersymmetry. This simple model is not satisfactory because the potential is roughly

$$V \approx \kappa^2 m_{\mathrm{SU}}^6 (S + S^*)^k \exp[-3(S + S^*)/|b_8|] . \qquad (18.8.9)$$

The power of $S + S^*$ comes from the Kähler potential for S and from the two-loop beta function. At small coupling (large S), where the calculation is valid, the potential has the qualitative form shown in figure 18.3 and there is no stable vacuum. Rather, the system rolls down the potential toward the point $\mathrm{Re}(S) = \infty$, where the theory is free and supersymmetric.

We will consider the problem of stabilizing the dilaton shortly, but for now let us see what happens if we assume that some higher correction, additional gauge group, or other modification gives rise to a stable supersymmetry-breaking vacuum at a point where S has roughly the value $8\pi^2/g_{\mathrm{YM}}^2 \approx 100$ found in simple grand unified models. The number 100 seems large, but noting that $|b_8| = 90$ this is actually the typical scale for the S-dependence.

Having broken supersymmetry, the next question is how this affects the masses of the ordinary quarks, leptons, gauge bosons, and their superpartners. The only tree-level coupling of the supersymmetry breaking field S to these fields is again through a gauge kinetic term, that of the Standard Model gauge fields. Thus F_S has a coupling of the same form as (18.8.5) but to the ordinary gauginos. Inserting the expectation value for F_S gives a gaugino mass term,

$$\kappa \langle F_S \rangle \bar{\lambda}\lambda \approx \kappa^2 m_{\mathrm{SU}}^3 \exp(-3\langle S \rangle / |b_8|)\bar{\lambda}\lambda \ . \tag{18.8.10}$$

The mass is

$$m_\lambda \approx \kappa^2 m_{\mathrm{SU}}^3 \exp(-3\langle S \rangle / |b_8|) \approx \exp(-3\langle S \rangle / |b_8|) \times 10^{18} \ \mathrm{GeV} \ . \tag{18.8.11}$$

To solve the Higgs naturalness problem the masses of the Standard Model superpartners must be of order 10^3 GeV or less. For the values $S \approx 100$ and $|b_8| = 90$ of this simple model this is not the case, but because these parameters appear in the exponent a modest ratio of parameters $S/|b| \approx 12$ would produce the observed large ratio of mass scales.

Once masses are generated for the Standard Model gauginos, loop corrections will give mass to the scalar partners of quarks and leptons. There is a simple reason why the (yet unseen) superpartners receive masses in this way while the quarks, leptons and gauge bosons do not: the latter masses are all forbidden by gauge invariance. Another feature to be understood is the negative mass-squared of the Higgs scalar, needed to break $SU(2) \times U(1)$, while the quark and lepton scalars must have positive masses-squared to avoid breaking baryon and lepton number. Again there is a simple general explanation, namely the one-loop correction to the Higgs potential coming from a top quark loop; the large top quark mass is just what is needed for this to work. The mass scale of the superpartners then determines the weak interaction scale.

The enormous ratio

$$\frac{m_{\mathrm{ew}}}{m_{\mathrm{grav}}} \approx 10^{-16} \tag{18.8.12}$$

thus arises ultimately from an exponent of order 10 in Λ_8, eq. (18.8.3). The renormalization group has this effect of amplifying modestly small couplings into large hierarchies. Thus, assuming the necessary stable vacuum, the enormous ratio of the weak and gravitational scales could emerge from a theory that has no free parameters.

We should point out that there is a distinction between the mass scale m_{sp} of the Standard Model superpartners and the scale $m_{\mathrm{SUSY}} = F_S^{1/2}$ of the supersymmetry-breaking expectation value. In fact,

$$m_{\mathrm{SUSY}}^2 \approx m_{\mathrm{sp}} m_{\mathrm{grav}} \ , \tag{18.8.13}$$

or

$$m_{\mathrm{sp}} \approx \kappa F_S \ . \tag{18.8.14}$$

This relation has a simple interpretation: the splittings in the Standard Model are given by the magnitude of the supersymmetry-breaking expectation value times the strength of the coupling between the Standard

Model and the supersymmetry breaking. There has also been much consideration of field theory models in which the two sectors couple more strongly, through gauge interactions, and m_{SUSY} is correspondingly lower. Such models could arise in string theory, in (0,2) vacua.

The form of supersymmetry breaking in this particular model, from $\langle F_S \rangle$, is known as *dilaton-mediated* supersymmetry breaking. Because the couplings of the dilaton are model-independent, the resulting pattern of superpartner masses is rather simple. In particular, the induced masses for the squarks and sleptons are to good accuracy the same for all three generations. This is important to account for the suppression of radiative corrections to rare decays (flavor changing neutral currents). More generally, radiative and other corrections can lead to a less universal pattern. Also, we have neglected all moduli other than the dilaton, but we will see below a simple model in which it is one of the Calabi–Yau moduli whose auxiliary field breaks supersymmetry.

The massless dilaton appears in the tree-level spectrum of every string theory, but not in nature: it would mediate a long-range scalar force of roughly gravitational strength. Measurements of the gravitational force at laboratory and greater scales restrict any force with a range greater than a few millimeters (corresponding to a mass of order 10^{-4} eV) to be several orders of magnitude *weaker* than gravity, ruling out a massless dilaton. We see from the present model that supersymmetry breaking can, and generically will, generate a potential for the dilaton. In this case there is no stable minimum, but the second derivative of the potential gives an indication of the typical mass

$$m_\Phi \approx m_{\mathrm{sp}} \; . \tag{18.8.15}$$

The superpotential (18.8.7) does not depend on any moduli other than S. This is because the scale Λ_8 is determined by the initial value of the gauge coupling, which at tree level depends only on S. We know that the one-loop correction to the gauge coupling depends on the other moduli, and this in turn induces a dependence in the superpotential. Thus if there is a stable minimum in the potential, generically all moduli will be massive.

Cosmological questions are outside our scope, but we note in passing that there is a potential cosmological problem with the moduli, in that their current energy density must not greatly exceed the critical density for closure of the universe. Typically the range of masses 10^{-7} GeV $< m < 10^4$ GeV is problematic. Below this, the mass is sufficiently small not to present a problem; above it, the decay rate of the particles is sufficiently great. Masses at either end of the range give interesting possibilities for dark matter.

Let us give an optimistic summary. Start with the simplest heterotic string vacuum with $N = 1$ supersymmetry, namely a (2,2) orbifold or

Calabi–Yau compactification. The result is a theory very much like the picture one obtains by starting from the Standard Model and trying to account for its patterns: gauge group E_6, chiral matter in the **27** representation, and a hidden sector that breaks supersymmetry (modulo the stabilization problem) and produces a realistic spectrum of superpartner masses. Of course, things may not work out so simply in detail; we know that the set of string vacua is vast, and we do not know any dynamical reason why these simple vacua should be preferred.

Another example

It is interesting to consider the following model,

$$K = -\ln(S + S^*) - 3\ln(T + T^*), \qquad (18.8.16a)$$

$$W = -w + \kappa m_{SU}^3 \exp(-3S/|b_8|). \qquad (18.8.16b)$$

The Kähler potential for T is based on the large-radius limit of Calabi–Yau compactification. Inclusion of a constant $-w$ in the tree-level superpotential is consistent with the scaling and PQ transformations. After some cancellation, the potential is proportional to a square,

$$V = \frac{(S + S^*)|W_{;S}|^2}{(T + T^*)^3}, \qquad (18.8.17a)$$

$$W_{;S} = \frac{w}{S + S^*} - \kappa m_{SU}^3 \exp(-3S/|b_8|)\left(\frac{3}{|b_8|} + \frac{1}{S + S^*}\right). \qquad (18.8.17b)$$

When $W_{;S} = 0$ the potential is minimized, and the value at the minimum is zero. Nevertheless supersymmetry is broken, as

$$W_{;T} = -\frac{3W}{T + T^*} \neq 0. \qquad (18.8.18)$$

This is intriguing: supersymmetry is broken nonperturbatively yet the vacuum energy is still zero. Also, the field T is undetermined, so there is a degenerate family of vacua with arbitrary supersymmetry-breaking scale $W_{;T}$. This is known as a *no-scale model*. The special properties of the potential depend on the detailed form of the Kähler potential and the superpotential, in particular the factor of 3 in the former and the fact that the latter is independent of T. Higher order effects will spoil this. For example, as we have noted above, threshold corrections will introduce a T-dependence into the superpotential.

Discussion

Since $S \propto g^{-2}$, the superpotential (18.8.7) is of order $\exp[-O(1/g^2)]$, which is characteristic of nonperturbative effects in field theory. It is not invariant under the PQ symmetry $S \to S + i\epsilon$ but transforms in a simple way. This can be related to the breaking of PQ invariance by instantons, but the argument is rather indirect and we will not pursue it.

It is interesting to consider at this point the order $\exp[-O(1/g)]$ stringy nonperturbative corrections deduced from the large order behavior of string perturbation theory. For the type II string we were able to relate these to D-instantons, but there is no analogous amplitude in the heterotic string. In the type II theory the D-instanton gives rise to an effect that does not occur in any order of perturbation theory, the nonconservation of the integrated R–R 1-form field strength. In the heterotic string it is unlikely that the stringy nonperturbative effects violate the perturbative nonrenormalization theorems. They would give rise to effects proportional to one of the forms

$$\exp(CS^{1/2}), \quad \exp[C(S + S^*)^{1/2}] \qquad (18.8.19)$$

with C a constant. The first form is holomorphic and the second is PQ-invariant. Corrections to the superpotential would have to be of the first form, but these have a complicated PQ transformation which is probably not allowed. In particular, it is believed that a discrete subgroup of the PQ symmetry is unbroken by anomalies; this would forbid the form $\exp(CS^{1/2})$. The nonperturbative effects could then only modify the Kähler potential, but this in any case receives corrections at all orders of perturbation theory.

Now we return to the stabilization of the dilaton. One possibility is that there are two competing strong gauge groups. In this case the dilaton potential can have a minimum, which for appropriate choices of the groups can be at the weak coupling $S \approx 100$ which is suggested by grand unification and needed for a large hierarchy. Another possibility is that a weak-coupling minimum can be produced by including the stringy nonperturbative corrections to the Kähler potential. It may seem odd that these corrections can be important at weak coupling, but it has been suggested that for the modestly small but not infinitesimal couplings of interest, the stringy nonperturbative effects can dominate the perturbative corrections. There may also be minima at very strong coupling, where the dual M-theory picture is more useful, or at couplings of order 1 which are close to neither limit.

Another idea would be that the potential really is as in figure 18.3 and that the dilaton is time-dependent, rolling toward large S. However, a brief calculation shows that these solutions cannot describe our universe: given the age of the universe, the supersymmetry breaking and gauge couplings would be far too small.

However, it is impossible to separate the stabilization of the dilaton from the cosmological constant problem. A generic potential on field space will have some number of local minima, but there is no reason that the value of the potential at any of the minima should vanish, either exactly or to the enormous accuracy required by the upper limit on the cosmological constant. So while the dilaton is stabilized, the metric is still 'unstable,' expanding exponentially, and the vacuum is not acceptable. The cosmological constant problem afflicts any theory of gravity, not just string theory. However, since predictive power in string theory is completely dependent on understanding the dynamics of the vacuum, any detailed discussion of the determination of the vacuum is likely to be premature until we understand why the cosmological constant is so small.

In any event, our current understanding would suggest that string theory has many stable vacua. Supersymmetry guarantees that the various moduli spaces with $N = 2$ and greater supersymmetry are exact solutions. In addition there are likely moduli spaces with $N = 1$ supersymmetry but no strong gauge groups and no breaking of supersymmetry. In addition there may be a number of isolated minima of approximate $N = 1$ supersymmetry, which are the ones we seek. There are also some string states of negative energy density. These are known to exist from one-loop calculations in nonsupersymmetric vacua with vanishing tree-level cosmological constant. The reader might worry that any vacuum with zero energy density will then be unstable. However, gravitational effects can completely forbid tunneling from a state of zero energy density to a state of negative energy density if the barrier between the two is sufficiently high. The conditions for this to occur are met rather generally in supersymmetric theories.

If there are many stable vacua, which of these the universe finds itself in would be a cosmological question, depending on the initial conditions, and the answer might be probabilistic rather than deterministic. This does not imply a lack of predictive power. Assuming that we eventually understand the dynamics well enough to determine the minima, there will likely be very few with such general features of the Standard Model as three generations. The key point is that because supersymmetry breaking leaves only isolated minima, there are no effective free parameters: the moduli are all determined by the dynamics.

This rather prosaic extrapolation is likely to be modified by new dynamical ideas. In particular, whatever principle is responsible for the suppression of the cosmological constant may radically change the rules of the game.

Exercises

18.1 Calculate the tree-level string amplitude with a model-independent axion and two gauge bosons.

18.2 Show from the explicit form of the string amplitudes that no scalar other than the model-independent axion has a tree-level coupling to $F_2 \wedge F_2$.

18.3 Derive the conditions cited at the end of section 18.6 for a heterotic string theory to have $N = 2$ and $N = 4$ spacetime supersymmetry.

18.4 Calculate the Zamolodchikov metric for two untwisted moduli of the $\mathbf{Z}_3$ orbifold and compare with the result obtained in chapter 16 by dimensional reduction.

18.5 Work out the one-loop vacuum amplitude for the twisted theory described at the beginning of section 18.7.

18.6 For the $SO(32)$ heterotic string on the $\mathbf{Z}_3$ orbifold, show that the gauge and mixed gauge–gravitational anomalies are nonzero. Show that they can be canceled by giving the superfield S the gauge transformation (18.7.6). Show that the resulting potential has supersymmetric minima.

18.7 If we integrate out the auxiliary field F_S, the couplings (18.8.5) lead to a tree-level interaction of four gauge fermions. Find this interaction using string perturbation theory. Note that it is independent of the compactification.

19

Advanced topics

In this final chapter we develop a number of intertwined ideas, concerning the perturbative and nonperturbative dynamics of the heterotic and type II theories. A common thread running through much of the chapter is world-sheet $N = 2$ superconformal symmetry, and we begin by developing this algebra in more detail. We then consider type II strings on Calabi–Yau and other (2,2) SCFTs, and heterotic strings on general (2,2) SCFTs. We next study string theories based on (2,2) minimal models, which leads us also to mirror symmetry. From there we move to some of the most interesting recent discoveries, phase transitions involving a change of topology of the compact space — the perturbative flop transition and the nonperturbative conifold transition. The final two sections deal with dualities of compactified theories, the first developing K3 compactification and the second the dualities of toroidally compactified heterotic strings.

19.1 The $N = 2$ superconformal algebra

The $N = 2$ superconformal algebra in operator product form, given in eq. (11.1.4), is repeated below:

$$T_B(z)T_F^\pm(0) \sim \frac{3}{2z^2}T_F^\pm(0) + \frac{1}{z}\partial T_F^\pm(0) \,, \tag{19.1.1a}$$

$$T_B(z)j(0) \sim \frac{1}{z^2}j(0) + \frac{1}{z}\partial j(0) \,, \tag{19.1.1b}$$

$$T_F^+(z)T_F^-(0) \sim \frac{2c}{3z^3} + \frac{2}{z^2}j(0) + \frac{2}{z}T_B(0) + \frac{1}{z}\partial j(0) \,, \tag{19.1.1c}$$

$$T_F^+(z)T_F^+(0) \sim T_F^-(z)T_F^-(0) \sim 0 \,, \tag{19.1.1d}$$

$$j(z)T_F^\pm(0) \sim \pm\frac{1}{z}T_F^\pm(0) \,, \tag{19.1.1e}$$

$$j(z)j(0) \sim \frac{c}{3z^2} \,. \tag{19.1.1f}$$

375

In the examples of interest the current j is single-valued with respect to all vertex operators. The Laurent expansions are then

$$T_B(z) = \sum_{n \in \mathbf{Z}} \frac{L_n}{z^{n+2}}, \qquad j(z) = \sum_{n \in \mathbf{Z}} \frac{J_n}{z^{n+1}}, \tag{19.1.2a}$$

$$T_F^+(z) = \sum_{r \in \mathbf{Z}+v} \frac{G_r^+}{z^{r+3/2}}, \qquad T_F^-(z) = \sum_{r \in \mathbf{Z}-v} \frac{G_r^-}{z^{r+3/2}}, \tag{19.1.2b}$$

where the shift v can take any real value. The OPEs (19.1.1) correspond to the $N = 2$ superconformal algebra

$$[L_m, G_r^\pm] = \left(\frac{m}{2} - r\right) G_{m+r}^\pm, \tag{19.1.3a}$$

$$[L_m, J_n] = -n J_{m+n}, \tag{19.1.3b}$$

$$\{G_r^+, G_s^-\} = 2L_{r+s} + (r - s)J_{r+s} + \frac{c}{3}\left(r^2 - \frac{1}{4}\right)\delta_{r,-s}, \tag{19.1.3c}$$

$$\{G_r^+, G_s^+\} = \{G_r^-, G_s^-\} = 0, \tag{19.1.3d}$$

$$[J_n, G_r^\pm] = \pm G_{r+n}^\pm, \tag{19.1.3e}$$

$$[J_m, J_n] = \frac{c}{3} m \delta_{m,-n}. \tag{19.1.3f}$$

It was shown in section 18.5 that every heterotic string theory with $d = 4$, $N = 1$ spacetime supersymmetry has a right-moving $N = 2$ superconformal algebra. In compactifications with the spin connection embedded in the gauge connection there is also a left-moving $N = 2$ algebra. Most of this final chapter deals with string theories having such (2,2) superconformal algebras. These are interesting for a number of reasons. First, they can also be taken as backgrounds for the type II string, where they lead to $d = 4$, $N = 2$ supersymmetry. This larger supersymmetry puts strong constraints on the dynamics, even nonperturbatively. Second, the large world-sheet superconformal algebra allows us to derive many general results concerning the low energy dynamics of heterotic string compactifications. Third, there are several additional constructions of (2,2) CFTs, and an interesting interplay between the different constructions. Finally, we have explained in the previous chapter that (0,2) CFTs have several phenomenological advantages over the more restricted (2,2) theories. However, many (0,2) theories are obtained from (2,2) theories by turning on Wilson lines or moduli. Also, many of the methods and constructions that we will develop for (2,2) theories can also be applied to the (0,2) case, though with more difficulty.

Heterotic string vertex operators

In this section we consider only the right-moving supersymmetry algebra, so that the results apply to all supersymmetric compactifications of the heterotic string. We take $\tilde{c} = 9$, as is relevant to four-dimensional theories. The local symmetry $\tilde{T}_F$ of the heterotic string is embedded in the $N = 2$ algebra as

$$\tilde{T}_F = \tilde{T}_F^+ + \tilde{T}_F^- . \qquad (19.1.4)$$

The separate generators $\tilde{T}_F^{\pm}$ must have the same periodicity as $\tilde{T}_F$: either NS ($v = \frac{1}{2}$) or R ($v = 0$). In addition to $N = 2$ superconformal symmetry, spacetime supersymmetry implies that all states have integer charge under the current (18.5.17). In a general vertex operator proportional to

$$\exp\left[l\tilde{\phi} + is_0\tilde{H}^0 + is_1\tilde{H}^1 + i\tilde{Q}(\tilde{H}/3^{1/2})\right] , \qquad (19.1.5)$$

it must then be the case that

$$l + s_0 + s_1 + \tilde{Q} \in 2\mathbf{Z} . \qquad (19.1.6)$$

Given the result $\tilde{j} = 3^{1/2}i\bar{\partial}\tilde{H}$ from section 18.5, it follows that $\tilde{Q}$ is the eigenvalue of $\tilde{J}_0$.

The vertex operators for the graviton, dilaton, and axion depend only on the noncompact coordinates and so are independent of compactification. For the remaining scalars, the weight $(1, \frac{1}{2})$ vertex operator in the -1 picture comes entirely from the compact CFT. The condition (19.1.6) in this case implies that $\tilde{Q}$ is an odd integer. The weight of $\exp(i\tilde{Q}\tilde{H}/3^{1/2})$ is $\tilde{h} = \tilde{Q}^2/6$, so the only possible values are $\tilde{Q} = \pm 1$ and the vertex operator takes one of the two forms

$$\mathscr{U} \exp(i\tilde{H}/3^{1/2}) , \quad \overline{\mathscr{U}} \exp(-i\tilde{H}/3^{1/2}) , \qquad (19.1.7)$$

with $\mathscr{U}$ having weight $(1, \frac{1}{3})$.

For fermions from the compact CFT, the internal part has weight $(1, \frac{3}{8})$. When the four-dimensional spinor is a **2**, then $s_0 + s_1$ is an odd integer and the allowed values of $\tilde{Q}$ are $\frac{3}{2}$ and $-\frac{1}{2}$, giving the vertex operators

$$j^a \exp(3^{1/2}i\tilde{H}/2) , \quad \mathscr{U} \exp[-i\tilde{H}/(2 \times 3^{1/2})] . \qquad (19.1.8)$$

For $\tilde{Q} = \frac{3}{2}$, the exponential saturates the right-moving weight $\frac{3}{8}$ and is identical to the compact part of the spacetime supercharge. The remaining factor j is a $(1, 0)$ current, so this state is a gaugino. For $\tilde{Q} = -\frac{1}{2}$, the remaining factor $\mathscr{U}$ is of weight $(1, \frac{1}{3})$, just as for the scalar. Because these theories have spacetime supersymmetry there is an isomorphism between the scalar and fermionic spectra. The OPE with the compact part $\exp(\pm 3^{1/2}i\tilde{H}/2)$ of the supercharge, which has $\tilde{Q} = \pm\frac{3}{2}$, relates the bosonic states with $\tilde{Q} = +1$ to the fermionic states with $\tilde{Q} = -\frac{1}{2}$. Similarly when the four-dimensional spinor is a **2'**, then $\tilde{Q}$ must be $\frac{1}{2}$ or $-\frac{3}{2}$, giving the vertex operators

$$\overline{\mathscr{U}} \exp[i\tilde{H}/(2 \times 3^{1/2})] , \quad j^a \exp(-3^{1/2}i\tilde{H}/2) . \qquad (19.1.9)$$

Chiral primary fields

The $N = 2$ superconformal algebra includes the anticommutators

$$\{\tilde{G}^+_{1/2}, \tilde{G}^-_{-1/2}\} = 2\tilde{L}_0 + \tilde{J}_0 , \tag{19.1.10a}$$

$$\{\tilde{G}^+_{-1/2}, \tilde{G}^-_{1/2}\} = 2\tilde{L}_0 - \tilde{J}_0 . \tag{19.1.10b}$$

We use the right-moving notation consistent with our convention for the heterotic string, but now allow arbitrary central charge. For central charge $\tilde{c}$, the bosonization of the $\tilde{j}\tilde{j}$ OPE implies that

$$\tilde{j} = i(\tilde{c}/3)^{1/2}\bar{\partial}\tilde{H} . \tag{19.1.11}$$

Taking the expectation values of the anticommutators (19.1.10) in any state, the left-hand side is nonnegative and so

$$2\tilde{h} \geq |\tilde{Q}| . \tag{19.1.12}$$

Let us consider an NS state $|c\rangle$ that saturates this inequality with $\tilde{Q} = 2\tilde{h}$. Such a state has the properties

$$\tilde{G}^\pm_r|c\rangle = 0 , \quad r > 0 , \tag{19.1.13a}$$

$$\tilde{L}_n|c\rangle = \tilde{J}_n|c\rangle = 0 , \quad n > 0 , \tag{19.1.13b}$$

$$\tilde{G}^+_{-1/2}|c\rangle = 0 . \tag{19.1.13c}$$

The first two lines state that $|c\rangle$ is annihilated by all of the lowering operators in the $N = 2$ algebra and so is an $N = 2$ superconformal primary field. The additional property of being annihilated by $\tilde{G}^+_{-1/2}$ defines a *chiral primary field*. To derive (19.1.13), note that all of the lowering operators except for $\tilde{G}^-_{1/2}$ take $|c\rangle$ into a state that would violate the inequality (19.1.12), and so must annihilate it. The expectation value of the anticommutator (19.1.10b) further implies that $\tilde{G}^-_{1/2}$ and $\tilde{G}^+_{-1/2}$ annihilate $|c\rangle$, giving the rest of eq. (19.1.13). A state with $\tilde{Q} = -2\tilde{h}$ is similarly a superconformal primary field that is also annihilated by $\tilde{G}^-_{-1/2}$, and is known as an *antichiral primary field*. The free boson $\tilde{H}$ contributes $3\tilde{Q}^2/2\tilde{c}$ to the weight of any state, so chiral primaries are possible only if

$$\frac{3\tilde{Q}^2}{2\tilde{c}} \leq \frac{|\tilde{Q}|}{2} \quad \Rightarrow \quad |\tilde{Q}| \leq \frac{\tilde{c}}{3} . \tag{19.1.14}$$

In particular the NS vertex operators (19.1.7), with $\tilde{Q} = \pm 1$ and $\tilde{h} = \frac{1}{2}$, are chiral and antichiral primaries. This property will be useful later. For the present we just use it to complete an argument from the previous chapter. We have seen that the -1 picture massless vertex operators have $U(1)$ charge $\tilde{Q} = \pm 1$. Acting with $G_{-1/2}$ to obtain the 0 picture operators could give $\tilde{Q} = \pm 2$ or 0. However, the chiral and antichiral properties imply that the terms with $\tilde{Q} = \pm 2$ vanish, so that the 0 picture operator must have $\tilde{Q} = 0$ and can depend on $\tilde{H}$ only through its derivative. Dimensionally it can then only be linear in $\bar{\partial}\tilde{H}$. Acting with $\tilde{J}_1$ picks out the coefficient of $\bar{\partial}\tilde{H}$,

$$\tilde{J}_1 \cdot \mathcal{V}^0 = \tilde{J}_1 G_{-1/2} \cdot \mathcal{V}^{-1} = (\tilde{G}_{-1/2}\tilde{J}_1 + \tilde{G}_{1/2}^+ - \tilde{G}_{1/2}^-) \cdot \mathcal{V}^{-1} = 0 \,, \quad (19.1.15)$$

the final equality holding because $\mathcal{V}^{-1}$ is primary. The 0 picture vertex operator is the change in the world-sheet action when a modulus is varied. We have established that this is independent of $\tilde{H}$, as needed above eq. (18.7.2).

Spectral flow

Suppose that we have a representation of the $N = 2$ algebra (19.1.3) with some periodicity v. Imagine shifting the $U(1)$ charge of every state by $-\tilde{c}\eta/3$, so that the free boson part of any vertex operator is shifted

$$\exp\left[i(3/\tilde{c})^{1/2}\tilde{Q}\tilde{H}\right] \to \exp\left[i(3/\tilde{c})^{1/2}\tilde{Q}\tilde{H} - i\eta(\tilde{c}/3)^{1/2}\tilde{H}\right] . \quad (19.1.16)$$

From their $U(1)$ charges we know that the $\tilde{T}_F^\pm$ depend on the free boson as $\exp[\pm i(3/\tilde{c})^{1/2}\tilde{H}]$. Then from the OPE of this factor with the exponential (19.1.16) it follows that the periodicity of $\tilde{T}_F^\pm$ with respect to any vertex operator shifts,

$$v \to v + \eta \,. \quad (19.1.17)$$

By this shift of the $U(1)$ charges, known as *spectral flow*, a representation with any periodicity can be converted to any other periodicity. The periodicities of the $U(1)$ current and energy-momentum tensor are unaffected. In the $d = 4$ heterotic string, the flow with $\eta = \frac{1}{2}$ converts a chiral primary into a $\tilde{Q} = -\frac{1}{2}$ R sector state, the flow with $\eta = -\frac{1}{2}$ converts an antichiral primary into a $\tilde{Q} = \frac{1}{2}$ R sector state, and vice versa: the superpartners are related to one another by spectral flow. The defining relations for chiral and antichiral primaries become

$$\tilde{G}_n^\pm|\psi\rangle = \tilde{L}_n|\psi\rangle = \tilde{J}_n|\psi\rangle = 0 \,, \quad n \geq 0 \,, \quad (19.1.18)$$

where $|\psi\rangle$ is the R sector state produced by the flow.

19.2 Type II strings on Calabi–Yau manifolds

Consider either type II string on a Calabi–Yau manifold. The compact CFT is the same as for the heterotic string, with the left-moving current algebra fermions λ^A for $A = 1, \ldots, 6$ replaced by fermions ψ^m and the remaining λ^A omitted. One can construct a right-moving spacetime supersymmetry precisely as in the heterotic string, and because the world-sheet theory is now the same on the right and left, there is a second spacetime supercharge from the left-movers. Thus either type II theory will have $d = 4$, $N = 2$ supersymmetry. The argument from section 18.5 shows further that this will be true for any compact CFT with (2,2) superconformal symmetry, provided it satisfies the generalized GSO projection (19.1.6) on both sides.

For the IIA string on a Calabi–Yau manifold, the massless fields come from the NS–NS fluctuations g_{MN}, b_{MN}, ϕ and the R–R fluctuations c_M and c_{MNP}. For any Calabi–Yau manifold these will include the four-dimensional metric $g_{\mu\nu}$, dilaton ϕ, and axion $b_{\mu\nu} \cong a$. The field c_μ is a massless vector. In addition, every Calabi–Yau manifold has exactly one (3,0)-form and one harmonic (0,3)-form, giving additional scalars from c_{ijk} and $c_{\bar{i}\bar{j}\bar{k}}$. For each harmonic (1,1)-form there is a scalar from $g_{i\bar{j}}$, another scalar from $b_{i\bar{j}}$, and a vector from $c_{\mu i\bar{j}}$. For each harmonic (2,1)-form there are scalars from g_{ij} and $g_{\bar{i}\bar{j}}$ just as for the heterotic string, and also scalars from $c_{ij\bar{k}}$ and $c_{\bar{i}\bar{j}k}$.

Let us see how these fit into multiplets of the $N = 2$ spacetime supersymmetry; the latter are summarized in section B.2. The metric $g_{\mu\nu}$ plus vector c_μ comprise the bosonic content of the supergravity multiplet. The remaining model-independent fields are four real scalars: ϕ, a, c_{ijk}, and $c_{\bar{i}\bar{j}\bar{k}}$. This is the bosonic content of one hypermultiplet. For each harmonic (1,1)-form there are two scalars and a vector, the bosonic content of a vector multiplet. For each harmonic (2,1)-form there are four scalars again forming a hypermultiplet. In all, there are

$$\text{IIA:} \quad h^{1,1} \text{ vector multiplets}, \quad h^{2,1} + 1 \text{ hypermultiplets}. \tag{19.2.1}$$

For the IIB string on a Calabi–Yau manifold, the massless fields come from the NS–NS fluctuations g_{MN}, b_{MN}, ϕ and the R–R fluctuations c, c_{MN}, and c_{MNPQ}. The model-independent fields are now the four-dimensional metric $g_{\mu\nu}$, dilaton ϕ, and axion $b_{\mu\nu} \cong a$, and also the scalar c, a second axion $c_{\mu\nu} \cong a'$, and a vector $c_{\mu ijk}$ from the (3,0)-form. For each harmonic (1,1)-form there is again a scalar from $g_{i\bar{j}}$ and one from $b_{i\bar{j}}$, and also one from $c_{i\bar{j}}$ and a fourth from the Poincaré dual of $c_{\mu\nu i\bar{j}}$. One might think that we should get additional scalars from $c_{ij\bar{k}\bar{l}}$ with the $h^{1,1}$ harmonic (2,2)-forms implied by the Hodge diamond (17.2.29), but because the 5-form field strength is self-dual these are actually identical to the states from $c_{\mu\nu i\bar{j}}$. For the same reason there is not an additional vector from $c_{\mu\bar{i}\bar{j}\bar{k}}$. For each harmonic (2,1)-form there are scalars from g_{ij} and $g_{\bar{i}\bar{j}}$ and a vector from $c_{\mu ij\bar{k}}$. Again the self-duality means that the vectors $c_{\mu\bar{i}\bar{j}k}$ give the same vector states. The massless IIB states form the $N = 2$ supergravity multiplet plus

$$\text{IIB:} \quad h^{2,1} \text{ vector multiplets}, \quad h^{1,1} + 1 \text{ hypermultiplets}. \tag{19.2.2}$$

Table 19.1. *Relations between Calabi–Yau moduli and supersymmetry multiplets in the two type II theories.*

	IIA	IIB
Kähler (1,1):	vector	hyper
complex structure (2,1):	hyper	vector

For convenient reference we have summarized the Calabi–Yau moduli of the type II theories in table 19.1.

Low energy actions

In section B.7 we describe the general low energy theory allowed by $N = 2$ supergravity. An important result is that the potential is determined entirely by the gauge interactions. Since the gauge fields in the type II compactifications all come from the R–R sector, all strings states are neutral and so the potential vanishes. Thus we can conclude that all the scalars found above are moduli. Moreover, because this is a consequence of symmetry it remains true to all orders in string and world-sheet perturbation theory, and even nonperturbatively. This is different from the $N = 1$ case, where we saw that nonperturbative effects could produce a potential.

The low energy action is then determined by supersymmetry in terms of the kinetic terms for the moduli — the metric on moduli space. Supersymmetry further implies that the kinetic terms for the hypermultiplet scalars are independent of the vector multiplet scalars and the kinetic terms for the vectors and their scalar partners are independent of the hypermultiplet scalars. In other words, the moduli space is a product. The vector multiplet moduli space is a special Kähler manifold and the hypermultiplet moduli space a quaternionic manifold, both defined in section B.7.

Now let us compare the IIA and IIB theories compactified on the same Calabi–Yau manifold. A hypermultiplet has twice as many scalars as a vector multiplet, so the IIA and IIB moduli spaces (19.2.1) and (19.2.2) do not in general even have the same dimension. However, they are related in interesting ways. If the R–R scalars are set to zero the tree-level IIA and IIB theories become identical, and indeed this removes two states from each hypermultiplet. Thus at string tree level, the R–R-vanishing subspace of each hypermultiplet moduli space should be a product of the dilaton–axion moduli space and a space identical to the vector multiplet moduli space of the *other* type II theory on the same Calabi–Yau manifold.

We can also go the other way, constructing the larger hypermultiplet moduli space from the smaller vector multiplet moduli space. Imagine compactifying one additional coordinate x^3 on a circle, going to $d = 3$. On a circle the IIA and IIB theories are T-dual, so the resulting moduli spaces should be identical. Indeed, each vector gives rise to two additional moduli, one from the vector component A_3 and one from the Poincaré dual of the $d = 3$ gauge field, so the dimensions are correct. Carrying out this reduction in detail gives the c-map from special Kähler manifolds to quaternionic manifolds. Since the hypermultiplet moduli spaces can be deduced in this way from the vector multiplet spaces, it follows that each can be characterized by a single holomorphic prepotential as in special Kähler geometry.

For the heterotic string we found a nonrenormalization theorem for the superpotential in world-sheet perturbation theory from the combination of holomorphicity and the symmetry $\delta T = i\epsilon$. It is interesting to apply these same constraints in the present case. Consider first a single Kähler modulus T representing the overall scale of the Calabi–Yau manifold. Just as for the heterotic string, eq. (17.5.4), one derives the Kähler potential

$$K = -3\ln(T + T^*) . \tag{19.2.3}$$

Up to a Kähler transformation, this is of the special geometry form (B.7.18),

$$K = -\ln \mathrm{Im}\left(\sum_I X^{I*}\partial_I F(X)\right) , \tag{19.2.4}$$

where

$$F(X) = \frac{(X^1)^3}{X^0} , \qquad T = \frac{iX^1}{X^0} . \tag{19.2.5}$$

The PQ symmetry $\delta T = i\epsilon$ is

$$\delta X^1 = \epsilon X^0 . \tag{19.2.6}$$

The function F is not invariant under this but changes by

$$\delta F = 3\epsilon(X^1)^2 . \tag{19.2.7}$$

The Kähler potential is then invariant; more generally, it is invariant provided that

$$\delta F = c_{IJ} X^I X^J \tag{19.2.8}$$

with real coefficients.

The function F must be of degree 2 in the X^I, and so an n-loop world-sheet correction would scale as $T^{3-n}(X^0)^2$. The only such correction that is allowed by the PQ symmetry and is not of the trivial form (19.2.8) is

$$\Delta F = i\lambda(X^0)^2 , \tag{19.2.9}$$

a three-loop correction to the leading interaction. This does in general appear, as we will note later. Further, in parallel to the heterotic string, nonperturbative world-sheet corrections to the Kähler moduli space are allowed by this argument but corrections to the complex structure moduli space are forbidden because the Kähler modulus T cannot couple to the complex structure moduli.

For more than one hypermultiplet, the PQ symmetries $\delta T^A = i\epsilon^A$ again greatly constrain the function F. It can be shown that any symmetry of the Kähler metric must be of the form

$$\delta X^I = \omega^{IJ} X^J , \tag{19.2.10}$$

so that up to a field redefinition we must have

$$T^A = \frac{iX^A}{X^0} , \quad A = 1,\ldots,n , \tag{19.2.11}$$

and $\omega^{A0} = \epsilon^A$. Requiring that F transform as in eq. (19.2.8) determines that it is of the form

$$F = \frac{d_{ABC} X^A X^B X^C}{X^0} + i\lambda (X^0)^2 . \tag{19.2.12}$$

This is consistent with the explicit results in section 17.5 for Calabi–Yau compactification, which were stated for the heterotic string but also apply to the type II theories. The coefficients d_{ABC} are the intersection numbers discussed there. This is the moduli space of vector multiplets in the IIA string, or the R–R-vanishing subspace of the IIB hypermultiplet moduli space. Since it is derived using the (1,1) PQ symmetry, this F receives world-sheet instanton corrections of order $\exp(-n_A T^A / 2\pi\alpha')$.

The complex structure moduli space must be a special Kähler manifold but is otherwise not restricted to a form as narrow as eq. (19.2.12). The one strong constraint is that the field-theory calculation of this moduli space receives no corrections from world-sheet interactions. The scale of the Calabi–Yau space, which governs these interactions, is a Kähler modulus. By the factorized property of the moduli space, it cannot appear in the complex structure metric. In section 19.6 we will describe the field theory calculation further.

The discussion of the moduli space metric thus far has been restricted to string tree level. For the potential, the $N = 2$ spacetime supersymmetry allowed us to draw strong conclusions that were valid even nonperturbatively. This is also the case for the metric: supersymmetry strongly constrains the form of possible string corrections, in the expansion parameter $g \sim e^{\Phi_4}$, as well as world-sheet corrections, in the expansion parameter $\alpha'/R_c^2 \sim 1/T$. The string coupling is governed by the dilaton, so any perturbative and nonperturbative corrections to the metric must depend on the dilaton. For both IIA and IIB compactifications, we have argued above that the dilaton is in a hypermultiplet. We arrived at this conclusion by counting states, but one can also show it directly (exercise 19.1). The low energy action for the vector multiplet cannot depend on the dilaton because of the product structure, and so receives no corrections from string interactions, either perturbative or nonperturbative.

Referring to table 19.1, one can conclude from the nonrenormalization theorems that of the four moduli spaces appearing in the type II theories, the IIB complex structure moduli space receives neither world-sheet nor string corrections. The tree-level result one obtains in the field theory approximation is exact. The other moduli spaces receive corrections of one or both kinds. Later we will see various extensions and applications of these results.

Chiral rings

As a final point, let us consider compactification on a general (2,2) SCFT. In parallel to the discussion for the heterotic string, the vertex operators for the NS–NS moduli must be of one of the forms

$$|c, \tilde{c}\rangle , \quad |c, \tilde{a}\rangle , \quad |a, \tilde{c}\rangle , \quad |a, \tilde{a}\rangle , \tag{19.2.13}$$

the states being chiral or antichiral primaries on each side. The corresponding operators are respectively denoted

$$\Phi^{++} , \quad \Phi^{+-} , \quad \Phi^{-+} , \quad \Phi^{--} . \tag{19.2.14}$$

Now consider a product of operators of the same type, for example chiral–chiral operators Φ^{++} and Ψ^{++}. The minimum weight for an operator in the OPE is

$$h \geq \frac{1}{2}(Q_\Phi + Q_\Psi) = h_\Phi + h_\Psi , \tag{19.2.15}$$

and similarly for $\tilde{h}$ and $\tilde{Q}$. The OPE is therefore nonsingular,

$$\Phi^{++}(z, \bar{z})\Psi^{++}(0,0) \sim (\Phi\Psi)^{++}(0,0) . \tag{19.2.16}$$

The operator $(\Phi\Psi)^{++}$ has $(h, \tilde{h}) = \frac{1}{2}(Q, \tilde{Q})$ and so is again chiral–chiral. The $(c, \tilde{c})$ operators thus form a multiplicative *chiral ring* (not a *group*, because an operator with $Q > 0$ has no inverse). The $(a, \tilde{a})$ operators form the conjugate ring, and the $(c, \tilde{a})$ and $(a, \tilde{c})$ operators form a different ring and its conjugate.

Let us connect this with the Calabi–Yau example. The (2,2) $U(1)$ currents are

$$j = \psi^i \psi^{\bar{i}} , \quad \tilde{j} = \tilde{\psi}^i \tilde{\psi}^{\bar{i}} . \tag{19.2.17}$$

From any harmonic (p, q)-form we can construct the operator

$$b_{i_1 \ldots i_p \bar{j}_1 \ldots \bar{j}_q}(X) \psi^{i_1} \ldots \psi^{i_p} \tilde{\psi}^{\bar{j}_1} \ldots \tilde{\psi}^{\bar{j}_q} . \tag{19.2.18}$$

This has charges and weights

$$Q = p , \quad h = \frac{p}{2} , \quad \tilde{Q} = -q , \quad \tilde{h} = \frac{q}{2} , \tag{19.2.19}$$

and so is a $(c, \tilde{a})$ chiral primary.[1] The weight comes entirely from the Fermi fields, because the form is harmonic. Naively multiplying two operators (19.2.18), the chiral ring is just the wedge product of the forms, which is the *cohomology ring*. This is correct at large radius, where the world-sheet interactions are weak, but the ring is corrected by world-sheet interactions. Note that the operator corresponding to a $(1, 1)$-form is just the vertex operator for the Kähler modulus. The product of three such operators is proportional to the corresponding Yukawa coupling.

Topological string theory

Notice that

$$(G_0^+)^2 = 0 . \tag{19.2.20}$$

This suggests that we think of G_0^+ as a BRST operator. The reader can show that the cohomology consists precisely of the chiral primary states, in the form (19.1.18), with vanishing spacetime momentum.

The operator G_0^+ is not conformally invariant, because the current T_F^+ has weight $(\frac{3}{2}, 0)$. Let us consider instead the energy-momentum tensor

$$T_B^{\text{top}} \equiv T_B + \frac{1}{2} \partial j . \tag{19.2.21}$$

The reader can verify the following properties:

$$T_B^{\text{top}}(z) T_B^{\text{top}}(0) \sim \frac{2}{z^2} T_B^{\text{top}}(z) + \frac{1}{z} \partial T_B^{\text{top}}(z) , \tag{19.2.22a}$$

$$T_B^{\text{top}}(z) T_F^+(0) \sim \frac{1}{z^2} T_F^+ + \frac{1}{z} \partial T_F^+ . \tag{19.2.22b}$$

This shows that T_B^{top} generates a conformal symmetry of central charge 0, and that under this symmetry T_F^+ has weight $(1, 0)$ and so G_0^+ is conformally invariant.

Starting with any (2,2) CFT, we can make a string theory by coupling the world-sheet metric to T_B^{top}. Because the central charge already vanishes, no additional ghosts are needed; the OPE

$$T_F^+(z) T_F^-(0) = \ldots + \frac{1}{z} T_B^{\text{top}}(0) + \ldots \tag{19.2.23}$$

---

[1] Often the sign convention for $\tilde{Q}$ is reversed so that the operator is $(c, \tilde{c})$.

shows that T_F^- plays the role of the b ghost. This theory has very few states because the cohomology is so small, and in particular has no dynamics because all physical states are time-independent. It is known as *topological string theory,* and its amplitudes are a special subset of the amplitudes of the related type II string theory.

19.3   Heterotic string theories with (2,2) SCFT

Now let us consider heterotic string theory with a general $c = \tilde{c} = 9$ (2,2) CFT. The remainder of the left-moving central charge for the compact theory comes from 26 free current algebra fermions. The noncompact fields are the usual X^μ and $\tilde{\psi}^\mu$. We continue to take the generalized GSO projection (19.1.6) on the right-moving side. On the left-moving side we will similarly generalize the GSO projection. We focus on the $E_8 \times E_8$ case. The current algebra fermions of interest will always be λ^A with $7 \le A \le 16$. For the second E_8, where $17 \le A \le 32$, we take the same GSO projection as in ten dimensions. The current algebra GSO projection then requires that the sum of the charge Q from the left-moving $N = 2$ SCFT and the charge for the current algebra number current

$$\sum_{K=4}^{8} \lambda^{K+}\lambda^{K-} \tag{19.3.1}$$

be an even integer.

From the current algebra fields and the free boson for the $U(1)$ of the left-moving superconformal algebra one can form the following (1,0) currents, all of which survive the GSO projection:

$$\lambda^A \lambda^B , \quad \Theta_{16}\exp(3^{1/2}iH/2) , \quad \Theta_{\overline{16}}\exp(-3^{1/2}iH/2) , \quad i\partial H . \tag{19.3.2}$$

Here Θ_{16} and $\Theta_{\overline{16}}$ are the R sector vertex operators for the current algebra fermions, with the subscript distinguishing the two spinor representations. These currents transform as

$$\mathbf{45 + 16 + \overline{16} + 1} \tag{19.3.3}$$

under the manifest $SO(10)$ current algebra. The gauge group must have an $SO(10)$ subgroup under which the adjoint representation decomposes in this way; this identifies it as E_6, whose adjoint is the **78**. In addition there is another E_8 from the second set of current algebra fermions. This $E_6 \times E_8$ is the full gauge symmetry of generic (2,2) compactifications. In special cases there are additional gauge symmetries, such as the $SU(3)$ of the $\mathbf{Z}_3$ orbifold.

To find the scalar spectrum, we start with the operator Φ^{++} for a state $|c, \tilde{c}\rangle$ with $Q = \tilde{Q} = 1$. On the right-moving side this is in the $-\frac{1}{2}$ picture, but on the left the superconformal symmetry is just a global symmetry and there are no pictures. Rather, we need total weight $h = 1$. We can obtain this and also satisfy the GSO projection with an additional λ^A excitation; the vertex operator is

$$\mathcal{V} = \lambda^A \Phi^{++} . \tag{19.3.4}$$

This is a **10** of the $SO(10)$ that acts on λ^A. By spectral flow on the left-moving part of Φ^{++} we also obtain

$$\Theta_{16} \Phi^{++} (1 \to -\tfrac{1}{2}) . \tag{19.3.5}$$

The notation indicates the charge Q after spectral flow. The charge is shifted by $-\tfrac{3}{2}$ units, which moves Φ^{++} from the NS to the R sector of the (2,2) CFT. The effect of spectral flow is to give $\Phi^{++}(Q \to Q')$ a weight

$$h = \frac{Q}{2} + \frac{Q'^2 - Q^2}{6} , \quad \tilde{h} = \frac{\tilde{Q}}{2} . \tag{19.3.6}$$

This is $(\tfrac{3}{8}, \tfrac{1}{2})$ in the present case. We have also included an R sector vertex operator Θ_{16} for the current algebra fermions, the subscript indicating its representation. The vertex operator then has the correct weight $(1, \tfrac{1}{2})$ and satisfies the GSO projection. Spectral flow also gives

$$\Phi^{++}(1 \to -2) . \tag{19.3.7}$$

This is now in the NS sector, with weight $(1, \tfrac{1}{2})$, and satisfies the GSO projection. These $SO(10)$ representations **10** + **16** + **1** add up to a **27** of E_6. As discussed in section 19.1, spectral flow on the right-moving side generates the fermionic partners of these scalars in the **2** of the four-dimensional Lorentz group.

There is one more massless scalar related to the above, with the weight $(1, \tfrac{1}{2})$ vertex operator

$$G^-_{-1/2} \cdot \Phi^{++} . \tag{19.3.8}$$

This is neutral under the gauge group. To see the significance of this state, consider using the same (2,2) CFT for compactification of one of the *type II* strings. In this case, Φ^{++} is the $(-1, -1)$ picture vertex operator for a modulus. The operator (19.3.8), which is in the heterotic -1 picture, is then identical to the zero-momentum vertex operator for the type II modulus in the $(0, -1)$ picture. Raising the right-moving picture in both theories, the 0 picture heterotic vertex operator is identical to the $(0, 0)$ picture type II vertex operator. These are the pictures that we add to the action when we turn on a background, so we conclude that we get the same CFT in the heterotic theory with a background of the scalar (19.3.8) as in the type II theories with a nonzero $(c, \tilde{c})$ modulus.

This further implies that the massless state (19.3.8) is a modulus, with vanishing potential. The argument is that we know from $N = 2$ spacetime supersymmetry that the corresponding type II state has no potential, and so the world-sheet theory with this background is an exact CFT whichever string theory we have. This kind of argument, using the larger supersymmetry of the type II theory to make arguments indirectly about the heterotic compactification, is very effective. It is important to note that it is valid only at string tree level: we have used the statement that CFTs correspond to tree-level backgrounds. At higher orders there is no relation between the two theories, because different states run around the loops. Some quantities that are not renormalized in the type II theory do get corrections in the less supersymmetric heterotic theory. For example, we argued that for the type II string, $N = 2$ spacetime supersymmetry implies that the flat directions are flat even nonperturbatively. In the heterotic string we know that gluino condensation and other effects can produce a potential.

Starting with a state $|a, \tilde{a}\rangle$ leads to the antiparticles of the above states. Starting with states $|c, \tilde{a}\rangle$ and $|a, \tilde{c}\rangle$ leads to a modulus plus a generation of the opposite chirality, the spacetime **2** being correlated with the gauge $\overline{\mathbf{27}}$. This pairing between generations and moduli of one type, and anti-generations and moduli of another type, generalizes the association with (1,1) and (2,1) forms found in Calabi–Yau compactification. In chapter 17 we argued that the moduli were exact by appealing to a result on the detailed form of instanton amplitudes, and now we have come to the same conclusion by appealing to results on the general $N = 2$ spacetime supersymmetric action. This second method is more general. For example, it also implies that the blowing-up modes for the fixed points of orbifolds are moduli, a result argued for in section 16.4 by citing detailed studies of twisted-state amplitudes.

In Calabi–Yau compactification we found additional E_6 singlets. In the abstract (2,2) description, these are states of weight $(1, \frac{1}{2})$ and $Q = 0$ that are $N = 2$ superconformal primary fields on the left-moving side. This is in contrast to the states (19.3.8), which are not annihilated by $G^+_{1/2}$.

We have used the relation between heterotic and type II compactifi-cations at string tree level, but let us note that any modular-invariant type II compactification also gives rise to a modular-invariant heterotic compactification. The modular transformation of the type II string theory mixes up the four sectors on each side, R vs NS and $\exp(\pi i F) = \pm 1$, in the same fashion as in the ten-dimensional theory in chapter 10. To make a heterotic theory we replace the two left-moving fermions $\psi^{2,3}$ with 26 left-moving current algebra fermions. The effect is independent of the (2,2) CFT and in particular is the same as in ten dimensions. Because the difference in the number of fermions is an odd multiple of eight, the signs in the type II and heterotic modular invariants differ (compare eq. (10.7.9) with eq. (11.2.13)), which is precisely as required by spacetime spin-statistics.

More on the low energy action

The argument that the scalars (19.3.8) are moduli is not self-contained, in that it uses results on $N = 2$ supergravity that we have not derived. To show these requires detailed analysis of the field theory actions and is beyond the scope of this book. One can also give a direct demonstration that the scalar (19.3.8) is a modulus, by extracting the effective action from an analysis of the heterotic string scattering amplitudes. The basic strategy is to consider a tree-level amplitude with any number of moduli (19.3.8), in any combination of the chiral and antichiral types. If the potential vanishes then this amplitude vanishes in the zero-momentum limit. Writing the operator $G_{-1/2}^-$ as a contour integral of T_F around Φ^{++}, one can deform the contour until it surrounds other vertex operators. It then takes one of the two forms

$$G_{-1/2}^- G_{-1/2}^- \cdot \Phi^{+\pm} = 0 \,, \tag{19.3.9a}$$

$$G_{-1/2}^- G_{-1/2}^+ \cdot \Phi^{-\pm} = (2L_{-1} - G_{-1/2}^+ G_{-1/2}^-) \cdot \Phi^{-\pm} = 2\partial \Phi^{-\pm} \,. \tag{19.3.9b}$$

We have used the relations

$$(G_{-1/2}^-)^2 = 0 \,, \quad G_{-1/2}^- \cdot \Phi^{-\pm} = 0 \,. \tag{19.3.10}$$

The final result is a total derivative and so should integrate to zero. To complete the argument one needs to show that there are no surface terms from vertex operators approaching one another; this uses the fact that the same structure appears on the right-moving side as on the left-moving one. Also, the fixed vertex operators require some additional bookkeeping. These details are left to the references. Below we cite further results that are found from a careful study of string amplitudes. These are obtained by the same approach, but the details are lengthy and again are left to the references.

In the previous section we discussed the constraints from $N = 2$ supergravity on the metrics for the type II moduli spaces, that is, on the kinetic terms for the moduli. We have argued that the CFT is the same for the type II and heterotic theories, and so the metric on moduli space should be the same in both string theories. In particular, the Zamolodchikov metric (18.6.10) gives the moduli space metric in terms of data from the CFT. This conclusion is confirmed by a study of moduli scattering amplitudes, which to order k^2 are the same in the type II and heterotic theories. Thus the (1,1) and (2,1) moduli spaces for the heterotic string each are special Kähler manifolds and are governed by a single holomorphic prepotential. This is in agreement with the explicit Calabi–Yau results in section 17.5. The analytic function $F_1(T)$ governing the Kähler moduli was there denoted $W(T)$, and the analytic function $F_2(Z)$ governing the complex structure moduli was there denoted $\mathscr{G}(Z)$.

For the **27**s and **$\overline{27}$**s, the model-dependent factors $\Phi^{\pm\pm'}$ in the vertex operators are the same as for the corresponding moduli. One would therefore expect that their amplitudes would be related to the amplitudes for the moduli in a model-independent way. Indeed, the low energy action is completely determined in terms of the holomorphic prepotentials F_1 and F_2 governing the (1,1) and (2,1) moduli, except for the extra E_6 singlets. The **$\overline{27}$** metric and superpotential are

$$G'_{A\bar{B}} = \exp[\kappa^2(K_2 - K_1)/3]G_{A\bar{B}} , \qquad (19.3.11a)$$

$$W(\phi) = \phi^A_{\ \bar{x}}\phi^B_{\ \bar{y}}\phi^C_{\ \bar{z}}d^{\bar{x}\bar{y}\bar{z}}\partial_A\partial_B\partial_C F_1(T) . \qquad (19.3.11b)$$

The **27** metric and superpotential are

$$G'_{a\bar{b}} = \exp[\kappa^2(K_1 - K_2)/3]G_{a\bar{b}} , \qquad (19.3.12a)$$

$$W(\chi) = \chi^a_{\ x}\chi^b_{\ y}\chi^c_{\ z}d^{xyz}\partial_a\partial_b\partial_c F_2(Z) . \qquad (19.3.12b)$$

Unlike earlier results, these cannot be derived from $N = 2$ supergravity, as the **27**s and **$\overline{27}$**s have no analogs in the type II theory. That the relations (19.3.11) and (19.3.12) are identical in form follows from the fact that the (1,1) and (2,1) states are essentially identical in CFT, differing only by a change in sign of the free scalar H from the superconformal algebra. The four-loop term (19.2.9) in F_1 does not affect $W(\phi)$.

These results generalize the Calabi–Yau results in section 17.5. We have also learned from the use of the PQ symmetries that the Kähler prepotential F_1 is of the form eq. (19.2.12) in world-sheet perturbation theory, and that F_2 cannot receive world-sheet corrections. Again we emphasize that the forms (19.3.11) and (19.3.12) are derived using CFT arguments and so are exact at string tree level, but that the relation between the different terms in the low energy action and the special form of the Kähler potential are not protected by the $N = 1$ supersymmetry of the heterotic string and so do not survive string loop corrections.

19.4 $N = 2$ **minimal models**

In chapter 15 we described the $N = 0$ and $N = 1$ minimal models. There is a similar family of solvable CFTs with $N = 2$ superconformal symmetry. It is interesting to consider heterotic string theories where the (2,2) CFT is a combination of these $N = 2$ minimal models, with total central charge

$(c, \tilde{c}) = 9$. This is another subject for which our treatment must be rather abbreviated. The full details of the constructions are lengthy and are left to the references.

A generalization of the method described in section 15.1 shows that unitary representations of the $N = 2$ superconformal algebra can exist only if $c \geq 3$, or at the discrete values

$$c = 3 - \frac{6}{k+2} = \frac{3k}{k+2}, \quad k = 0, 1, \ldots . \tag{19.4.1}$$

For the discrete theories, the allowed weights and $U(1)$ charges are

$$\text{NS:} \quad h = \frac{l(l+2) - q^2}{4(k+2)}, \quad Q = \frac{q}{k+2}, \tag{19.4.2a}$$

$$\text{R:} \quad h = \frac{l(l+2) - (q \pm 1)^2}{4(k+2)} + \frac{1}{8}, \quad Q = \frac{q \pm 1}{k+2} \mp \frac{1}{2}, \tag{19.4.2b}$$

where $0 \leq l \leq k$ and $-l \leq q \leq l$.

We showed that the $N = 0$ minimal models could be constructed as cosets starting from $SU(2)$ current algebras. There is a similar relation here. The central charge (19.4.1) is precisely the central charge of the $SU(2)$ current algebra at level k. The connection is as follows. Recall from section 15.5 that we can represent one current, say j^3, in terms of a free boson $i(k/2)^{1/2}\partial H$, and the CFT then separates into the free boson CFT and a so-called parafermionic theory. All other operators separate, for example

$$j^+ = \psi_1 \exp\left[i\left(\frac{2}{k}\right)^{1/2} H\right], \quad j^- = \psi_1^\dagger \exp\left[-i\left(\frac{2}{k}\right)^{1/2} H\right]. \tag{19.4.3}$$

Now define

$$T_F^+ = \psi_1 \exp\left[i\left(\frac{k+2}{k}\right)^{1/2} H\right], \quad T_F^- = \psi_1^\dagger \exp\left[-i\left(\frac{k+2}{k}\right)^{1/2} H\right]. \tag{19.4.4}$$

These operators have conformal weight

$$1 - \frac{1}{2}\left(\frac{2}{k}\right) + \frac{1}{2}\left(\frac{k+2}{k}\right) = \frac{3}{2}, \tag{19.4.5}$$

and one can show that they satisfy the $N = 2$ superconformal OPE. The parafermionic plus free-boson central charge remains at its original value.

Similarly, the current algebra primary fields factorize

$$\mathcal{O}_m^j = \psi_m^j \exp\left[im\left(\frac{2}{k}\right)^{1/2} H\right]. \tag{19.4.6}$$

Define now

$$\mathcal{O}_m'^j = \psi_m^j \exp\left[i\frac{2m}{k^{1/2}(k+2)^{1/2}} H\right]. \tag{19.4.7}$$

Relative to the current algebra primary, the exponent in $\mathcal{O}_m'^j$ is multiplied by $[2/(k+2)]^{1/2}$. The exponent in T_F is multiplied by the reciprocal factor relative to j^3, so the leading singularity z^{-1} in the current–primary OPE remains the same, and the operators $\mathcal{O}_m'^j$ are NS primaries under the $N=2$ algebra. Subtracting and adding the free-boson contributions, the weight of $\mathcal{O}_m'^j$ is

$$h = \frac{j(j+1)}{k+2} - \frac{m^2}{k} + \frac{2m^2}{k(k+2)} = \frac{j(j+1)-m^2}{k+2} . \tag{19.4.8}$$

This matches the weight of the NS primary (19.4.2a), with the identification $l=2j$ and $q=2m$. The ranges of l and q then match the ranges of the current algebra primaries. With the properly normalized $N=2$ current

$$j = i[k/(k+2)]^{1/2}\partial H , \tag{19.4.9}$$

the charge $Q = 2m/(k+2)$ also matches that of the current algebra primary. Similarly, the fields

$$\psi_m^j \exp\left[i\frac{2m \pm k/2}{k^{1/2}(k+2)^{1/2}}H\right] \tag{19.4.10}$$

have an additional factor $z^{\pm 1/2}$ in their OPEs with the currents. They are therefore primary fields in the R sector and are also annihilated by $G_0^{\pm}$, the sign correlating with that in the exponential. The weight and $U(1)$ charge agree with eq. (19.4.2b).

Landau–Ginzburg models

We now give a Lagrangian representation of the minimal models, the *Landau–Ginzburg* description. The rigid subgroup of the $(2,2)$ superconformal algebra is $(2,2)$ world-sheet supersymmetry. Having four supercharges, this is the dimensional reduction of $d=4$, $N=1$ supersymmetry. Any $d=4$, $N=1$ theory becomes a $(2,2)$ world-sheet theory by dimensional reduction, requiring the fields to be independent of $x^{2,3}$.

In particular, let us take a single chiral superfield with superpotential

$$W(\Phi) = \Phi^{k+2} . \tag{19.4.11}$$

Consider a scale transformation

$$\sigma \rightarrow \lambda\sigma , \quad \phi = \lambda^\omega\phi , \tag{19.4.12a}$$

$$\psi \rightarrow \lambda^{\omega-1/2}\psi , \quad F = \lambda^{\omega-1}F , \tag{19.4.12b}$$

with ω as yet unspecified. The relation between the scaling of the various components of the superfield is determined by the fact that the supersymmetry transformation squares to a translation. Including the scaling of $d^2\sigma$, the terms in the action (B.2.16) that are linear in W scale as

$$\lambda^{2-1+(k+2)\omega}, \qquad (19.4.13)$$

and so are invariant if $\omega = -1/(k + 2)$. With this value for ω, the kinetic terms scale as $\lambda^{-2/(k+2)}$ and are less important at long distance (large λ). Thus the theory at long distance is scale-invariant, and so also conformally invariant by the discussion of the c-theorem. Normally one must worry about quantum corrections to scaling, but not here because the superpotential is not renormalized. In this case the nonrenormalization theorem can be understood from symmetry. The theory with superpotential (19.4.11) has an R symmetry (defined in eq. (B.2.21)) under which ϕ has charge $2/(k + 2)$. This allows no corrections to the superpotential. If we began with a superpotential which also had higher powers of Φ, their effect would scale away at long distance.

The combination of conformal invariance and rigid supersymmetry generates the full (2,2) superconformal theory. Thus, the long distance limit of the theory has this symmetry, and it is this limiting *critical* theory that can be used as a string compactification. Equivalently, but more in the language of renormalization, we can hold the distance fixed but take to zero the 'cutoff' length at which the original field theory is defined.

We expect the critical theory to be a minimal model. The chiral superfield without a superpotential is the usual $c = 3$ free field representation. As in the discussion of $N = 0$ Landau–Ginzburg theories in chapter 15, the superpotential should reduce the effective number of degrees of freedom and so reduce the central charge. To see which minimal model we have, let us note that the field ϕ is a $(c, \tilde{c})$ primary. Its supersymmetry transformation (B.2.14) contains a projection operator P_+ onto four-dimensional spinors with $s_0 + s_1 = \pm 1$. The value of s_0 determines which of $P^0 \pm P^1$ the supersymmetry squares to, and so whether it is left- or right-moving. The projection P_+ thus implies that one rigid supersymmetry on each side annihilates ϕ; by convention we call these $G^-_{-1/2}$ and $\tilde{G}^-_{-1/2}$, so ϕ is $(c, \tilde{c})$. The chiral–chiral property is also consistent with the weight and charge. The scale transformation (19.4.12) implies that $h_\phi + \tilde{h}_\phi = -\omega$, and ϕ is spinless so

$$h_\phi = \tilde{h}_\phi = \frac{1}{2(k + 2)}. \qquad (19.4.14)$$

The R symmetry, under which ϕ has charge $2/(k + 2)$, acts on all components of the supercharge and so is equal to $Q + \tilde{Q}$. Thus

$$Q_\phi = \tilde{Q}_\phi = \frac{1}{k + 2}, \qquad (19.4.15)$$

and ϕ satisfies $Q = 2h$ and $\tilde{Q} = 2\tilde{h}$.

We can now identify the Landau–Ginzburg theory (19.4.11) with the minimal model at the same k. The minimal model field $\mathcal{O}'^{1/2}_{1/2}$ is a chiral primary, as one sees from the relation $Q = 2h$, and its weight agrees with that of ϕ at the same k. Also, by the chiral ring argument, we can make further primaries as powers ϕ^l. These correspond to $\mathcal{O}'^{l/2}_{l/2}$. However, the process terminates, because the equation of motion

$$\partial_\phi W(\phi) = (k + 2)\phi^{k+1} = 0 \qquad (19.4.16)$$

implies that $l \leq k$. This matches the minimal model bound on l, as well as the general bound that the maximum charge of a chiral primary is

$$Q = \frac{c}{3} = \frac{k}{k + 2} . \qquad (19.4.17)$$

An important role is played by the $\mathbf{Z}_{k+2}$ symmetry of the Landau–Ginzburg theory,

$$\Phi \to \exp\left(\frac{2\pi i}{k + 2}\right)\Phi . \qquad (19.4.18)$$

This acts in the same way on all components of the superfield and leaves the superpotential invariant. It is a discrete subgroup of the superconformal $U(1)$ generated by

$$\exp(2\pi i Q) ; \qquad (19.4.19)$$

this operator acts on ϕ as in (19.4.18), and it commutes with $T_F^{\pm}$ and so acts in the same way on all components of a world-sheet superfield. The operator $\exp(2\pi i \tilde{Q})$ is not an independent symmetry, because all fields in the Landau–Ginzburg theory are invariant under $\exp[2\pi i(Q - \tilde{Q})]$.

The Landau–Ginzburg theory is strongly interacting at long distance (since the interaction dominates the kinetic term) and so cannot be solved explicitly. Nevertheless, as in the examples we have seen, most of the quantities of interest in the low energy limit of string theory can be determined using constraints from supersymmetry. Much of the physics can then be rather directly understood from this representation, as opposed to the more abstract CFT construction of the minimal models. Landau–Ginzburg theories can be generalized to multiple superfields, where the classification of superpotentials uses methods from singularity theory. There are also more general current algebra constructions.

19.5 Gepner models

Now we wish to use the exact CFTs from the previous section to construct string theories. In order to obtain central charge (9,9) we need several minimal/Landau–Ginzburg models, with

$$\sum_i \frac{k_i}{k_i + 2} = 3 . \tag{19.5.1}$$

There are many combinations that satisfy this. Now consider the product of the Landau–Ginzburg path integrals, where we sum over common periodic or antiperiodic boundary conditions on all the fermions ψ_i and $\tilde{\psi}_i$ at once. The result is modular-invariant: the modular transformations mix the path integral sectors in the usual way, and the left–right symmetry guarantees the absence of anomalous phases. In terms of the abstract CFT description this is the diagonal invariant, taking the same $N = 2$ representation on the left and right and summing over representations; to be precise, one separates each representation into two halves according to $\exp(\pi i F)$ before combining left and right.

This is a consistent CFT for either the type II or heterotic string, but it is not yet spacetime supersymmetric. We must now impose the GSO projection (19.1.6), namely

$$l + s_0 + s_1 + Q \in 2\mathbf{Z} . \tag{19.5.2}$$

Normally this is imposed as a $\mathbf{Z}_2$ projection, beginning with a spectrum for which the combination $l + s_0 + s_1 + Q$ takes only integer values. It is therefore necessary first to twist by the group generated by

$$g_q = \exp(\pi i s + 2\pi i Q) = \exp(\pi i s) \prod_i \exp(2\pi i Q_i) , \tag{19.5.3}$$

where we define s to be even in the NS sector and odd in the R sector. The extra factor of $exp(\pi i s)$ is needed because $l + s_0 + s_1$ is integer in the NS sector but half-integer in the R sector. The operator (19.5.3) contains the product of the $\mathbf{Z}_{k_i+2}$ generators for the separate minimal model factors and so generates $\mathbf{Z}_p$, where p is the least common multiple of the $k_i + 2$.

There are two possible subtleties. First, since the projection (19.5.3) is not left–right symmetric, modular invariance is not guaranteed. The issue is the same as for the orbifold, discussed in section 16.1, and the necessary and sufficient condition is level matching just as in that case. Second, the phase of the operator (19.5.3) is determined by level matching and may not be that which we wanted. In the references it is shown that under rather general conditions, which include the case at hand, these subtleties do not arise and so the resulting theory is consistent and supersymmetric. This argument also applies to a more general set of $(2,2)$ CFTs known as Kazama–Suzuki theories, which are also constructed from current algebras.

Let us illustrate these general results for the notationally simple case of N minimal model factors having equal levels k. The central charge condition

$$\frac{Nk}{k+2} = 3 \qquad (19.5.4)$$

has integer solutions

$$k^N = 1^9,\ 2^6,\ 3^5,\ 6^4 . \qquad (19.5.5)$$

Before twisting, the NS–NS primaries are

$$\prod_{i=1}^{N} \psi_{m_i}^{j_i} \tilde{\psi}_{m_i}^{j_i} \exp\left[i\frac{2m(H_i + \tilde{H}_i)}{k^{1/2}(k+2)^{1/2}}\right] . \qquad (19.5.6)$$

We continue to use the $SU(2)$ notation, though the common notation in the literature on this subject is to use integer-valued labels $l = 2j$ and $q = 2m$ (or, confusingly, m equal to twice its $SU(2)$ value).

Now twist by g_q. An operator of charge $Q_i = l$ depends on the free scalar H_i from the ith factor as

$$\exp\left[il\left(\frac{k+2}{k}\right)^{1/2}H_i\right] . \qquad (19.5.7)$$

This picks up an extra phase $\exp(2\pi inl)$ when transported around a vertex operator in a sector twisted by g_q^n. It follows that the vertex operators in that sector contain an additional factor

$$\exp\left[in\left(\frac{k}{k+2}\right)^{1/2}H_i\right] . \qquad (19.5.8)$$

Thus the untwisted vertex operator (19.5.6) becomes

$$\prod_{i=1}^{N} \psi_{m_i}^{j_i} \tilde{\psi}_{m_i}^{j_i} \exp\left[i\frac{(2m_i + nk)H_i + 2m_i\tilde{H}_i}{k^{1/2}(k+2)^{1/2}}\right] . \qquad (19.5.9)$$

Using eq. (19.5.4), this has total $U(1)$ charge

$$Q = \frac{1}{k+2}\sum_{i=1}^{N}(2m_i + nk) = 3n + \frac{2}{k+2}\sum_{i=1}^{N}m_i . \qquad (19.5.10)$$

The level mismatch is

$$L_0 - \tilde{L}_0 = \frac{1}{2k(k+2)}\sum_{i=1}^{N}\left\{(2m_i + nk)^2 - (2m_i)^2\right\}$$

$$= \frac{3n^2}{2} + \frac{2n}{k+2}\sum_{i=1}^{N}m_i . \qquad (19.5.11)$$

Thus, requiring the charge (19.5.10) to be an integer implies that the level mismatch is a multiple of $\frac{1}{2}$, which is the appropriate result for the NS–NS sector before GSO projecting. The other sectors work as well.

Now let us look for $(c, \tilde{c})$ states. On the right-moving side, the charge and weight are given by the untwisted values (19.4.2a), so

$$\tilde{h} - \frac{\tilde{Q}}{2} = \sum_{i=1}^{N} \frac{j_i(j_i + 1) - m_i(m_i + 1)}{k + 2} . \qquad (19.5.12)$$

The chiral primaries have $m_i = j_i$ for all i. In the untwisted sector, $n = 0$, these are paired with chiral primaries on the left. The number of such states having $Q = 1$ is given by all sets of j_i such that

$$\sum_i j_i = \frac{k + 2}{2} , \quad |j_i| \le \frac{k}{2} . \qquad (19.5.13)$$

From the structure of the $N = 2$ superconformal representations one can show that there are no $(c, \tilde{c})$ states in the twisted sectors; these moduli, or the **27**s in the heterotic string, come entirely from the untwisted sector.

The $(a, \tilde{c})$ states, or $\overline{\mathbf{27}}$s, come from primaries with opposite m on the right and left,

$$m_i = -\tilde{m}_i = -j_i , \quad \text{all } i . \qquad (19.5.14)$$

These must come from the twisted sectors. Note that the states (19.5.9) are in general excited states in their representations, and the $(a, \tilde{c})$ states are obtained with lowering operators. A little thought shows that $(a, \tilde{c})$ states can arise only if $\tilde{m}_i$ is independent of i; one such state is consistent with the conditions (19.5.13) for the 3^5 and 6^4 cases. For example, the $(a, \tilde{c})$ state in the 3^5 model is obtained from the state with $m_i = \frac{1}{2}$ and $n = 1$ by acting with $G_{-1/2}^-$ in each of the five factors. The reader can check that this has the correct weight and charge, and that the OPE implies that it is nonzero.

In summary, the numbers $(n_{\mathbf{27}}, n_{\overline{\mathbf{27}}})$ for the k^N models are

$$1^9 : (84, 0) , \quad 2^6 : (90, 0) , \quad 3^5 : (101, 1) , \quad 6^4 : (149, 1) . \qquad (19.5.15)$$

Connection to Calabi–Yau compactification

An interesting point about the Gepner models is that most are in the same moduli space as Calabi–Yau compactifications. The simplest example is 3^5, five copies of the $k = 3$ model. The discrete symmetry is

$$S_5 \ltimes \mathbf{Z}_5^4 , \qquad (19.5.16)$$

where the permutation group S_5 interchanges the various factors. The $\mathbf{Z}_5$s come from the separate minimal model factors,

$$\exp(2\pi i Q_i) , \quad i = 2, 3, 4, 5 . \qquad (19.5.17)$$

The symmetry $\exp(2\pi i Q_1)$ is not independent because the projection (19.5.3) relates it to the others.

Now consider our simple example of a Calabi–Yau model, the quintic in CP^4, for the special polynomial

$$G(z) = z_1^5 + z_2^5 + z_3^5 + z_4^5 + z_5^5 \ . \tag{19.5.18}$$

This is invariant under the same discrete symmetry (19.5.16); the permutation acts on the z_i, and the four $\mathbf{Z}_5$s are

$$z_i \rightarrow \exp(2\pi i n_i/5)z_i \ , \quad n_1 = 0 \ . \tag{19.5.19}$$

We can set $n_1 = 0$ because an overall phase rotation of the z_i is trivial by the projective equivalence. Further, this Gepner model has 101 **27**s and one $\overline{\mathbf{27}}$, the same as the Calabi–Yau theory. The generations in each theory can be shown to fall into the same representations of the discrete symmetry. The only difference in the massless spectrum is that the Gepner model has four extra $U(1)$ gauge symmetries. These come from the currents ∂H_i for the separate factors, minus one linear combination that is already part of E_6. Such enhancements are common at special points of moduli space, as for toroidal compactification at the self-dual point. There are also extra E_6 singlet $U(1)$ charged states. All the extra states become massive by the Higgs mechanism as we move away from the Gepner point.

This is strong evidence that the 3^5 Gepner model is the same theory as the quintic (19.5.18). The same is true of other Gepner models, though in many cases one needs a Calabi–Yau manifold constructed from *weighted projective space*, where the projection (17.2.34) that defines CP^n is generalized to allow different scalings for the different z_i. To understand the connection in more detail, note the suggestive fact that the total Landau–Ginzburg superpotential

$$\sum_{i=1}^{5} \Phi_i^5 \tag{19.5.20}$$

is the same as the defining polynomial (19.5.18) of the Calabi–Yau manifold.

To make this observation more precise we generalize the previous Landau–Ginzburg construction, starting again with a theory of (2,2) rigid supersymmetry obtained by dimensional reduction from a $d = 4$, $N = 1$ theory. We take the five superfields Φ_i and an additional superfield P, as well as a $U(1)$ gauge field. The superpotential is

$$W = PG(\Phi) \ , \tag{19.5.21}$$

where we take an arbitrary quintic polynomial as in the Calabi–Yau case. This is gauge-invariant with $U(1)$ gauge charges

$$q_\Phi = 1 \ , \quad q_P = -5 \ . \tag{19.5.22}$$

The gauge coupling is e, and if we start from the general four-dimensional action (B.2.16) then there is one more parameter at our disposal, a $U(1)$ Fayet–Iliopoulos term which we will denote $\xi = -r/2$.

The potential energy for this *linear sigma model* is

$$U = |G(\phi)|^2 + |p|^2 \sum_{i=1}^{5} \left| \frac{\partial G}{\partial \phi_i} \right| + \frac{e^2}{2} \left(r + 5|p|^2 - \sum_{i=1}^{5} |\phi_i|^2 \right)$$

$$+ (A_2^2 + A_3^2) \left(25|p|^2 + \sum_{i=1}^{5} |\phi_i|^2 \right), \qquad (19.5.23)$$

coming from the F-terms, the D-terms, and the dimensional reduction of the kinetic terms. We use lower case letters for the scalar components of superfields. We are interested in the low energy dynamics of this field theory, and so in those points in field space where the potential vanishes.

Let us first restrict attention to polynomials that are *transverse*, meaning that the five equations

$$\frac{\partial G}{\partial \phi_i} = 0 \qquad (19.5.24)$$

have no simultaneous solutions except at $\phi = 0$. The reason for imposing this condition is that we are going to make contact with the Calabi–Yau manifold defined by the embedding $G(\phi) = 0$. If the gradient vanishes at any point, the condition $G(\phi) = 0$ degenerates and does not define a smooth manifold (if the gradient vanishes at some point ϕ_i, this point automatically lies on the submanifold $G = 0$ because $\phi_i \partial_i G = 5G$). These are actually five equations for four independent unknowns because of the projective equivalence (homogeneity of G). They therefore generically have no solutions other than $\phi = 0$; the case in which they do is very interesting and will be discussed in section 19.7.

Let us first consider the case $r > 0$. Transversality implies that the second term in the potential vanishes only if p vanishes and/or all the ϕ_i vanish. Combined with the vanishing of the third term this implies that

$$p = 0, \quad \sum_{i=1}^{5} |\phi_i|^2 = r. \qquad (19.5.25)$$

The fourth term forces $A_2 = A_3 = 0$, so finally we are left with

$$G(\phi) = 0. \qquad (19.5.26)$$

The manifold of vacua is identical to the Calabi–Yau manifold defined by $G = 0$ in CP^4. The condition (19.5.25) on ϕ can be regarded as a partial fixing of the projective invariance. The remaining invariance, a common phase rotation of the ϕ_i, is the $U(1)$ gauge invariance. The metric on this space is induced by the flat metric in the kinetic term $|\partial_a \phi_i|^2$. In particular, the size of the Calabi–Yau manifold is

$$R_c^2 \propto r \ . \tag{19.5.27}$$

One can show that this classical analysis becomes quantitatively accurate for large r.

Now consider $r < 0$. The unique zero-energy point is

$$|p|^2 = \frac{r}{5} \ , \quad \phi_i = 0 \ , \quad A_2 = A_3 = 0 \ . \tag{19.5.28}$$

Although this is an isolated zero of the potential, the fields ϕ_i are massless because their potential is of order $|\phi|^8$. In fact, they are described by a generalized Landau–Ginzburg theory, with superpotential

$$W = \langle p \rangle \, G(\Phi) \ ; \tag{19.5.29}$$

we can replace p with its mean value because the fluctuations are massive. This superpotential produces a nontrivial critical theory, by a generalization of the earlier argument.

We have seen that for positive values the parameter r has an interpretation as a modulus. It is the only Kähler modulus for this CFT. The complex structure moduli are the parameters in the polynomial G. Thus we conclude that the Landau–Ginzburg theories represent a different region in the same moduli space. The identification (19.5.27) would suggest that they correspond to unphysical negative values of R_c^2, but that identification is valid only at large r. There is an important distinction between the $r \to +\infty$ and $r \to -\infty$ limits. The former really represents an infinite distance in moduli space, corresponding to the fact that the Calabi–Yau space is becoming very big. As $r \to -\infty$, however, the low energy critical theory is determined by the superpotential (19.5.29). This depends on r through $\langle p \rangle$, but that can be absorbed in a rescaling of the fields. It follows that the low energy theory becomes independent of r as $r \to -\infty$. This point is actually at finite distance in moduli space, and the region of moduli space described by the Landau–Ginzburg theory is in the interior.

Recall that to construct a string theory from the Landau–Ginzburg theory we had to twist by the Z_5 symmetry g_q. It is interesting to see how this arises in the present construction. The expectation value of p breaks the $U(1)$ gauge symmetry, but a discrete subgroup

$$p \to p \ , \quad \phi_i \to \exp(2\pi i/5)\phi_i \ , \tag{19.5.30}$$

remains as an unbroken gauge symmetry of the low energy theory. As discussed in section 8.5, gauging of a discrete symmetry is one way to think about the twisting construction.

The analysis above breaks down at $r = 0$. The potential requires p and ϕ_i to vanish, and the fields $A_{2,3}$ then have no potential. This infinite volume in field space could produce a singularity that prevents continuation from positive to negative r. In fact, there is a singularity (to jump ahead a little, it is the mirror of the conifold singularity in the complex structure), but it does not prevent continuation between the two regions. The point is that the Kähler modulus is a complex field, and we have identified only its real part. To find the imaginary part, the $B_{i\bar{\jmath}}$ background, recall that this gives a total derivative on the world-sheet. There is one natural total derivative to add to the present theory, namely

$$i\frac{\theta}{2\pi} \int F_2 \,, \tag{19.5.31}$$

where F_2 is the $U(1)$ field strength 2-form. This does indeed correspond to the imaginary part of the modulus. In the Calabi–Yau phase one can use the equation of motion for the gauge field A_1 to show this.

At $r = 0$ but with θ nonzero, the world-sheet theory is nonsingular and so one can continue past the $r = \theta = 0$ singularity. The θ parameter in two dimensions has been extensively discussed in field theory, in part as a model for the instanton θ parameter in four dimensions. It does not change the equations of motion but changes the boundary conditions, so that there is a fractional electric flux

$$F_{12} = \frac{\theta}{2\pi} \,. \tag{19.5.32}$$

This flux will produce a nonzero energy density unless it is screened. A fractional flux cannot be screened by massive integer charged quanta. It can be screened by massless integer charges in two dimensions, or by the condensate if the $U(1)$ symmetry is spontaneously broken. In the present case a charged field, either p or ϕ_i, has an expectation value when r is nonzero and then the θ parameter has no effect. When r vanishes the $U(1)$ is unbroken. If $A_{2,3}$ are nonzero then all charged fields are massive and there is an energy density. Only at the point where all the fields vanish does the energy density go to zero, so the field space is effectively compact at low energy and the theory is nonsingular.

Thus the two parts of moduli space are smoothly connected. The term *phases* is often used to describe the two regions. Like the water/steam case, the two phases are continuously connected but display qualitatively different physics.

We have focused on the simplest example, but there are clearly many possible generalizations. It is interesting to note the following point. In the (2,2) algebra there are left- and right-moving $U(1)$s, under which the fields that move in the opposite direction are neutral. In the Landau–Ginzburg theory we identified the sum of these charges as an R symmetry.

To identify the separate symmetries we need also the rotation S_1 in the (2,3) plane, which in the dimensionally reduced theory becomes an internal symmetry. Since a four-dimensional Weyl spinor has $s_0 = -s_1$, the charge S_1 is correlated with the direction of motion, while the R charge is independent of it. By forming linear combinations of R and S_1 we can obtain symmetries Q and $\tilde{Q}$ under which either the left- or right-movers are neutral. To do this simultaneously for all fields, we must assign a common R-charge to all superfields. Since there is a gauge field we must be concerned about a possible anomaly in a current that acts on Fermi fields moving in one direction. The anomaly comes from a current–current OPE, as in section 12.2. By the above construction Q and $\tilde{Q}$ are the same for all superfields and so their anomalies are proportional to the sum of the gauge $U(1)$ charges. This is $-5 + (5 \times 1) = 0$ for the model at hand, so the anomalies vanish. If there were an anomaly, then one would not expect to have independent conserved Q and $\tilde{Q}$ and so there could be no (2,2) superconformal algebra. In fact one finds in this case quantum corrections that invalidate the classical analysis used above. In more general models, the anomaly cancellation condition turns out to be equivalent to the condition that in the Calabi–Yau phase the first Chern class vanishes, which was a necessary condition for conformal invariance.

19.6 Mirror symmetry and applications

In CFT it is arbitrary which states we call $(c, \tilde{c})$ and which $(a, \tilde{c})$. These just differ by a redefinition $H \to -H$ of the free scalar for the left-moving $U(1)$ current. However, for CFTs obtained from Calabi–Yau compactification these have very different geometric interpretations, in terms of the Kähler and complex structure moduli respectively. This suggests that Calabi–Yau manifolds might exist in *mirror pairs* $\mathcal{M}$ and $\mathcal{W}$, where

$$(h^{1,1}, h^{2,1})_{\mathcal{M}} = (h^{2,1}, h^{1,1})_{\mathcal{W}} , \tag{19.6.1}$$

and where the two CFTs are isomorphic, being related by $H \to -H$.

We can illustrate this for the analog of Calabi–Yau compactification with two compact dimensions. The holonomy is in $SU(1)$, which is trivial, so the compact dimensions must be a 2-torus. Calling the compact directions $x^{8,9}$, act with T-duality in the 9-direction. This flips the sign of $X_L^9(z)$ and so that of $\psi^9(z)$. Therefore it also flips the $U(1)$ current $i\psi^8\psi^9$. The 2-torus is thus its own mirror, but with different values of the moduli. Referring back to the discussion at the end of section 8.4, we noted there that T-duality on one axis interchanged the Kähler modulus ρ with the complex structure modulus τ. When the Kähler modulus ρ is large, the 2-torus is large; when the complex structure modulus τ is large, the 2-torus is long and thin.

The T-duality between the IIA and IIB theories implies that the IIA string on one 2-torus is the same as the IIB string on its mirror. This last will also be true for six-dimensional Calabi–Yau manifolds: the reversal of the $U(1)$ charge on one side also reverses the GSO projection on that side, interchanging the two type II strings. This is consistent with our results (19.2.1) and (19.2.2) for the moduli spaces. If we put the IIA theory on $\mathcal{M}$ and the IIB theory on $\mathcal{W}$, the number of vector multiplets $h^{1,1}_{\mathcal{M}} = h^{2,1}_{\mathcal{W}}$ is the same, and similarly the number of hypermultiplets.

The explicit construction of the mirror transformation for six-dimensional Calabi–Yau manifolds is less straightforward. Circumstantial evidence for the existence of mirror pairs was found when the $(h^{1,1}, h^{2,1})$ values were plotted for large classes of Calabi–Yau manifolds: if a given point was present, then a manifold with reversed Hodge numbers $(h^{2,1}, h^{1,1})$ usually also existed. This does not prove that the manifolds are mirrors, because the Hodge numbers do not determine the full CFT, but it is suggestive.

There is one class of Calabi–Yau manifolds where the mirror can be constructed explicitly, the ones that are related to Gepner models. Consider our usual example 3^5. The subgroup of the global $\mathbf{Z}_5^4$ symmetry that commutes with the spacetime supersymmetry is the group $\Gamma = \mathbf{Z}_5^3$ with elements

$$\exp\left\{2\pi i[r(Q_2 - Q_3) + s(Q_3 - Q_4) + t(Q_4 - Q_5)]\right\} \qquad (19.6.2)$$

for integer r, s, and t. We claim that if the theory is twisted by Γ then something simple happens. Consider first a single periodic scalar X compactified at radius $R = (\alpha' n)^{1/2}$ for some integer n. The translation

$$X \to X + 2\pi(\alpha'/n)^{1/2} \qquad (19.6.3)$$

generates a $\mathbf{Z}_n$. If we twist by this $\mathbf{Z}_n$ then we obtain the scalar at radius

$$R' = (\alpha'/n)^{1/2} . \qquad (19.6.4)$$

This is T-dual to the original radius, so the result is isomorphic to the original CFT, differing only by $X_L(z) \to -X_L(z)$. We leave it to the reader to show that the twist by Γ has the same effect in the 3^5 model, turning the Gepner CFT into one that is isomorphic under $H(z) \to -H(z)$. The point of this exercise is that we now have a geometric relation between the original theory and its mirror. This Gepner model maps to the quintic, which we will denote $\mathcal{M}$. The group Γ acts on the CP^4 coordinates in the Calabi–Yau description as

$$(z_1, z_2, z_3, z_4, z_5) \to (z_1, \alpha^r z_2, \alpha^{s-r} z_3, \alpha^{t-s} z_4, \alpha^s z_5) , \qquad (19.6.5)$$

where $\alpha = \exp(2\pi i/5)$. Twisting by Γ produces the coset space

$$\mathscr{W} = \mathscr{M}/\Gamma \,. \tag{19.6.6}$$

Some of the transformations have fixed points, so the space $\mathscr{W}$ is not a manifold but has orbifold singularities. These can be blown up, and the resulting smooth manifold indeed has Hodge numbers $(h^{1,1}, h^{2,1}) = (101, 1)$, the reverse of the $(1, 101)$ of the quintic.

The explicit twist can be carried out only at the Gepner point in moduli space, but the existence of mirror symmetry at this point is sufficient to imply it for the whole moduli space. The point is that the isomorphism of CFTs implies a one-to-one mapping of moduli, so the effect of turning on a modulus in one theory is identical to that of the equivalent modulus in the dual theory.

There have been many attempts to derive mirror symmetry in a more general way, with partial success. *Toric geometry* is a generalization of the projective identification that defines CP^n corresponding to the most general linear sigma model. It provides a framework for constructing many Calabi–Yau manifolds and their mirrors. In another direction, one might wonder whether a connection can be made to T-duality, as in the case of the 2-torus. Indeed, this has been done as follows. Put the IIA string on a Calabi–Yau manifold $\mathscr{M}$, and consider the manifold of states of a D0-brane: this is just the Calabi–Yau manifold itself, since the D0-brane can be anywhere. In the IIB string on the mirror manifold $\mathscr{W}$, BPS states come from Dp-branes with p odd, wrapped on nontrivial cycles of the mirror. Since $b_1 = b_5 = 0$, we must have $p = 3$. This immediately suggests a T-duality on three axes. Three of the coordinates of the D0-brane map to internal Wilson lines on the D3-brane, which therefore must be topologically a 3-torus. By following this line of argument one can show that $\mathscr{W}$ is a T^3 *fibration*. That is, it is locally a product $T^3 \times X$ with X a three-manifold, but with the shape of the T^3 *fiber* varying over X. The mirror transformation is T-duality on the three axes of T^3, and $\mathscr{M}$ is also a T^3 fibration. Any Calabi–Yau space with a Calabi–Yau mirror must be such a fibration; this property is not uncommon.

Moduli spaces

An important consequence of mirror symmetry is that it allows the full low energy field theory to be obtained at string tree level but exactly in world-sheet perturbation theory. We have argued that the field theory calculation of the complex structure moduli space is exact, but now we can also obtain the Kähler moduli space from the complex structure moduli space of the mirror.

Let us explain further how this works, taking our usual example of the quintic. We focus on the Kähler moduli space, which has a single modulus T. The general polynomial invariant under the $\mathbf{Z}_5^3$ twist (19.6.5) is

$$G(z) = z_1^5 + z_2^5 + z_3^5 + z_4^5 + z_5^5 - 5\psi z_1 z_2 z_3 z_4 z_5 . \tag{19.6.7}$$

This polynomial is parameterized by one complex parameter ψ, which survives as the sole complex structure modulus of the mirror.

The low energy action for the complex structure modulus of the mirror can be obtained as described in section 17.4. The special coordinates and periods are defined by the integrals of the harmonic $(3,0)$-form Ω over closed cycles,

$$Z^I = \int_{A^I} \Omega_{3,0} , \quad \mathscr{G}_I(Z) = \int_{B_I} \Omega_{3,0} . \tag{19.6.8}$$

The range of I is from 1 to $h^{2,1} + 1$, which is 2 in this example. For this construction the cycles and $\Omega_{3,0}$ can be given explicitly and the integrals evaluated. The result is that Z^a and $\mathscr{G}_I$ are hypergeometric functions of ψ. These in turn determine the prepotential $\mathscr{G} = \frac{1}{2} Z^I \mathscr{G}_I$, and so the low energy action for ψ.

There are three special points in this space, $\psi = 0, 1$, and ∞. The Gepner point $\psi = 0$ is where the theory can be described by a product of minimal models. The *conifold* point $\psi = 1$ is a singular Calabi–Yau space. The singularity is very interesting, and will be described in detail in the next section. The *large complex structure limit* is $\psi = \infty$. It is the only point at infinite distance in the moduli space metric, and so must be related by mirror symmetry to the large-radius limit $T = \infty$.

To exploit mirror symmetry we need the precise mapping between ψ and T. As in section 19.2, T is related to the special coordinates on Kähler moduli space by $T = iX^1/X^0$. The Z^I are special coordinates on the complex structure moduli space. Special geometry allows only a symplectic transformation (B.7.20) between different sets of special coordinates. The precise form of the symplectic transformation (which depends on the basis of cycles used in eq. (19.6.8)) can be found by comparing the exact prepotential as $\psi \to \infty$ with the large-radius limit of the Kähler prepotential. To leading approximation at large radius the result is

$$T \approx \frac{5}{2\pi} \ln(5\psi) . \tag{19.6.9}$$

The full mapping gives the exact prepotential for T and so the low energy action. Expanded around large T it agrees with the general form in section 19.2,

$$F = (X^0)^2 \left[\frac{5i}{6} T^3 - \frac{25i}{2\pi^3} \zeta(3) + \sum_{k=1}^{\infty} C_k \exp(-2\pi k T) \right] . \tag{19.6.10}$$

The first term is the tree-level interaction, the normalization agreeing with the expression (17.4.16) in terms of the intersection number. The second term is the three-loop correction, related to the R^4 term in the effective action which has the distinctive coefficient $\zeta(3)$. The final term represents a sum over instantons, where k is the total winding number of the world-sheet over the nontrivial 2-cycle of the Calabi–Yau manifold. The contribution of each instanton is rather simple, so the numerical constant just counts the number n_k of instantons (holomorphic curves) of given winding number, up to some simple factors. Expanding out the result from the mirror map gives immediately the number of such curves, which grows rapidly:

$$n_k = 2875,\ 609250,\ 317206375,\ 242467530000,\ \ldots\ . \tag{19.6.11}$$

The direct geometric determination of n_k is much more involved. Initially only the first few values were known, but now the full series has been determined, in agreement with the mirror symmetry prediction.

All of the above applies to string tree level. The string corrections depend on which string theory is put on the Calabi–Yau space. For the IIA string, the Kähler moduli are in vector multiplets and their low energy action receives no corrections. For the IIB string the low energy action for the complex structure moduli receives no corrections. For the heterotic or type I string there is only $d = 4$, $N = 1$ supersymmetry and so both moduli spaces may be corrected, while the superpotential may receive nonperturbative corrections.

The flop

The integral of the Kähler form over a 2-cycle is

$$\mathrm{Re}(T^A) = \int_{N^A} J_{1,1} = \int_{N^A} d^2w\, G_{i\bar{j}} \frac{\partial X^i}{\partial w} \frac{\partial X^{\bar{j}}}{\partial \bar{w}} > 0\,. \tag{19.6.12}$$

This must be positive for every 2-cycle, and similarly for the integral of $J_{1,1} \wedge J_{1,1}$ over any 4-cycle and of $J_{1,1} \wedge J_{1,1} \wedge J_{1,1}$ over the whole Calabi–Yau space. These conditions define the Kähler moduli space as a cone in the space parameterized by T^A.

In combination with mirror symmetry, this presents a puzzle. The boundary of the cone has codimension 1, since $\mathrm{Re}(T^A) = 0$ is a single real condition on the geometry. This must agree with the structure of the complex structure moduli space of the mirror manifold. The puzzle is that such boundaries do not appear in the complex structure moduli space. All special points in the latter are determined by complex equations, and so lie on manifolds of even codimension; an example is the point $\psi = 1$ of the quintic.

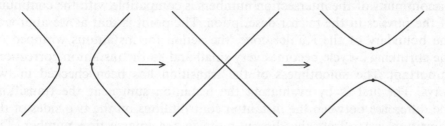

Fig. 19.1. The flop transition, projected onto the $\text{Re}(\phi_1)$–$\text{Re}(\rho_1)$ plane. The dots indicate the intersection of the minimal 2-spheres with this plane. The conifold transition is similar, but with a 3-sphere before the transition and a 2-sphere after.

The resolution of this puzzle is suggested by geometry. The integral of the Kähler form represents the minimum volume of a 2-sphere in the given homology class. This goes to zero at the boundary and would be negative beyond it. There is a sense in which the geometry can be continued to 'negative volumes.' A model for the region of the small sphere is given in terms of four complex scalars ϕ_1, ϕ_2, ρ_1, and ρ_2. To make a six-dimensional manifold we impose the condition

$$\phi^* \cdot \phi - \rho^* \cdot \rho - r = 0 \qquad (19.6.13)$$

for some real parameter r, and also the identification

$$(\phi_i, \rho_i) \cong (e^{i\lambda}\phi_i, e^{-i\lambda}\rho_i) . \qquad (19.6.14)$$

Let r first be positive, and look at the space parameterized by ϕ when $\rho_i = 0$. The condition (19.6.13) defines a 3-sphere and the identification (19.6.14) reduces this to a 2-sphere, with volume $4\pi r$. As ρ_i varies, the size of this 2-sphere grows, so $4\pi r$ is the minimum volume. For $r = 0$ the volume is zero and the space singular, but for $r < 0$ the space is perfectly smooth: the previous picture goes through with ϕ_i and ρ_i interchanged. The smallest 2-sphere has volume $4\pi|r|$, but it is a different 2-sphere from the one considered at positive r. This is shown schematically in figure 19.1. The transition from positive to negative r is known as a *flop*.

The mirror symmetry argument strongly suggests that the CFT at the $r = 0$ point is nonsingular, and that one can pass smoothly through it. One can check this in various ways. The flop transition does not change the Hodge numbers $h^{1,1}$ and $h^{2,1}$; this is consistent with the fact that nothing is happening in the mirror description. It does change the topology, however, as measured for example by the intersection numbers of various 2-cycles. These intersection numbers determine the low energy interactions of the Kähler moduli in the field-theory limit, so we need to understand how the discontinuity of the intersection numbers is compatible with the continuity of the physics in the mirror description. The point is that as we approach the boundary of the Kähler cone, the action for instantons wrapped on the shrinking 2-cycle becomes very small and so the instanton corrections important. The smoothness of the transition has been checked in two ways. The first is by evaluating the instanton sum near the transition: the difference between the instanton contributions on the two sides of the transition just offsets the discontinuity in the intersection number. The second is by looking at points on either side of the transition but far from it: the calculation in the mirror is then found to reduce to the appropriate intersection number in the various limits. As a final argument for the smoothness of the transition we can use a linear sigma model. In fact, if we take four chiral superfields and gauge the $U(1)$ symmetry (19.6.14), the D-term condition and gauge equivalence just reproduce the above model of the flop. As in the earlier application of the linear sigma model, we can interpolate from positive r to negative r along a path of nonzero θ.

The full picture is that in each moduli space the only singularities are of codimension at least 2. In the complex structure description the topology is the same throughout. In the mirror-equivalent Kähler description the cones for the different topologies join smoothly. However, smooth Calabi–Yau manifolds do not cover the whole Kähler moduli space. Some regions have a description in terms of orbifolds of Calabi–Yau manifolds, or Landau–Ginzburg models, or a hybrid of the two.

We cannot illustrate the flop transition with the quintic. This has only one Kähler modulus, and when it vanishes the volume of the whole Calabi–Yau manifold goes to zero. Incidentally, the moduli space of the quintic (in either the Kähler or complex structure description) is multiply connected. One nontrivial path runs from ψ to $\exp(2\pi i/5)\psi$; these points are equivalent with the coordinate change $z_1 \to \exp(-2\pi i/5)z_1$. The Gepner model is a fixed point for this operation. A second nontrivial path circles the conifold point $\psi = 1$. Together these generate the full modular group. This acts in a complicated way in terms of the variable T, but has a fundamental region with $\mathrm{Re}(T)$ positive.

We could consider a situation in which the moduli are time-dependent, moving from one Kähler cone to another. From the four-dimensional point of view, this is just the smooth evolution of a scalar field. If we consider the same process with the radius of the manifold blown up to macroscopic scales, we would see a region of the compact space pinch down and then expand in a topologically distinct way.

In general relativity the geometry of spacetime is dynamical, but it is an old question as to whether the topology is as well: spacetime can bend, but can it break? String theory, as a complete theory of quantum gravity, should answer this, and it does. At least in the limited way considered here, and in the somewhat more drastic way that we are about to consider, topology can change. It remains to understand the full extent of this and to learn what ideas are to replace geometry and topology as the foundation of our understanding of spacetime.

19.7 The conifold

Following eq. (19.5.24) we have discussed the requirement that the polynomial defining the embedding of the Calabi–Yau space in CP^4 be transverse, its gradient nonvanishing. As explained there, the vanishing of the gradient gives five conditions for four unknowns and generically has no solutions. If we allow the complex structure moduli to vary we get additional unknowns, and there will in general be solutions having complex codimension 1 (real codimension 2). The conifold is a realization of this. The vanishing of the gradient implies that

$$z_i^5 = \psi z_1 z_2 z_3 z_4 z_5 , \quad i = 1,\ldots,5 . \tag{19.7.1}$$

Multiplying these five equations together implies either that all the z_i vanish (which point is excluded from CP^4 by definition) or that

$$\psi^5 = 1 . \tag{19.7.2}$$

This has isolated solutions in the complex plane, consistent with the counting. We have noted above that ψ and $\exp(2\pi i/5)\psi$ are equivalent, so there is one possible singular point, $\psi = 1$. The singular manifold is known as a *conifold*, with $\psi = 1$ the conifold point in moduli space. The singularity, or node, on the manifold itself is at the point $z_1 = z_2 = z_3 = z_4 = z_5$.

Let us see the nature of the singularity. Generically, and in this example, the matrix of second derivatives of G is nonvanishing. We can then find complex coordinates $w = (w_1,\ldots,w_4)$ such that near the singularity the manifold is of the form

$$\sum_i w_i^2 = 0 , \tag{19.7.3}$$

the gradient of the left-hand side vanishing at the point $w = 0$. These are ordinary, not projective coordinates: one can fix the projective invariance by $z_1 = 1$, and the w_i are linear functions of $z_2,\ldots,z_5$. This equation then defines a space of $4 - 1$ complex dimensions as it should. The space is a cone, meaning that if w is on it then so is aw for any real a. To see the cross-section of the cone, consider the intersection with the 3-sphere

$$\sum_i |w_i^2| = 2\rho^2 . \tag{19.7.4}$$

Separating w_i into real and imaginary parts $w_i = x_i + iy_i$, this becomes

$$x \cdot x = \rho^2 , \quad y \cdot y = \rho^2 , \quad x \cdot y = 0 . \tag{19.7.5}$$

That is, x lies on a 3-sphere, and for given x the coordinate y lies on a 2-sphere. This is in fact a direct product, so the whole geometry near the singularity is

$$S^3 \times S^2 \times \mathbf{R}^+ . \tag{19.7.6}$$

If the complex structure is deformed away from the singular value then the embedding equation becomes

$$\sum_i w_i^2 = \psi - 1 . \tag{19.7.7}$$

There is now a minimum 3-sphere of radius $|\psi - 1|^{1/2}$. For example, taking $\psi - 1$ to be real, this would be given by

$$x \cdot x = |\psi - 1| , \quad y = 0 . \tag{19.7.8}$$

We have seen that for manifolds with orbifold or flop singularities, the CFT and associated string theory remain perfectly well-behaved. This is not the case at a conifold singularity. The exact calculation described in the previous section shows that there is a singularity at the conifold point in moduli space. Specifically, let us take the A^1 cycle to be the 3-sphere that is contracting to zero size at $\psi = 1$. The special coordinate Z^1 is defined by an integral (19.6.8) over A^1, so it must be that in terms of this coordinate the conifold singularity is at $Z^1 = 0$. The result of the exact calculation is then that the period has a singularity

$$\mathcal{G}_1 = \frac{1}{2\pi i} Z^1 \ln Z^1 + \text{holomorphic terms} . \tag{19.7.9}$$

This implies in turn a logarithmic singularity in the metric $G_{1\bar{1}}$ on the moduli space. The singularity (19.7.9) can be understood as follows. Observe that if Z^1 is taken once around the origin then the period is multivalued:

$$\mathcal{G}_1 \rightarrow \mathcal{G}_1 + Z^1 . \tag{19.7.10}$$

Now, this period is defined by an integral (19.6.8) over a cycle B_1 that intersects the shrinking cycle A^1 once. This does not define B_1 uniquely, and it is a general result that if we take a surface in the topological class of B_1 and follow it as we deform the complex structure through a cycle around the conifold point, then it ends up as a cycle topologically equivalent to $B_1 + A^1$, which also intersects A^1 once. This *monodromy* of the cycles translates into the monodromy (19.7.10) of the period.

We wish to understand the meaning of this singularity. We focus on the IIB string, where the issue is particularly sharp. In this case the complex structure moduli are in vector multiplets, and so the low energy action does not receive quantum corrections. The conifold singularity is then a property of the exact low energy field theory.

A general physical principle, which seems to hold true even in light of all recent discoveries about dynamics, is that singularities in low energy actions are IR effects, arising because one or more particles is becoming massless. For a nonsingular description of the physics we must keep these extra massless particles explicitly in the effective theory. We then need to understand why a particle would become massless at the conifold point in moduli space. The fact that this point is associated with a 3-cycle shrinking to zero size suggests a natural mechanism. A 3-brane wrapped around this surface would at least classically have a mass proportional to its area, and so become massless at $Z^1 = 0$.

This classical reasoning could be invalidated by quantum corrections, which might add a zero-point energy to the mass of the soliton. This does not happen for the following reason. The vector multiplet Z^1 comes from the (2,1)-form ω_1 that has unit integral over the cycle A^1 and zero integral over the other basis 3-cycles. In other words, the R–R 4-form potential is

$$c_{\mu npq}(x, y) = c_{\mu}^1(x)\omega_{1npq}(y) , \qquad (19.7.11)$$

with c_{μ}^1 the four-dimensional gauge field. For a D3-brane whose world-volume D is the product of the cycle A^1 on which the brane is wrapped and a path P in the noncompact dimensions, the coupling to the R–R 4-form is

$$\int_D c_4 = \int_P c_1^1 . \qquad (19.7.12)$$

The D3-brane thus has unit charge under the $U(1)$ gauge symmetry associated with the vector multiplet of Z^1. There is a BPS bound that the mass of any state with $U(1)$ charge is at least the charge times $|Z_1|$, times an additional nonzero factor. A BPS state, which attains the bound, thus has a mass that vanishes at $Z_1 = 0$. For the wrapped D3-brane, an analysis far from the conifold point in moduli space, where it is large and its world-volume theory weakly coupled, shows that it has one hypermultiplet of BPS states. This is also consistent with the low energy supersymmetry algebra, which allows a mass term (B.7.11) proportional to $|Z^1|$ for a charged hypermultiplet.

Finally, the logarithm in the low energy effective action arises from loops of the light charged particles. By a standard field theory calculation a hypermultiplet of unit charge and mass M contributes

$$-\frac{1}{32\pi^2} \ln(\Lambda^2/M^2)F_{\mu\nu}F^{\mu\nu} \qquad (19.7.13)$$

to the effective Lagrangian density. Here Λ is the effective cutoff on the momentum integral. This is in precise agreement with the singularity (19.7.9): $N = 2$ supergravity implies that the gauge kinetic term is proportional to

$$\frac{1}{8\pi}\text{Re}(i\partial_1\mathscr{G}_1)\,. \tag{19.7.14}$$

Thus the singular interactions at the conifold point, though they are found in a tree-level string calculation, can only be understood in terms of the full nonperturbative spectrum of the theory. This is another indication of the tight structure of nonperturbative string theory. We call the D-brane nonperturbative because it is not part of the ordinary string spectrum, and because at any fixed Z^1 the ratio of its mass to the masses of the string states goes to infinity as g is taken to zero. Note that in four dimensions the scale $g^{-1}\alpha'^{-1/2}$ of its mass is the four-dimensional Planck mass up to numerical factors. This is consistent with the fact that the BPS bound is derived in supergravity using only the gravitational and gauge part of the action, and so when written in units of the Planck scale cannot depend on the dilaton.

In some early papers the state that is becoming massless is referred to as a black hole. As discussed in section 14.8, the black hole and D-brane pictures apply in different regimes. In the present case the particle is singly charged and so the D-brane picture is the relevant one in the string perturbative regime $g < 1$. For $g > 1$ we would have to use a dual description of the IIB string; in this description the D-brane picture is again the relevant one.

Previously we encountered D-branes as large, essentially classical objects. It is not clear in what regimes it is sensible to sum over virtual D-branes, but clearly here where a D-brane becomes a light particle it is necessary to do so. One might think that as $g = e^\Phi$ goes to zero, the D-brane would have to decouple because it becomes very massive, meaning that its effect would go to zero. However, the complex structure action is independent of the dilaton Φ. Evidently we must take the upper cutoff Λ in the loop amplitude (19.7.13) also to scale as $1/g$ as compared to the string mass. This is another indication of the existence of distances shorter than the string scale.

The conifold transition

We should consider the possibility that at the point where the D-brane hypermultiplet becomes massless, there is another branch of moduli space where it acquires an expectation value. This does not happen in the example above because there is a quartic potential. In the notation of section B.7, where the two scalars in the hypermultiplet are denoted Φ_α, the condition that the potential vanish is

$$\Phi_\alpha^\dagger \sigma_{\alpha\beta}^A \Phi_\beta = 0 \,, \quad A = 1, 2, 3 \,. \tag{19.7.15}$$

It is easy to show that the only solution is $\Phi_1 = \Phi_2 = 0$. However, in more intricate examples a new branch of moduli space does emerge from a conifold point.

Let us consider a different point in the complex structure moduli space of the quintic, where the embedding equation is

$$z^1 H_1(z) + z^2 H_2(z) = 0 \,, \tag{19.7.16}$$

with H_1 and H_2 generic quartic polynomials in the z^i. This has singular points when

$$z^1 = z^2 = H_1(z) = H_2(z) = 0 \,. \tag{19.7.17}$$

The simultaneous quartic equations generically have 16 solutions, so this is the number of singular points on the Calabi–Yau manifold. Sixteen 3-spheres have shrunk to zero size.

The new feature of this example is that the shrinking 3-spheres are not all topologically distinct. Their sum is trivial in homology, which is to say that there is a four-dimensional surface whose boundary consists of these sixteen 3-spheres. Thus there are only fifteen distinct homology cycles and so fifteen associated $U(1)$ gauge groups. However, there are sixteen charged hypermultiplets that become massless at the point (19.7.16) in moduli space, since a D3-brane can wrap each small 3-sphere. The fact that the sum of the cycles is trivial translates into the statement that the sum of the charges of the sixteen light hypermultiplets is zero. Labeling the hypermultiplets by $i = 1, \ldots, 16$, we can take a basis $I = 1, \ldots, 15$ for the $U(1)$s such that the charges are

$$q^I_i = \delta^I_i \,, \quad i = 1, \ldots, 15 \,, \quad q^I_{16} = -1 \,. \tag{19.7.18}$$

The condition that the potential for the charged hypermultiplets vanish is then

$$\Phi_{i\alpha}^\dagger \sigma_{\alpha\beta}^A \Phi_{i\beta} - \Phi_{16\alpha}^\dagger \sigma_{\alpha\beta}^A \Phi_{16\beta} = 0 \,, \quad A = 1, 2, 3 \,, \quad i = 1, \ldots, 15 \,. \tag{19.7.19}$$

This has nonzero solutions, namely

$$\Phi_{i\alpha} = \Phi_{16\alpha} \,, \quad i = 1, \ldots, 15 \,. \tag{19.7.20}$$

Thus there is a new branch of moduli space. The fifteen $U(1)$s are spontaneously broken, so the number of vector multiplets is reduced from 101 to 86, while the potential leaves one additional hypermultiplet modulus (19.7.20) for a total of two.

As with the flop transition, this stringy phenomenon is already hinted at in geometry. We have discussed blowing up the 3-sphere at the apex

of the cone (19.7.6), but it is also possible to blow up the 2-sphere.
There are certain global obstructions to how this can be done; it cannot
be done for the simple singularity at $\psi = 1$ but it can be done in the
case (19.7.16). In fact these obstructions just coincide with the condition
that the hypermultiplet potential has flat directions. The resulting Calabi–
Yau manifold has just the Hodge numbers that would be deduced from
the low energy field theory, which are

$$(h^{1,1}, h^{2,1}) = (2, 86) \qquad\qquad (19.7.21)$$

in the present case. Thus the condensate of D-branes has a classical
interpretation in terms of a change of the topology of the manifold.

This change of topology is more radical than the flop, in that the
Hodge numbers change, and in particular the Euler number $\chi = 2(h^{1,1} - h^{2,1})$ changes. This is another example of the phenomenon, illustrated in
figure 14.4, that the more we understand string dynamics the more we
find that all theories and vacua are connected to one another. It appears
that all Calabi–Yau vacua may be connected by conifold transitions.

The conifold transition is also more radical in that it is nonperturbative
while the flop occurs in CFT, at string tree level. In fact the Euler number
cannot change in CFT. One way to see this is by considering the dynamics
of type II strings on the Calabi–Yau manifold. To have a potential that
can give mass to some moduli, we need charged matter as above. However,
the low energy gauge fields are all from the R–R sector and do not couple
to ordinary strings. We can also see it by putting the heterotic string on the
same space. At tree level the only way generations and antigenerations
could become massive is in pairs, through a coupling $\mathbf{1 \cdot 27 \cdot \overline{27}}$ when
a singlet acquires an expectation value. This leaves the Euler number
unchanged.

For a different string theory on the same Calabi–Yau manifold, the
nonperturbative physics will be different. For the IIB theory there are no
3-branes that could become massless at the conifold point. Also in this case
the complex structure moduli are in hypermultiplets, so the low energy
effective action can receive string corrections. It is then possible and in fact
likely that these corrections remove the singularity present in the tree-level
action. This is similar to the way that world-sheet instantons remove the
singularity at the edge of the Kähler cone in the flop transition. On the
other hand, mirror symmetry relates the conifold singularity in complex
structure moduli space to a singularity in the Kähler moduli space of the
mirror. The IIB theory has the same behavior at this singularity as the
IIA theory at the singularity in complex structure moduli space.

The two heterotic theories and the type I theory on a Calabi–Yau
manifold have only $d = 4$, $N = 1$ supersymmetry, so there is less control
over their nonperturbative behavior. One might think that the argument

above about generations and antigenerations becoming massive in pairs would exclude any Euler number changing conifold transition in these cases. However, one of the things that has been learned from the recent study of nonperturbative dynamics in field and string theory is that at a nontrivial fixed point (meaning that the interactions remain nontrivial to arbitrarily long distances) one can have phase transitions that cannot be described by any classical Lagrangian. We will illustrate one such transition in the next section. There is no physical principle (such as an index theorem) to exclude the possibility that as one passes through such a fixed point to a new branch of moduli space, unpaired generations become massive due to strong interaction effects. It has been argued that this does actually occur, though in a somewhat different situation.

19.8 String theories on K3

A Calabi–Yau manifold of $2n$ real dimensions has $SU(n)$ holonomy. The number of six-dimensional Calabi–Yau manifolds is large, but in the discussion of mirror symmetry we saw that there is a unique two-dimensional example T^2. In four dimensions there are exactly two Calabi–Yau manifolds, the flat T^4 and the manifold K3, which has nontrivial $SU(2)$ holonomy. Compactification on K3 down to six noncompact dimensions is of interest for a number of reasons. The resulting six-dimensional theories have interesting dynamics but are highly constrained by Lorentz invariance and supersymmetry. Also, compactification on K3 often appears as an intermediate step to a four-dimensional theory, where the compact space is locally the product of K3 and a 2-manifold.

Compactification on K3 breaks half of the supersymmetry of the original theory. Under $SO(9,1) \to SO(5,1) \times SO(4)$, the ten-dimensional spinors decompose

$$\mathbf{16} \to (\mathbf{4}, \mathbf{2}) + (\mathbf{4'}, \mathbf{2'}) \,, \tag{19.8.1a}$$

$$\mathbf{16'} \to (\mathbf{4}, \mathbf{2'}) + (\mathbf{4'}, \mathbf{2}) \,. \tag{19.8.1b}$$

Under $SO(4) \to SU(2) \times SU(2)$, the $\mathbf{2}$ transforms under the first $SU(2)$ and the $\mathbf{2'}$ under the second, so if the holonomy lies in the first $SU(2)$ then a constant $\mathbf{2'}$ spinor is also covariantly constant and gives rise to an unbroken supersymmetry. The smallest $d = 6$ supersymmetry algebra (reviewed in section B.7) has eight supercharges, so each ten-dimensional supersymmetry gives rise to one six-dimensional supersymmetry. The decompositions (19.8.1) determine the chiralities: the IIA theory on K3 has nonchiral $d = 6$ (1,1) supersymmetry, the IIB theory has chiral (2,0) supersymmetry, and the heterotic or type I theory has (1,0) supersymmetry.

The Hodge diamond of K3 is

$$
\begin{matrix}
& & h^{2,2} & & \\
& h^{2,1} & & h^{1,2} & \\
h^{2,0} & & h^{1,1} & & h^{0,2} \\
& h^{1,0} & & h^{0,1} & \\
& & h^{0,0} & &
\end{matrix}
\quad = \quad
\begin{matrix}
& & 1 & & \\
& 0 & & 0 & \\
1 & & 20 & & 1 \\
& 0 & & 0 & \\
& & 1 & &
\end{matrix}
\quad . \tag{19.8.2}
$$

In four spatial dimensions, the Poincaré dual squares to one, $** = 1$, so we can define self-dual or anti-self-dual 2-forms,

$$
* \omega_2 = \pm \omega_2 \ . \tag{19.8.3}
$$

On K3, 19 of the (1,1)-forms are self-dual, and the remaining (1,1)-form and the (2,0) and (0,2)-forms are anti-self-dual.

For the IIA string, the fluctuations without internal indices are $g_{\mu\nu}$, $b_{\mu\nu}$, ϕ, c_μ, and $c_{\mu\nu\rho}$, the last being related by Poincaré duality to a second vector c'_μ. Each (1,1)-form gives rise to a Kähler modulus $g_{i\bar{j}}$ and an axion $b_{i\bar{j}}$. An additional scalar arises from each of b_{ij} and $b_{\bar{i}\bar{j}}$. The complex structure moduli arise from (1,1)-forms by using the (2,0)-form $\Omega_{2,0}$, in parallel to their connection with (2,1)-forms in four-dimensional theories:

$$
g_{ij} = \Omega_{[i}{}^{\bar{k}} \omega_{j]\bar{k}} \ , \tag{19.8.4}
$$

and similarly for $g_{\bar{i}\bar{j}}$. This vanishes when ω_2 is the Kähler form so there are a total of $19 + 19 = 38$ complex structure moduli. The total number of moduli for the K3 surface is then 80, of which 58 parameterize the metric and 22 the antisymmetric tensor background. Finally, $c_{\mu np}$ gives 22 vectors, one for each 2-form. In all the spectrum consists of the (1,1) supergravity multiplet (B.6.7) and 20 vector multiplets (B.6.8).

For the IIB string, there are the same NS–NS fluctuations $g_{\mu\nu}$, $b_{\mu\nu}$, ϕ, g_{mn} and b_{mn}. There is also an R–R scalar c, another from the dual of $c_{\mu\nu\rho\sigma}$, and another antisymmetric tensor $c_{\mu\nu}$. The components c_{mn} give an additional 22 scalars from the 2-forms, while $c_{\mu\nu pq}$ give 22 tensors. For the latter we must be careful about the duality properties. The ten-dimensional field strength is

$$
H_{\mu\nu\sigma pq} = H_{\mu\nu\sigma} \omega_{pq} \ . \tag{19.8.5}
$$

The ten-dimensional $*$ factorizes

$$
*_{10} = *_4 \, *_6 \ . \tag{19.8.6}
$$

Since the ten-dimensional field strength is self-dual in the IIB string, the four-dimensional field strength transforms in the same way as the internal form ω_2. The tensors $b_{\mu\nu}$ and $c_{\mu\nu}$ have both self-dual and anti-self-dual parts, so the total spectrum contains 21 self-dual tensors and 5 anti-self-dual tensors. The bosonic fields add up to the (2,0) supergravity multiplet (B.6.10) and 21 tensor multiplets (B.6.11).

It is interesting that the properties of the cohomology of K3 can be deduced entirely from physical considerations. The (2,0) supergravity theory is chiral and so potentially anomalous. We leave the discussion of anomalies in six dimensions to the references, but the result is that the anomaly from the supergravity multiplet can only be canceled if there are exactly 21 tensor multiplets. This determines the cohomology, and so indirectly the spectrum of the nonchiral IIA theory on the same manifold.

Before going on to the heterotic string, let us note some further properties of K3 and the associated CFT. First, there are various orbifold limits. Two were developed in the exercises to chapter 16, namely $T^4/\mathbf{Z}_2$ and $T^4/\mathbf{Z}_3$. The spectra of the type II theories on each of these orbifolds are the same as those that we have just found. Second, the manifold K3 is hyper-Kähler. In section B.7 hyper-Kähler geometry is defined in the context of field space, but the idea also applies to the spacetime geometry: the holonomy $SU(2) \subset SO(4)$ is the case $m = 1$ of the discussion in the appendix. In fact, spacetime and moduli space are not so distinct. Consider a Dp-brane for $p < 5$, oriented so that it is extended in the noncompact directions and at a point in the K3. We leave it to the reader to show that this breaks half the supersymmetries of the type II theory on K3, leaving eight unbroken. The four collective coordinates for the motion of the Dp-brane within K3 lie in a hypermultiplet and so their moduli space geometry is hyper-Kähler. However, the moduli space of the collective coordinates is just the space in which the Dp-brane moves, K3. This is an elementary example of a very fruitful idea, the interrelation between spacetime geometry and the moduli spaces of quantum field theories on branes. Third, the 80-dimensional moduli space of the NS–NS fields on K3 is guaranteed by supersymmetry to be of the form (B.6.1), namely

$$\frac{SO(20,4,\mathbf{R})}{SO(20,\mathbf{R}) \times SO(4,\mathbf{R})} ,$$ (19.8.7)

up to a right identification under some discrete T-duality group.

Finally, the CFT of the string on K3 has (4,4) world-sheet superconformal invariance. This is closely related to the condition for $d = 4$, $N = 2$ supersymmetry cited in section 18.5. In geometric terms it comes about as follows. The basic world-sheet supercurrent is

$$T_F = i\psi_m \partial X^m = i\psi_r e^r_m \partial X^m ,$$ (19.8.8)

and similarly for right-movers. We have used the tetrad e^r_m to convert the index on ψ to tangent space. This tangent space index transforms as a $\mathbf{4} = (\mathbf{2}, \mathbf{2})$ of $SO(4) = SU(2) \times SU(2)$. The curvature of K3 lies entirely within the first $SU(2)$, so rotations of ψ_r in the second $SU(2)$ leave the action invariant. However, they do not leave the supercurrent invariant, and the three infinitesimal $SU(2)$ rotations generate three additional conserved supercurrents.

For the heterotic string we need to specify the gauge background. We start by embedding the spin connection in the gauge connection. This breaks the gauge symmetry to $E_7 \times E_8$ or $SO(28) \times SU(2)$. The bosonic spectrum includes the same states $g_{\mu\nu}$, $b_{\mu\nu}$, ϕ, g_{mn}, and b_{mn} found in the NS–NS spectrum of the type II theories. These comprise the bosonic content of the $d = 6$, $N = 1$ supergravity multiplet, one tensor multiplet, and 20 hypermultiplets. In addition there are vector multiplets in the adjoint of the gauge group. Finally there are additional hypermultiplets not related to the cohomology, which come from varying the gauge connection so that it is no longer equal to the spin connection. For the $E_8 \times E_8$ theory these hypermultiplets lie in the representations

$$(\mathbf{56}, \mathbf{1})^{10} + (\mathbf{1}, \mathbf{1})^{65} \tag{19.8.9}$$

of $E_7 \times E_8$. For the $SO(32)$ theory they lie in

$$(\mathbf{28}, \mathbf{2})^{10} + (\mathbf{1}, \mathbf{1})^{65} \tag{19.8.10}$$

of $SO(28) \times SU(2)$. Let us mention another result from the analysis of anomalies. A necessary condition for anomaly cancellation is that the numbers of hyper, tensor, and vector multiplets satisfy

$$n_H + 29n_T - n_V = 273 . \tag{19.8.11}$$

In both of the present theories this is $625 + 29 - 381 = 273$. The full story of anomaly cancellation is more involved, because of the possibility of multiple tensors, and is left to the references.

The potential for the charged hypermultiplets has flat directions, and there is a nice geometric description of the resulting moduli space. The conditions (17.1.12), namely $F_{ij} = F_{\bar{i}\bar{j}} = F_i^i = 0$, translate for four compact dimensions into the statement that the field strength is self-dual,

$$F = *F . \tag{19.8.12}$$

This is the condition that defines instantons in Yang–Mills theory; K3 is a four-dimensional Euclidean manifold, which is the usual setting for Yang–Mills instantons. The integral of the Bianchi identity (17.1.13),

$$\int_{\text{K3}} \text{tr}(R_2 \wedge R_2) = \int_{\text{K3}} \text{Tr}_v(F_2 \wedge F_2) , \tag{19.8.13}$$

determines the instanton charge: the number works out to 24. Thus the moduli space parameterized by the charged hypermultiplets is the space of gauge fields of instanton number 24 on K3. Supersymmetry guarantees that these lowest order solutions are exact. One can think of the moduli as representing the sizes of the instantons, their positions on K3, and their orientations within the gauge group; these parameters are not completely independent for the different instantons because there are constraints in order for the gauge field to be globally well defined. With spin connection equal to gauge connection all instantons are in the same $SU(2)$ subgroup and the unbroken symmetry is rather large. Generically they have various gauge orientations and the unbroken symmetry is smaller, E_8 for the $E_8 \times E_8$ theory and $SO(8)$ for the $SO(32)$ theory. For the $E_8 \times E_8$ theory, the gauge field with spin connection equal to gauge connection lies entirely in one E_8. By varying the moduli one can break the first E_8 entirely but the gauge field in the second E_8 remains zero. This is because the instanton numbers in the respective groups start at $(n_1, n_2) = (24, 0)$ and cannot change continuously. There are other branches of moduli space with different values of (n_1, n_2) such that $n_1 + n_2 = 24$.

Finally, it is very interesting to consider what happens when one or more instantons shrink to zero size. Note that all of these instantons are 5-branes, in that they are localized on K3 but the fields are independent of the six noncompact dimensions. We have discussed small instantons for the type I string in section 14.3: a new $SU(2)$ gauge symmetry appears on the 5-brane. The type I theory is the dual of the $SO(32)$ heterotic theory so the same must happen in the latter case. The gauge symmetry in the core cannot change as we go from weak to strong coupling by varying the neutral dilaton. Independent of duality, some of the arguments that were used in the type I case to derive the existence of the $SU(2)$ gauge symmetry apply also in the heterotic case — the ones based on the instanton moduli space and on the need for complete hypermultiplet representations. The group grows to $Sp(m)$ for m coincident zero-size instantons.

For the $E_8 \times E_8$ theory the same analysis leads to a very different result. To understand what happens, let us remember that the $E_8 \times E_8$ heterotic string is M-theory compactified on a segment of length $\alpha'^{1/2} g$. The eleven-dimensional spacetime is bounded by two ten-dimensional walls, with one E_8 living in each wall. The claim is that when an instanton in one of the walls shrinks to zero size, it can detach from the wall and move into the eleven-dimensional bulk. It remains extended in the noncompact directions so must be some 5-brane; there is a natural candidate, the 5-brane of M-theory discussed in section 14.4.

We have seen a similar phenomenon in section 13.6, where an instanton constructed from the gauge fields on a D4-brane could be contracted to a point and then detached from the D4-brane as a D0-brane. In fact, the present situation is dual to this, as shown in figure 19.2. If we compactify one of the noncompact dimensions with a small radius and regard this as the eleventh direction, we get the IIA string compactified to five dimensions on $\text{K3} \times S_1 / \mathbf{Z}_2$. The gauge fields live on D8-branes, and the instanton detaches as a D4-brane; this is T-dual to the D4–D0 system.

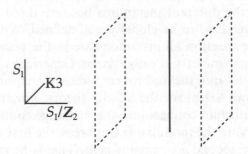

Fig. 19.2.   A schematic picture of M-theory on $K3 \times (S_1/\mathbf{Z}_2) \times S_1$. The $4+1$ noncompact dimensions are suppressed, and K3 is represented by a single dimension. An M5-brane, extended in the noncompact and S_1 directions, is shown. When the S_1 is small, this is the IIA theory with a D4-brane. When the $S_1/\mathbf{Z}_2$ is small it is the $E_8 \times E_8$ heterotic string with a detached M5-brane. The M5-brane can move to either boundary and become an instanton in one of the E_8s.

In all, there can be some number n_5 of M5-branes, and this and the instanton numbers now satisfy

$$n_1 + n_2 + n_5 = 24 \ . \tag{19.8.14}$$

The different (n_1, n_2) moduli spaces discussed above are now connected, as an instanton can detach from one wall, move across the bulk, and attach to the other. In chapter 14 we argued that the world-volume of the M5-brane includes a massless tensor and five scalars. Here the M5-brane is extended in the noncompact dimensions, so these become massless fields in the six-dimensional low energy field theory. Four of the scalars, forming a hypermultiplet, represent the position of the brane within K3. The fifth scalar, in a tensor multiplet, represents the position in the $S_1/\mathbf{Z}_2$ direction. The total number of tensor multiplets is $n_T = n_5 + 1$.

The instanton and M5-brane branches meet at a point, and the nature of the transition is quite interesting. As the vacuum moves onto the M5-brane branch, the number n_T of tensor multiplets increases by one. The anomaly cancellation condition $n_H + 29n_T - n_V = 273$ requires a compensating change in the number of hyper or vector multiplets. Typically, the number of hypermultiplets associated with the gauge background decreases by 30 when the instanton number goes down by one, offsetting the contribution of the tensor and hypermultiplets on the M5-brane.

The ordinary Higgs mechanism preserves the anomaly cancellation by giving mass to a vector and hypermultiplet. For the Higgs mechanism there is a familiar classical Lagrangian description. There is no classical Lagrangian that exhibits this new phase transition where a tensor multiplet becomes massless and a net of 29 hypermultiplets massive, or the reverse. In this respect it is like the generation-changing transition discussed in the previous section, and so would until recently have been considered impossible. We now understand that such transitions can occur at nontrivial fixed points. In fact, the Δn_T transition point is similar to the tensionless string theory that arises on coincident M5-branes. We have not discussed in detail the boundary conditions on the ends of the S_1/Z_2, but an M2-brane can end on them, as well as on an M5-brane as before (using duality, the reader can derive this fact in various ways). An M2-brane stretched between an M5-brane and the wall is a string with tension proportional to the separation, becoming tensionless when the M5-brane reaches the wall.

19.9 String duality below ten dimensions

In chapter 14 we focused on the nonperturbative dynamics of string theories in ten dimensions, and in a few toroidal compactifications. In this chapter we have seen some further phenomena that arise in compactified theories, in particular the conifold transition and the instanton/5-brane transition.

We should emphasize that many things that are impossible in CFT (string tree level) can happen nonperturbatively. One is the conifold transition itself, as we have explained. Another is heterotic string theory with $n_T > 1$, which we have just found. To get a massless tensor from a perturbative string state requires exciting a right-moving vector oscillator and a left-moving vector oscillator, and there is exactly one way to do this. The vacua with $n_5 > 0$ then do not have a perturbative string description. A third concerns the maximum rank of the gauge group in the heterotic string. Focusing on the maximal commuting subgroup $U(1)^r$, each $U(1)$ contributes 1 to the central charge, or $\frac{3}{2}$ for a right-mover, for a maximum of $r = 16 + 2k$, where k is the number of compactified dimensions. On the other hand, the $SO(32)$ theory in the limit that all instantons are pointlike has gauge group $SO(32) \times Sp(1)^{24}$, or as large as $SO(32) \times Sp(24)$ if the instantons are coincident. Each of these has rank 40, exceeding the 24 allowed in CFT. A fourth is the no-go theorem for the Standard Model in type II theory. This was proved in section 18.2, but the possibility of nonperturbative breakdown was also discussed.

This does not mean that the various results obtained in CFT are valueless. First, an understanding of the tree-level spectrum is a necessary step toward determining the nonperturbative dynamics. Second, Calabi–Yau compactification of the weakly coupled $E_8 \times E_8$ string resembles

the grand unified Standard Model sufficiently closely to suggest that our
vacuum may be of this type, at least approximately.

The conifold and small instanton transitions both occurred in theories with eight supersymmetries; such theories can have rich dynamics.
Presumably theories with four and fewer supersymmetries have dynamics
that is at least as rich, though the understanding of these is less complete. However, even with 16 supersymmetries there are some important
phenomena that come with compactification. In particular, the toroidal
compactifications of the heterotic string have this supersymmetry, and
these are the main subject of this final section.

Heterotic strings in $7 \leq d \leq 9$

We would like to determine the strong-coupling behavior of the heterotic
string compactified on T^k. The answer would seem to be obvious, because
we know the duals in ten dimensions and we can just compactify these. To
see what the issue is, recall the $SO(32)$ heterotic–type I relations (14.3.4),

$$G_{I\mu\nu} = g_{\mathrm{h}}^{-1} G_{\mathrm{h}\mu\nu} \,, \tag{19.9.1a}$$

$$g_I = g_{\mathrm{h}}^{-1} \,. \tag{19.9.1b}$$

This symmetry acts locally on the fields, and so should take a given
spacetime into the same spacetime in the dual theory. However, the metric
is rescaled; therefore, for toroidal compactification, the radii are rescaled

$$R_{mI} \propto g_{\mathrm{h}}^{-1/2} R_{m\mathrm{h}} \,. \tag{19.9.2}$$

As the heterotic coupling becomes large the k-torus in the type I theory
becomes small. As usual, we seek a description where the compact manifold is fixed in size or large, because g is not an accurate measure of
the effective coupling with a very small compact manifold. Thus we will
follow a succession of dualities, as we did in section 14.5 in deducing the
dual of the $E_8 \times E_8$ heterotic string.

The obvious next step is T-duality. This gives

$$g' \propto V_I^{-1} g_I \propto V_{\mathrm{h}}^{-1} g_{\mathrm{h}}^{(k-2)/2} \,, \tag{19.9.3a}$$

$$R'_m \propto R_{mI}^{-1} \propto g_{\mathrm{h}}^{1/2} R_{m\mathrm{h}}^{-1} \,. \tag{19.9.3b}$$

We have defined the volume $V = \prod_{m=10-k}^{9} (2\pi R_m)$ in each theory. The
compact space is now an orientifold as discussed in section 13.2,

$$T^k/\mathbf{Z}_2 \,, \quad \mathbf{Z}_2 = \{1, \Omega\hat{\beta}\} \,. \tag{19.9.4}$$

Here $\hat{\beta}$ is essentially a reflection in the compact directions, to be studied
in more detail below.

At strong heterotic coupling the compact space is now large, while the coupling is proportional to $g_h^{(k-2)/2}$. For $k = 1$ the theory that we have arrived at is weakly coupled, but even here there is a subtlety. If we begin with a compactification that has vanishing Wilson lines, we know from the discussion in chapter 8 that in the T-dual theory the 16 $D(9-k)$-branes will be at a single fixed point. The R–R and dilaton charges of the fixed points and D-branes cancel globally, but not locally. The dilaton, and therefore the effective coupling, is position-dependent. It diverges at the fixed points without D-branes when $k \geq 2$, and even for $k = 1$ it will diverge if the dual spacetime is too large. To keep things simple we will always start with a configuration of Wilson lines such that the D-branes are distributed equally among the fixed points. The number of fixed points is 2^k, so that by using half-D-branes we can do this for k as large as 5.

For $k = 2$ the coupling in the dual theory is V_h^{-1}. If the original 2-torus is larger than the string scale then we have reached a weakly coupled description, and if it is smaller then we simply start with an additional T-duality. If it is of order the string scale then the coupling g' is of order 1 and this is the simplest description that we can reach.

For $k = 3$ the coupling g' is strong, suggesting a further weak–strong duality. The bulk physics for k odd is that of the IIA theory, so strong coupling gives an eleven-dimensional theory. The necessary transformations (12.1.9) were obtained from the dimensional reduction of $d = 11$ supergravity, giving

$$R_{10M} \propto g'^{2/3} \propto g_h^{1/3} V_h^{-2/3} \,, \tag{19.9.5a}$$

$$R_{mM} \propto g'^{-1/3} R'_m \propto g_h^{1/3} R_h^{-1} V_h^{1/3} \,. \tag{19.9.5b}$$

All the radii grow with g_h, so the strongly coupled theory is eleven-dimensional. We will make some further remarks about the $k = 3$ case after the discussion of $k = 4$.

Heterotic–type IIA duality in six dimensions

The case $k = 4$ is interesting for a number of reasons, and we will discuss it in some detail. The description (19.9.3) is strongly coupled and the bulk physics is described by the IIB string, so we make a further IIB weak–strong transformation to obtain

$$g'' \propto g'^{-1} \propto g_h^{-1} V_h \,, \tag{19.9.6a}$$

$$R''_m \propto g'^{-1/2} R'_m \propto R_{mh}^{-1} V_h^{1/2} \,. \tag{19.9.6b}$$

We now have a weakly coupled description on a space of fixed volume as g_h becomes large.

To be more precise about the nature of the dual theory we must determine the $\mathbf{Z}_2$ identification. This is related to the Ω of the type I theory by a T-duality and then a IIB S-duality. The T-duality is a redefinition by β_R, a reflection in the compact directions acting only on the right-movers. Then $\Omega\hat{\beta}$ is the image of Ω under this,

$$\Omega\hat{\beta} = \beta_R^{-1}\Omega\beta_R = \Omega\beta_L^{-1}\beta_R \ . \tag{19.9.7}$$

For $k = 2n$ even, β has a convenient definition $\exp(\pi iJ)$ as a rotation by π in n planes. Then

$$\hat{\beta} = \beta_L^{-1}\beta_R = \beta_L^{-2}\beta_L\beta_R = \exp(-2\pi iJ_L)\beta = \exp(\pi inF_L)\beta \ . \tag{19.9.8}$$

In other words, this differs from the simple parity operation β by an extra $(-1)^n$ in the left-moving R sector. For T^4 this is simply β, and the $\mathbf{Z}_2$ is $\Omega\beta$. We must now consider the effect of the IIB weak–strong duality. The image of β is β, because duality commutes with the Lorentz group. To determine the image of Ω let us note its effect on the massless fields of the IIB theory, as discussed in section 10.6,

$$G_{\mu\nu}+\ , \quad B_{\mu\nu}-\ , \quad \Phi+\ , \quad C-\ , \quad C_{\mu\nu}+\ , \quad C_{\mu\nu\rho\sigma}-\ . \tag{19.9.9}$$

The weak–strong duality interchanges $B_{\mu\nu}$ and $C_{\mu\nu}$ and inverts $e^{-\Phi} + iC$. Conjugating the operation (19.9.9) by this results in

$$G_{\mu\nu}+\ , \quad B_{\mu\nu}+\ , \quad \Phi+\ , \quad C-\ , \quad C_{\mu\nu}-\ , \quad C_{\mu\nu\rho\sigma}-\ . \tag{19.9.10}$$

This acts as $+1$ on NS–NS fields and -1 on R–R fields. This identifies it as $\exp(i\pi F_L)$ (or $\exp(i\pi F_R)$ — which one we choose is arbitrary). As another check, Ω commutes with one of the two supercharges (the sum of the left- and right-movers), as does $\exp(\pi iF_L)$ (the supercharge in the NS–R sector).

Thus our dual to the heterotic theory on T^4 is the IIB theory on

$$T^k/\mathbf{Z}_2\ , \quad \mathbf{Z}_2 = \{1, \exp(\pi iF_L)\beta\}. \tag{19.9.11}$$

We can bring this to a more familiar form by a further T-duality transformation on a single coordinate, say X^9; since the radii are independent of the heterotic coupling this still defines a good dual. This gives the IIA theory with

$$g_A = g''R_9''^{-1} = g_h^{-1}R_{9h}V_h^{1/2}\ , \tag{19.9.12a}$$

$$R_{9A} = R_9''^{-1} = V_h^{-1/2}R_{9h}\ , \tag{19.9.12b}$$

$$R_{mA} = R_m'' = V_h^{1/2}R_{mh}^{-1}\ , \quad m = 6, 7, 8\ . \tag{19.9.12c}$$

The T-duality adds or deletes a 9-index on each R–R field so that β will act with the opposite sign. This cancels the action of $\exp(\pi iF_L)$, so the

IIA image of the space (19.9.11) is the ordinary orbifold

$$T^4/\mathbf{Z}_2 , \quad \mathbf{Z}_2 = \{1, \beta\} . \tag{19.9.13}$$

This orbifold is a special case of K3.

Thus the dual of the heterotic string on T^4 is the IIA string on K3; this is often termed *string–string duality*. In particular we have found that a special configuration of Wilson lines in the heterotic theory maps to an orbifold K3, but since the duality at this point implies an isomorphism between the respective moduli we can in the usual way extend this to the full moduli space. Indeed, the moduli space (19.8.7) of the IIA string on K3 is identical to the Narain moduli space (11.6.14) of the heterotic string. The coset structure is just a consequence of the 16 supercharges, but the number 20 in each case is a nontrivial check. Also, a careful analysis of the discrete T-duality of the K3 CFT has shown that it is identical to that of the heterotic theory on T^4.

In perturbation theory the gauge group of the IIA string on K3 is $U(1)^{24}$. This is also the gauge group at generic points in heterotic moduli space. At special points non-Abelian symmetries appear, the low energy physics being the usual Higgs mechanism. These same symmetries must appear on the IIA side. The $U(1)$s all come from the R–R sector, so the charged gauge bosons must arise from D-branes. In particular, the gauge fields associated with 2-forms couple to D2-branes wrapped around the corresponding 2-cycles. These must become massless at the enhanced symmetry points, and we know from the conifold example that this can occur if one or more 2-cycles shrinks to zero size. Indeed, the possible singularities of K3 are known to have an A–D–E classification, meaning that they are associated with the Dynkin diagrams of the simply-laced Lie algebras. At such a singularity the charges of the massless D-brane states are the roots of the associated algebra. Thus nonperturbative string theory provides a connection between the A–D–E classification of singularities and the corresponding algebra.

A single collapsed 2-sphere gives a $\mathbf{Z}_2$ orbifold singularity. The orbifold CFT is solvable and nonsingular. One expects that if a CFT is nonsingular then string perturbation theory should be a good description at weak coupling, meaning that there should not be massless nonperturbative states. This seems to contradict the argument that the collapsed 2-sphere gives rise to massless wrapped D-brane states. In fact, the massless D-brane should appear only when both the real and imaginary parts of the Kähler modulus $T = v + ib$ for the 2-sphere vanish. A careful analysis shows that the solvable theory is the orbifold limit with $T = i\pi$. The modulus b is a twisted state in the orbifold theory, so to reach the point of enhanced symmetry one must turn on a twisted state background and the CFT is no longer solvable.

Define the six-dimensional dilaton by

$$e^{-2\Phi_6} = V e^{-2\Phi} \,. \tag{19.9.14}$$

Tracing through the various dualities, the map between heterotic and IIA fields is

$$\Phi_6 \to -\Phi_6 \,, \qquad G_{\mu\nu} \to e^{-2\Phi_6} G_{\mu\nu} \,, \tag{19.9.15a}$$

$$\tilde{H}_3 \to e^{-2\Phi_6} *_6 \tilde{H}_3 \,, \qquad F_2^a \to F_2^a \,. \tag{19.9.15b}$$

The transformation takes the same form in both directions, heterotic $\to$ IIA and IIA $\to$ heterotic. The tensors and forms (19.9.15) are all in the noncompact directions. In the special case of a $\mathbf{Z}_2$ orbifold, the mapping of the moduli is given in eq. (19.9.12). The dimensionally reduced six-dimensional action for the fields (19.9.15) in the heterotic string is

$$S_{\text{het}} = \frac{1}{2\kappa_6^2} \int d^6x \, (-G_6)^{1/2} e^{-2\Phi_6} \left(R + 4\partial_\mu \Phi_6 \partial^\mu \Phi_6 \right.$$
$$\left. - \frac{1}{2}|\tilde{H}_3|^2 - \frac{\kappa_6^2}{2g_6^2}|F_2|^2 \right) \,. \tag{19.9.16}$$

The same action for the IIA theory is

$$S_{\text{IIA}} = \frac{1}{2\kappa_6^2} \int d^6x \, (-G_6)^{1/2} \left(e^{-2\Phi_6} R + 4e^{-2\Phi_6} \partial_\mu \Phi_6 \partial^\mu \Phi_6 \right.$$
$$\left. - \frac{1}{2}|\tilde{H}_3|^2 - \frac{\kappa_6^2}{2g_6^2}e^{-2\Phi_6}|F_2|^2 \right) \,. \tag{19.9.17}$$

We have omitted the kinetic terms for the moduli and the dependence of g_6 on the moduli; it is left to the reader to include these. The transformation (19.9.15) converts one theory to the other.

We should mention that the strategy that we used to find the dual of the ten-dimensional type I and IIB theories, following the D-string to strong coupling, was first applied to the six-dimensional heterotic–IIA duality. Consider the IIA NS5-brane, with four of its dimensions wrapped around K3. This is extended in one noncompact direction, and so is a string. A study of its fluctuations shows that they are the same as those of a heterotic string. The ratio of the tensions of the solitonic and fundamental strings is g^{-2}, as compared to the g^{-1} of the D-string. This again becomes small at strong coupling, so we can make the same duality argument as for the D-string. Similarly the fluctuations of the heterotic NS5-brane wrapped on T^4 are the same as those of the fundamental IIA string, and so this argument yields an element of the U-duality group.

Let us return to the case $k = 3$. To deduce the spacetime geometry, we need to understand how the $\mathbf{Z}_2$ identification acts on the M-theory circle. Again the $\mathbf{Z}_2$ arises via T-duality from the Ω projection of the type I theory. Recall from section 10.6 that the field $C_{\mu789}$ is odd under the latter. In the T-dual description the field C_μ is then odd. Since this couples to 10-momentum, it must be that the $\mathbf{Z}_2$ reflects the M-theory circle as well as the original T^3. Thus, the $d = 7$ heterotic string is dual to M-theory on

$$T^4/\mathbf{Z}_2 = \mathrm{K3}. \tag{19.9.18}$$

Recall from the Narain description that the moduli space of this heterotic compactification, including the dilaton, is locally

$$\frac{SO(19,3,\mathbf{R})}{SO(19,\mathbf{R}) \times SO(3,\mathbf{R})} \times \mathbf{R}^+ . \tag{19.9.19}$$

This 58-parameter space is identical to the space of metrics on K3. This is different from string theory on K3: M-theory has no 2-form field, so there are fewer moduli. The enhanced gauge symmetries on the heterotic side come from M2-branes wrapped on collapsed cycles of the K3.

Heterotic S-duality in four dimensions

The six-dimensional duality that we have just found can be used to find duals of four-dimensional theories. Let us consider the most supersymmetric case, compactification on a further 2-torus, to give the heterotic string on T^6 and the IIA string on $T^2 \times \mathrm{K3}$. The four-dimensional dilaton

$$e^{-2\Phi_4} \propto R_5 R_6 e^{-2\Phi_6} \tag{19.9.20}$$

transforms as

$$e^{\Phi_4} \to (R_5 R_6)^{-1/2}, \tag{19.9.21}$$

where again the transformation is the same in both directions. The 3-form field strength transforms as

$$*_4\tilde{H} \to e^{-2\Phi_4} dB_{56} , \tag{19.9.22}$$

but also in each theory this field strength is related to the axion by

$$*_4\tilde{H} \propto e^{-2\Phi_4} da . \tag{19.9.23}$$

It follows that the dilaton–axion field $S = e^{-2\Phi_4} + ia$ is related to the scalar $\rho \sim B_{56} + iR_5 R_6$ introduced in section 8.4 by

$$S \to i\rho^* . \tag{19.9.24}$$

From this we learn something interesting. The T-duality in the (5,6)-directions acts by the usual $SL(2,\mathbf{Z})$ transformation on ρ in each theory. It follows from the duality (19.9.24) that in each theory there is also an $SL(2,\mathbf{Z})$ acting on S (and hence called S-duality). This includes a weak–strong duality $S \to 1/S$, as well as discrete shifts of the axion. Thus we have deduced the strong-coupling dual of the heterotic string on T^4: itself. We see that the heterotic string compactified on tori has a complicated but consistent pattern of duals in different dimensions. In no case does one find two *different* weakly coupled duals of the same theory: that would be a contradiction.

In the heterotic theory on T^6, the interactions at energies far below the Planck scale reduce to $d = 4$, $N = 4$ gauge theory, and the $SL(2, \mathbf{Z})$ reduces to the Montonen–Olive symmetry of gauge theory, discussed in section 14.1. In both theories the full moduli space is

$$\frac{SU(1, 1)}{U(1) \times SU(1, 1, \mathbf{Z})} \times \frac{O(22, 6, \mathbf{R})}{O(22, \mathbf{R}) \times O(6, \mathbf{R}) \times O(22, 6, \mathbf{Z})} . \qquad (19.9.25)$$

As usual, the continuous identifications act on the left and the discrete ones on the right. In the heterotic string the first factor is from the dilaton superfield and the second from the moduli of Narain compactification. As is usually the case, the integer subgroup of the symmetry in the numerator of each factor is a symmetry of the full theory. In the IIA theory the first factor is from the ρ field, while the dilaton–axion field, the K3 moduli, and additional moduli from the T^2 compactification combine to give a single coset. The $O(22, 6, \mathbf{Z})$ duality then includes the perturbative duality of the K3, the S-duality of the dilaton–axion field, and U-dualities that mix these.

The six-dimensional duality is also useful in constructing dual pairs with less supersymmetry. Many Calabi–Yau manifolds are *K3 fibrations,* locally a product of K3 with a two-dimensional manifold. Applying the heterotic–IIA duality locally, the IIA theory on such a space is dual to the heterotic string on the corresponding T^4 fibration. For heterotic string compactifications with $d = 4$, $N = 2$ supersymmetry, the dilaton is in a vector multiplet. To see this, note that the dilaton is obtained by exciting one left- and one right-moving oscillator and so is of the form $|1, -1\rangle$ or $|-1, 1\rangle$, where the notation refers to the helicity s_1 carried on each side. Spacetime supersymmetry acts only on the right, generating a multiplet of four states. A helicity $\pm\frac{3}{2}$ on one side is not possible at the massless level, as the conformal weight would be at least $\frac{9}{8}$. The supermultiplet must then consist of

$$|1, -1\rangle, \quad |1, -\tfrac{1}{2}\rangle^2, \quad |1, 0\rangle \qquad (19.9.26)$$

and the CPT conjugates. This is the helicity content of a vector multiplet. It follows that the hypermultiplet moduli space does not have string loop or nonperturbative corrections in $d = 4$, $N = 2$ compactifications of the heterotic string, just as the vector multiplet moduli space does not have such corrections in the dual type II theory. This is analogous to the constraints from mirror symmetry, but for string rather than world-sheet corrections. In some cases one can combine mirror symmetry and string duality to determine the exact low energy action for a $d = 4$, $N = 2$ compactification. As with mirror symmetry, comparing the exact result in one theory with the loop and instanton corrections in its dual leads to unexpected mathematical connections.

19.10 Conclusion

Especially in this final chapter, we have only been able to scratch the surface of many important and beautiful ideas. String theory is a rich structure, whose full form is not yet understood. It is a mathematical structure, but deeply grounded in physics. It incorporates and unifies the central principles of physics: quantum mechanics, gauge symmetry, and general relativity, as well as anticipated new principles: supersymmetry, grand unification, and Kaluza–Klein theory. Undoubtedly there are many remarkable discoveries still to be made.

Exercises

19.1 Verify directly that the type II dilaton is in a hypermultiplet, by the method of eq. (19.9.26).

19.2 Fill in the details of the counting of $(a, \tilde{c})$ states in the Gepner models, as discussed below eq. (19.5.14).

19.3 Show explicitly that the net effect of the twist (19.6.2) on the spectrum is to reverse the sign of the left-moving $U(1)$ charge.

19.4 For compactification of the type I string on T^k for $k \leq 5$, give explicitly the Wilson line configuration such that in the T-dual theory there is an equal number of D-branes coincident with each orientifold fixed plane. What is the unbroken gauge group in each case?

19.5 By composing S, T, S, and T dualities as discussed in section 19.9, show that in both directions the string–string duality transformation takes the form (19.9.15). Show that this transforms the heterotic action into the IIA action. Find the action for the moduli R_m and show that it is invariant.

Appendix B

Spinors and supersymmetry
in various dimensions

Results about spinors and supersymmetry in various spacetime dimensions are used throughout this volume. This appendix provides an introduction to these subjects. The appropriate sections of the appendix should be read as noted at various points in the text.

B.1   Spinors in various dimensions

We develop first the Dirac matrices, which represent the Clifford algebra

$$\{\Gamma^\mu, \Gamma^\nu\} = 2\eta^{\mu\nu} . \tag{B.1.1}$$

We then go on to representations of the Lorentz group. To be specific we will take signature $(d-1, 1)$, so that $\eta^{\mu\nu} = \mathrm{diag}(-1, +1, \ldots, +1)$. The extension to signature $(d, 0)$ (and to more than one timelike dimension) will be indicated later. Throughout this appendix the dimensionality of spacetime is denoted by d; we generally reserve D to designate the total spacetime dimensionality of a string theory.

We begin with an even dimension $d = 2k + 2$. Group the Γ^μ into $k + 1$ sets of anticommuting raising and lowering operators,

$$\Gamma^{0\pm} = \frac{1}{2}(\pm\Gamma^0 + \Gamma^1) , \tag{B.1.2a}$$

$$\Gamma^{a\pm} = \frac{1}{2}(\Gamma^{2a} \pm i\Gamma^{2a+1}) , \quad a = 1, \ldots, k . \tag{B.1.2b}$$

These satisfy

$$\{\Gamma^{a+}, \Gamma^{b-}\} = \delta^{ab} , \tag{B.1.3a}$$

$$\{\Gamma^{a+}, \Gamma^{b+}\} = \{\Gamma^{a-}, \Gamma^{b-}\} = 0 . \tag{B.1.3b}$$

In particular, $(\Gamma^{a+})^2 = (\Gamma^{a-})^2 = 0$. It follows that by acting repeatedly with the Γ^{a-} we can reach a spinor annihilated by all the Γ^{a-},

$$\Gamma^{a-}\zeta = 0 \quad \text{for all } a \,. \tag{B.1.4}$$

Starting from ζ one obtains a representation of dimension 2^{k+1} by acting in all possible ways with the Γ^{a+}, at most once each. We will label these by with $\mathbf{s} \equiv (s_0, s_1, \ldots, s_k)$, where each of the s_a is $\pm\frac{1}{2}$:

$$\zeta^{(\mathbf{s})} \equiv (\Gamma^{k+})^{s_k+1/2} \ldots (\Gamma^{0+})^{s_0+1/2}\zeta \,. \tag{B.1.5}$$

In particular, the original ζ corresponds to all $s_a = -\frac{1}{2}$.

Taking the $\zeta^{(\mathbf{s})}$ as a basis, the matrix elements of Γ^μ can be derived from the definitions and the anticommutation relations. Increasing d by two doubles the size of the Dirac matrices, so we can give an iterative expression starting in $d = 2$, where

$$\Gamma^0 = \begin{bmatrix} 0 & 1 \\ -1 & 0 \end{bmatrix} , \quad \Gamma^1 = \begin{bmatrix} 0 & 1 \\ 1 & 0 \end{bmatrix} . \tag{B.1.6}$$

Then in $d = 2k + 2$,

$$\Gamma^\mu = \gamma^\mu \otimes \begin{bmatrix} -1 & 0 \\ 0 & 1 \end{bmatrix} , \quad \mu = 0, \ldots, d-3 \,, \tag{B.1.7a}$$

$$\Gamma^{d-2} = I \otimes \begin{bmatrix} 0 & 1 \\ 1 & 0 \end{bmatrix} , \quad \Gamma^{d-1} = I \otimes \begin{bmatrix} 0 & -i \\ i & 0 \end{bmatrix} , \tag{B.1.7b}$$

with γ^μ the $2^k \times 2^k$ Dirac matrices in $d - 2$ dimensions and I the $2^k \times 2^k$ identity. The 2×2 matrices act on the index s_k, which is added in going from $2k$ to $2k + 2$ dimensions.

The notation $\mathbf{s}$ reflects the Lorentz properties of the spinors. The Lorentz generators

$$\Sigma^{\mu\nu} = -\frac{i}{4}[\Gamma^\mu, \Gamma^\nu] \tag{B.1.8}$$

satisfy the $SO(d-1, 1)$ algebra

$$i[\Sigma^{\mu\nu}, \Sigma^{\sigma\rho}] = \eta^{\nu\sigma}\Sigma^{\mu\rho} + \eta^{\mu\rho}\Sigma^{\nu\sigma} - \eta^{\nu\rho}\Sigma^{\mu\sigma} - \eta^{\mu\sigma}\Sigma^{\nu\rho} \,. \tag{B.1.9}$$

The generators $\Sigma^{2a,2a+1}$ commute and can be simultaneously diagonalized. In terms of the raising and lowering operators,

$$S_a \equiv i^{\delta_{a,0}}\Sigma^{2a,2a+1} = \Gamma^{a+}\Gamma^{a-} - \frac{1}{2} \tag{B.1.10}$$

so $\zeta^{(\mathbf{s})}$ is a simultaneous eigenstate of the S_a with eigenvalues s_a. The half-integer values show that this is a spinor representation. The spinors form the 2^{k+1}-dimensional *Dirac* representation of the Lorentz algebra $SO(2k + 1, 1)$.

The Dirac representation is reducible as a representation of the Lorentz algebra. Because $\Sigma^{\mu\nu}$ is quadratic in the Γ matrices, the $\zeta^{(\mathbf{s})}$ with even and odd numbers of $+\frac{1}{2}$s do not mix. Define

$$\Gamma = i^{-k}\Gamma^0\Gamma^1 \ldots \Gamma^{d-1} \, , \tag{B.1.11}$$

which has the properties

$$(\Gamma)^2 = 1 \, , \quad \{\Gamma, \Gamma^\mu\} = 0 \, , \quad [\Gamma, \Sigma^{\mu\nu}] = 0 \, . \tag{B.1.12}$$

The eigenvalues of Γ are ± 1. The conventional notation for Γ in $d = 4$ is Γ_5, but this is inconvenient in general d. Noting that

$$\Gamma = 2^{k+1}S_0 S_1 \ldots S_k \, , \tag{B.1.13}$$

we see that $\Gamma_{ss'}$ is diagonal, taking the value $+1$ when the s_a include an even number of $-\frac{1}{2}$s and -1 for an odd number of $-\frac{1}{2}$s. The 2^k states with Γ eigenvalue (*chirality*) $+1$ form a *Weyl* representation of the Lorentz algebra, and the 2^k states with eigenvalue -1 form a second, inequivalent, Weyl representation. For $d = 4$, the Dirac representation is the familiar four-dimensional one, which separates into 2 two-dimensional Weyl representations,

$$\mathbf{4}_{\text{Dirac}} = \mathbf{2} + \mathbf{2}' \, . \tag{B.1.14}$$

Here we have used a common notation, labeling a representation by its dimension (in boldface). In $d = 10$ the representations are

$$\mathbf{32}_{\text{Dirac}} = \mathbf{16} + \mathbf{16}' \, . \tag{B.1.15}$$

For an odd dimension $d = 2k + 3$, simply add $\Gamma^d = \Gamma$ or $\Gamma^d = -\Gamma$ to the Γ matrices for $d = 2k + 2$. This is now an irreducible representation of the Lorentz algebra, because $\Sigma^{\mu d}$ anticommutes with Γ. Thus there is a single spinor representation of $SO(2k + 2, 1)$, which has dimension 2^{k+1}.

Majorana spinors

The above construction of the irreducible representation of the Γ matrices shows that in even dimensions $d = 2k + 2$ it is unique up to a change of basis. The matrices $\Gamma^{\mu*}$ and $-\Gamma^{\mu*}$ satisfy the same Clifford algebra as Γ^μ, and so must be related to Γ^μ by a similarity transformation. In the basis **s**, the matrix elements of $\Gamma^{a\pm}$ are real, so it follows from the definition (B.1.2) that $\Gamma^3, \Gamma^5, \ldots, \Gamma^{d-1}$ are imaginary and the remaining Γ^μ real. This is also consistent with the explicit expression (B.1.7). Defining

$$B_1 = \Gamma^3\Gamma^5 \ldots \Gamma^{d-1} \, , \quad B_2 = \Gamma B_1 \, , \tag{B.1.16}$$

one finds by anticommutation that

$$B_1\Gamma^\mu B_1^{-1} = (-1)^k\Gamma^{\mu*} \, , \quad B_2\Gamma^\mu B_2^{-1} = (-1)^{k+1}\Gamma^{\mu*} \, . \tag{B.1.17}$$

For either B_1 or B_2 (and only for these two matrices),

$$B\Sigma^{\mu\nu}B^{-1} = -\Sigma^{\mu\nu*} \, . \tag{B.1.18}$$

It follows from eq. (B.1.18) that the spinors ζ and $B^{-1}\zeta^*$ transform in the same way under the Lorentz group, so the Dirac representation is its own conjugate. Acting on the chirality matrix Γ, one finds

$$B_1 \Gamma B_1^{-1} = B_2 \Gamma B_2^{-1} = (-1)^k \Gamma^* , \tag{B.1.19}$$

so that either form for B will change the eigenvalue of Γ when k is odd and not when it is even. For k even ($d = 2$ mod 4) each Weyl representation is its own conjugate. For k odd ($d = 0$ mod 4) each Weyl representation is conjugate to the other. Thus in $d = 4$ we can designate the representations as $\mathbf{2}$ and $\bar{\mathbf{2}}$ rather than $\mathbf{2}$ and $\mathbf{2}'$, but in $d = 10$, only as $\mathbf{16}$ and $\mathbf{16}'$

Just as the gravitational and gauge fields are real, various spinor fields satisfy a *Majorana* condition, which relates ζ^* to ζ. This condition must be consistent with Lorentz transformations and so must have the form

$$\zeta^* = B\zeta \tag{B.1.20}$$

with B satisfying (B.1.18). Taking the conjugate gives $\zeta = B^*\zeta^* = B^*B\zeta$, so such a condition is consistent if and only if $B^*B = 1$. Using the reality and anticommutation properties of the Γ-matrices one finds

$$B_1^*B_1 = (-1)^{k(k+1)/2} , \quad B_2^*B_2 = (-1)^{k(k-1)/2} . \tag{B.1.21}$$

A Majorana condition using B_1 is therefore possible only if $k = 0$ or 3 (mod 4), and using B_2 only if $k = 0$ or 1 (mod 4). If $k = 0$ both conditions are possible but they are physically equivalent, being related by a similarity transformation.

A Majorana condition can be imposed on a Weyl spinor only if $B^*B = 1$ *and* the Weyl representation is conjugate to itself. For k odd, which is $d = 0$ or 4 (mod 8), it is therefore not possible to impose both the Majorana and Weyl conditions on a spinor: one can impose one or the other. Precisely for $k = 0$ mod 4, which is $d = 2$ (mod 8), a spinor can simultaneously satisfy the Majorana and Weyl conditions. Majorana–Weyl spinors in $d = 10$ play a key role in the spacetime theory of the superstring, and Majorana–Weyl spinors in $d = 2$ (ψ^μ and $\tilde{\psi}^\mu$) play a key role on the world-sheet.

Extending to odd dimensions, $\Gamma^d = \pm\Gamma$, and so the conjugation (B.1.19) of Γ^d is compatible with the conjugation (B.1.17) of the other Γ^μ only for B_1, so that $k = 0$ or 3 (mod 4). In all, a Majorana condition is possible if $d = 0, 1, 2, 3,$ or 4 (mod 8). When the Majorana condition is allowed, there is a basis in which B is either 1 or Γ and so commutes with all the $\Sigma^{\mu\nu}$. In this basis the $\Sigma^{\mu\nu}$ are imaginary.

All these results are summarized in the table B.1. The number of real parameters in the smallest representation is indicated in each case. This is twice the dimension of the Dirac representation, reduced by a factor of 2 for a Weyl condition and 2 for a Majorana condition. The derivation

Table B.1. *Dimensions in which various conditions are allowed for SO(d − 1,1) spinors. A dash indicates that the condition cannot be imposed. For the Weyl representation, it is indicated whether these are conjugate to themselves or to each other (complex). The final column lists the smallest representation in each dimension, counting the number of real components. Except for the final column the properties depend only on d mod 8.*

d	Majorana	Weyl	Majorana–Weyl	min. rep.
2	yes	self	yes	1
3	yes	-	-	2
4	yes	complex	-	4
5	-	-	-	8
6	-	self	-	8
7	-	-	-	16
8	yes	complex	-	16
9	yes	-	-	16
10=2+8	yes	self	yes	16
11=3+8	yes	-	-	32
12=4+8	yes	complex	-	64

implies that the properties are periodic in d with period 8, except the dimension of the representation which increases by a factor of 16.

For d a multiple of 4, a spinor may have the Majorana or Weyl property but not both: conjugation changes one Weyl representation into the other. In fact, the two cases are physically identical, there being a one-to-one mapping between them. Define the chirality projection operators

$$P_\pm = \frac{1 \pm \Gamma}{2} .$$ (B.1.22)

Given a Majorana spinor ζ or a Weyl spinor χ, the maps

$$\zeta \to P_+\zeta , \quad \chi \to \chi + B\chi^* $$ (B.1.23)

give a spinor of the other type, and these maps are inverse to one another.

The matrices $-\Gamma^{\mu T}$ also satisfy the Clifford algebra. The *charge conjugation* matrix has the property

$$C\Gamma^\mu C^{-1} = -\Gamma^{\mu T} .$$ (B.1.24)

Using the hermiticity property

$$\Gamma^{\mu\dagger} = \Gamma_\mu = -\Gamma^0\Gamma^\mu(\Gamma^0)^{-1} ,$$ (B.1.25)

this implies that

$$C\Gamma^0\Gamma^\mu(C\Gamma^0)^{-1} = \Gamma^{\mu*} .$$ (B.1.26)

Then for even d,

$$C = B_1\Gamma^0 \, , \, d = 2 \bmod 4 \; ; \quad C = B_2\Gamma^0 \, , \, d = 4 \bmod 4 \, . \qquad \text{(B.1.27)}$$

For odd $d = 2k + 3$, again only $C = B_1\Gamma^0$ acts uniformly on Γ^μ for all μ; with this definition $C\Gamma^\mu C^{-1} = (-1)^{k+1}\Gamma^{\mu T}$. In all cases,

$$C\Sigma^{\mu\nu}C^{-1} = -\Sigma^{\mu\nu T} \, . \qquad \text{(B.1.28)}$$

Additional properties of the matrices B and C are developed in exercise B.1.

Product representations

We now wish to develop the decomposition of a product of spinor representations. A product of spinors ζ and χ will have integer spins and so can be decomposed into tensor representations. Recall the standard spinor invariant

$$\bar\zeta\chi = \zeta^\dagger\Gamma^0\chi \, . \qquad \text{(B.1.29)}$$

Similarly

$$\bar\zeta\Gamma^{\mu_1}\Gamma^{\mu_2}\ldots\Gamma^{\mu_m}\chi \qquad \text{(B.1.30)}$$

transforms as the indicated tensor. However, this involves conjugation of the spinor ζ. From the properties of C it follows that $\zeta^T C$ transforms in the same way as $\bar\zeta$, so for the product of spinors without conjugation

$$\zeta^T C\Gamma^{\mu_1}\Gamma^{\mu_2}\ldots\Gamma^{\mu_m}\chi \qquad \text{(B.1.31)}$$

transforms as a tensor.

Starting now with the case of $d = 2k + 3$ odd, we claim that

$$\zeta^T C\Gamma^{\mu_1\mu_2\cdots\mu_m}\chi \qquad \text{(B.1.32)}$$

for $m \le k + 1$ comprise a complete set of independent tensors. Here

$$\Gamma^{\mu_1\mu_2\cdots\mu_m} = \Gamma^{[\mu_1}\Gamma^{\mu_2}\ldots\Gamma^{\mu_m]} \qquad \text{(B.1.33)}$$

is the completely antisymmetrized product. Without the antisymmetry these would not be independent, as the anticommutation relation would allow a pair of Γ matrices to be removed. The restriction $m \le k + 1$ comes about as follows. The definition of Γ implies in even dimensions that

$$\Gamma^{\mu_1\cdots\mu_s}\Gamma = -\frac{i^{-k+s(s+1)}}{(d-s)!}\epsilon^{\mu_1\cdots\mu_d}\Gamma_{\mu_{s+1}\cdots\mu_d} \, . \qquad \text{(B.1.34)}$$

In odd dimensions, where $\Gamma^d = \pm\Gamma$, it follows that the antisymmetrized products (B.1.33) for m and $d-m$ are linearly related. There are no further restrictions, and the dimensions agree: $2^{k+1} \cdot 2^{k+1}$ in the product of spinors and 2^{2k+2} from the binomial expansion. Thus

$$2^{k+1} \times 2^{k+1} = [0] + [1] + \ldots + [k+1] \,, \qquad (B.1.35)$$

where $[m]$ denotes the antisymmetric m-tensor.

For even $d = 2k + 2$, the products of m and $d - m$ Γ matrices are independent, and the same construction leads to

$$
\begin{aligned}
2^{k+1}_{\mathrm{Dirac}} \times 2^{k+1}_{\mathrm{Dirac}} &= [0] + [1] + \ldots + [2k+2] \\
&= [0]^2 + [1]^2 + \ldots + [k]^2 + [k+1] \,. \qquad (B.1.36)
\end{aligned}
$$

In the second line we have used the equivalence $[m] = [d - m]$ from contraction with the ϵ-tensor. Again the dimensionality is correct.

To find the products of the separate Weyl representations, use

$$\zeta^T C \Gamma^{\mu_1 \mu_2 \cdots \mu_m} \Gamma \chi = (-1)^{k+m+1} (\Gamma \zeta)^T C \Gamma^{\mu_1 \mu_2 \cdots \mu_m} \chi \,, \qquad (B.1.37)$$

as follows from the definition of C. The tensor (B.1.32) is then nonvanishing if $k + m$ is odd and the chiralities of ζ and χ are the same, or if $k + m$ is even and the chiralities are opposite. This allows us to separate the product (B.1.36):

$$
2^k \times 2^k = \begin{cases} [1] + [3] + \ldots + [k+1]_+ \,, & k \text{ even} \,, \\ [0] + [2] + \ldots + [k+1]_+ \,, & k \text{ odd} \,, \end{cases} \qquad (B.1.38a)
$$

$$
2^{k\prime} \times 2^{k\prime} = \begin{cases} [1] + [3] + \ldots + [k+1]_- \,, & k \text{ even} \,, \\ [0] + [2] + \ldots + [k+1]_- \,, & k \text{ odd} \,, \end{cases} \qquad (B.1.38b)
$$

$$
2^k \times 2^{k\prime} = \begin{cases} [0] + [2] + \ldots + [k] \,, & k \text{ even} \,, \\ [1] + [3] + \ldots + [k] \,, & k \text{ odd} \,. \end{cases} \qquad (B.1.38c)
$$

The relation (B.1.34) implies that the tensors of rank $k + 1 = d/2$ satisfy a self-duality condition with a sign that depends on the chirality of the spinor. A self-dual tensor representation can only be real for k even.

Some of the facts that we have deduced can also be verified quickly by considering the eigenvalues s_a. Consider the reality properties of the Weyl spinors. Conjugation flips the rotation eigenvalues $s_1, \ldots, s_k$ but not the boost eigenvalue s_0. For k even, this is an even number of flips and gives a state of the same chirality; for k odd it reverses the chirality. This is consistent with the third column of table B.1. For the tensor products of Weyl representations, note that the even-rank tensors $[2n]$ (e.g. the invariant $[0]$) always contain a component with eigenvalues $s_a = (0, 0, \ldots, 0)$, while the odd-rank tensors do not. This would be obtained, for example, from the product of spinor components $s_a = (\frac{1}{2}, \frac{1}{2}, \ldots, \frac{1}{2})$ and $s_a = (-\frac{1}{2}, -\frac{1}{2}, \ldots, -\frac{1}{2})$. For k even these have opposite chirality, as in

Table B.2. *Dimensions in which various conditions are allowed for $SO(N)$ spinors.*

N mod 8	real	Weyl	real and Weyl
0	yes	self	yes
1	yes	-	-
2	yes	complex	-
3	pseudo	-	-
4	pseudo	self	-
5	pseudo	-	-
6	yes	complex	-
7	yes	-	-

the product (B.1.38c). For k odd they have the same chirality, as in the products (B.1.38a) and (B.1.38b).

Spinors of $SO(N)$

For $SO(N)$ the analysis is quite parallel. For $N = 2l$, there is a 2^l-dimensional representation of the Γ-matrices which reduces to two 2^{l-1}-dimensional spinor representations of $SO(2l)$, while for $SO(2l + 1)$ there is a single representation of dimension 2^l. The reality properties can be analyzed as in the Minkowski case. Essentially one ignores $\mu = 0, 1$, so $SO(N)$ is analogous to $SO(N + 1, 1)$, with the results shown in table B.2. Here *real* means the algebra can be written in terms of purely imaginary matrices. The term *pseudoreal* is often used for $N - 3, 4, 5$ mod 8, where the representation is conjugate to itself but cannot be written in terms of imaginary matrices.

The familiar case of a pseudoreal representation is the **2** of $SO(3)$. This is conjugate to itself because it is the only two-dimensional representation, but it must act on a complex doublet. It should be noted, however, that two wrongs make a right — the product of two pseudoreal representations is real. Let the indices on u_{ij} both be $SU(2)$ doublets, either of the same or different $SU(2)$s. Then the reality condition

$$u_{ij}^* = \epsilon_{ii'}\epsilon_{jj'}u_{i'j'} \tag{B.1.39}$$

is invariant. With just a single index, the analogous condition $u_i^* = \epsilon_{ii'}u_{i'}$ would force u to vanish. Incidentally, one can impose a Majorana condition on the **2** of $SO(2, 1)$, consistent with table B.1. A real basis for the Γ-matrices is

$$\Gamma^0 = i\sigma^2, \quad \Gamma^1 = \sigma^1, \quad \Gamma^2 = \sigma^3. \tag{B.1.40}$$

Product representations are obtained as in the Minkowski case, with the result in $N = 2l$

$$2^{l-1} \times 2^{l-1} = \begin{cases} [0] + [2] + \ldots + [l]_+ \,, & l \text{ even}, \\ [1] + [3] + \ldots + [l]_+ \,, & l \text{ odd}, \end{cases} \tag{B.1.41a}$$

$$2^{l-1'} \times 2^{l-1'} = \begin{cases} [0] + [2] + \ldots + [l]_- \,, & l \text{ even}, \\ [1] + [3] + \ldots + [l]_- \,, & l \text{ odd}, \end{cases} \tag{B.1.41b}$$

$$2^{l-1} \times 2^{l-1'} = \begin{cases} [1] + [3] + \ldots + [l-1] \,, & l \text{ even}, \\ [0] + [2] + \ldots + [l-1] \,, & l \text{ odd}. \end{cases} \tag{B.1.41c}$$

For more than one timelike dimension, the analog of table B.1 or B.2 depends on the *difference* of the number of spacelike and timelike dimensions.

Decomposition under subgroups

We frequently consider subgroups such as

$$SO(9,1) \to SO(3,1) \times SO(6) \,. \tag{B.1.42}$$

We can directly match representations by comparing the eigenvalues of S_a. In particular, for the case in which all the dimensions are even,

$$SO(2k+1,1) \to SO(2l+1,1) \times SO(2k-2l) \,, \tag{B.1.43}$$

the Weyl spinors decompose

$$2^k \to (2^l, 2^{k-l-1}) + (2^{l'}, 2^{k-l-1'}) \,, \tag{B.1.44a}$$
$$2^{k'} \to (2^{l'}, 2^{k-l-1}) + (2^l, 2^{k-l-1'}) \,. \tag{B.1.44b}$$

Another subgroup that has particular relevance for the superstring is

$$SO(2n) \to SU(n) \times U(1) \,. \tag{B.1.45}$$

To describe this subgroup, consider again the complex linear combinations (B.1.2) of Γ-matrices, where $a = 1, \ldots, n$. A general $SO(2n)$ rotation will mix the Γ^{a+} both among themselves and with the Γ^{a-}. The subgroup that mixes the Γ^{a+} only among themselves is $U(n) = SU(n) \times U(1)$. Now let us consider how the spinor representation decomposes. Again we start with the spinor ζ annihilated by all the Γ^{a-}. This condition $\Gamma^{a-}\zeta = 0$ is invariant under $U(n)$ rotations so that ζ rotates at most by a phase. Thus

$$\zeta \in \mathbf{1}_{-n} \,, \tag{B.1.46}$$

where the $U(n)$ charge, indicated by the subscript, has been normalized to $2\sum S_a$. Acting with a raising operator adds an $SU(n)$ index and increases the $U(1)$ charge by 2, giving

$$2^n \to [0]_{-n} + [1]_{2-n} + [2]_{4-n} + \ldots + [n]_n \,, \tag{B.1.47}$$

where $[k]$ refers to the k-times antisymmetrized $\mathbf{n}$ of $SU(n)$. The completely antisymmetrized $[n]$ is the same as $[0] = \mathbf{1}$, while $[n-1] = \overline{[1]} = \bar{\mathbf{n}}$, and so on. Decomposing further into the Weyl representations, the last term $[0]_n$ is in the $\mathbf{2^{n-1}}$, and the successive terms alternate. Thus in particular for

$$SO(6) \rightarrow SU(3) \times U(1) \,, \tag{B.1.48}$$

we have

$$\mathbf{4} \rightarrow \mathbf{1}_3 + \mathbf{3}_{-1} \,, \tag{B.1.49a}$$
$$\bar{\mathbf{4}} \rightarrow \mathbf{1}_{-3} + \bar{\mathbf{3}}_1 \,. \tag{B.1.49b}$$

A relation that arises often is

$$SO(4) = SU(2) \times SU(2) \,. \tag{B.1.50}$$

To see this, combine the four components of a vector into a 2×2 matrix

$$x = x^4 I + ix^i \sigma^i \,, \ i = 1, 2, 3 \,; \quad \det x = \sum_{m=1}^{4} (x^m)^2 \,. \tag{B.1.51}$$

The length of x is invariant under independent left- and right-hand $SU(2)$ rotations

$$x' = g_1 x g_2^{-1} \,, \tag{B.1.52}$$

giving the decomposition (B.1.50). Then

$$\mathbf{4} = (\mathbf{2}, \mathbf{2}) \,, \tag{B.1.53a}$$
$$\mathbf{2} = (\mathbf{2}, \mathbf{1}) \,, \tag{B.1.53b}$$
$$\mathbf{2'} = (\mathbf{1}, \mathbf{2}) \,. \tag{B.1.53c}$$

The decomposition of the vector is just eq. (B.1.52), while those of the spinors can be derived in various ways.

B.2 Introduction to supersymmetry: $d = 4$

The familiar conserved quantities, such as energy-momentum, angular momentum, and charge, transform as vectors, tensors, and scalars under the Lorentz group. It is also possible for a conserved quantity to transform as a spinor. Such a *supersymmetry (SUSY)* will relate the properties of fermions to those of bosons. Supersymmetry is a feature of all consistent string theories. Further, as discussed in section 16.2, there is good reason to expect that it will be found with particle accelerators.

In this appendix we summarize the various results that will be needed in the text. We are interested in the algebras, their representations, the transformations of the fields, and the invariant actions. The reader should be able to follow the derivation of the various representations (massless, standard massive, and BPS massive). However, the transformations and actions require detailed calculation, and so for these we simply cite for reference some of the key results.

$d = 4$, $N = 1$ *supersymmetry*

According to table B.1, the smallest spinor in four dimensions has four real degrees of freedom. As shown in eq. (B.1.23) this can be described either as a Weyl spinor, with two complex components, or as a Majorana spinor, with four components satisfying a reality condition.

The smallest $d = 4$ supersymmetry algebra would have one Weyl or Majorana spinor of supercharges. Again these are identical, the same four linearly independent supercharges described in two different notations; we will use the Majorana description. A more general supersymmetry algebra in $d = 4$ would have $4N$ supercharges. For $N > 1$ this is known as *extended supersymmetry*. In any number of dimensions the ratio of the number of supercharges to the smallest spinor representation is denoted by N. However, the structure of the theory depends more on the actual number of supercharges than on the ratio N, so subsequent sections are organized according to this total number. For pedagogic purposes we find it convenient in this section to start with the smallest algebra and build up, but later we will start with the largest algebra and work downwards, from 32 to 16 to 8. The number of supercharges need not be a power of 2, but in the great majority of examples it is and so these are the cases on which we focus.

The $N = 1$ supersymmetry algebra is uniquely determined to be

$$\{Q_\alpha, \bar{Q}_\beta\} = -2P_\mu \Gamma^\mu_{\alpha\beta} \,, \tag{B.2.1a}$$

$$[P^\mu, Q_\alpha] = 0 \,, \tag{B.2.1b}$$

where P_μ is the spacetime momentum. The minus sign is due to our metric signature $(-+\ldots+)$. Recall that from the Majorana property, $\bar{Q} \equiv Q^\dagger \Gamma^0 = Q^T C$.

It is easy to work out the representations of this algebra. The massless and massive representations differ, and we consider the former first. For massless states choose a frame in which $k_1 = k_0$. The supersymmetry algebra becomes

$$\{Q_\alpha, Q^\dagger_\beta\} = 2k^0(1 + \Gamma^0\Gamma^1)_{\alpha\beta} = 2k^0(1 + 2S_0)_{\alpha\beta} \,. \tag{B.2.2}$$

In the s-basis, the Majorana condition becomes $Q^\dagger_{s_0 s_1} = Q_{s_0, -s_1}$ and the anticommutator becomes

$$\{Q_{s'_0 s'_1}, Q^\dagger_{s_0 s_1}\} = 4k^0 \delta_{s_0, 1/2} \delta_{ss'} \,. \tag{B.2.3}$$

The matrix elements of $Q_{-1/2,s_1}$ must vanish in these momentum eigenstates because

$$0 = \langle\psi|\{Q_{-1/2,s_1}, Q^\dagger_{-1/2,s_1}\}|\psi\rangle$$
$$= \|Q_{-1/2,s_1}|\psi\rangle\|^2 + \|Q^\dagger_{-1/2,s_1}|\psi\rangle\|^2 . \qquad \text{(B.2.4)}$$

The remaining supercharges form a fermionic oscillator algebra. Defining

$$b = (4k^0)^{-1/2}Q_{1/2,-1/2} , \quad b^\dagger = (4k^0)^{-1/2}Q_{1/2,1/2} , \qquad \text{(B.2.5)}$$

the supersymmetry algebra becomes

$$\{b, b^\dagger\} = 1 , \quad b^2 = b^{\dagger 2} = 0 . \qquad \text{(B.2.6)}$$

Starting from a state $|\lambda\rangle$ such that

$$S_1|\lambda\rangle = \lambda|\lambda\rangle , \quad b|\lambda\rangle = 0 , \qquad \text{(B.2.7)}$$

the algebra generates exactly one additional state

$$b^\dagger|\lambda\rangle = |\lambda + \tfrac{1}{2}\rangle , \quad S_1|\lambda + \tfrac{1}{2}\rangle = (\lambda + \tfrac{1}{2})|\lambda + \tfrac{1}{2}\rangle . \qquad \text{(B.2.8)}$$

The massless irreducible multiplets thus each consist of two states with helicities differing by $\tfrac{1}{2}$: one state in each multiplet is a fermion and one a boson. These are also representations of Poincaré symmetry. However, CPT, which appears to be an exact symmetry of string theory as it is of field theory, requires that each multiplet be accompanied by its conjugate with opposite helicities and quantum numbers. Thus we have the following $(\lambda, \lambda + \tfrac{1}{2})$ multiplets:

- The *chiral multiplet* consists of a $(0, \tfrac{1}{2})$ multiplet and its CPT conjugate $(-\tfrac{1}{2}, 0)$, corresponding to a Weyl fermion and a complex scalar.

- The *vector multiplet* $(\tfrac{1}{2}, 1)$ plus $(-1, -\tfrac{1}{2})$ contains a gauge boson and a Weyl fermion, both necessarily in the adjoint of the gauge group.

- The *gravitino multiplet* $(1, \tfrac{3}{2})$ plus $(-\tfrac{3}{2}, -1)$ contains an additional spin-$\tfrac{3}{2}$ gravitino and so is not relevant since there is only one supersymmetry and so only the gravitino in the graviton multiplet. This multiplet would be relevant if we had a larger supersymmetry and decomposed it into $N = 1$ representations.

- The *graviton multiplet* $(\tfrac{3}{2}, 2)$ plus $(-2, -\tfrac{3}{2})$ contains the graviton and gravitino.

- Massless particles with helicities greater than 2 are believed to be impossible to couple to gravity, and have not arisen in string theory.

In an $N = 1$ supersymmetric extension of the Standard Model, the Higgs boson and spin-$\frac{1}{2}$ fermions are in chiral multiplets. The Standard Model fermions cannot be in vector multiplets because the latter must be in the adjoint representation.

For massive representations, the anticommutator in the rest frame is

$$\{Q_{s_0's_1'}, Q^\dagger_{s_0s_1}\} = 2m\delta_{ss'} . \tag{B.2.9}$$

This is now *two* copies of the fermionic oscillator algebra,

$$b_1 = (2m)^{-1/2}Q_{1/2,-1/2} , \quad b_2 = (2m)^{-1/2}Q_{-1/2,-1/2} , \tag{B.2.10a}$$

$$\{b_i, b_j^\dagger\} = \delta_{ij} , \quad \{b_i, b_j\} = \{b_i^\dagger, b_j^\dagger\} = 0 . \tag{B.2.10b}$$

Starting again from a state

$$S_1|\lambda\rangle = \lambda|\lambda\rangle , \quad b_i|\lambda\rangle = 0 , \tag{B.2.11}$$

the algebra generates the additional three states

$$b_1^\dagger|\lambda\rangle, \ b_2^\dagger|\lambda\rangle, \ b_1^\dagger b_2^\dagger|\lambda\rangle , \quad S_1 = \lambda + \tfrac{1}{2}, \ \lambda + \tfrac{1}{2}, \ \lambda + 1 . \tag{B.2.12}$$

For example, the massive chiral multiplet is $\lambda = -\frac{1}{2}, 0, 0, \frac{1}{2}$, the same as the *CPT*-extended massless multiplet. The multiplet $\lambda = 0, \frac{1}{2}, \frac{1}{2}, 1$ is incomplete, even without *CPT*, because massive states must be a representation of the rotation group $SU(2)$. Adding in $\lambda = -1, -\frac{1}{2}, -\frac{1}{2}, 0$, we obtain a spin-1, two spin-$\frac{1}{2}$, and one spin-0 particle. These are the same states as a massless vector plus chiral multiplet, and can be obtained from them via the Higgs mechanism.

Actions with $d = 4$, $N = 1$ SUSY

From section 16.4 on, we need some results about $d = 4$, $N = 1$ super-symmetry transformations and invariant actions. We collect these here, without derivation. A general renormalizable theory will contain a number of massless chiral and vector multiplets; the larger massive multiplets can always be decomposed into these. The particle content of the massless chiral multiplet corresponds to a complex scalar field ϕ and a Majorana (or Weyl) spinor ψ. That of a massless vector multiplet corresponds to a gauge field A_μ and a Majorana (or Weyl) spinor λ. In each case it is useful, though not essential, to add a complex auxiliary field, a complex field F in the chiral multiplet and a real field D in the vector multiplet. We then have the following *superfields*

$$\Phi^i: \ \phi^i , \ \psi^i , \ F^i , \tag{B.2.13a}$$

$$V^a: \ A_\mu^a , \ \lambda^a , \ D^a . \tag{B.2.13b}$$

These have the supersymmetry transformations

$$\delta\phi^i/2^{1/2} = i\bar{\zeta}P_+\psi^i = i\bar{\psi}^iP_+\zeta , \tag{B.2.14a}$$

$$\delta(P_+\psi^i)/2^{1/2} = P_+\zeta F^i + \Gamma^\mu P_-\zeta D_\mu\phi^i , \tag{B.2.14b}$$

$$\delta F^i/2^{1/2} = -i\bar{\zeta}\Gamma^\mu D_\mu P_+\psi^i , \tag{B.2.14c}$$

and

$$\delta A_\mu^a = -i\bar{\zeta}\Gamma_\mu\lambda^a , \tag{B.2.15a}$$

$$\delta\lambda^a = \frac{1}{2}\Gamma^{\mu\nu}\zeta F_{\mu\nu}^a + i\Gamma\zeta D^a , \tag{B.2.15b}$$

$$\delta D^a = -\bar{\zeta}\Gamma\Gamma^\mu D_\mu\lambda^a , \tag{B.2.15c}$$

in terms of a Majorana SUSY parameter ζ.

The most general renormalizable action is determined by the gauge couplings g_a (which of course must be equal within each simple group) and the *superpotential* $W(\Phi)$, which is a *holomorphic* function of the superfields. Also, for each $U(1)$ gauge group there is an additional parameter ξ_a, the *Fayet–Iliopoulos term*. The Lagrangian density is

$$\mathscr{L} = \mathscr{L}_1 + \mathscr{L}_2 , \tag{B.2.16}$$

where

$$\mathscr{L}_1 = -D_\mu\phi^{i*}D^\mu\phi^i - \frac{i}{2}\bar{\psi}^i\Gamma^\mu D_\mu\psi^i - \frac{1}{4g_a^2}F_{\mu\nu}^a F^{a\mu\nu} - \frac{i}{2g_a^2}\bar{\lambda}^a\Gamma^\mu D_\mu\lambda^a$$

$$- \frac{1}{2}\Big[iW_{,ij}(\phi)\bar{\psi}^iP_+\psi^j + 2^{1/2}\phi^{i*}t_{ij}^a\bar{\lambda}^aP_+\psi^j\Big] + \text{c.c.} , \tag{B.2.17}$$

and

$$\mathscr{L}_2 = F^{i*}F^i + \frac{1}{2g_a^2}D^{a2} + W_{,i}(\phi)F^i + \text{c.c.} + \frac{1}{2}D^a(2\xi_a + \phi^{i*}t_{ij}^a\phi^j) . \tag{B.2.18}$$

In $\mathscr{L}_1$ are the kinetic terms, fermion masses and Yukawa couplings, while in $\mathscr{L}_2$ are all terms involving the auxiliary fields. The t_{ij}^a are the gauge group representation matrices. Renormalizability requires the superpotential W to be at most cubic in the fields. Carrying out the Gaussian path integration over the auxiliary fields gives a scalar potential

$$-\mathscr{L}_2' = V = |F^i(\phi)|^2 + \frac{1}{2g_a^2}[D^a(\phi,\phi^*)]^2 , \tag{B.2.19}$$

where

$$F^i(\phi) = -W_{,i}(\phi)^* , \tag{B.2.20a}$$

$$D^a(\phi,\phi^*) = -\frac{g_a^2}{2}(2\xi_a + \phi^{i*}t_{ij}^a\phi^j) . \tag{B.2.20b}$$

The two terms in the potential are known respectively as the F-term, from the superpotential, and the D-term, from the gauge interaction.

An important nonrenormalization theorem states that the tree-level superpotential does not receive perturbative corrections. It is also important that this is only a perturbative statement, and that there can be nonperturbative corrections to the superpotential. An example of this arises in chapter 18.

Two kinds of internal symmetry are possible in supersymmetry. The first is a unitary rotation U_{ij} acting uniformly on all fields ϕ^i, $P_+\psi^i$ and F^i in a given chiral multiplet. This is a symmetry if W is invariant. The gauge fields A_μ^a couple to such a symmetry. The second, known as an R *symmetry*, acts differently on different components:

$$\phi^i \to \exp(iq_i\alpha)\phi^i , \quad P_+\psi^i \to \exp[i(q_i-1)\alpha]P_+\psi^i , \quad \text{(B.2.21a)}$$

$$F^i \to \exp[i(q_i-2)\alpha]F^i , \quad \text{(B.2.21b)}$$

$$A_\mu^a \to A_\mu^a , \quad P_+\lambda^a \to \exp(i\alpha)P_+\lambda^a , \quad D^a \to D^a . \quad \text{(B.2.21c)}$$

Examining the action, for example the Yukawa terms, one sees that this is a symmetry provided the superpotential transforms as

$$W(\phi) \to \exp(2i\alpha)W(\phi) . \quad \text{(B.2.22)}$$

In addition, the R symmetry must commute with the gauge symmetry.

Spontaneous supersymmetry breaking

As with an ordinary internal symmetry, spontaneous breaking of supersymmetry is signified by certain nonvanishing vacuum expectation values. In particular, consider

$$\langle 0|\{Q_\alpha, \chi_\beta\}|0\rangle , \quad \text{(B.2.23)}$$

where Q_α is some component of the supercharge. We can assume the operator χ_β to be fermionic; otherwise, the expectation value vanishes automatically by Lorentz invariance. If supersymmetry is unbroken, $Q_\alpha|0\rangle = \langle 0|Q_\alpha = 0$ and all such vacuum expectation values vanish. Classically the condition for unbroken supersymmetry becomes

$$\delta\psi^i = \delta\lambda^a = 0 . \quad \text{(B.2.24)}$$

From the variations (B.2.14) and (B.2.15), it follows that a configuration is supersymmetric if the fields are position-independent, the gauge field is zero, and

$$F^i(\phi) = D^a(\phi) = 0 . \quad \text{(B.2.25)}$$

Moreover, we see from the potential (B.2.19) that if such a configuration exists it will be a minimum of the energy. Supersymmetry will be spontaneously broken if there are no solutions to eqs. (B.2.25). The simplest example of a system with broken supersymmetry is a single superfield with superpotential

$$W = f\phi_1 \,, \tag{B.2.26}$$

with f a nonzero constant; then $F = -f^* \neq 0$. This is rather trivial as it stands, but by coupling ϕ_1 appropriately to other fields, for example

$$W = f\phi_1 + m\phi_2\phi_3 + g\phi_1\phi_2^2 \,, \tag{B.2.27}$$

one obtains a theory with a nonsupersymmetric spectrum.

Higher corrections and supergravity

In the usual power counting in four dimensions, the scalar field and vector potential have dimension l^{-1} and the spinors dimension $l^{-3/2}$, l being length. These are determined by the kinetic terms. It follows from the transformations (B.2.14) and (B.2.15) that the supersymmetry parameter ζ has dimension $l^{1/2}$, consistent with the product of two supersymmetry transformations being a translation. Also, the auxiliary fields F^i and D^a have dimension l^{-2}. Including the l^4 from d^4x, the renormalizable action retains all terms that are relevant at long distance, that is, all terms of dimension l^n with $n \geq 0$.

Power counting in renormalization theory is based on the scaling of the quantum fluctuations of the fields. However, in string theory we have encountered the phenomenon of moduli, scalar fields with flat potentials. These can have large *classical* values. In order to write an effective Lagrangian valid in all of moduli space,[1] we need a different power counting that assigns scalars scaling l^0. Supersymmetry then assigns their fermionic partners scaling $l^{-1/2}$. We wish to keep all terms of the same order as the kinetic terms for these fields, and therefore all terms in the Lagrangian density having dimension l^m with $m \geq -2$. In order to keep the kinetic terms for the gauge multiplet, assign A_μ scaling l^0 and λ scaling $l^{-1/2}$. Finally, we assign the metric scaling l^0, since it has a classical expectation value. Incidentally, this 'moduli space' power counting is the same in all dimensions, whereas the renormalization power counting is dimension-dependent.

---

[1] To be precise, the effective Lagrangian will still break down at particular points in moduli space, namely those points where extra massless fields occur. In the neighborhood of such a point, one needs an effective Lagrangian which includes these additional fields.

In this approximation, the low energy effective action includes all the earlier terms plus additional ones. It depends now on three functions:

- The superpotential $W(\Phi)$, which is still holomorphic but need no longer be cubic.

- An arbitrary holomorphic function $f_{ab}(\Phi)$ replacing the gauge coupling g_a^{-2}.

- The *Kähler* potential $K(\Phi, \Phi^*)$, which is a general function of the superfields.

Again there is a Fayet–Iliopoulos parameter ξ_a for each $U(1)$. The full Lagrangian density is quite lengthy, so we give only the purely bosonic terms,

$$\frac{\mathcal{L}_{\text{bos}}}{(-G)^{1/2}} = \frac{1}{2\kappa^2} R - K_{,\bar{\imath}j} D_\mu \phi^{i*} D^\mu \phi^j - \frac{1}{4} \text{Re}(f_{ab}(\phi)) F^a_{\mu\nu} F^{b\mu\nu}$$

$$- \frac{1}{8} \text{Im}(f_{ab}(\phi)) \epsilon^{\mu\nu\sigma\rho} F^a_{\mu\nu} F^b_{\sigma\rho} - V(\phi, \phi^*) \,. \qquad \text{(B.2.28)}$$

The potential is

$$V(\phi, \phi^*) = \exp(\kappa^2 K)(K^{\bar{\imath}j} W^*_{;i} W_{;j} - 3\kappa^2 W^* W) + \frac{1}{2} f_{ab} D^a D^b \,. \qquad \text{(B.2.29)}$$

Here $K^{\bar{\imath}j}$ is the inverse matrix to $\partial_j \partial_{\bar{k}} K$ and

$$W_{;i} = \partial_i W + \kappa^2 \partial_i K W \qquad \text{(B.2.30a)}$$

$$\text{Re}(f_{ab}(\phi)) D^b = -2\xi_a - K_{,i} t^a_{ij} \phi^j \,. \qquad \text{(B.2.30b)}$$

The negative term proportional to κ^2 is a supergravity effect. The other terms generalize the earlier potential (B.2.19).

The kinetic term for the scalars is now field-dependent. The second derivative

$$K_{,\bar{\imath}j} = \frac{\partial^2 K(\phi, \phi^*)}{\partial \phi^{i*} \partial \phi^j} \qquad \text{(B.2.31)}$$

plays the role of a metric for the space of scalar fields, generalizing the flat metric $\delta_{\bar{\imath}j}$ of the renormalizable theory. The flat metric is the special case $K = \phi^{i*} \phi^i$. A metric of the form (B.2.31) is known as a *Kähler metric*. In a similar way, the function $f_{ab}(\phi)$ gives rise to a field-dependent (*nonminimal*) kinetic term for the gauge fields, as well as a field-dependent $F_2 \wedge F_2$ coupling. The metric (B.2.31) is invariant under *Kähler transformations*,

$$K(\phi, \phi^*) \rightarrow K(\phi, \phi^*) + f(\phi) + f(\phi)^* \,. \qquad \text{(B.2.32)}$$

This is an invariance of the whole action provided also that the superpotential transforms as

$$W(\phi) \to \exp[-\kappa^2 f(\phi)] W(\phi) . \tag{B.2.33}$$

This is important because in interesting examples the field space has a nontrivial topology and the Kähler potential is not globally defined.

The supersymmetry transformations of the fermions are

$$\delta P_+ \psi^i / 2^{1/2} = -K^{i\bar{j}} W^*_{;\bar{j}} P_+ \zeta + \Gamma^\mu P_- \zeta D_\mu \phi^i , \tag{B.2.34a}$$

$$\delta \lambda^a = \frac{1}{2} \Gamma^{\mu\nu} \zeta F^a_{\mu\nu} + i \Gamma \zeta D^a , \tag{B.2.34b}$$

$$\delta \psi_\mu = D_\mu \zeta + \frac{1}{2} \Gamma_\mu \zeta \exp(\kappa^2 K / 2) W . \tag{B.2.34c}$$

Here ψ_μ is the gravitino. The covariant derivative of the spinor ζ includes the spin connection. The variations (B.2.34) all vanish if the metric is flat, the gauge field zero, the scalars and ζ constant, and $\partial_i W = D^a = W = 0$.

Extended supersymmetry in $d = 4$

With several supersymmetries Q^A_α for $A = 1, \ldots, N$, the straightforward generalization of the earlier algebra is

$$\{ Q^A_\alpha, \bar{Q}^B_\beta \} = -2\delta^{AB} P_\mu \Gamma^\mu_{\alpha\beta} , \qquad [P^\mu, Q^A_\alpha] = 0 . \tag{B.2.35}$$

This is not the most general algebra, but we analyze it first. For massless particles, the earlier fermionic oscillator is replaced by N oscillators b_A. These generate 2^N states in a binomial distribution from helicity λ to helicity $\lambda + \frac{1}{2}N$. For example, for $N = 2$ the following massless multiplets are important:

hypermultiplet: $(-\frac{1}{2}, 0^2, \frac{1}{2}) + (-\frac{1}{2}, 0^2, \frac{1}{2})$,

vector multiplet: $(-1, -\frac{1}{2}^2, 0) + (0, \frac{1}{2}^2, 1)$,

supergravity multiplet: $(-2, -\frac{3}{2}^2, -1) + (1, \frac{3}{2}^2, 2)$.

In each case there are two SUSY multiplets, related by CPT. For states that are their own CPT conjugates, a half-hypermultiplet is allowed.

Let us note an important feature of these multiplets. If we just look at the SUSY multiplets, not making use of CPT, then all states in the multiplet have the same gauge quantum numbers because the supersymmetry charges commute with the gauge symmetries.[2] It follows that an $N = 2$ theory cannot have chiral gauge interactions. The only SUSY multiplets

---

[2] There is an exception to this known as gauged supergravity, but it is not relevant to the gauge interactions of the Standard Model.

with spin-$\frac{1}{2}$ states are the half-hypermultiplet and the vector multiplet. The former contains states of helicities $\pm\frac{1}{2}$ with the same gauge quantum numbers and so is nonchiral. The latter is necessarily in the (real) adjoint representation and so is also nonchiral.

For $N = 4$ the multiplets are larger:

$$\text{vector multiplet:} \qquad (-1, -\tfrac{1}{2}^4, 0^6, \tfrac{1}{2}^4, 1) \; ;$$
$$\text{supergravity multiplet:} \quad (-2, -\tfrac{3}{2}^4, -1^6, -\tfrac{1}{2}^4, 0) + (0, \tfrac{1}{2}^4, 1^6, \tfrac{3}{2}^4, 2) \; .$$

Finally, for $N = 8$ there is only a single possible representation:

$$\text{supergravity multiplet:} \quad (-2, -\tfrac{3}{2}^8, -1^{28}, -\tfrac{1}{2}^{56}, 0^{70}, \tfrac{1}{2}^{56}, 1^{28}, \tfrac{3}{2}^8, 2) \; .$$

Larger algebras would require helicities greater than 2, which is believed to be impossible (there are some uninteresting exceptions, such as free field theories). String theory has several times turned up loopholes in such statements, but not yet here.

Massive representations of extended supersymmetry similarly contain 2^{2N} states generated by b_{1A} and b_{2A}.

The most general extended supersymmetry algebra allowed by Lorentz invariance is

$$\{Q_\alpha^A, \bar{Q}_\beta^B\} = -2\delta^{AB} P_\mu \Gamma_{\alpha\beta}^\mu - 2i Z^{AB} \delta_{\alpha\beta} \; , \tag{B.2.36a}$$

$$[P^\mu, Q_\alpha^A] = [Z^{AB}, Q_\alpha^C] = [Z^{AB}, P_\mu] = [Z^{AB}, Z^{CD}] = 0 \; . \tag{B.2.36b}$$

Here Z^{AB} is some set of conserved charges. It must be antisymmetric in AB due to the Majorana property and the antisymmetry of the charge conjugation matrix C.

To be precise, this is the most general algebra if we include only charges that can be carried by point particles. Including charges that can be carried by *extended objects,* additional terms appear. Rather than explain this here, we introduce it in its natural physical context: first in section 11.6, and then in more variety in chapter 13. The same caveat applies to the higher-dimensional algebras to be introduced later in this appendix.

To see the effect of the additional term consider a particle in its rest frame, for which the algebra becomes

$$\{Q_\alpha^A, Q_\beta^{B\dagger}\} = 2m\delta^{AB}\delta_{\alpha\beta} + 2i Z^{AB} \Gamma_{\alpha\beta}^0 \; . \tag{B.2.37}$$

Taking an eigenstate of the charges Z^{AB}, we can go to a basis in which

$$
Z^{AB} = \begin{bmatrix}
0 & q_1 & 0 & 0 \\
-q_1 & 0 & 0 & 0 & \cdots \\
0 & 0 & 0 & q_2 \\
0 & 0 & -q_2 & 0 \\
\vdots & & & & \ddots
\end{bmatrix},
\tag{B.2.38}
$$

with $q_i \geq 0$. The left-hand side of the algebra (B.2.37) is nonnegative as a matrix in $(A\alpha, B\beta)$. The eigenvalues $2(m \pm q_i)$ on the right-hand side must therefore also be nonnegative, implying the *Bogomolnyi–Prasad–Sommerfield (BPS) bound*

$$
m \geq q_i .
\tag{B.2.39}
$$

Thus the mass is bounded below by the charges, and in particular massless states must be neutral. If m is strictly greater than all the q_i, the massive representations are unaffected and contain 2^{2N} states. If the largest k q_is are equal to one another and to m, the algebra requires $2k$ pairs of fermionic oscillators to annihilate the states, just as half the oscillators do for a massless representation. This gives a *short* or *BPS* representation with $2^{2(N-k)}$ states. If all the q_i are equal to one another and to m, the result is an *ultrashort* representation of dimension 2^N (for N even), the same as the massless representation.

B.3 Supersymmetry in $d = 2$

In this section we briefly make the connection with the world-sheet algebras of string theory. The smallest spinor representation in two dimensions is Majorana–Weyl and has one Hermitean component. The general $(N, \tilde{N})$ algebra would have N Hermitean left-moving supercharges Q_L^A and $\tilde{N}$ Hermitean right-moving supercharges Q_R^A. The algebra is

$$
\{Q_L^A, Q_L^B\} = \delta^{AB}(P^0 - P^1), \quad \{Q_R^A, Q_R^B\} = \delta^{AB}(P^0 + P^1),
\tag{B.3.1a}
$$
$$
\{Q_L^A, Q_R^B\} = Z^{AB},
\tag{B.3.1b}
$$

where now Z^{AB} need have no special symmetry. The superconformal generators G_0 and $\tilde{G}_0$ satisfy this algebra. Thus the R sector of the $(N, \tilde{N})$ superconformal theory contains the $(N, \tilde{N})$ supersymmetry algebra. In fact, this was one of several independent routes by which supersymmetry was first discovered.

The dimensional reduction of the $d = 4$, $N = 1$ supersymmetry algebra gives the $d = 2$ (2,2) algebra.

B.4   Differential forms and generalized gauge fields

Various antisymmetric tensor fields appear in supergravity and string theory. Differential forms are a convenient notation to minimize the bookkeeping of indices and combinatoric factors. A *p-form* A is simply a completely antisymmetric p-index tensor $A_{\mu_1 \ldots \mu_p}$ with the indices omitted. Because we encounter many different forms, we will denote the rank of any form by an italicized subscript, A_p. The product of a p-form A_p and a q-form B_q is written $A_p \wedge B_q$ or simply $A_p B_q$, and is defined

$$(A_p \wedge B_q)_{\mu_1 \ldots \mu_{p+q}} = \frac{(p+q)!}{p! q!} A_{[\mu_1 \ldots \mu_p} B_{\mu_{p+1} \ldots \mu_{p+q}]} . \qquad (B.4.1)$$

Again, [ ] denotes antisymmetrization, averaging over permutations with a ± 1 for odd permutations. The wedge product of a p-form A and q-form B has the property

$$A_p \wedge B_q = (-1)^{pq} B_q \wedge A_p . \qquad (B.4.2)$$

The exterior derivative d takes a p-form into a $(p+1)$-form:

$$(dA_p)_{\mu_1 \ldots \mu_{p+1}} = (p+1)\partial_{[\mu_1} A_{\mu_2 \ldots \mu_{p+1}]} . \qquad (B.4.3)$$

It has the important property $d^2 = 0$.

The integral of a d-form is coordinate-invariant,

$$\int d^d x \, A_{01 \ldots d-1} \equiv \int A_d , \qquad (B.4.4)$$

the transformation of the tensor offsetting that of the measure. Because of the antisymmetry, one must specify an orientation. Similarly, a p-form can be integrated over any p-dimensional submanifold. For a manifold with boundary one has Stokes's theorem,

$$\int_{\mathcal{M}} dA_{p-1} = \int_{\partial \mathcal{M}} A_{p-1} \qquad (B.4.5)$$

where $\mathcal{M}$ is p-dimensional.

None of the above constructions requires a metric. In particular d contains only the ordinary derivative, but it is invariant due to the antisymmetry. One construction that does require a metric is the Poincaré dual, or, more properly, Hodge star. It is defined as

$$*A_{\mu_1 \ldots \mu_{d-p}} = \frac{1}{p!} \epsilon_{\mu_1 \ldots \mu_{d-p}}{}^{\nu_1 \ldots \nu_p} A_{\nu_1 \ldots \nu_p} . \qquad (B.4.6)$$

The Levi–Civita symbol $\epsilon_{\mu_1 \ldots \mu_d}$ is defined to transform as a tensor. Thus with all lower indices its components are $\pm(-G)^{1/2}$ and 0, while with all upper indices its components are $\pm(-G)^{-1/2}$ and 0. One can check that on a p-form,

$$** = (-1)^{p(d-p)+1} \,, \tag{B.4.7}$$

the $+1$ coming from the Minkowski signature.

One can also represent the above by introducing an algebra of d anticommuting *differentials* dx^μ, writing

$$A_p = \frac{1}{p!} A_{\mu_1 \ldots \mu_p} dx^{\mu_1} \ldots dx^{\mu_p} \,. \tag{B.4.8}$$

The factorial just offsets the sum over permutations so that each independent component appears once. The product of a p-form A and a q-form B is then (B.4.1), and the exterior derivative is $d = dx^\nu \partial_\nu$.

In this notation, an Abelian field strength, vector potential, and gauge transformation are written

$$F_2 = dA_1 \,, \quad \delta A_1 = d\lambda \,. \tag{B.4.9}$$

In the non-Abelian case, writing the fields as matrices, these become

$$F_2 = dA_1 - iA_1 \wedge A_1 \equiv dA_1 - iA_1^2 \,, \quad \delta A_1 = d\lambda - iA_1 \lambda + i\lambda A_1 \,. \tag{B.4.10}$$

In the Abelian case there is a straightforward generalization to a p-form gauge transformation

$$F_{p+2} = dA_{p+1} \,, \quad \delta A_{p+1} = d\lambda_p \,. \tag{B.4.11}$$

The action is

$$-\frac{1}{2} \int d^d x \, (-G)^{1/2} |F_{p+2}|^2 = -\frac{1}{2} \int d^d x \, \frac{(-G)^{1/2}}{(p+2)!} F_{\mu_1 \ldots \mu_{p+2}} F^{\mu_1 \ldots \mu_{p+2}} \,. \tag{B.4.12}$$

A given component, say $A_{1\ldots p+1}$, then appears with the canonical normalization for a real scalar, $-\frac{1}{2} \partial_\mu A_{1\ldots p+1} \partial^\mu A_{1\ldots p+1}$. There is no straightforward non-Abelian generalization. For $p = -1$, the gauge invariance is trivial and this describes a massless scalar.

Using the gauge invariance (B.4.11), we can set $n^\mu A_{\mu \nu_1 \ldots \nu_p} = 0$. The field equation then also implies $k^\mu A_{\mu \nu_1 \ldots \nu_p} = 0$ and $k^2 = 0$. The potential A_{p+1} thus gives rise to a massless particle in the representation $[p+1]$ of the spin $SO(d-2)$.

Since $[p+1] = [d-p-3]$ for $SO(d-2)$, a $(p+1)$-form potential and a $(d-p-3)$-form potential describe the same particle states. For $d = 4$ and $p = 1$, this is the familiar fact that $B_{\mu\nu}$ describes the axion. We can also show this at the level of the fields. The Bianchi identity from $F_{p+2} = dA_{p+1}$ and the equation of motion from the action (B.4.12) are

$$dF_{p+2} = 0 \,, \quad d*F_{p+2} = 0 \,. \tag{B.4.13}$$

There is an obvious symmetry here: defining

$$F'_{d-p-2} = *F_{p+2} \tag{B.4.14}$$

simply switches the field equation and Bianchi identity, and in particular one can solve the new Bianchi identity in terms of a new potential A'_{d-p-3}, where $dA'_{d-p-3} = F'_{d-p-2}$. These theories are therefore equivalent, and one need consider only potentials of rank up to $\frac{1}{2}d - 1$. Again note that it is the field strength, not the potential, that is dualized. One can also see the equivalence in the action:

$$-\frac{1}{2}\int d^d x \, (-G)^{1/2}|dA_{p+1}|^2 \;\rightarrow\; -\frac{1}{2}\int d^d x \, (-G)^{1/2}|F_{p+2}|^2$$
$$+ \int A'_{d-p-3} \wedge dF_{p+2}$$
$$\rightarrow\; -\frac{1}{2}\int d^d x \, (-G)^{1/2}|dA'_{d-p-3}|^2 \;. \quad \text{(B.4.15)}$$

In the first action the potential A_{p+1} is the variable of integration. In the second, F_{p+2} is the variable of integration; the Bianchi identity is no longer automatic so a Lagrange multiplier A'_{d-p-3} has been introduced to enforce it. In the final form the original F_{p+2} has been integrated out, leaving a gauge action for A'_{d-p-3}. In $d = 4$, this is electric–magnetic duality of Maxwell's equations. In $d = 3$, it implies that a vector potential is equivalent to a massless scalar. In $d = 2$ a massless scalar is equivalent to a dual scalar; in fact, this is equivalent to the world-sheet T-duality $X \rightarrow X'$. Again, note that it is the field strength to which the Poincaré duality is applied, not the potential.

For $d = 2$ mod 4, where $*^2 = 1$ on $(d/2)$-forms, it is consistent with the field equation and Bianchi identity to impose one of

$$F_{d/2} = \pm *F_{d/2} \;. \quad \text{(B.4.16)}$$

These are consistent theories with half as many components. In $d = 2$ they correspond to the left- or right-moving parts of a massless scalar. The action (B.4.12) no longer gives the field equation, as

$$|F_{d/2}|^2 = \pm F_{d/2} \wedge F_{d/2} = 0 \quad \text{(B.4.17)}$$

vanishes. There are more complicated actions which are not manifestly covariant.

B.5   Thirty-two supersymmetries

We now begin a survey of some of the supersymmetric theories that arise as low energy limits in string theory. A more complete treatment can be found in the references.

$d = 11$ *supergravity*

In four dimensions the largest supersymmetry algebra, $N = 8$, contains 32 supercharges. This same limit holds in higher dimensions, since we could reduce to four by compactifying on tori. Table B.1 then implies that $d = 11$ is the maximum in which supersymmetry can exist,[3] since the spinor representations are too large for $d \geq 12$. Although this exceeds by one the critical dimension of superstring theory, we will start with this case.

The Majorana spinor supercharge again satisfies the algebra

$$\{Q_\alpha, \overline{Q}_\beta\} = -2P_\mu \Gamma^\mu_{\alpha\beta} \ . \tag{B.5.1}$$

The massless irreducible representation contains $2^8 = 256$ states, half fermions and half bosons. By calculating the spins $S_1, \ldots, S_4$ one finds that the graviton multiplet contains two bosonic representations of $SO(9)$: a traceless symmetric tensor (the graviton) with $\frac{1}{2} \times 9 \times 10 - 1 = 44$ components and a completely antisymmetric three-index tensor with $9 \times 8 \times 7/3! = 84$ components for 128 in all. There is a single fermionic vector-spinor representation. The spinor index takes 16 values and the vector 9 values; 16 components vanish by a trace condition as in eq. (10.5.19), leaving $16 \times 9 - 16 = 128$ fermionic components.

With two or fewer derivatives there is a unique supersymmetric action, whose bosonic part is

$$S_{11} = \frac{1}{2\kappa^2} \int d^{11}x \, (-G)^{1/2} \left(R - \frac{1}{2}|F_4|^2 \right) - \frac{1}{12\kappa^2} \int A_3 \wedge F_4 \wedge F_4 \tag{B.5.2}$$

with A_3 a 3-form potential and F_4 its 4-form field strength. The final *Chern–Simons* term is gauge-invariant in spite of the explicit appearance of A_3 because the term from the variation $\delta A_3 = d\lambda_2$ vanishes by parts.

$d = 10$ *IIA supergravity*

By compactifying the $d = 11$ theory on a torus and keeping only the massless fields (*dimensional reduction*), we obtain a $d = 10$ theory with 32 supercharges. The $d = 11$ Majorana spinor becomes a $d = 10$ Majorana spinor, which reduces to one Majorana–Weyl spinor of each chirality,

$$Q^1_\alpha \in \mathbf{16} \, , \quad Q^2_\alpha \in \mathbf{16}' \, . \tag{B.5.3}$$

The product of two spinors of the same chirality contains a vector, while the product of spinors of opposite chirality contains a scalar (eq. (B.1.38)),

---

[3] With *two* timelike dimensions a Majorana–Weyl spinor with 32 components is allowed at $d = 12$, but we will not try to figure out what this might mean.

so from the eleven-dimensional algebra we deduce

$$\{Q_\alpha^1, \overline{Q}_\beta^1\} = -2P_\mu(P_+\Gamma^\mu)_{\alpha\beta} \ , \quad \{Q_\alpha^2, \overline{Q}_\beta^2\} = -2P_\mu(P_-\Gamma^\mu)_{\alpha\beta} \ , \quad \text{(B.5.4a)}$$

$$\{Q_\alpha^1, \overline{Q}_\beta^2\} = -2P_{10}(P_+\Gamma)_{\alpha\beta} \ . \quad\quad\quad\quad\quad\quad\quad\quad \text{(B.5.4b)}$$

Here $\Gamma = \Gamma^{10}$ is from the toroidal dimension. A notable feature is the appearance of a central charge proportional to the Kaluza–Klein momentum. This is one of the ways that a central charge in the supersymmetry algebra can arise; additional central charges carried by extended objects are introduced in their physical context in section 13.2.

The dimensional reduction of the $d = 11$ theory leaves a scalar from $G_{10\,10}$, a Kaluza–Klein vector from $G_{\mu 10}$, a 2-form potential from $B_{\mu\nu 10}$ and a 3-form from $B_{\mu\nu\sigma}$. This is the same as the massless content of the IIA superstring, the scalar dilaton and the 2-form being from the NS–NS sector and the 1- and 3-forms from the R–R sector. This is no surprise because the large amount of supersymmetry determines the massless particle content completely. What is a surprise is that there really is an eleventh dimension hidden in the IIA string, invisible in perturbation theory but visible at strong coupling. This is discussed in chapter 14.

The action can be obtained by dimensional reduction; further details are given in section 12.1.

$d = 10$ *IIB supergravity*

There is another ten-dimensional supergravity, which is not obtained by compactifying an eleven-dimensional theory. This has two supercharges of the same chirality, which we can define to be **16**. The algebra is

$$\{Q_\alpha^A, \overline{Q}_\beta^B\} = -2\delta^{AB}P_\mu(P_+\Gamma^\mu)_{\alpha\beta} \ . \quad\quad \text{(B.5.5)}$$

The graviton multiplet contains two scalars, the traceless symmetric graviton, two antisymmetric 2-forms, and a 4-form with self-dual field strength, for

$$2 + 35 + 28 + 28 + 35 = 128 \quad\quad \text{(B.5.6)}$$

bosonic states in all. This is the same as the massless content of the IIB superstring. More details are given in chapters 10 and 12.

$d < 10$ *supergravity*

The supergravities with 32 supercharges in $d < 10$ can be obtained by dimensional reduction of the IIA string, or equivalently of $d = 11$ supergravity. In this section we discuss some of the main features; this subject is relevant in particular to section 14.2. We need not consider the

Table B.3. *Supergravities with 32 supercharges. The group G is a symmetry of the low energy supergravity theory, and the moduli space is locally G/H.*

d	scalars	vectors	G	H
10A	1	1	$SO(1,1,\mathbf{R})$	-
10B	2	0	$SL(2,\mathbf{R})$	$SO(2,\mathbf{R})$
9	3	3	$SL(2,\mathbf{R}) \times SO(1,1,\mathbf{R})$	$SO(2,\mathbf{R})$
8	7	6	$SL(2,\mathbf{R}) \times SL(3,\mathbf{R})$	$SO(2,\mathbf{R}) \times SO(3,\mathbf{R})$
7	14	10	$SL(5,\mathbf{R})$	$SO(5,\mathbf{R})$
6	25	16	$SO(5,5,\mathbf{R})$	$SO(5,\mathbf{R}) \times SO(5,\mathbf{R})$
5	42	27	$E_{6(6)}(\mathbf{R})$	$USp(8)$
4	70	28	$E_{7(7)}(\mathbf{R})$	$SU(8)$
3	128	-	$E_{8(8)}(\mathbf{R})$	$SO(16,\mathbf{R})$

IIB string separately below $d = 10$, because after compactification on a circle it is T-dual to the IIA string (chapter 13).

The first issue we wish to consider is the number of scalars. Compactifying k of the dimensions of $d = 11$ supergravity, there are

$$\frac{1}{2}k(k+1) \tag{B.5.7}$$

scalars from G_{mn} and

$$\frac{1}{3!}k(k-1)(k-2) \tag{B.5.8}$$

from B_{mnp}. Again m, n, p are compactified and μ, ν are noncompact. In addition, in $d = 5$, the Poincaré dual $*(H_{\mu\nu\rho\sigma})$ gives the field strength (gradient) for an extra scalar, as discussed at the end of section B.4. In $d = 4$, $*(H_{\mu\nu\rho m})$ gives 7 extra scalars. In $d = 3$, $*(H_{\mu\nu mn})$ gives $\frac{1}{2}8 \times 7 = 28$ extra scalars. Also in $d = 3$ the duals of the 8 Kaluza–Klein vectors give additional scalars. The total number is indicated in table B.3.

The second issue is the number of vectors: k from $G_{\mu n}$ and $\frac{1}{2}k(k-1)$ from $B_{\mu mn}$. In addition there is one in $d = 6$ from $*(H_{\mu\nu\rho\sigma})$ and six in $d = 5$ from $*(H_{\mu\nu\rho m})$. In $d = 4$, $*(H_{\mu\nu mn})$ is just the magnetic description of the $B_{\mu mn}$ vectors, and there are no vectors in $d = 3$ because we have converted them all to scalars by Poincaré duality. The results are summarized in the second column of the table. The gauge group is $U(1)^{n_V}$.

Third, there is no potential — the scalars are moduli — and the moduli space metric is completely determined by symmetry. The moduli spaces are cosets G/H, as listed in the table. The structure is the same as in the toroidal example in section 8.4, a coset of a noncompact group by a compact group. In the string case there was a further identification by the discrete T-duality group. This discrete identification does not affect the local structure of moduli space, and in particular not the effective action, and so it is not determined at this point by supersymmetry. Rather, it is determined by short-distance physics as is described in chapter 14. In the bosonic case the dilaton was decoupled, giving a separate space $SO(1,1,\mathbf{R}) = \mathbf{R}$. Below $d = 9$ in table B.3 it combines with other moduli into a larger homogeneous space.

In each case, the noncompact group in the numerator is a global symmetry of the supergravity theory, and the compact group in the denominator is the unbroken symmetry at any point in moduli space. The notation $E_{n(n)}(\mathbf{R})$ refers to an exceptional group with some sign changes in the algebra to make it noncompact, just as $SO(n,m,\mathbf{R})$ is related to $SO(n+m,\mathbf{R})$ and $SL(n,\mathbf{R})$ to $SU(n)$. The details of table B.3 are not at this point important, but it is interesting to see in chapter 14 how the structure fits into string theory.

For $d = 4$, the count in the table agrees with the $N = 8$ multiplet. To dimensionally reduce the supersymmetry algebra, separate the 11-dimensional 32-valued spinor index into a 4-valued $SO(3,1)$ index α and an 8-valued $SO(7)$ index A. The 11-dimensional algebra (B.5.1) becomes

$$\{Q_\alpha^A, \overline{Q}_\beta^B\} = -2P_\mu \delta^{AB}\Gamma_{\alpha\beta}^\mu - 2P_m \Gamma_{(7)}^{mAB}\Gamma_{(4)\alpha\beta} \ . \tag{B.5.9}$$

Here Γ^m factors into $\Gamma_{(7)}^{mAB}\Gamma_{(4)\alpha\beta}$ with $\Gamma_{(7)}^{mAB}$ being $SO(7)$ Γ matrices. The factor of

$$\Gamma_{(4)\alpha\beta} = i(\Gamma^0\Gamma^1\Gamma^2\Gamma^3)_{\alpha\beta} \tag{B.5.10}$$

must appear because Γ^m anticommutes with Γ^μ. Again, a central charge has arisen from the compact momenta. Since

$$(P_m\Gamma^m P_n\Gamma^n)^{AB} = \delta^{AB} P_m P^m \ , \tag{B.5.11}$$

the eigenvalues q_i of the central charge are all equal and any BPS multiplet will be ultrashort, with the same 256 states as a massless multiplet. In this case there is a simple explanation. The BPS condition is $-P_\mu P^\mu = P_m P^m$, so a BPS multiplet is actually a massless multiplet from the higher-dimensional point of view.

The $d = 4$, $N = 8$ theory has 28 gauge bosons, but only the 7 Kaluza–Klein charges appear in the dimensionally reduced algebra (B.5.9). In fact, the full algebra contains all 28 gauge charges, the remainder arising from the extended-object charges in higher dimensions. The antisymmetric matrix Z^{AB} has precisely 28 components, so in general all are nonzero. The gauge charges can be organized into an antisymmetric matrix P^{AB} so that the algebra (after a chirality rotation to remove $\Gamma_{(4)}$) is

$$\{Q_\alpha^A, \overline{Q}_\beta^B\} = -2P_\mu \delta^{AB}\Gamma_{\alpha\beta}^\mu - 2P^{AB}\delta_{\alpha\beta} \ . \tag{B.5.12}$$

It should be noted that the compact momenta depend on the moduli, for example $p = n/R$ for compactification on a single circle. When the central charge is written in terms of integer charges such as n, it has explicit dependence on the moduli.

B.6   Sixteen supersymmetries

$d = 10$, $N = 1$ *(type I) supergravity*

This algebra has a single Majorana–Weyl $\mathbf{16}$ supercharge. The massless vector representation has 16 states, $\mathbf{8}_v + \mathbf{8}'$ under the $SO(8)$ little group. The supergravity multiplet is $\mathbf{8}_v \times (\mathbf{8}_v + \mathbf{8}')$ as found in the bosonic and type I strings. The bosonic content is then a graviton, an antisymmetric tensor, and a dilaton from the supergravity multiplet plus $\dim g$ vectors from the gauge multiplets, g being the gauge group. The bosonic action is given in section 12.1. The action is classically invariant for any g, but as discussed in section 12.2 there are anomalies unless $g = SO(32)$ or $E_8 \times E_8$.

$d < 10$ *supergravity*

Toroidal compactification of k dimensions gives supergravity with 16 supersymmetries in $d = 10 - k$. There are a total of $k(k + r) + 1$ moduli, r being the rank of the ten-dimensional gauge group g. The metric gives rise to $\frac{1}{2}k(k+1)$ moduli, the antisymmetric tensor to $\frac{1}{2}k(k-1)$, the Wilson lines to kr, and the original ten-dimensional dilaton to the final one. Of course g is $SO(32)$ or $E_8 \times E_8$, both having $r = 16$, in a consistent ten-dimensional theory, but here we are just using this as a trick to generate theories in lower dimensions. The reduced theories are parity-symmetric and have no anomalies, and so can have any g. In fact, various $r < 16$ theories can be obtained in string theory by slightly more complicated compactifications. The moduli space is as given explicitly by the Narain compactification of the heterotic string,

$$\frac{SO(1, 1, \mathbf{R}) \times SO(k + r, k, \mathbf{R})}{SO(k + r, \mathbf{R}) \times SO(k, \mathbf{R})} , \tag{B.6.1}$$

the $SO(1, 1, \mathbf{R})$ being from the dilaton. In $d = 4$ the antisymmetric tensor gives another scalar, the axion, via Poincaré duality; this combines with the dilaton to form $SL(2, \mathbf{R})/SO(2, \mathbf{R})$. In $d = 3$ $(k = 7)$, the Poincaré duals of the $14 + r$ vectors combine with the dilaton and the other moduli to enlarge the moduli space (B.6.1) to

$$\frac{SO(8 + r, 8, \mathbf{R})}{SO(8 + r, \mathbf{R}) \times SO(8, \mathbf{R})} . \tag{B.6.2}$$

Toroidal compactification gives gauge group $U(1)^{2k+r}$ at generic points in moduli space, the original gauge group being broken to $U(1)^r$ by Wilson lines. At special points it will be enhanced to various non-Abelian groups; the rank remains $2k + r$.

$$d = 6, \; N = 2 \; supersymmetry$$

Under $SO(9,1) \rightarrow SO(5,1) \times SO(4)$, the ten-dimensional $N = 1$ supersymmetry decomposes

$$\mathbf{16} \rightarrow (\mathbf{4,2}) + (\mathbf{4',2'}) \,. \tag{B.6.3}$$

The $(\mathbf{4,2})$ has eight real components, forming a single complex $\mathbf{4}$. The $\mathbf{4}$ cannot have a Majorana condition imposed so the complex $\mathbf{4}$ is the smallest algebra in $d = 6$. The dimensionally reduced algebra is $d = 6$ $(1,1)$ supersymmetry, one supercharge in the $\mathbf{4}$ and one in the $\mathbf{4'}$. The only representation with spins ≤ 1 is the vector, which is the dimensional reduction of the $d = 10$ vector and so consists of one vector and four scalars.

Decomposing $SO(5,1)$ into the $SO(1,1)$ of the $(0,1)$-plane and the transverse $SO(4)$, the Weyl spinor supercharges decompose

$$\mathbf{4} \rightarrow (+\tfrac{1}{2}, \mathbf{2}) + (-\tfrac{1}{2}, \mathbf{2'}) = (+\tfrac{1}{2}, \tfrac{1}{2}, 0) + (-\tfrac{1}{2}, 0, \tfrac{1}{2}) \,, \tag{B.6.4a}$$

$$\mathbf{4'} \rightarrow (+\tfrac{1}{2}, \mathbf{2'}) + (-\tfrac{1}{2}, \mathbf{2}) = (+\tfrac{1}{2}, 0, \tfrac{1}{2}) + (-\tfrac{1}{2}, \tfrac{1}{2}, 0) \,. \tag{B.6.4b}$$

These are complex representations, so their adjoints are independent operators. The representations of $SO(1,1)$ are all one-dimensional and are labeled by the helicity S_0. In the second equality of each line we have used the relation $SO(4) = SU(2) \times SU(2)$ and labeled the $SU(2)$ representations by their spin j, so the notation is (s_0, j_1, j_2). As in section B.1, the $s_0 = -\tfrac{1}{2}$ generators annihilate the massless states. The latter then form a representation of the generators with $s_0 = \tfrac{1}{2}$, these being Q_α and Q'_β in the $(\tfrac{1}{2}, 0)$ and $(0, \tfrac{1}{2})$ of $SU(2) \times SU(2)$. Treating these as lowering operators and their adjoints as raising operators, by taking all combinations of raising operators one obtains the representations

$$r = (\tfrac{1}{2}, \tfrac{1}{2}) + (\tfrac{1}{2}, 0)^2 + (0, \tfrac{1}{2})^2 + (0, 0)^4 \,. \tag{B.6.5}$$

Starting from an $SU(2) \times SU(2)$ multiplet $|j_1, j_2\rangle$ annihilated by the lowering operators, the raising operators generate the representations

$$r \times |j_1, j_2\rangle \,. \tag{B.6.6}$$

The supergravity multiplet is built on $|\tfrac{1}{2}, \tfrac{1}{2}\rangle$, giving the states

$$|1,1\rangle + |1,0\rangle + |0,1\rangle + |0,0\rangle + |\tfrac{1}{2}, \tfrac{1}{2}\rangle^4$$
$$+ |1, \tfrac{1}{2}\rangle^2 + |\tfrac{1}{2}, 1\rangle^2 + |0, \tfrac{1}{2}\rangle^2 + |\tfrac{1}{2}, 0\rangle^2 \,. \tag{B.6.7}$$

The bosonic content, in the first line, is a graviton, an antisymmetric tensor, a scalar, and four vectors. The vector multiplet is built on $|0,0\rangle$, giving

$$|\tfrac{1}{2},\tfrac{1}{2}\rangle + |0,0\rangle^4 + |\tfrac{1}{2},0\rangle^2 + |0,\tfrac{1}{2}\rangle^2 . \tag{B.6.8}$$

There is a second $d = 6$ algebra with 16 supercharges, the $(2,0)$ algebra with two complex **4** supercharges. The raising operators now form the representations

$$r' = (1,0) + (\tfrac{1}{2},0)^4 + (0,0)^5 . \tag{B.6.9}$$

Acting on $|0,1\rangle$, these produce the supergravity multiplet

$$|1,1\rangle + |\tfrac{1}{2},1\rangle^4 + |0,1\rangle^5 , \tag{B.6.10}$$

whose bosonic content is a graviton and five anti-self-dual antisymmetric tensors. Acting on $|0,0\rangle$ they produce the tensor multiplet

$$|1,0\rangle + |\tfrac{1}{2},0\rangle^4 + |0,0\rangle^5 , \tag{B.6.11}$$

with self-dual antisymmetric tensor and five scalars.

$$d = 4,\ N = 4\ \textit{gauge theory}$$

The four-dimensional $N = 4$ algebra is

$$\{Q_\alpha^A, \bar{Q}_\beta^B\} = -2P_\mu \delta^{AB}\Gamma^\mu_{\alpha\beta} - 2P_{Rm}\Gamma^{mAB}\delta_{\alpha\beta} . \tag{B.6.12}$$

In this case only six of the gauge charges appear; in the heterotic string these are the ones coming from right-moving currents.

We now consider the effective renormalizable theory near a point of non-Abelian symmetry h. It will be useful to derive the full action and SUSY transformation by dimensional reduction of ten-dimensional supersymmetric Yang–Mills theory, whose Lagrangian density is

$$-\frac{1}{4g^2}\mathrm{Tr}(F_{MN}F^{MN}) - \frac{i}{2g^2}\mathrm{Tr}(\bar{\lambda}\Gamma^M D_M \lambda) . \tag{B.6.13}$$

The gauge field and the gaugino λ (a Majorana–Weyl **16**) are written in matrix notation, and M, N run from 0 to 9. The supersymmetry transformation is

$$\delta A_M = -i\bar{\zeta}\Gamma_M\lambda ,\qquad\qquad\text{(B.6.14a)}$$

$$\delta\lambda = \frac{1}{2}F_{MN}\Gamma^{MN}\zeta .\qquad\qquad\text{(B.6.14b)}$$

Reducing $M \to \mu, m$, the Lagrangian density becomes

$$-\frac{1}{4g^2}\text{Tr}\Big(F_{\mu\nu}F^{\mu\nu} + 2D_\mu A_m D^\mu A_m - [A_m, A_n]^2\Big)$$

$$-\frac{i}{2g^2}\text{Tr}(\bar{\lambda}\Gamma^\mu D_\mu\lambda + i\bar{\lambda}\Gamma_m[A_m, \lambda]) .\qquad\text{(B.6.15)}$$

The six compact components of the gauge field become the six scalars A_m of the $N = 4$ vector multiplet. The **16** index separates into $(\mathbf{2}, \mathbf{4}) + (\bar{\mathbf{2}}, \bar{\mathbf{4}})$ under $SO(3, 1) \times SO(6)$, so the ten-dimensional spinor becomes four Weyl spinors. Similarly the transformation laws reduce to

$$\delta A_\mu = -i\bar{\zeta}\Gamma_\mu\lambda\qquad\qquad\text{(B.6.16a)}$$

$$\delta A_m = -i\bar{\zeta}\Gamma_m\lambda\qquad\qquad\text{(B.6.16b)}$$

$$\delta\lambda = \Big(\frac{1}{2}F_{\mu\nu}\Gamma^{\mu\nu} + D_\mu\phi_n\Gamma^{\mu n} + \frac{i}{2}[A_m, A_n]\Gamma^{mn}\Big)\zeta .\qquad\text{(B.6.16c)}$$

The potential

$$V = -\frac{1}{4g^2}\text{Tr}\left([A_m, A_n]^2\right)\qquad\qquad\text{(B.6.17)}$$

is nonnegative and vanishes only if $[A_m, A_n] = 0$ for all m, n. Thus, in the flat directions the A_m can be taken simultaneously diagonal, and the moduli are just the $6\,\text{rank}(h)$ eigenvalues. At generic points the group is broken to $U(1)^{\text{rank}(h)}$. We have seen this potential before, in eq. (8.7.11) for the D-brane moduli. This is no accident, as the T-duality that produces the D-brane has the effect of dimensionally reducing the open string Yang–Mills action.

Eq. (B.6.15) is the most general renormalizable action consistent with $N = 4$ global supersymmetry. It remains the most general action if we adopt the looser moduli space power counting described in section B.2, which would have allowed field-dependent kinetic terms. In other words, $N = 4$ global supersymmetry requires the moduli space to be flat. This is no contradiction with the curved moduli space (B.6.1) found in supergravity. The only scale there is the Planck scale, so the dimensionless variable is κA and the nonlinearities vanish in the limit $\kappa \to 0$ where we ignore gravity.

The $N = 4$ Yang–Mills theory has a number of interesting properties, the first being that its beta function vanishes identically — the coupling does not run. Unlike most gauge theories, different values of g really give different theories, rather than being transmuted to a change of scale. It is easy to prove this statement, not just to all orders of perturbation theory but exactly. Consider for simplicity $h = SU(2)$. Generically, $A_m \propto \sigma^3$ breaks $SU(2)$ to $U(1)$ at a scale

$$v = (A_m A_m)^{1/2} \equiv A \,, \tag{B.6.18}$$

so the massless theory contains only an Abelian vector multiplet. Consider the gauge field kinetic term in the effective action. Its coefficient is $-1/4g^2$, but if the coupling runs in $SU(2)$ we must figure out at what scale to evaluate g. The answer is v, because this is where $SU(2)$ breaks and the coupling stops running. The scale v depends on the massless moduli, so what we really have is an effective Lagrangian density

$$-\frac{1}{4g^2(A)} F_{\mu\nu} F^{\mu\nu} \,. \tag{B.6.19}$$

However, this is a field-dependent kinetic term, which we have just stated is inconsistent with $N = 4$ supersymmetry — unless in fact the coupling is independent of scale as claimed. This argument is typical of the recent analysis of supersymmetric gauge theories, but is particularly simple because of the large amount of supersymmetry.

B.7 Eight supersymmetries

$d = 6$, $N = 1$ *supersymmetry*

We start in six dimensions, the maximum in which a spinor with eight components is allowed according to table B.1. We obtain the massless representations as in eq. (B.6.6), where now

$$r'' = (\tfrac{1}{2}, 0) + (0, 0)^2 \,. \tag{B.7.1}$$

The supergravity multiplet, built on $|\tfrac{1}{2}, 1\rangle$, is

$$|1, 1\rangle + |\tfrac{1}{2}, 1\rangle^2 + |0, 1\rangle \tag{B.7.2}$$

containing the graviton, gravitino (which requires two copies of $|\tfrac{1}{2}, 1\rangle$), and the $(0, 1)$ which is an anti-self-dual 2-form. The other relevant multiplets are built on $|0, 0\rangle$, $|0, \tfrac{1}{2}\rangle$, and $|\tfrac{1}{2}, 0\rangle$, giving

half-hypermultiplet:	$|\tfrac{1}{2}, 0\rangle$, $|0, 0\rangle^2$,
vector multiplet:	$|\tfrac{1}{2}, \tfrac{1}{2}\rangle$, $|0, \tfrac{1}{2}\rangle^2$,
tensor multiplet:	$|1, 0\rangle$, $|\tfrac{1}{2}, 0\rangle^2$, $|0, 0\rangle$.

The respective bosonic content is: two scalars; a vector; a self-dual tensor plus scalar.

A general theory will have some number of vector, hyper-, and tensor multiplets. We describe the general bosonic Lagrangian, first in the most restrictive form of keeping only terms that would be renormalizable when reduced to four dimensions. The Lagrangian consists of gauge-invariant kinetic terms for the various fields and a potential for the hypermultiplets. The tensors must be neutral under the gauge group, so in this limit the tensor representation is decoupled from all other fields. To write the potential, we collect two half-hypermultiplets into a complex doublet of scalars Φ_α^i, with α the doublet index and i labeling the hypermultiplets.[4] The $N = 2$ D-term is

$$D^{Aa} = \frac{g^2}{2} \Phi_\alpha^{i*} \sigma_{\alpha\beta}^A t_{ij}^a \Phi_\beta^j ,$$
                                                            (B.7.3)

with $\sigma_{\alpha\beta}^A$ the Pauli matrices ($A = 1, 2, 3$) and t_{ij}^a the group representation. The potential, determined entirely by the gauge symmetry, is

$$\frac{1}{2g^2} D^{Aa} D^{Aa} .$$
                                                            (B.7.4)

The interactions, incidentally, are nonrenormalizable in six dimensions.

Now consider the less restrictive moduli space action, where field-dependent kinetic terms are included, but with gravity still decoupled. Supersymmetry does not allow the gauge field kinetic term to depend on the hypermultiplet moduli, and it is allowed to depend on the scalar t in the tensor multiplets only in the precise form

$$t \text{Tr}(F_{\mu\nu} F^{\mu\nu}) .$$
                                                            (B.7.5)

The linear dependence is fixed because this term is related by supersymmetry to a coupling of the self-dual tensor,

$$B_2 \text{Tr}(F_2 \wedge F_2) ,$$
                                                            (B.7.6)

where the tensor gauge invariance allows only the linear coupling. The term (B.7.6) is needed to cancel anomalies in a six-dimensional version of the Green–Schwarz mechanism, so these terms can arise only at exactly one loop, with a coefficient that is determined by the gauge quantum numbers of the hyper- and vector multiplets.

The hypermultiplet kinetic term may depend on the hypermultiplet moduli but not the tensor moduli. Representing the moduli by real fields ϕ^r, one has

$$G_{rs}(\phi) \partial_\mu \phi^r \partial^\mu \phi^s .$$
                                                            (B.7.7)

---

[4] If the scalars are in a pseudoreal representation of the gauge group, meaning that the conjugation matrix C_{ij} is antisymmetric, one can reduce to a half-hypermultiplet by the reality condition $\Phi_\alpha^{i*} = \epsilon_{\alpha\beta} C_{ij} \Phi_\beta^j$.

For the case of $d = 4$, $N = 1$ supersymmetry, we have explained in section B.2 that the moduli space must be Kähler, meaning that the $2n$ real moduli can be grouped into n complex fields ϕ^i with the metric $G_{\bar{i}j} = K_{\bar{i}j}$. Here the supersymmetry is doubled and the metric correspondingly more restricted: it must be *hyper-Kähler*. This means that there are three different complex Kähler structures, three different ways to group the real moduli into complex fields, each giving a Kähler metric. Further there is a relation between the three complex structures. Given any complex structure, a set of complex coordinates ϕ^i and ϕ^{i*}, we can define the tensor J by

$$J^i_j = i\delta^i_j, \quad J^{\bar{i}}_{\bar{j}} = -i\delta^{\bar{i}}_{\bar{j}}, \quad J^i_{\bar{j}} = J^{\bar{i}}_j = 0 . \tag{B.7.8}$$

This tensor can be defined for any complex manifold and is also known as the *complex structure*. We have defined it in a particular coordinate system but now can translate it to arbitrary coordinates. It satisfies $J^2 = -1$, a frame-independent statement. The three complex structures of the hyper-Kähler space are required to satisfy

$$J^A J^B = -\delta^{AB} + \epsilon^{ABC} J^C . \tag{B.7.9}$$

These properties require the number of real moduli to be a multiple of 4.

An alternative characterization is as follows. There are $4m$ moduli, so a general metric (B.7.7) would have holonomy $SO(4m)$. That is, parallel transport of a vector around a loop in moduli space brings it back to itself rotated by a general element of $SO(4m)$. These ideas are familiar from general relativity, in the context of the spacetime manifold, but we emphasize that the manifold in question here is field space. Now consider the following $SU(2)$ subgroup of $SO(4m)$. We know that $SO(4) = SU(2) \times SU(2)$. Take the first $SU(2)$ and replace the elements with $m \times m$ identity matrices to make a subgroup of $SO(4m)$. The subgroup of $SO(4m)$ that commutes with this $SU(2)$ is $Sp(m)$. Then a hyper-Kähler manifold is one for which the holonomy lies in this $Sp(m)$ subgroup, the Js being the $SU(2)$ generators.

$d = 4$, $N = 2$ *supersymmetry*

Let us first consider the reduction of the self-dual tensor multiplet from $d = 6$ to $d = 5$. The components $B_{\mu 5}$ become a vector. The dual $*H_{\sigma\rho\omega}$ would give the field strength of a second vector if the tensor were unconstrained, but due to the self-duality this is the same as $H_{\mu\nu 5}$. Thus one has in all a vector and a scalar. This is the same as the content of the vector multiplet, where the scalar comes from the reduction of A_5, so these multiplets are identical in $d = 5$ and consequently in $d = 4$.

Thus in $d = 4$ we need consider only the hypermultiplet with four scalars, and the vector multiplet with two scalars A_4, A_5 from the six-dimensional vector. First in the renormalizable limit, the action includes the bosonic terms discussed in $d = 6$, the gauge-invariant kinetic terms and the potential (B.7.4). The potential has additional terms

$$-\frac{1}{4g^2}\text{Tr}([A_4, A_5]^2) \qquad (B.7.10)$$

from reduction of the field strength and

$$\Phi_\alpha^{i\dagger}(M_4^2 + M_5^2)_{ij}\Phi_\alpha^j \,, \qquad (B.7.11)$$

where

$$M_{mij} = A_m^a t_{ij}^a + q_{mij} \,. \qquad (B.7.12)$$

The first term in M is from reduction of the covariant derivative; the parameters q_{mij} are allowed by supersymmetry and can be thought of as arising from a 'dummy' gauge field.

The potential has various flat directions. We discuss first the moduli space approximation with gravity still decoupled. Supersymmetry requires the kinetic term for the vector multiplet to depend only on the vector multiplet moduli and the kinetic term for the hypermultiplet to depend only on the hypermultiplet moduli. The latter is required to be a hyper-Kähler space just as in $d = 6$. The vector moduli space is also a Kähler metric with extra conditions. Namely, forming the complex scalars $A^a = A_4^a + iA_5^a$ with a indexing the gauge generators, the Kähler potential must be of the form

$$K(A, A^*) = \text{Im}\left(\sum_a A^{a*}\partial_a F(A)\right) \qquad (B.7.13)$$

for some holomorphic *prepotential* $F(A)$. The metric on moduli space is then

$$G_{a\bar{b}} = \text{Im}(\partial_a\partial_b F) \,. \qquad (B.7.14)$$

This is known as a *rigid special Kähler metric*.

Turning on $N = 2$ supergravity, the moduli space acquires additional curvature as it did for $N = 4$, but it remains a direct product of hyper-multiplet and vector multiplet moduli spaces. The hyper-Kähler metrics are replaced by *quaternionic* metrics, where the $SU(2)$ holonomy is no longer zero but has a definite curvature of order κ^2. The vector moduli space becomes a *special Kähler space*. These spaces are also relevant to $N = 1$ compactifications of the heterotic string, so we describe them in some detail. For n hypermultiplets, it is useful to begin with $n+1$ complex coordinates X^I with the projective identification

$$(X^0, X^1, \ldots, X^n) \cong (\lambda X^0, \lambda X^1, \ldots, \lambda X^n) \qquad \text{(B.7.15)}$$

for any nonzero complex λ. One can also introduce invariant coordinates; for example, away from the subspace $X^0 = 0$ the set

$$T^A = \frac{X^A}{X^0}, \quad A = 1, \ldots, n. \qquad \text{(B.7.16)}$$

The low energy action is determined by a single complex function $F(X)$, which must be homogeneous of degree 2 under the identification (B.7.15),

$$F(\lambda X) = \lambda^2 F(X). \qquad \text{(B.7.17)}$$

The Kähler potential is then

$$K = -\ln \operatorname{Im} \left(\sum_I X^{I*} \partial_I F(X) \right). \qquad \text{(B.7.18)}$$

Under a projective transformation (B.7.15),

$$K \to K - \ln \lambda - \ln \lambda^*. \qquad \text{(B.7.19)}$$

This is a Kähler transformation (B.2.32), so the metric is well defined on the projective space produced by the identification. The number of vectors is $n+1$, including the one from the supergravity multiplet, so the fields A_μ^I for $I = 0, \ldots, n$ are independent. Their kinetic term is again determined by F and depends only on the vector multiplet moduli; its explicit form is left to the references.

The forms (B.7.15) and (B.7.18) are not invariant under arbitrary changes of coordinates: the coordinates X^I are known as *special coordinates*. The forms are clearly invariant under linear redefinitions of the special coordinates, but there is in fact a larger set of transformations that preserves the form, namely

$$\begin{bmatrix} X'^I \\ \partial_{I'} F' \end{bmatrix} = S \begin{bmatrix} X^I \\ \partial_I F \end{bmatrix} \qquad \text{(B.7.20)}$$

for S a $2(n+1) \times 2(n+1)$ real symplectic matrix. As a final comment, in recent literature it has been noted that in special cases the symplectic transformation (B.7.20) gives a would-be gradient $\partial_{I'} F'$ whose curl is actually nonvanishing. For these the definition of special geometry needs to be generalized.

Appendix B

Exercises

B.1 In section B.1 we defined B and C in a particular basis. The properties (B.1.17) and (B.1.24) define them in general. Under a change of basis

$$\Gamma^\mu \to U\Gamma^\mu U^{-1}$$

for unitary U, find the transformations of B and C. Show that the properties (B.1.18), (B.1.19), (B.1.21), (B.1.25), and (B.1.27) are independent of such a change of basis. Determine the relations between B and B^T and between C and C^T, and show that these are independent of basis.

B.2 Extend the decomposition (B.1.44) to the general $SO(d-1,1) \to SO(d'-1,1) \times SO(d-d')$, where some of the dimensions are odd.

B.3 Work out the details of the reduction of the $d = 4$, $N = 1$ supersymmetry algebra to the $d = 2$ (2,2) algebra. Identify the central charges.

B.4 Verify eq. (B.4.7) for $*^2$ and derive the corresponding result for Euclidean space.

B.5 List the helicities (s_1, s_2, s_3, s_4) for the massless $\mathbf{8}_v + \mathbf{8}$ open string states and show that these constitute a representation of the type I supersymmetry algebra.

References

The literature on the subjects covered in these two volumes easily exceeds 5000 papers, and probably approaches twice that number. To represent accurately the contributions of all who have worked in these many areas is an undertaking beyond the scope of this book. Rather, I have tried to assemble a list that is short enough to be useful in developing specific subjects and in giving points of entry to the literature. The resulting list includes many review articles, plus papers from the original literature where I felt they were needed to supplement the treatment in the text.

Papers listed as 'e-print: hep-th/yymmnnn' are available electronically at the Los Alamos physics e-print archive,

$$\text{http://xxx.lanl.gov/abs/hep-th/yymmnnn}.$$

A list of corrections to the text is maintained at

$$\text{http://www.itp.ucsb.edu/~joep/errata.html}.$$

General references

Other books and lectures covering material in volume two include Green, Schwarz, & Witten (1987) (henceforth denoted GSW), Peskin (1987), Lüst & Theisen (1989), Alvarez-Gaumé & Vazquez-Mozo (1995), D'Hoker (1993), Ooguri & Yin (1997), and Kiritsis (1997). A number of the review articles cited are collected in Efthimiou & Greene (1997). Many of the key papers up to 1985 can be found in Schwarz (1985), and in the reference section of GSW.

Chapter 10

Many of the topics of this chapter are covered in the general references. In addition, Friedan, Martinec, & Shenker (1986) cover superconformal field

467

theory, bosonization, and vertex operators. The original article Gliozzi, Scherk, & Olive (1977) is quite readable. The review Schwarz (1982) covers many of the topics in this chapter. The cancellation of divergences in the $SO(32)$ type I theory is discussed in Green & Schwarz (1985); whereas we consider the vacuum amplitude, they study the *planar amplitude,* where there are vertex operators but on one boundary only.

Chapter 11

The classification of superconformal algebras is from Ademollo *et al.* (1976) and Sevrin, Troost, & Van Proeyen (1988). Strings based on higher spin (W) algebras are reviewed in Pope (1995). Strings based on fractional spin algebras are reviewed in Tye (1995). Strings based on the $N = 2$ superconformal algebra are reviewed and a spacetime interpretation given in Ooguri & Vafa (1991). The covariant manifestly spacetime-supersymmetric string is introduced in Green & Schwarz (1984a). Topological string theory is introduced in Witten (1988).

The ten-dimensional supersymmetric heterotic string is introduced in Gross, Harvey, Martinec, & Rohm (1985). Our treatment of the nonsupersymmetric theories is similar to that in Kawai, Lewellen, & Tye (1986b), which also has references to earlier constructions of the various models. Level matching and discrete torsion are discussed in Vafa (1986).

Wybourne (1974) and Georgi (1982) are introductions to Lie algebras and groups. A recent reference covering both Lie and current algebras is Fuchs & Schweigert (1997). Many useful facts are tabulated in Slansky (1981). Goddard & Olive (1986) is a thorough review of current algebra. For more on the Sugawara construction see Goddard & Olive (1985); further current algebra references are given for chapter 15. The algebra of Chan–Paton factors is analyzed in Marcus & Sagnotti (1982).

For more on toroidal compactification of the heterotic string, see Narain (1986) and Narain, Sarmadi, & Witten (1987). The heterotic string as a BPS state is discussed in Dabholkar, Gibbons, Harvey, & Ruiz Ruiz (1990). Gauntlett, Harvey, & Liu (1993) is a reference/review on magnetic monopoles in toroidally compactified heterotic string theory.

Chapter 12

For further discussion and references on supergravity actions see chapter 13 of GSW and Townsend (1996). Gravitational anomalies are discussed in detail in Alvarez-Gaumé & Witten (1983). Anomaly cancellation is discussed in Green & Schwarz (1984b) and chapters 10 and 13 of GSW.

Superspace and picture-changing are discussed in Friedan, Martinec, & Shenker (1986). The superspace formalism for nonlinear sigma models with various supersymmetries is reviewed in Roček (1993). The connection between the world-sheet anomaly and the spacetime Chern–Simons action is discussed in Hull & Witten (1985).

Our treatment of superstring perturbation theory is similar to that in Friedan, Martinec, & Shenker (1986), Verlinde & Verlinde (1987), Martinec (1987), Alvarez-Gaumé *et al.* (1988), and Giddings (1992). See also Atick, Moore, & Sen (1988), La & Nelson (1989), and Aoki, D'Hoker, & Phong (1990). Light-cone methods are developed in chapters 7–11 of GSW.

The various tree amplitudes are treated in Schwarz (1982), Gross, Harvey, Martinec, & Rohm (1986), Kawai, Lewellen, & Tye (1986a), and chapter 7 of GSW. The extraction of higher dimension corrections to the low energy action is discussed in Grisaru, van de Ven, & Zanon (1986) and Gross & Sloan (1987). One-loop calculations in the light-cone gauge are in chapters 8–10 of GSW. The explicit one-loop calculations in section 12.6 are based on Lerche, Nilsson, Schellekens, & Warner (1988) and Abe, Kubota, & Sakai (1988). Some related higher-loop amplitudes that can be evaluated in closed form are discussed in Bershadsky, Cecotti, Ooguri, & Vafa (1994) and Antoniadis, Gava, Narain, & Taylor (1994). Non-renormalization theorems are derived by world-sheet contour arguments in Martinec (1986).

Chapter 13

Most of the subjects in this chapter are covered in the review by Polchinski (1997). For more on *T*-duality see the review by Giveon, Porrati, & Rabinovici (1994). Much of the discussion in the first three sections follows Polchinski (1995). The $e^{-O(1/g)}$ effects are discussed in Shenker (1991). Their connection with D-instantons is discussed in Polchinski (1994); Green & Gutperle (1997) give a detailed treatment of D-instanton effects. A recent discussion of the Born–Infeld action appears in Tseytlin (1997).

The discussion of branes at angles is similar to that in Berkooz, Douglas, & Leigh (1996). The quartic identity of Riemann and other theta function identities are in Mumford (1983). For more on D-brane scattering see Bachas (1996), Lifschytz (1996), and Douglas, Kabat, Pouliot, & Shenker (1997) and references therein.

Non-Abelian D-brane dynamics and F–D bound states are discussed in Witten (1996a). For more on D0–D0 bound states see Sen (1996a) and Sethi & Stern (1997). For more on D0–D4 bound states see Sen (1996b) and Vafa (1996a). For more on the connection between D-branes and instantons see Witten (1996b), Vafa (1996b), and Douglas (1996).

Chapter 14

A number of string duality conjectures have been put forward over the years, but the coherent picture presented in this chapter took shape with the work of Hull & Townsend (1995), Townsend (1995), & Witten (1995). Some reviews are Townsend (1996), Sen (1997,1998), and Schwarz (1997). For more on $SO(32)$ type I/heterotic duality see Polchinski & Witten (1996). For more on the strongly coupled $E_8 \times E_8$ theory see Hořava & Witten (1996). Some tests of string dualities based on perturbative and nonperturbative string amplitudes are discussed in Tseytlin (1995) and Green (1997).

Callan, Harvey, & Strominger (1992) review extended objects with NS–NS charges. For black p-branes see Horowitz & Strominger (1991). Townsend (1996), Duff (1997), and Stelle (1998) review the various extended objects that play a role in string duality and the connection between extended objects in M-theory and in IIA string theory. Strominger (1996) discusses extended objects ending on other objects. Harvey (1997) reviews magnetic monopoles and Montonen–Olive duality. The discussion of type I D5-branes follows Gimon & Polchinski (1996) and Witten (1996b).

Matrix theory is introduced in Banks, Fischler, Shenker, & Susskind (1997). Banks (1997) and Bigatti & Susskind (1997) are reviews.

For background on black hole thermodynamics see Carter (1979) and Wald (1997). The entropy calculation in the chapter is similar to that in Strominger & Vafa (1996). Horowitz (1997), Peet (1997), and Maldacena (1998) review D-brane calculations of black hole entropies and other properties. The correspondence principle is discussed in Horowitz & Polchinski (1997). Page (1994) gives a review of the black hole information problem. Susskind (1995) discusses the possible breakdown of locality in string theory.

The connection between branes and gauge theory dynamics is reviewed in Giveon & Kutasov (1998). For very recent progress see Maldacena (1997).

Chapter 15

Much of the first three sections is based on Belavin, Polyakov, & Zamolodchikov (1984). The review by Ginsparg (1990) covers many of the subjects in this chapter. Many of the relevant papers are collected in Goddard & Olive (1988) or in Itzykson, Saleur, & Zuber (1988).

The unitary representations of the Virasora algebra are discussed by Friedan, Qiu, & Shenker (1984). Thorn (1984) gives a stringy derivation of the Kac formula. Cardy (1986) discusses general aspects of modular invariance, and Cappelli, Itzykson, & Zuber (1987) give the modular-invariant partition functions for the minimal models and $SU(2)$ current algebras. For the solution of minimal models using the Feigin–Fuchs representation see Dotsenko & Fateev (1984, 1985).

The exact solution of current algebra CFTs is described in Knizhnik & Zamolodchikov (1984) and Gepner & Witten (1986). The nonlinear sigma model interpretation is in Witten (1984). Free-field representations of current algebras are obtained in Bershadsky & Ooguri (1989). The coset construction is developed in Goddard, Kent, & Olive (1986). Parafermionic theories are described in Zamolodchikov & Fateev (1985). W-algebras are reviewed in Bouwknegt & Schoutens (1993) and de Boer, Harmsze, & Tjin (1996). Our discussion of rational CFT is largely based on Vafa (1988). Moore & Seiberg (1989) give a systematic treatment of the monodromy and other constraints. Irrational CFT is reviewed in Halpern, Kiritsis, Obers, & Clubok (1996).

Many of the subjects in the final two sections are developed in the review by Cardy (1990). For more on the c-theorem see Zamolodchikov (1986b), and for more on Landau–Ginzburg models see Zamolodchikov (1986a).

Chapter 16

Dixon, Harvey, Vafa, & Witten (1985, 1986) develop the general framework for strings on orbifolds. Modular invariance is discussed in Vafa (1986). Orbifold vertex operators and interactions are treated in Dixon, Friedan, Martinec, & Shenker (1987), Hamidi & Vafa (1987), and the review by Dixon (1988). These papers also discuss the blowing up of the fixed points; our discussion is similar to that in Hamidi & Vafa.

Asymmetric orbifolds are developed in Narain, Sarmadi, & Vafa (1987). Antoniadis, Bachas, & Kounnas (1987) and Kawai, Lewellen, & Tye (1987) develop general free-fermion models. A generalized free-boson construction appears in Lerche, Schellekens, & Warner (1989).

Two (of the many) discussions of the motivation for spacetime super-symmetry and of general aspects of supersymmetric model building are Witten (1981) and Dine (1997). Ross (1984) is an introduction to grand unification.

Font, Ibàñez, Quevedo, & Sierra (1990) is a review of three gener-ation orbifold models; the model (16.3.32) appears in section 4.2. A much-vamped three-generation free-fermion model appears in Antoniadis,

Ellis, Hagelin, & Nanopoulos (1989). Kakushadze, Shiu, Tye, & Vtorov-Karevsky (1997) is a recent review of free-field models with particular attention to higher level three-generation models, which can have ordinary grand unified symmetry breaking.

The discussion of the action for untwisted moduli is patterned on Witten (1985). The general expression for the one-loop threshold correction is obtained in Kaplunovsky (1988); the lectures by Kiritsis (1997) give a thorough treatment. The evaluation of Δ_a for orbifold models is in Dixon, Kaplunovsky, & Louis (1991). The paper by Ibànez & Lüst (1992) reviews many aspects of the low energy physics of orbifolds, especially those connected with T-duality and with threshold corrections. Quevedo (1996) is a review of low energy string physics.

Chapter 17

The necessary geometric background is given in more detail in chapter 15 of GSW and in Candelas (1988). Hübsch (1992) is a full length treatment at a more advanced level. Calabi–Yau compactification is developed in Candelas, Horowitz, Strominger, & Witten (1985) and in chapter 16 of GSW. Strominger & Witten (1985) discuss various aspects of the low energy physics. For more on the low energy action see Candelas & de la Ossa (1991). The nonrenormalization theorem is from Witten (1986), who also discusses (0,2) compactifications. World-sheet instantons are discussed in Dine, Seiberg, Wen, & Witten (1986, 1987). An analysis of the field equations without the vanishing torsion assumption is in Strominger (1986).

Chapter 18

Continuous symmetries are discussed in Banks & Dixon (1988). Dine (1995) discusses discrete symmetries and the strong CP problem in string theory.

Closed string gauge couplings are discussed in Ginsparg (1987). Constraints on right-moving and type II gauge symmetries are in Dixon, Kaplunovsky, & Vafa (1987). Dienes (1997) is an extensive review of coupling constant unification in string theory. The argument in figure 18.1 for the proximity of the compactification and string scales is based on Kaplunovsky (1985). The discussion of the effect of an extra dimension in figure 18.2 is based on Witten (1996c). The derivation of the moduli independence of $\sin^2 \theta_{\mathrm{w}}$ follows Banks, Dixon, Friedan, & Martinec (1988). The unification of the couplings in supersymmetric theories is reviewed in Dimopoulos, Raby, & Wilczek (1991). The discussion of fractional charges is taken from Schellekens (1990). For more on proton stability in supersymmetric and string theories see Ibàñez & Ross (1992), Pati (1996).

The general argument that spacetime supersymmetry requires $N = 2$ world-sheet supersymmetry is from Banks, Dixon, Friedan, & Martinec (1988). The analysis for extended supersymmetry is in Banks & Dixon (1988). The world-sheet argument that supersymmetry breaking cannot be turned on continuously is also in that paper; the spacetime derivation of the same result is in Dine & Seiberg (1988). The use of PQ symmetry and the scaling of S to derive nonrenormalization theorems is in Dine & Seiberg (1986). Derivation of nonrenormalization theorems from the structure of string perturbation theory is in Martinec (1986). The reader will note that the spacetime derivations are generally shorter and less intricate, and can in some cases give nonperturbative information as well. Generation of D-terms by string loops is discussed in Dine, Seiberg, & Witten (1987). The reviews by Quevedo (1996) and Dine (1997) discuss nonperturbative supersymmetry breaking in more detail, with extensive references. The cosmological constant problem is reviewed in Weinberg (1989).

Chapter 19

Many of the subjects in this chapter are covered in the review by Greene (1997).

For more on chiral rings see Lerche, Vafa, & Warner (1989). For type II strings on Calabi–Yau manifolds and their low energy actions, see Cecotti, Ferrara, & Girardello (1989). The world-sheet argument for the vanishing of the potential for the moduli is given in more detail in Dixon (1988). For a systematic derivation of the constraints from (2,2) superconformal symmetry, derived from analysis of string scattering amplitudes, see Dixon, Kaplunovsky, & Louis (1990). For arguments using the relation between type II and heterotic compactification see Dine & Seiberg (1988).

For more on $N = 2$ minimal models see Boucher, Friedan, & Kent (1986); for more on their connection with $SU(2)$ current algebra see Zamolodchikov & Fateev (1986) and Qiu (1987). For more on $N = 2$ Landau–Ginzburg models and singularity theory see Martinec (1989) and Vafa & Warner (1989). Gepner models are constructed in Gepner (1988). Our discussion is based on Vafa (1989); our discussion of the connection to Calabi–Yau compactification is based on Witten (1993).

Numerical evidence for mirror symmetry appears in Candelas, Lynker, & Schimmrigk (1990). The construction via twisted Gepner models is in Greene & Plesser (1990). Strominger, Yau, & Zaslow (1996) obtain the connection to T-duality. Toric geometry and other advanced ideas are covered in the review by Greene (1997). The use of mirror symmetry to obtain the exact low energy action is in Candelas, de la Ossa, Green, & Parkes (1991). The flop transition is described in Aspinwall, Greene, & Morrison (1993) and Witten (1993). Cox & Katz (1998) is a recent treatment of mathematics and mirror symmetry.

The interpretation of the conifold singularity in terms of a light black hole/D-brane is in Strominger (1995). Shenker (1995) discusses the short-distance cutoff on the loop graph. Greene, Morrison, & Strominger (1995) show that condensation of these states leads to topology change, providing a physical interpretation for the geometric observations of Candelas, Green, & Hübsch (1989).

The basic features of string theories on K3 are described in Seiberg (1988) (the discussion of Calabi–Yau moduli space in that paper has been superceded by later references). Aspinwall (1997) gives an extended review of this subject. The lectures by Sagnotti (1997) and Schwarz (1997) also cover various six-dimensional string theories, discussing in particular anomaly cancellation. The tensionless string phase transition is described in Seiberg & Witten (1996) and Ganor & Hanany (1996).

For more on the duals of toroidally compactified heterotic strings see Hull & Townsend (1995) and Witten (1995). Our discussion is similar to that in Sen (1997). Sen (1994) is a review of $SL(2, \mathbf{Z})$ duality of the heterotic string on T^6. F-theory is introduced in Vafa (1996c). Kachru & Silverstein (1997) apply F-theory to find heterotic phase transitions that change generation number, and give further references. There is a growing literature on duals of theories with $N = 1$ and $N = 2$ supersymmetry. Vafa & Witten (1995) and Ferrara, Harvey, Strominger, & Vafa (1995) give some relatively simple examples.

Appendix

Our treatment of the spinor representations of $SO(D - 1, 1)$ and $SO(n)$ follows the treatment for $SO(n)$ in Georgi (1982). Sohnius (1985) also discusses spinors in general dimensions.

Two references on $d = 4$, $N = 1$ supersymmetry are Sohnius (1985) and Wess & Bagger (1992); the former also has some discussion of extended supersymmetry and higher-dimensional theories. The general $d = 4$, $N = 1$ supergravity action is given in Cremmer, Ferrara, Girardello, & Van Proeyen (1983). The significance of the BPS property is developed in Witten & Olive (1978).

The $d = 11$ supergravity theory appears in Cremmer, Julia, & Scherk (1978). Table B.3 (with a misprint corrected) is taken from Hull & Townsend (1995), who give original references. Chapter 13 of GSW and

Townsend (1996) have more on supergravity actions in $d = 11$ and $d = 10$; Townsend also discusses the central charges in the supersymmetry algebra. Salam & Sezgin (1989) is a collection of many relevant papers.

The general $d = 4$, $N = 4$ supergravity theory is obtained in de Roo (1985). The general $d = 4$, $N = 2$ supergravity theory is obtained in Andrianopoli *et al.* (1996). The hypermultiplet moduli space is described in Bagger & Witten (1983) and Hitchin, Karlhede, Lindstrom, & Roček (1987). The vector multiplet moduli space is described in de Wit, Lauwers, & Van Proeyen (1985). Seiberg & Witten (1994) give a review of the global supersymmetry limit.

References

Abe, M., Kubota, H., & Sakai, N., (1988). Loop corrections to the $E_8 \times E_8$ heterotic string effective Lagrangian. *Nuclear Physics,* **B306**, 405.

Ademollo, M., Brink, L., D'Adda, A., D'Auria, R., Napolitano, E., Sciuto, S., Del Guidice, E., Di Vecchia, P., Ferrara, S., Gliozzi, F., Musto, R., Pettorino, R., & Schwarz, J. H. (1976). Dual string with $U(1)$ color symmetry. *Nuclear Physics,* **B111**, 77.

Alvarez-Gaumé, L., Nelson, P., Gomez, C., Sierra, G., & Vafa, C. (1988). Fermionic strings in the operator formalism. *Nuclear Physics,* **B311**, 333.

Alvarez-Gaumé, L., & Vazquez-Mozo M. A. (1995). Topics in string theory and quantum gravity. In *Les Houches Summer School on Gravitation and Quantizations, 1992* eds. J. Zinn-Justin & B. Julia, pp. 481–636. Amsterdam: North-Holland. E-print hep-th/9212006.

Alvarez-Gaumé, L., & Witten, E. (1983). Gravitational anomalies. *Nuclear Physics,* **B234**, 269.

Andrianopoli, L., Bertolini, M., Ceresole, A., D'Auria, R., Ferrara, S., & Fré, P. (1996). General matter coupled $N = 2$ supergravity. *Nuclear Physics,* **B476**, 397. E-print hep-th/9603004.

Antoniadis, I., Bachas, C. P., & Kounnas, C. (1987). Four-dimensional superstrings. *Nuclear Physics,* **B289**, 87.

Antoniadis, I., Ellis, J., Hagelin, J. S., & Nanopoulos, D. V. (1989). The flipped $SU(5) \times U(1)$ string model revamped. *Physics Letters,* **B231**, 65.

Antoniadis, I., Gava, E., Narain, K. S., & Taylor, T. R. (1994). Topological amplitudes in string theory. *Nuclear Physics,* **B413**, 162. E-print hep-th/9307158.

Aoki, K., D'Hoker, E., & Phong, D. H. (1990). Unitarity of closed superstring perturbation theory. *Nuclear Physics,* **B342**, 149.

Aspinwall, P. S. (1997). K3 surfaces and string duality. In *Fields, Strings, and Duality, TASI 1996,* eds. C. Efthimiou & B. Greene, pp. 421–540. Singapore: World Scientific. E-print hep-th/9611137.

Aspinwall, P. S., Greene, B. R., & Morrison, D. R. (1993). Multiple mirror manifolds and topology change in string theory. *Physics Letters,* **B303,** 249. E-print hep-th/9301043.

Atick, J. J., Moore, G., & Sen A. (1988). Catoptric tadpoles. *Nuclear Physics,* **B307,** 221.

Bachas, C. (1996). D-brane dynamics. *Physics Letters,* **B374,** 37. E-print hep-th/9511043.

Bagger, J., & Witten, E. (1983). Matter couplings in $N = 2$ supergravity. *Nuclear Physics,* **B222,** 1.

Banks, T. (1997). Matrix theory. E-print hep-th/9710231.

Banks, T., & Dixon, L. J. (1988). Constraints on string vacua with spacetime supersymmetry. *Nuclear Physics,* **B307,** 93.

Banks, T., Dixon, L. J., Friedan, D., & Martinec, E. (1988). Phenomenology and conformal field theory or can string theory predict the weak mixing angle? *Nuclear Physics,* **B299,** 613.

Banks, T., Fischler, W., Shenker, S. H., & Susskind, L. (1997). M theory as a matrix model: a conjecture. *Physical Review,* **D55,** 5112. E-print hep-th/9610043.

Belavin, A. A., Polyakov, A. M., & Zamolodchikov, A. B. (1984). Infinite conformal symmetry in two-dimensional quantum field theory. *Nuclear Physics,* **B241,** 333. Also in Goddard & Olive (1988), Itzykson *et al.* (1988).

Berkooz, M., Douglas, M. R., & Leigh, R. G. (1996). Branes intersecting at angles. *Nuclear Physics,* **B480,** 265. E-print hep-th/9606139.

Bershadsky, M., Cecotti, S., Ooguri, H., & Vafa, C. (1994). Kodaira–Spencer theory of gravity and exact results for quantum string amplitudes. *Communications in Mathematical Physics,* **165,** 311. E-print hep-th/9309140.

Bershadsky, M., & Ooguri, H. (1989). Hidden $SL(n)$ symmetry in conformal field theories. *Communications in Mathematical Physics,* **126,** 49.

Bigatti, D., & Susskind, L. (1997). Review of matrix theory. E-print hep-th/9712072.

Boucher, W., Friedan, D., & Kent, A. (1986). Determinant formulae and unitarity for the $N = 2$ superconformal algebras in two dimensions or exact results on string compactification. *Physics Letters,* **B172,** 316.

Bouwknegt, P., & Schoutens, K. (1993). W-symmetry in conformal field theory. *Physics Reports,* **223,** 183. E-print hep-th/9210010.

Callan, C. G., Harvey, J. A., & Strominger, A. (1992). Supersymmetric string solitons. In *String Theory and Quantum Gravity, Trieste 1991,* eds. J. A. Harvey *et al.,* pp. 208–244. Singapore: World Scientific. E-print hep-th/9112030.

Candelas, P. (1988). Lectures on complex manifolds. In *Superstrings '87,* eds. L. Alvarez-Gaumé *et al.,* pp. 1–88. Singapore: World Scientific.

Candelas, P., & de la Ossa, X. C. (1991). Moduli space of Calabi–Yau manifolds. *Nuclear Physics*, **B355**, 455.

Candelas, P., de la Ossa, X. C., Green, P. S., & Parkes, L. (1991). A pair of Calabi–Yau manifolds as an exactly soluble superconformal theory. *Nuclear Physics*, **B359**, 21.

Candelas, P., Green, P. S., & Hübsch, T. (1989). Rolling among Calabi–Yau vacua. *Nuclear Physics*, **B330**, 49.

Candelas, P., Horowitz, G. T., Strominger, A., & Witten, E. (1985). Vacuum configurations for superstrings. *Nuclear Physics*, **B258**, 46.

Candelas, P., Lynker, M., & Schimmrigk, R. (1990). Calabi–Yau manifolds in weighted P_4. *Nuclear Physics*, **B341**, 383.

Cappelli, A., Itzykson, C., & Zuber, J.-B. (1987). Modular-invariant partition functions in two dimensions. *Nuclear Physics*, **B280**, 445. Also in Itzykson *et al.* (1988).

Cardy, J. L. (1986). Operator content of two-dimensional conformally invariant theories. *Nuclear Physics*, **B270**, 186. Also in Goddard & Olive (1988), Itzykson *et al.* (1988).

Cardy, J. L. (1990). Conformal invariance and statistical mechanics. In *Fields, Strings, and Critical Phenomena, Les Houches 1988*, eds. E. Brezin & J. Zinn-Justin, pp. 169–246. Amsterdam: North-Holland.

Carter, B. (1979). The general theory of the mechanical, electromagnetic, and thermodynamic properties of black holes. In *General Relativity. An Einstein Centennial Survey*, eds. S. W. Hawking & W. Israel. Cambridge: Cambridge University Press.

Cecotti, S., Ferrara, S., & Girardello, L. (1989). Geometry of type II superstrings and the moduli of superconformal field theories. *International Journal of Modern Physics*, **A4**, 2475.

Cox, D. A., & Katz, S. (1998). *Mirror Symmetry and Algebraic Geometry*, American Mathematical Society, in press.

Cremmer, E., Ferrara, S., Girardello, L., & Van Proeyen, A. (1983). Yang–Mills theories with local supersymmetry: Lagrangian, transformation laws, and super-Higgs effect. *Nuclear Physics*, **B212**, 413.

Cremmer, E., Julia, B., & Scherk, J. (1978). Supergravity theory in eleven dimensions. *Physics Letters*, **B76**, 409.

Dabholkar, A., Gibbons, G., Harvey, J. A., & Ruiz Ruiz, F. (1990). Superstrings and solitons. *Nuclear Physics*, **B340**, 33.

de Boer, J., Harmsze, F., & Tjin, T. (1996). Non-linear finite W-symmetries and applications in elementary systems. *Physics Reports*, **272**, 139. E-print hep-th/9503161.

de Roo, M. (1985). Matter coupling in $N = 4$ supergravity. *Nuclear Physics*, **B255**, 515.

de Wit, B., Lauwers, P. G., & Van Proeyen, A. (1985). Lagrangians of $N = 2$ supergravity-matter systems. *Nuclear Physics,* **B255**, 569.

D'Hoker, E. (1993). TASI Lectures on critical string theory. In *Recent Directions in Particle Theory, TASI 1992,* eds. J. Harvey & J. Polchinski, pp. 1–100. Singapore: World Scientific, 1993.

Dienes, K. R. (1997). String theory and the path to unification: a review of recent developments. *Physics Reports,* **287**, 447. E-print hep-th/9602045 .

Dimopoulos, S., Raby, S. A., & Wilczek, F. (1991) Unification of couplings. *Physics Today,* **44**, No. 10, 25.

Dine, M. (1995). Topics in string phenomenology. In *Proceedings of Strings '93,* eds. M. B. Halpern, G. Rivlis, & A. Sevrin, pp. 268–284. Singapore: World Scientific. E-print hep-ph/9309319.

Dine, M. (1997). Supersymmetry phenomenology (with a broad brush). In *Fields, Strings, and Duality, TASI 1996,* eds. C. Efthimiou & B. Greene, pp. 813–882. Singapore: World Scientific. E-print hep-ph/9612389.

Dine, M., & Seiberg, N. (1986). Nonrenormalization theorems in superstring theory. *Physical Review Letters,* **57**, 2625.

Dine, M., & Seiberg, N. (1988). Microscopic knowledge from macroscopic physics in string theory. *Nuclear Physics,* **B301**, 357.

Dine, M., Seiberg, N., Wen, X.-G., & Witten, E. (1986). Nonperturbative effects on the string world sheet. *Nuclear Physics,* **B278**, 769.

Dine, M., Seiberg, N., Wen, X.-G., & Witten, E. (1987). Nonperturbative effects on the string world sheet. II. *Nuclear Physics,* **B289**, 319.

Dine, M., Seiberg, N., & Witten, E. (1987). Fayet–Iliopoulos terms in string theory. *Nuclear Physics,* **B289**, 589.

Dixon, L. J. (1988). Some world sheet properties of superstring compactifications, on orbifolds and otherwise. In *Superstrings, Unified Theories, and Cosmology, 1987,* eds. G. Furlan *et al.,* pp. 67–126. Singapore: World Scientific.

Dixon, L. J., Friedan, D., Martinec, E., & Shenker, S. (1987). The conformal field theory of orbifolds. *Nuclear Physics,* **B282**, 13.

Dixon, L. J., Harvey, J. A., Vafa, C., & Witten, E. (1985). Strings on orbifolds. *Nuclear Physics,* **B261**, 678.

Dixon, L. J., Harvey, J. A., Vafa, C., & Witten, E. (1986). Strings on orbifolds. II. *Nuclear Physics,* **B274**, 285.

Dixon, L. J., Kaplunovsky, V. S., & Louis, J. (1990). On effective field theories describing (2,2) vacua of the heterotic string. *Nuclear Physics,* **B329**, 27.

Dixon, L. J., Kaplunovsky, V. S., & Louis, J. (1991). Moduli dependence of string loop corrections to gauge coupling constants. *Nuclear Physics,* **B355**, 649.

Dixon, L. J., Kaplunovsky, V. S., & Vafa, C. (1987). On four-dimensional gauge theories from type II superstrings. *Nuclear Physics,* **B294**, 43.

Dotsenko, V. S., & Fateev, V. A. (1984). Conformal algebra and multipoint correlation functions in two-dimensional statistical models. *Nuclear Physics,* **B240**, 312. Also in Itzykson *et al.* (1988).

Dotsenko, V. S., & Fateev, V. A. (1985). Four-point correlation functions and the operator algebra in two-dimensional conformal-invariant theories with central charge $c \leq 1$. *Nuclear Physics,* **B251**, 691.

Douglas, M. R. (1996). Gauge fields and D-branes. E-print hep-th/9604198.

Douglas, M. R., Kabat, D., Pouliot, P., & Shenker, S. H. (1997). D-branes and short distances in string theory. *Nuclear Physics,* **B485**, 85. E-print hep-th/9608024.

Duff, M. (1997). Supermembranes. In *Fields, Strings, and Duality, TASI 1996,* eds. C. Efthimiou & B. Greene, pp. 219–290. Singapore: World Scientific. E-print hep-th/9611203.

Efthimiou, C., & Greene, B., eds. (1997). *Fields, Strings, and Duality, TASI 1996,* Singapore: World Scientific.

Ferrara, S., Harvey, J. A., Strominger, A., & Vafa, C. (1995). Second-quantized mirror symmetry. *Physics Letters,* **B361**, 59. E-print hep-th/9505162.

Font, A., Ibáñez, L. E., Quevedo, F., & Sierra, A. (1990). The construction of 'realistic' four-dimensional strings through orbifolds. *Nuclear Physics,* **B331**, 421.

Friedan, D., Martinec, E., & Shenker, S. (1986). Conformal invariance, supersymmetry, and string theory. *Nuclear Physics,* **B271**, 93.

Friedan, D., Qiu, Z., & Shenker, S. (1984). Conformal invariance, unitarity and critical exponents in two dimensions. *Physical Review Letters,* **52**, 1575. Also in Goddard & Olive (1988), Itzykson *et al.* (1988).

Fuchs, J., & Schweigert, C. (1997). *Symmetries, Lie Algebras and Representations.* Cambridge: Cambridge University Press.

Ganor, O. J., & Hanany, A. (1996). Small E_8 instantons and tensionless noncritical strings. *Nuclear Physics,* **B474**, 122. E-print hep-th/9602120.

Gauntlett, J. P., Harvey, J. A., & Liu, J. T. (1993). Magnetic monopoles in string theory. *Nuclear Physics,* **B409**, 363. E-print hep-th/9211056.

Georgi, H. (1982). *Lie Algebras in Particle Physics.* Reading: Benjamin-Cummings.

Gepner, D. (1988). Spacetime supersymmetry in compactified string theory and superconformal models. *Nuclear Physics,* **B296**, 757.

Gepner, D., & Witten, E. (1986). String theory on group manifolds. *Nuclear Physics,* **B278**, 493.

Giddings, S. B. (1992). Punctures on super-Riemann surfaces. *Communications in Mathematical Physics,* **143**, 355.

Gimon, E. G., & Polchinski, J. (1996). Consistency conditions for orientifolds and D-manifolds. *Physical Review,* **D54**, 1667. E-print hep-th/9601038.

Ginsparg, P. (1987). Gauge and gravitational couplings in four-dimensional string theories. *Physics Letters,* **197**, 139.

Ginsparg, P. (1990). Applied conformal field theory. In *Fields, Strings, and Critical Phenomena, Les Houches 1988,* eds. E. Brezin & J. Zinn-Justin, pp. 1–168. Amsterdam: North-Holland.

Giveon, A., Porrati, M., & Rabinovici, E. (1994). Target space duality in string theory. *Physics Reports,* **244**, 77. E-print hep-th/9401139.

Giveon, A., & Kutasov, D. (1998). Brane dynamics and gauge theory. E-print hep-th/9802067.

Gliozzi, F., Scherk, J., & Olive, D. (1977). Supersymmetry, supergravity theories, and the dual spinor model. *Nuclear Physics,* **B122**, 253.

Goddard, P., Kent, A., & Olive, D. (1986). Unitary representations of the Virasoro and super-Virasoro algebras. *Communications in Mathematical Physics,* **103**, 105. Also in Goddard & Olive (1988), Itzykson *et al.* (1988).

Goddard, P., & Olive, D. (1985). Kac–Moody algebras, conformal symmetry, and critical exponents. *Nuclear Physics,* **B257**, 226.

Goddard, P., & Olive, D. (1986). Kac–Moody and Virasoro algebras in relation to quantum physics. *International Journal of Modern Physics,* **A1**, 303. Also in Goddard & Olive (1988).

Goddard, P., & Olive, D., eds. (1988). *Kac–Moody and Virasoro Algebras.* Singapore: World Scientific.

Green, M. B. (1997). Connections between M-theory and superstrings. E-print hep-th/9712195.

Green, M. B., & Gutperle, M. (1997). Effects of D-instantons. *Nuclear Physics,* **B498**, 195. E-print hep-th/9701093.

Green, M. B., & Schwarz, J. H. (1984a). Covariant description of superstrings. *Physics Letters,* **B136**, 367.

Green, M. B., & Schwarz, J. H. (1984b). Anomaly cancellations in supersymmetric $d = 10$ gauge theory and superstring theory. *Physics Letters,* **B149**, 117.

Green, M. B., & Schwarz, J. H. (1985). Infinity cancellations in $SO(32)$ superstring theory. *Physics Letters,* **B151**, 21.

Green, M. B., Schwarz, J. H., & Witten, E. (1987). *Superstring Theory,* in two volumes. Cambridge: Cambridge University Press.

Greene, B. R. (1997). String theory on Calabi–Yau manifolds. In *Fields, Strings, and Duality, TASI 1996,* eds. C. Efthimiou & B. Greene, pp. 543–726. Singapore: World Scientific. E-print hep-th/9702155.

Greene, B. R., & Plesser, M. R. (1990). Duality in Calabi–Yau moduli space. *Nuclear Physics,* **B338**, 15.

Greene, B. R., Morrison, D. R., & Strominger, A. (1995). Black hole condensation and the unification of string vacua. *Nuclear Physics,* **B451**, 109. E-print hep-th/9504145.

Grisaru, M. T., van de Ven, A. E. M., & Zanon, D. (1986). Two-dimensional supersymmetric sigma models on Ricci-flat Kähler manifolds are not finite. *Nuclear Physics,* **B277**, 388.

Gross, D. J., Harvey, J. A., Martinec, E., & Rohm, R. (1985). Heterotic string theory (I). The free heterotic string. *Nuclear Physics,* **B256**, 253.

Gross, D. J., Harvey, J. A., Martinec, E., & Rohm, R. (1986). Heterotic string theory (II). The interacting heterotic string. *Nuclear Physics,* **B267**, 75.

Gross, D. J., & Sloan, J. H. (1987). The quartic effective action for the heterotic string. *Nuclear Physics,* **B291**, 41.

Halpern, M. B., Kiritsis, E., Obers, N., & Clubok, K. (1996). Irrational conformal field theory. *Physics Reports,* **265**, 1. E-print hep-th/9501144.

Hamidi, S., & Vafa, C. (1987). Interactions on orbifolds. *Nuclear Physics,* **B279**, 465.

Harvey, J. A. (1997). Magnetic monopoles, duality, and supersymmetry. In *Fields, Strings, and Duality, TASI 1996*, eds. C. Efthimiou & B. Greene, pp. 157–216. Singapore: World Scientific. E-print hep-th/9603086.

Hitchin, N. J., Karlhede, A., Lindström, U., & Roček, M. (1987). Hyperkähler metrics and supersymmetry. *Communications in Mathematical Physics,* **108**, 535.

Hořava, P., & Witten, E. (1996). Heterotic and type I string dynamics from eleven dimensions. *Nuclear Physics,* **B460**, 506. E-print hep-th/9510209.

Horowitz, G. T. (1997). Quantum states of black holes. E-print gr-qc/9704072.

Horowitz, G. T., & Polchinski, J. (1997). Correspondence principle for black holes and strings. *Physical Review,* **D55**, 6189. E-print hep-th/9612146.

Horowitz, G. T., & Strominger, A. (1991). Black strings and p-branes. *Nuclear Physics,* **B360**, 197.

Hübsch, T. (1992). *Calabi–Yau Manifolds a Bestiary for Physicists.* Singapore: World Scientific.

Hull, C. M., & Townsend, P. K. (1995). Unity of superstring dualities. *Nuclear Physics,* **B438**, 109. E-print hep-th/9410167.

Hull, C. M., & Witten, E. (1985). Supersymmetric sigma models and the heterotic string. *Physics Letters,* **160B**, 398.

Ibáñez, L. E., & Lüst, D. (1992). Duality-anomaly cancellation, minimal string unification, and the effective low-energy Lagrangian of 4-D strings. *Nuclear Physics,* **B382**, 305. E-print hep-th/9202046.

Ibáñez, L. E., & Ross, G. G. (1992). Discrete gauge symmetries and the origin of baryon and lepton number conservation in supersymmetric versions of the standard model. *Nuclear Physics,* **B368**, 3.

Itzykson, C., Saleur, H., & Zuber, J.-B., eds. (1988). *Conformal Invariance and Applications to Statistical Mechanics.* Singapore: World Scientific.

Kachru, S., & Silverstein, E. (1997). Chirality changing phase transitions in 4d string vacua. *Nuclear Physics,* **B504**, 272. E-print hep-th/9704185.

Kakushadze, Z., Shiu, G., Tye, S. H. H., & Vtorov-Karevsky, Y. (1997). A review of three family grand unified string models. E-print hep-th/9710149.

Kaplunovsky, V. S. (1985). Mass scales of string unification. *Physical Review Letters,* **55**, 1036.

Kaplunovsky, V. S. (1988). One-loop threshold effects in string unification. *Nuclear Physics,* **B307**, 145; erratum *ibid.* **B382**, 436. E-print hep-th/9205068.

Kawai, H., Lewellen D. C., & Tye, S.-H. H. (1986a). A relation between tree amplitudes of closed and open strings. *Nuclear Physics,* **B269**, 1.

Kawai, H., Lewellen D. C., & Tye, S.-H. H. (1986b). Classification of closed-fermionic-string models. *Physical Review,* **D34**, 3794.

Kawai, H., Lewellen D. C., & Tye, S.-H. H. (1987). Construction of fermionic string models in four dimensions. *Nuclear Physics,* **B288**, 1.

Kiritsis, E. (1997). *Introduction to Superstring Theory.* Leuven: Leuven University, in press. E-print hep-th/9709062.

Knizhnik, V. G., & Zamolodchikov A. B. (1984). *Nuclear Physics,* **B247**, 83. Also in Goddard & Olive (1988), Itzykson *et al.* (1988).

La, H., & Nelson, P. (1989). Unambiguous fermionic string amplitudes. *Physical Review Letters,* **63**, 24.

Lerche, W., Nilsson, B. E. W., Schellekens, A. N., & Warner, N. P. (1988). Anomaly cancelling terms from the elliptic genus. *Nuclear Physics,* **B299**, 91.

Lerche, W., Schellekens, A. N., & Warner, N. P. (1989). Lattices and strings. *Physics Reports,* **177**, 1.

Lerche, W., Vafa, C., & Warner, N. P. (1989). Chiral rings in $N = 2$ superconformal theories. *Nuclear Physics,* **B324**, 427.

Lifschytz, G. (1996). Comparing D-branes to black-branes. *Physics Letters,* **B388**, 720. E-print hep-th/9604156.

Lüst, D., & Theisen, S. (1989). *Lectures on String Theory.* Berlin: Springer-Verlag.

Maldacena, J. M. (1997). The large-N limit of superconformal field theories and supergravity. E-print hep-th/9711200.

Maldacena, J. M. (1998). Black holes and D-branes. *Nuclear Physics Proceedings Supplement,* **61A**, 111. E-print hep-th/9705078.

Marcus, N., & Sagnotti, A. (1982). Tree-level constraints on gauge groups for type I superstrings. *Physics Letters,* **119B**, 97.

Martinec, E. (1986). Nonrenormalization theorems and fermionic string finiteness. *Physics Letters,* **B171**, 189.

Martinec, E. (1987). Conformal field theory on a (super-)Riemann surface. *Nuclear Physics,* **B281**, 157.

Martinec, E. (1989). Algebraic geometry and effective Lagrangians. *Physics Letters,* **B217**, 431.

Misner, C. W., Thorne, K. S., & Wheeler, J. A. (1973). *Gravitation.* San Francisco: Freeman.

Moore, G., & Seiberg, N. (1989). Classical and quantum conformal field theory. *Communications in Mathematical Physics,* **123**, 177.

Mumford, D. (1983). *Tata Lectures on Theta,* volume one. Boston: Birkhauser.

Narain, K. S. (1986). New heterotic string theories in uncompactified dimensions < 10. *Physics Letters,* **169B**, 41.

Narain, K. S., Sarmadi, M. H., & Vafa, C. (1987). Asymmetric orbifolds. *Nuclear Physics,* **B288**, 551.

Narain, K. S., Sarmadi, M. H., & Witten, E. (1987). A note on toroidal compactification of heterotic string theory. *Nuclear Physics,* **B279**, 369.

Ooguri, H., & Vafa, C. (1991). Geometry of $N = 2$ strings. *Nuclear Physics,* **B361**, 469.

Ooguri, H., & Yin, Z. (1997). Lectures on perturbative string theories. In *Fields, Strings, and Duality, TASI 1996,* eds. C. Efthimiou & B. Greene, pp. 5–82. Singapore: World Scientific. E-print hep-th/9612254.

Page, D. N. (1994). Black hole information. In *Proceedings of the 5th Canadian Conference on General Relativity and Relativistic Astrophysics,* eds. R. B. Mann & R. G. McLenaghan, pp. 1–41. Singapore: World Scientific. E-print hep-th/9305040.

Pati, J. C. (1996). The essential role of string-derived symmetries in ensuring proton-stability and light neutrino masses. *Physics Letters,* **B388**, 532.

Peet, A. W. (1997). The Bekenstein formula and string theory. E-print hep-th/9712253.

Peskin, M. (1987). Introduction to string and superstring theory. II. In *From the Planck Scale to the Weak Scale, TASI 1986,* ed. H. Haber, pp. 277–408. Singapore: World Scientific.

Polchinski, J. (1994). Combinatorics of boundaries in string theory. *Physical Review,* **D50**, 6041. E-print hep-th/9407031.

Polchinski, J. (1995). Dirichlet branes and Ramond–Ramond charges. *Physical Review Letters,* **75**, 4724. E-print hep-th/9510017.

Polchinski, J. (1997). TASI lectures on D-branes. In *Fields, Strings, and Duality, TASI 1996,* eds. C. Efthimiou & B. Greene, pp. 293–356. Singapore: World Scientific. E-print hep-th/9611050.

Polchinski, J., & Witten, E. (1996). Evidence for heterotic–type I string duality. *Nuclear Physics,* **B460**, 525. E-print hep-th/9510169.

Pope, C. N. (1995). *W* strings '93. In *Proceedings of Strings '93,* eds. M. B. Halpern, G. Rivlis, & A. Sevrin, pp. 439–450. Singapore: World Scientific. E-print hep-th/9309125.

Qiu, Z. (1987). Nonlocal current algebra and $N = 2$ superconformal field theory in two dimensions. *Physics Letters,* **B188**, 207.

Quevedo, F. (1996). Lectures on superstring phenomenology. In *Workshops on Particles and Fields and Phenomenology of Fundamental Interactions, Puebla 1995,* eds. J. C. D'Olivo, A. Fernánadez, & M. A. Pérez, pp. 202–242. Woodbury, NY: American Institute of Physics. E-print hep-th/9603074.

Roček, M. (1993). Introduction to supersymmetry. In *Recent Directions in Particle Theory, TASI 1992,* eds. J. Harvey & J. Polchinski, pp. 101–140. Singapore: World Scientific, 1993.

Ross, G. G. (1984). *Grand Unified Theories.* Menlo Park: Benjamin/Cummings.

Sagnotti, A. (1997). Surprises in open string perturbation theory. *Nuclear Physics Proceedings Supplement,* **56B**, 332. E-print hep-th/9702093.

Salam, A., & Sezgin, E., eds. (1989). *Supergravities in Diverse Dimensions.* New York, NY: Elsevier.

Schellekens, A. N. (1990). Electric charge quantization in string theory. *Physics Letters,* **B237**, 363.

Schwarz, J. H. (1982). Superstring theory. *Physics Reports,* **89**, 223.

Schwarz, J. H., ed. (1985). *Superstrings. The First 15 Years of Superstring Theory,* in two volumes. Singapore: World Scientific.

Schwarz, J. H. (1997). Lectures on superstring and M theory dualities. In *Fields, Strings, and Duality, TASI 1996,* eds. C. Efthimiou & B. Greene, pp. 359–418. Singapore: World Scientific. E-print hep-th/9607201.

Seiberg, N. (1988). Observations on the moduli space of superconformal field theories. *Nuclear Physics,* **B303**, 286.

Seiberg, N., & Witten, E. (1994). Electric–magnetic duality, monopole condensation, and confinement in $N = 2$ supersymmetric Yang–Mills theory. *Nuclear Physics,* **B426**, 19. E-print hep-th/9407087.

Seiberg, N., & Witten, E. (1996). Comments on string dynamics in six dimensions. *Nuclear Physics,* **B471**, 121. E-print hep-th/9603003.

Sen, A. (1994). Strong–weak coupling duality in four-dimensional string theory. *International Journal of Modern Physics,* **A9**, 3707. E-print hep-th/9402002.

Sen, A. (1996a). A note on marginally stable bound states in type II string theory. *Physical Review,* **D54**, 2964. E-print hep-th/9510229.

Sen, A. (1996b). *U*-duality and intersecting D-branes. *Physical Review,* **D53**, 2874. E-print hep-th/9511026.

Sen, A. (1997). Unification of string dualities. *Nuclear Physics Proceedings Supplement,* **58**, 5. E-print hep-th/9609176.

Sen, A. (1998). An introduction to nonperturbative string theory. E-print hep-th/9802051.

Sethi, S., & Stern, M. (1997). D-brane bound states redux. E-print hep-th/9705046.

Sevrin, A., Troost, W., & Van Proeyen, A. (1988). Superconformal algebras in two dimensions with $N = 4$. *Physics Letters,* **208B**, 447.

Shenker, S. (1991). The strength of nonperturbative effects in string theory. In *Random Surfaces and Quantum Gravity*, eds. O. Alvarez, E. Marinari, & P. Windey, pp. 191–200. New York: Plenum.

Shenker, S. (1995). Another length scale in string theory? E-print hep-th/9509132.

Slansky, R. (1981). Group theory for unified model building. *Physics Reports,* **79**, 1.

Sohnius, M. F. (1985). Introducing supersymmetry. *Physics Reports,* **128**, 39.

Stelle, K. S. (1998). BPS branes in supergravity. E-print hep-th/9803116.

Strominger, A. (1986). Superstrings with torsion. *Nuclear Physics,* **B274**, 253.

Strominger, A. (1995). Massless black holes and conifolds in string theory. *Nuclear Physics,* **B451**, 96. E-print hep-th/9504090.

Strominger, A. (1996). Open *p*-branes. *Physics Letters,* **B383**, 44. E-print hep-th/9512059.

Strominger, A., & Vafa, C. (1996). Microscopic origin of the Bekenstein–Hawking entropy. *Physics Letters,* **B379**, 99. E-print hep-th/9601029.

Strominger, A., & Witten, E. (1985). New manifolds for superstring compactification. *Communications in Mathematical Physics,* **101**, 341.

Strominger, A., Yau, S.-T., & Zaslow, E. (1996). Mirror symmetry is *T*-duality. *Nuclear Physics,* **B479**, 243. E-print hep-th/9606040.

Susskind, L. (1995). The world as a hologram. *Journal of Mathematical Physics,* **36**, 6377. E-print hep-th/9409089.

Thorn, C. B. (1984). Computing the Kac determinant using dual model techniques and more about the no-ghost theorem. *Nuclear Physics,* **B248**, 551.

Townsend, P. K. (1995). The eleven-dimensional membrane revisited. *Physics Letters,* **B350**, 184. E-print hep-th/9501068.

Townsend, P. K. (1996). Four lectures on M-theory. In *High-Energy Physics and Cosmology, Trieste 1996*, eds. E. Gava *et al.*, pp. 385–438. Singapore: World Scientific. E-print hep-th/9612121.

Tseytlin, A. A. (1996). On $SO(32)$ heterotic–type I superstring duality in ten dimensions. *Physics Letters*, **B367**, 84. E-print hep-th/9510173.

Tseytlin, A. A. (1997). On non-Abelian generalisation of Born–Infeld action in string theory. *Nuclear Physics*, **B501**, 41. E-orint hep-th/9701125.

Tye, S. H. H. (1995). Status of fractional superstrings. In *Proceedings of Strings '93*, eds. M. B. Halpern, G. Rivlis, & A. Sevrin, pp. 364–378. Singapore: World Scientific. E-print hep-th/9311021.

Vafa, C. (1986). Modular invariance and discrete torsion on orbifolds. *Nuclear Physics*, **B273**, 592.

Vafa, C. (1988). Toward classification of conformal theories. *Physics Letters*, **206B**, 421.

Vafa, C. (1989). String vacua and orbifoldized LG models. *Modern Physics Letters*, **A4**, 1169.

Vafa, C. (1996a). Gas of D-branes and Hagedorn density of BPS states. *Nuclear Physics*, **B463**, 415. E-print hep-th/9511088.

Vafa, C. (1996b). Instantons on D-branes. *Nuclear Physics*, **B463**, 435. E-print hep-th/9512078.

Vafa, C. (1996c). Evidence for F-theory. *Nuclear Physics*, **469**, 403. E-print hep-th/9602022.

Vafa, C., & Warner, N. (1989). Catastrophes and the classification of conformal theories. *Physics Letters*, **B218**, 51.

Vafa, C., & Witten, E. (1995). Dual string pairs with $N = 1$ and $N = 2$ supersymmetry in four dimensions. E-print hep-th/9507050.

Verlinde, E., & Verlinde, H. (1987). Multiloop calculations in covariant superstring theory. *Physics Letters*, **B192**, 95.

Wald, R. M. (1997). Black holes and thermodynamics. E-print gr-qc/9702022.

Weinberg, S. (1989). The cosmological constant problem. *Reviews of Modern Physics*, **61**, 1.

Wess, J., & Bagger, J. (1992). *Supersymmetry and Supergravity*. Princeton: Princeton University Press.

Witten, E. (1981). Dynamical breaking of supersymmetry. *Nuclear Physics*, **B188**, 513.

Witten, E. (1984). Non-Abelian bosonization in two dimensions. *Communications in Mathematical Physics*, **92**, 455. Also in Goddard & Olive (1988).

Witten, E. (1985). Dimensional reduction of superstring models. *Physics Letters*, **155B**, 151.

Witten, E. (1986). New issues in manifolds of $SU(3)$ holonomy. *Nuclear Physics*, **B268**, 79.

Witten, E. (1988). Topological sigma models. *Communications in Mathematical Physics,* **118**, 411.

Witten, E. (1993). Phases of $N = 2$ theories in two dimensions. *Nuclear Physics,* **B403**, 159. E-print hep-th/9301042.

Witten, E. (1995). String theory dynamics in various dimensions. *Nuclear Physics,* **B443**, 85. E-print hep-th/9503124.

Witten, E. (1996a). Bound states of strings and p-branes. *Nuclear Physics,* **B460**, 335. E-print hep-th/9510135.

Witten, E. (1996b). Small instantons in string theory. *Nuclear Physics,* **B460**, 541. E-print hep-th/9511030.

Witten, E. (1996c). Strong coupling expansion of Calabi–Yau compactification. *Nuclear Physics,* **B471**, 135. E-print hep-th/9602070.

Witten, E., & Olive, D. (1978). Supersymmetry algebras that include topological charges. *Physics Letters,* **B78**, 97.

Wybourne, B. G. (1974). *Classical Groups for Physicists.* New York: Wiley.

Zamolodchikov, A. B. (1986a). Conformal symmetry and multicritical points in two-dimensional quantum field theory. *Soviet Journal of Nuclear Physics,* **44**, 529. Also in Itzykson *et al.* (1988).

Zamolodchikov, A. B. (1986b). 'Irreversibility' of the flux of the renormalization group in a 2D field theory. *JETP Letters,* **43**, 730. Also in Itzykson *et al.* (1988).

Zamolodchikov, A. B., & Fateev, V. A. (1985). Nonlocal (parafermion) currents in two-dimensional conformal quantum field theory and self-dual critical points in Z_N-symmetric statistical mechanics. *Soviet Physics JETP,* **62**, 215. Also in Itzykson *et al.* (1988).

Zamolodchikov, A. B., & Fateev, V. A. (1986). Disorder fields in two-dimensional conformal quantum field theory and $N = 2$ extended supersymmetry. *Soviet Physics JETP,* **63**, 913.

Glossary

(0, 2) compactification a generic heterotic string vacuum having $d = 4$, $N = 1$ supersymmetry. The world-sheet CFT has $N = 2$ right-moving superconformal invariance.

(2, 0) theory in current usage, this refers to a family of nontrivial fixed point theories with tensionless strings and $d = 6$ $(2, 0)$ supersymmetry. These arise on coincident M5-branes and IIA NS5-branes, and on the IIB theory at an A–D–E singularity.

(2, 2) compactification one of a special subset of $d = 4$, $N = 1$ heterotic string vacua, which includes the Calabi–Yau compactifications. The world-sheet CFT has both right-moving and left-moving $N = 2$ superconformal invariance. In the type II string theories, these CFTs give vacua with $d = 4$, $N = 2$ supersymmetry.

A–D–E singularity a singularity of a four-(real)-dimensional complex manifold, resulting from the collapse of one or more two-spheres to zero volume. The terminology A–D–E refers to the Dynkin diagrams of the simply-laced Lie algebras, which describe the intersection numbers of the collapsed spheres.

Abelian differential a globally defined holomorphic (1,0)-form on a Riemann surface.

abstruse identity one of a set of quartic theta function identities due to Jacobi, it implies the degeneracy of bosons and fermions in GSO-projected string theories as required by supersymmetry.

affine Lie algebra see *current algebra*.

anomaly the violation of a classical symmetry by quantum effects. A *gravitational anomaly* is an anomaly in coordinate invariance. A *global anomaly* is an anomaly in a large symmetry transformation (one not continuously connected to the identity).

anomaly polynomial a formal $(d + 2)$-form in d-dimensions, which encodes the gauge and gravitational anomalies.

asymptotically locally Euclidean (ALE) space a space which at long distance approaches flat Euclidean space identified under a discrete group. This is the geometry in the neighborhood of an orbifold fixed point (or blown-up fixed point).

Atiyah–Drinfeld–Hitchin–Manin (ADHM) construction a method for the construction of all Yang–Mills field configurations having self-dual field strength.

auxiliary field a nonpropagating field, one whose field equation is algebraic rather than differential. In many supersymmetric theories, the transformations can be simplified by introducing such fields.

axion a Goldstone boson associated with spontaneously broken PQ symmetry. The *model-independent axion* appears in every perturbative string theory, and is closely related to the graviton and dilaton.

bc **CFT** a free CFT of anticommuting fields with an action of first order in derivatives. There is a family of such CFTs, parameterized by the weight $h_b = 1 - h_c$. For $h_b = 2$ this CFT describes the Faddeev–Popov ghosts associated with conformal invariance.

$\beta\gamma$ **CFT** a free CFT of commuting fields with an action of first order in derivatives. There is a family of such CFTs, parameterized by the weight $h_\beta = 1 - h_\gamma$. For $h_\beta = \frac{3}{2}$ this CFT describes the Faddeev–Popov ghosts associated with superconformal invariance.

Batalin–Vilkovisky formalism an extension of the BRST formalism, for quantizing more general theories with constraints. This has been useful in string field theory.

Becchi–Rouet–Stora–Tyutin (BRST) invariance a nilpotent symmetry of Faddeev–Popov gauge-fixed theories, which encodes the information contained in the original gauge symmetry.

Beltrami differential the derivative with respect to the moduli of the complex structure of a Riemann surface.

Berezin integration a linear operation taking functions of Grassmann variables to complex numbers, with many of the key properties of ordinary integration.

beta function 1. in quantum field theory, the derivative of the effective strength of an interaction with respect to length scale; 2. a special function involving a ratio of gamma functions, which appears in the Veneziano amplitude.

Betti numbers the number of nontrivial p-forms in de Rham cohomology, denoted B_p.

black hole entropy a quantity S proportional to the area of the horizon of a black hole, $S = 2\pi A/\kappa^2$. This has the properties of a thermodynamic entropy: it is nondecreasing in classical general relativity, and the sum of the black hole entropy and the ordinary entropy is nondecreasing even with the inclusion of Hawking radiation. To find a statistical mechanical derivation of this entropy has been a major goal, partly realized in recent work.

black hole evaporation the emission of thermal (Hawking) radiation by a black hole, due to pair production near the horizon.

black hole information paradox a conflict between quantum mechanics and general relativity. Information falling into a black hole is lost and does not reappear when the black hole evaporates; this is inconsistent with ordinary quantum mechanical evolution. It apparently requires either a significant modification of quantum mechanics, or a significant breakdown of the usual understanding of locality.

black p-brane a p-dimensional extended object with an event horizon: a space that is translationally invariant in p directions and has a black hole geometry in the remaining directions.

blow up to deform a singular manifold into a smooth manifold.

Bogomolnyi–Prasad–Sommerfield (BPS) state a state that is invariant under a nontrivial subalgebra of the full supersymmetry algebra. Such states always carry conserved charges, and the supersymmetry algebra determines the mass of the state exactly in terms of its charges. BPS states lie in smaller supersymmetry representations than non-BPS states, so-called *short representations*. When there are short representations of different sizes, one also distinguishes *ultrashort representations,* which are the smallest possible (generally their dimension is the square root of the non-BPS dimension).

Borel summation a method of defining the sum of a divergent series. This has been used as a means of studying nonperturbative effects in field and string theories, but it should be understood that most nonperturbative effects are not usefully studied in terms of the perturbation series.

Born–Infeld action a generalization of the usual gauge field action which is nonpolynomial in the gauge field strength. This was originally proposed as a possible short-distance modification of electromagnetism. It arises as the low energy effective action of the gauge fields on D-branes.

bosonization the exact equivalence of a theory of fermionic fields and a theory of bosonic fields, possible in two dimensions. The boson is a fermion–antifermion pair; the fermion is a coherent state of bosons.

c-**map** a method for constructing the hypermultiplet moduli space of a type II string theory compactified on a Calabi–Yau three-fold from the vector multiplet moduli space of the other type II theory on the same three-fold.

c-**theorem** the existence, in unitary CFTs in two dimensions, of a positive quantity c that is monotonically nonincreasing with increasing length scale, and which at fixed points is stationary and equal to the central charge. This is a strong constraint on the global form of the renormalization group flow; no simple analog seems to exist in higher dimensions. Also known as the *Zamolodchikov c-theorem.*

CPT symmetry the combined operation of parity-reversal, time-reversal, and charge conjugation, which is a symmetry of all Lorentz-invariant local quantum field theories.

Calabi–Yau manifold a Kähler manifold with vanishing first Chern class. A *Calabi–Yau n-fold* has $2n$ real $= n$ complex coordinates. Yau's theorem guarantees the existence of a Ricci-flat metric of $SU(n)$ holonomy.

canceled propagator argument a general principle implying, under broad conditions, the vanishing of surface terms on the moduli space of Riemann surfaces and therefore the decoupling of unphysical states in string amplitudes. Such amplitudes are defined by analytic continuation from a regime where the integrand falls rapidly at the boundary and the surface term is identically zero; its continuation is therefore also identically zero.

Casimir energy a shift in the ground state energy of a quantum field theory due to boundary conditions on the fields.

center-of-mass mode the zeroth spatial Fourier component of a quantum field.

central charge an operator (which might be a constant) that appears on the right-hand side of a Lie algebra and commutes with all operators in the algebra. Prominent examples include the constant term in the Virasoro algebra and the charges appearing on the right-hand sides of many supersymmetry algebras.

Chan–Paton degrees of freedom degrees of freedom localized at the endpoints of open strings. These are now interpreted as designating the D-brane on which the string ends.

Chan–Paton factor the vertex operator factor for the state of the Chan–Paton degrees of freedom.

Chern–Simons term a term in the action which involves p-form potentials as well as field strengths. Such a term is gauge-invariant as a consequence of the Bianchi identity and/or the modification of the p-form gauge transformation. These terms usually have a close connection to topology and to anomalies.

chiral 1. acting in a parity asymmetric fashion; see *chiral multiplet, chiral symmetry, chiral theory, chirality, extended chiral algebra*; 2. invariant under part of the supersymmetry algebra; see *chiral field, chiral multiplet, chiral primary, chiral ring*.

chiral field in supersymmetry, a local operator that is invariant under part of the algebra: the operator analog of a *BPS state*.

chiral multiplet the multiplet of $d = 4$, $N = 1$ supersymmetry with two real scalars. The quarks and leptons are contained in such multiplets. This multiplet is connected to both senses of *chiral*: it contains a fermion with chiral couplings, and the integral of the associated superfield is invariant under half of the supersymmetry algebra.

chiral primary in an $N = 2$ SCFT, a primary field that is also annihilated by one of the rigid supersymmetries $G^{\pm}_{-1/2}$.

chiral ring the closed OPE algebra of chiral fields.

chiral symmetry a symmetry whose action on spinor fields is not parity-symmetric.

chiral theory a gauge theory in which the gauge couplings are not parity-symmetric.

chirality in $d = 2k$ dimensions, the eigenvalue of the operator Γ^d which anti-commutes with the Γ^μ. This eigenvalue distinguishes the two Weyl representations, which are related to one another by parity.

Christoffel connection in general relativity, the connection that is constructed from the metric.

critical behavior the behavior of a quantum field theory at an IR fixed point with massless fields, and the approach to this behavior.

closed string a string with the topology of a circle.

cocycle in a vertex operator, an operator-valued phase factor which multiplies the creation–annihilation normal ordered exponential. This in needed in some cases in order to give the operator the correct commuting or anticommuting property.

coefficient functions the position-dependent coefficients of local operators appearing in the expansion of an operator product.

cohomology in any vector space with a nilpotent operator Q (one such that $Q^2 = 0$), the kernel of Q modulo the image of Q. That is, the space of *closed* states (those annihilated by Q) with the *exact* states (those of the form $Q\psi$) defined to be equivalent to zero. *De Rham cohomology* is the cohomology of the exterior derivative d acting on differential forms. On a complex manifold, *Dolbeault cohomology* is the cohomology of ∂ and $\bar\partial$ (the $(1,0)$ and $(0,1)$ parts of d) on (p,q)-forms. *Homology* is the cohomology of the boundary operator. *BRST cohomology* is the cohomology of the BRST operator, and defines the physical space of a gauge-invariant theory.

Coleman–Weinberg formula the expression for the vacuum energy density of a free quantum field, from the renormalized sum of the zero-point energies of its modes.

collapsing cycle a cycle whose volume vanishes in a limit, usually giving rise to a singular manifold.

collective coordinate in quantizing a soliton or other extended object, the degrees of freedom corresponding to its position or configuration.

compact CFT a CFT in which the number of states with energy less than any given value is finite. This is defined by analogy with the spectrum of a differential operator on a compact space.

compactification scale the characteristic mass scale of states whose wavefunctions have a nontrivial dependence on the compact dimensions.

compactify to consider a field theory or string theory in a spacetime, some of whose spatial dimensions are compact.

complex manifold a manifold with an assigned system of complex coordinates, modulo holomorphic reparameterizations of these coordinates.

complex structure an equivalence class of complex coordinates. A given differentiable manifold may have many inequivalent complex structures.

complex structure moduli the moduli that parameterize the inequivalent complex structures on a manifold. In compactification on a Calabi–Yau 3-fold, these are associated with $(2, 1)$-forms.

conformal block in CFT, the contribution of a single conformal family to a sum over states.

conformal bootstrap the partially successful program to construct all CFTs by using only symmetry and consistency conditions.

conformal family the set of states obtained by acting on a highest weight state with Virasoro raising generators in all inequivalent ways; or, the corresponding set of local operators. A *degenerate conformal family* contains null states, which are orthogonal to all states in the family.

conformal field theory (CFT) a conformally invariant quantum field theory.

conformal gauge a choice of coordinates in two dimensions, such that the metric is proportional to the unit metric.

conformal Killing vector a globally defined infinitesimal diff×Weyl transformation that leaves the metric invariant.

conformal transformation a mapping of Euclidean or Minkowski space to itself that leaves the flat metric invariant up to a position-dependent rescaling; equivalently, the subgroup of diff × Weyl that leaves invariant the flat metric. In $d \geq 3$ dimensions this has $\frac{1}{2}(d + 1)(d + 2)$ parameters. In two dimensions it is the set of all holomorphic maps. Finite transformations require the inclusion of points at infinity, as in the case of the Möbius transformations of the sphere.
This usage has become standard in string theory and quantum field theory, but in general relativity *conformal transformation* is defined to be any position-dependent rescaling of the metric, now called a *Weyl transformation* in string theory.

conifold a Calabi–Yau manifold with a singular complex structure, corresponding to the collapse of a three-cycle. The string theory on this space is singular; this is now understood to be due to the quantum effects of a massless 3-brane wrapped on the cycle.

conifold transition a change of topology due to condensation of massless 3-brane fields. Under appropriate conditions, the potential for the massless 3-brane fields on a conifold with multiple collapsed cycles has a flat direction for these fields; this corresponds to a change of topology, blowing up a 2-cycle rather than the collapsed 3-cycle.

constraint a symmetry generator whose matrix elements are required to vanish in physical states, either in the BRST or OCQ sense. These can usually be understood as arising from a gauge symmetry (for so-called *first class constraints,* which are all that we consider), and consist of those gauge symmetry generators that do not vanish by the equations of motion.

coset CFT a CFT constructed as one of the factors of a known CFT, when the energy-momentum tensor of the latter can be written as a sum of commuting pieces. In the classic example the full CFT G is a current algebra, as is one of the factors (H). This can also be thought of as gauging the symmetry H.

cosmological constant the energy density of the vacuum. In a nonsupersymmetric quantum theory (including one with spontaneously broken supersymmetry), there are many effects that give rise to such an energy density. The *cosmological constant problem* is the problem that the cosmological constant in nature is many orders of magnitude smaller than known nonzero effects.

critical dimension the dimension in which a perturbative string theory is consistent in flat Minkowski spacetime; the value is 26 for the bosonic string and 10 for the supersymmetric string theories.

current algebra in quantum field theory, the algebra of the currents associated with a continuous symmetry group g (or of their Fourier modes). As used here, it is the specific algebra that occurs in two-dimensional CFTs, with the energy-momentum tensor defined to be of Sugawara form. The terms *affine Lie algebra* and *affine Kac–Moody algebra* are also used for this algebra, though like *current algebra* they both have broader definitions as well. The term *affine* refers to the c-number (Schwinger) term. An *untwisted* current algebra is the algebra of periodic currents, with integer modes. An algebra can be *twisted* by any automorphism of g.

cycle a topologically nontrivial submanifold (in the sense of homology); a *p-cycle* is p-dimensional. The *A-* and *B-cycles* are a standard basis for the nontrivial one-cycles on a Riemann surface.

D-brane in the type I, IIA, and IIB string theories, a dynamical object on which strings can end. The term is a contraction of *Dirichlet brane.* The coordinates of the attached strings satisfy Dirichlet boundary conditions in the directions normal to the brane and Neumann conditions in the directions tangent to the brane. A *Dp-brane* is p-dimensional, with p taking any even value in the IIA theory, any odd value in the IIB theory, and the values 1, 5, and 9 in the type I theory; a D9-brane fills space and so corresponds to an ordinary Neumann boundary condition. The Dp-brane is a source for the $(p + 1)$-form R–R gauge field. The mass or tension of a D-brane is intermediate between that of an ordinary quantum or a fundamental string and that of a *soliton.* The low energy fluctuations of D-branes are described by supersymmetric gauge theory, which is non-Abelian for coincident branes.

D-instanton an object localized in (Euclidean) time as well as space, defined by Dirichlet conditions on all coordinates of attached strings. This is similar to a field-theoretic instanton, corresponding to a tunneling process that changes the value of an R–R field strength. More generally, the $(p + 1)$-dimensional world-volume of a Dp-brane, when localized in time and wrapped on a $(p + 1)$-cycle in space, has similar effects.

D-string a D1-brane, in the type I and IIB string theories.

D-term 1. in gauge theories with four or eight supersymmetries, the auxiliary field in the gauge multiplet; 2 the potential term proportional to the square of this auxiliary field, which depends only on the gauge couplings and, in the $U(1)$ case, the value of the Fayet–Iliopoulos parameter.

Del Guidice–Di Vecchia–Fubini (DDF) operators operators satisfying an oscillator algebra, which create a complete set of physical states in OCQ.

descendant a state obtained by acting on a highest weight state with Virasoro raising generators.

diagonal modular invariant a modular-invariant CFT formed by imposing common boundary conditions on the left- and right-moving fields.

diff invariance general coordinate (reparameterization) invariance, usually applied to the world-sheet coordinates.

dilaton the massless scalar with gravitational-strength couplings, found in all perturbative string theories. An exactly massless dilaton would violate limits on nongravitational interactions, but a mass for the dilaton is not forbidden by any symmetry and so dynamical effects will generate one in vacua with broken supersymmetry (the same holds for other moduli). The string coupling constant is determined by the value of the dilaton field.

dimensional reduction in the simplest cases, toroidal compactification retaining only the states of zero compact momentum. More generally (and less physically) the construction of a lower-dimensional field theory by requiring all fields to be invariant under a set of symmetries; this may have no interpretation in terms of compactification.

Dirac quantization condition for an electric and a magnetic charge, the condition that the product be quantized, $\mu_e \mu'_m = 2\pi n$. For two dyons, which have both charges, the condition is $\mu_e \mu'_m - \mu_m \mu'_e = 2\pi n$. These conditions generalize to objects of dimension p and $d - p - 4$ in d dimensions, where one is the source of a $(p + 2)$-form field strength and the other of the Poincaré dual $(d - p + 2)$-form field strength.

Dirac spinor the unique irreducible representation of the algebra of Dirac matrices, which is also known as a Clifford algebra. This is also a representation of the Lorentz group; in even dimensions it is reducible to two Weyl representations of the Lorentz group. The Dirac spinor is complex; in certain dimensions a further Majorana (reality) condition is compatible with Lorentz invariance.

Dirichlet boundary condition the condition that the value of a field be fixed at a boundary. This is the relevant usage in string theory, but in other contexts only the tangent derivative need be fixed.

discrete torsion in forming a *twisted CFT,* a change in the phases of the path integral sectors and therefore in the projection on the Hilbert space.

doubling trick the representation of holomorphic and antiholomorphic fields on a manifold with boundary by holomorphic fields alone, on the doubled copy of the manifold obtained by reflecting through the boundary.

dual resonance model a phenomenological model of the strong interaction, which developed into string theory. The *dual* refers here to *world-sheet duality.*

duality the equivalence of seemingly distinct physical systems. Such an equivalence often arises when a single quantum theory has distinct classical limits. One classic example is *particle–wave duality,* wherein a quantum field theory has one limit described by classical field theory and another described by classical particle mechanics. Another is the high-temperature–low-temperature duality of the Ising model. Here, low temperature is the statistical mechanical analog of the classical limit, the Boltzmann sum being dominated by the configurations of lowest energy. See *Montonen–Olive duality, S-duality, string–string duality, T-duality, U-duality, world-sheet duality.*

effective field theory the description of a physical system below a given energy scale (or equivalently, above a given length scale).

Einstein metric the metric whose leading low energy action is the *Hilbert action*

$$\frac{1}{2\kappa^2} \int d^d x \, (-G)^{1/2} R \; ;$$

this is independent of other fields. Here κ is the gravitational coupling, related to the Planck length by $\kappa = (8\pi)^{1/2} L_{\mathrm{P}}$. This metric is related to other metrics such as the *sigma-model metric* by a field-dependent Weyl transformation. The existence of distinct metrics would appear to violate the equivalence principle, but when the dilaton and other moduli are massive the distinction disappears.

electroweak scale the mass scale of electroweak symmetry breaking, roughly 10^2 GeV.

enhanced gauge symmetry a gauge symmetry appearing at special points in moduli space, which is not evident in the original formulation of a theory. The classic examples are the gauge symmetries that arise at special radii of toroidal compactification, whose gauge bosons are winding states. Many other mechanisms are now known: D-branes and black p-branes wrapped on collapsing cycles, F-strings or D-strings stretched between various branes in the limit that the latter become coincident, and the gauge symmetry appearing on a zero size $SO(32)$ instanton.

Euclidean having a metric of strictly positive signature. The original connotation that the metric be flat is somewhat disregarded; thus one refers to *Euclidean quantum gravity,* a conjectured analytic continuation of the Minkowskian theory. For the metric itself, the term *Riemannian* for a curved metric of Euclidean signature is more precise.

Euclidean adjoint in a Euclidean quantum theory, the Hermitean adjoint combined with time-reversal. The latter operation undoes the time-reversing effect of the adjoint, so that the combined operation is local.

Euler number the topological invariant

$$\chi = \sum_{p=0}^{d} (-1)^p B_p \, ,$$

where B_p is the pth Betti number. It is equal to $2(1 - g)$ for a Riemann surface of genus g. More properly, the *Euler characteristic*.

expectation value a path integral with specified insertions. This is the term that we have chosen to use, but *correlation function* and *correlator* are also in common usage.

extended chiral algebra the full set of holomorphic operators in a CFT.

extended supersymmetry a supersymmetry algebra in which the supercharges comprise more than one copy of the smallest spinor representation for the given spacetime dimension.

***F*-term** 1. the auxiliary field in the chiral multiplet of $d = 4$, $N = 1$ supersymmetry; 2. the potential term proportional to the square of this field.

F theory 1. a systematic description of IIB superstring states with nontrivial dilaton and R–R scalar backgrounds, which relates these fields to the modulus τ of an auxiliary two-torus; 2. a conjectured twelve-dimensional quantum theory underlying the IIB string. The name is inspired by M-theory, with F for father.

Faddeev–Popov determinant the Jacobian determinant arising from the reduction of a gauge-invariant functional integral to an integral over a gauge slice.

Faddeev–Popov ghosts the wrong-statistics quantum fields used to give a functional integral representation of the Faddeev–Popov determinant.

Fayet–Iliopoulos term in $U(1)$ gauge theories with four or eight supersymmetries, a term in the action which is linear in the auxiliary D-term field.

Feigin–Fuchs representation a representation of minimal model expectation values in terms of free fields.

fibration a space which is locally the product of a fiber F and a base B. The geometry of the fiber varies as one moves over the base, and may become singular. Typical fibers are tori and the $K3$ manifold.

first Chern class on a complex manifold, the Dolbeault cohomology class of the Ricci form $R_{i\bar{j}} dz^i d\bar{z}^{\bar{j}}$.

first-quantized description the representation of a quantized particle theory as a sum over particle paths, or of a string theory as a sum over worldsheets. *Second-quantized* refers to the representation in terms of a functional integral over ordinary or string fields. The term second-quantized implies the reinterpretation of the first-quantized *wavefunction* as a *field operator*. This terminology is in common usage, but it has been argued that it is unsatisfactory, in that it implies a deep principle where none may exist. Since the sum over world-sheets is itself a quantum field theory, one can equally well call it second-quantized, in which case string field theory is *third-quantized*. Third quantization of an ordinary field theory would describe operators that create and destroy universes, a concept which may or may not be useful.

Fischler–Susskind mechanism the cancellation of divergences and anomalies in the world-sheet quantum field theory against divergences and anomalies from integration over small topological features at higher orders of string perturbation theory. This is needed for the consistency of string perturbation theory in a quantum-corrected background.

fixed point 1. (*in geometry*) a point left invariant by a given symmetry transformation. This becomes a boundary point or a singularity if the space is identified under the transformation; 2. (*in quantum field theory*) a quantum theory whose physics is independent of length scale (scale-invariant). Usually such a theory is conformally invariant as well. A *UV fixed point* is the theory governing the short-distance physics of a quantum field theory; an *IR fixed point* is the theory governing the long-distance physics of a quantum field theory. A *trivial IR fixed point* has no massless fields. A *nontrivial IR fixed point* has massless fields with nonvanishing interactions. A theory whose IR limit is a massless free field theory is therefore described by neither of these terms; it is a *noninteracting IR fixed point*.

flat direction in scalar field space, a line of degenerate local minima. The field corresponding to this direction is a *modulus*.

flop a change of topology which can occur in weakly coupled string theory, where a two-cycle collapses and then a different two-cycle blows up.

fractional charge an unconfined particle whose electric charge is not a multiple of that of the electron. These exist in most $d = 4$ string theories, although in many cases all are superheavy.

fractional string theory a proposed generalization of string theory having constraints whose spin is not a multiple of $\frac{1}{2}$. No complete construction exists.

Fuchsian group a discrete subgroup Γ of the $SL(2, \mathbf{R})$ Möbius transformations of the complex upper-half-plane H, with additional conditions such that H/Γ is a manifold and in particular a Riemann surface.

functional integral in our usage, synonymous with *path integral*.

fundamental region in relation to a coset space M/Γ where Γ is a discrete group, a region F such that every point in M is identified with exactly one point in the interior of F or with one or more points on the boundary of F.

fundamental string (F-string) the original string whose quantization defines a weakly coupled string theory, as distinguished from D-strings and solitonic strings.

fusion rule the specification of which conformal families appear in the operator product of any two primary fields in a given CFT.

gauge-fixing the reduction of a redundant (gauge-invariant) description of a quantum theory to a description with a single representative from each equivalence class.

gaugino a spin-$\frac{1}{2}$ fermion in the same supersymmetry multiplet as a gauge boson.

gaugino condensation a strong coupling effect where a product of gaugino fields acquires a vacuum expectation value. This generally breaks a chiral symmetry but does not directly break supersymmetry; however, in combination with other fields it often induces supersymmetry breaking.

generation a family of quarks and leptons, described by spinor fields in a chiral but anomaly-free set of $SU(3) \times SU(2) \times U(1)$ representations. In $SU(5)$ grand unification these become a **5** + **$\overline{10}$**, in $SO(10)$ they are contained in a **16**, and in E_6 they are contained in a **27**. An *antigeneration* is the conjugate representation; the distinction between generation and antigeneration is a matter of convention.

genus the number g of handles on a closed oriented Riemann surface: $g = 0$ is a sphere, $g = 1$ is a torus, and so on.

Gepner model a string model based on $N = 2$ *minimal model* CFTs.

ghosts see *Faddeev–Popov ghosts*.

Gliozzi–Scherk–Olive (GSO) projection a construction of modular-invariant string theories by summing over R and NS boundary conditions on the fermion fields and projecting onto states of definite world-sheet fermion number. In supersymmetric string theories there are independent GSO projections on the left-movers and right-movers. The diagonal projection, which acts simultaneously on both sides, produces a nonsupersymmetric theory.

goldstino the massless spin-$\frac{1}{2}$ Goldstone fermion associated with spontaneously broken supersymmetry. In supergravity it combines with the gravitino to form a massive fermion.

Goldstone boson the massless scalar corresponding to fluctuations of the direction of spontaneous symmetry breaking.

grand unification the unification of the $SU(3) \times SU(2) \times U(1)$ gauge symmetries in a simple group.

grand unification scale the mass scale of spontaneous breaking of the grand unified group. Proton stability and the unification of the couplings require that it be within two or three orders of magnitude of the gravitational scale.

Grassmann variable the elements θ_i of an algebra with the relation $\theta_i \theta_j = -\theta_j \theta_i$. These are used to give a path integral representation of fermionic fields, and to define *superspace*. They are also called *anticommuting c-numbers*.

gravitational scale the mass scale at which the dimensionless gravitational coupling becomes of order 1, $m_{\text{grav}} = \kappa^{-1} = 2.4 \times 10^{18}$ GeV; this is $(8\pi)^{-1/2}$ times the Planck mass.

gravitino a spin-$\frac{3}{2}$ fermion in the same supersymmetry multiplet as the graviton.

Green–Schwarz mechanism the cancellation of an anomaly by the modified transformation law of a *p*-form potential in a *Chern–Simons term*.

Green–Schwarz superstring a manifestly supersymmetric formulation of the supersymmetric string theories, with a spacetime-fermionic gauge invariance known as *κ symmetry*. There is no simple covariant gauge fixing.

H-monopole a monopole carrying the magnetic charge of the antisymmetric tensor gauge field $B_{\mu n}$.

heterotic 5-brane the 5-brane carrying the magnetic charge of the massless heterotic string 2-form potential. It is obtained as the limit of a zero size instanton in the heterotic string gauge fields. The instanton configuration is localized in four spatial dimensions, and is therefore a 5-brane in nine spatial dimensions.

heterotic string a string with different constraint algebras acting on the left- and right-moving fields. The case of phenomenological interest has a $(0, 1)$ superconformal constraint algebra, with spacetime supersymmetry acting only on the right-movers and with gauge group $E_8 \times E_8$ or $SO(32)$.

Hagedorn temperature the temperature at which the thermal partition function of free strings diverges, due to the exponential growth of the density of states of highly excited strings.

hidden sector the fields that couple to the Standard Model only through gravitational-strength interactions. In *hidden sector models,* these include the fields responsible for supersymmetry breaking.

highest weight state in CFT, a state annihilated by all Virasoro lowering operators, or more generally by all lowering operators in a given algebra.

Hodge number the number of nontrivial (p, q)-forms in Dolbeault cohomology, denoted $h^{p,q}$.

holographic principle the conjecture that the states of quantum gravity in d dimensions have a natural description in terms of a $(d - 1)$-dimensional theory. This radical departure from local field theory was motivated by the black hole information problem, and has played a role in attempts to formulate M-theory.

holomorphic analytic, as used in the theory of complex variables. The Minkowskian continuation is *left-moving*. An *antiholomorphic* field is analytic in the conjugate variable, and its continuation is *right-moving*.

holomorphic quadratic differential a globally defined holomorphic $(2, 0)$-form on a Riemann surface.

holomorphic vector field a globally defined holomorphic $(-1, 0)$-form on a Riemann surface.

holonomy consider the parallel transport of a vector around a closed loop on a d-dimensional manifold: it returns to an $O(n)$ rotation of its original value. The set of all rotations that are obtained in this way for a given manifold is a subgroup of $O(n)$; this is the *holonomy group*.

homology see *cohomology*.

hyperelliptic surface a Riemann surface with a Z_2 symmetry. This can be represented as a two-sheeted cover of the sphere with branch cuts.

hyper-Kähler manifold a $4k$-dimensional manifold of holonomy $Sp(k) \subset SO(4k)$. This is the geometry of the moduli space of hypermultiplets in $d = 6$, $N = 1$ or $d = 4$, $N = 2$ supersymmetry, in the limit in which gravity decouples.

hypermultiplet in $d = 6$, $N = 1$ or $d = 4$, $N = 2$ supersymmetry, the multiplet whose bosonic content is four real massless scalars.

identify to define two points (or other objects) to be equivalent, thus producing a coset space.

infrared (IR) divergence a divergence arising from long distances in spacetime, usually signifying that one has calculated the wrong thing.

inheritance principle in twisted (orbifold) theories, the principle that the tree-level amplitudes of untwisted states are the same as in the untwisted theory.

insertion the integrand of a path integral, excluding the weight $\exp(iS)$ or $\exp(-S)$.

instanton in a Euclidean path integral, a nonconstant configuration that is a local but not a global minimum of the action. Such configurations are usually localized in spacetime, are usually topologically nontrivial, and are of interest when they give rise to effects such as tunneling that are not obtained from small fluctuations around a constant configuration. *Spacetime instantons* are instantons in the effective field theory in spacetime. *World-sheet instantons* are instantons in the world-sheet quantum field theory, and correspond to world-sheets wrapping around nontrivial two-cycles of spacetime.

intersection number the number of points at which a set of surfaces intersect, weighted by the orientation of the intersection.

irrelevant interaction an interaction whose dimensionless strength decreases with increasing length scale. In perturbation theory, this is equivalent to a *non-renormalizable* interaction.

K3 manifold the unique nontrivial Calabi–Yau manifold of four (real) dimensions. To be precise, it is topologically unique, but possesses complex structure and Kähler moduli. Its holonomy is $SU(2)$, so that half of the supersymmetries of a theory are broken upon compactification on K3.

Kac determinant the determinant of the matrix of inner products of states at a given L_0 level of a Verma module.

Kac–Moody algebra see *current algebra*.

Kähler form the $(1, 1)$-form $G_{i\bar{j}}dz^i d\bar{z}^{\bar{j}}$, formed from a Kähler metric on a complex manifold.

Kähler manifold a complex manifold of $U(n)$ holonomy in n complex dimensions.

Kähler moduli the moduli parameterizing the Kähler form.

Kähler potential the potential $K(z, \bar{z})$, in terms of which the metric of a Kähler manifold is determined, $G_{i\bar{j}} = \partial_i \partial_{\bar{j}} K$. This is not globally defined, being determined only up to a *Kähler transformation* $K(z, \bar{z}) \to K(z, \bar{z}) + f(z) + f(z)^*$.

Kaluza–Klein gauge field in a compactified theory, a gauge field originating from the metric of the higher-dimensional theory. The gauge group is the isometry group of the compact space.

Kaluza–Klein monopole a monopole carrying the magnetic charge of a $U(1)$ Kaluza–Klein gauge symmetry. The monopole configuration is the smooth *Taub–NUT* spacetime. It is localized in three spatial dimensions, and is therefore a 6-brane in nine spatial dimensions.

Kaluza–Klein states states with nonzero momentum in a compact spatial direction.

Knizhnik–Zamolodchikov (KZ) equation the differential equation determining the expectation values of the primary fields of a current algebra.

Landau–Ginzburg model a scalar field theory which has a nontrivial IR fixed point when the potential is appropriately tuned. In particular, this gives a Lagrangian representation of the minimal model CFTs.

large coordinate transformation a coordinate transformation that is not continuously connected to the identity.

lattice the set Γ of integer linear combinations of n linearly independent basis vectors in n dimensions. Given a Euclidean or Lorentzian metric, an *even lattice* is one whose points have even length-squared. The dual lattice Γ^* is the set of points v such that $v \cdot w \in \mathbf{Z}$ for all $w \in \Gamma$. The *root lattice* is the set of integer linear combinations of the roots of a Lie algebra.

level 1. the quantized c-number term in a current algebra, also known as the *Schwinger term*; 2. the total oscillator excitation number in free field theory; 3. in a conformal family, the difference between the L_0 eigenvalue of a given state and that of the highest weight state.

level-matching the modular invariance condition that $L_0 - \tilde{L}_0 \in \mathbf{Z}$.

Lie algebra an algebra with an antisymmetric product that satisfies the Jacobi identity. A *simple* Lie algebra has no subalgebra that commutes with its complement. A *simply-laced* Lie algebra has all roots of equal length. A *graded Lie algebra* has odd and even elements, with a symmetric product between odd elements.

light-cone gauge in string theory, the choice of world-sheet time coordinate to coincide with a particular spacetime null coordinate. In theories with local symmetries, a gauge choice such that the connection in a given null direction vanishes.

linear dilaton theory a scalar CFT in which the energy-momentum tensor includes a term proportional to the second derivative of the scalar. This arises in string theory when the dilaton is a linear function of position.

linear sigma model a scalar field theory whose kinetic term is field independent, but whose long-distance physics is governed by a nonlinear sigma model.

Liouville field theory the CFT of a scalar field with an exponential interaction. This arises in various situations, including the *noncritical string*. It corresponds to bosonic string theory in a linear dilaton plus exponential tachyon background.

little string theory one of several interacting string theories without gravity, notably found on NS5-branes in the limit of zero string coupling.

loop expansion in quantum field theory, the Feynman graph expansion, which is equivalent to the expansion in powers of $\hbar$. The *string loop expansion* is the sum over Riemann surfaces, with dimensionless string coupling g. The *world-sheet loop expansion* is the nonlinear sigma model perturbation expansion, in powers of α'/R_c^2 with R_c the compactification radius.

Lorentzian having a mixed signature $(-, \ldots, -, +, \ldots, +)$.

lowering operator operators that reduce the energy of a given state. In CFT, operators that reduce the Virasoro generator L_0 (or $\tilde{L}_0$) by n units carry a grading (subscript) n.

M-theory 1. (*narrow*) the limit of strongly coupled IIA theory with eleven-dimensional Poincaré invariance; 2. (*broad; most common usage*) the entire quantum theory whose limits include the various weakly coupled string theories as well as M-theory in the narrow sense. The name is deliberately ambiguous, reflecting the unknown nature of the theory; M has variously been suggested to stand for membrane, matrix, mother, and mystery.

M2-brane the 2-brane of M-theory, which couples to the potential A_3 of eleven-dimensional supergravity.

M5-brane the 5-brane of M-theory, which carries the magnetic charge of the potential A_3 of eleven-dimensional supergravity.

macroscopic string a string whose length is much greater than the characteristic string length scale. In particular, it is sometimes useful to consider an infinite string stretching across spacetime.

Majorana condition a Lorentz-invariant reality condition on a spinor field. This can be imposed only if the spacetime dimension is 1, 2, 3, 4, or 8 (mod 8).

marginal interaction an interaction whose dimensionless strength is independent of the length scale. In general this might hold only to first order in the coupling of the interaction; a *truly marginal* interaction is one that remains marginal even with finite coupling.

matrix models quantum mechanical systems with matrix degrees of freedom, with critical points governed by noncritical string theories.

matrix theory a quantum mechanical system with matrix degrees of freedom and 32 supercharges, obtained by dimensional reduction of $d = 10$ supersymmetric $U(n)$ Yang–Mills theory. In the large-n limit this is conjectured to define M-theory (in the broad sense).

minimal models several families of solvable CFTs, in which every conformal family is degenerate. There are infinite series of unitary minimal models having $N = 0$, $N = 1$, and $N = 2$ superconformal symmetries, which converge from below on the central charges 1, $\frac{3}{2}$, and 3 respectively.

Minkowskian having a signature $(-, +, +, \ldots, +)$.

mirror symmetry an equivalence between string theories compactified on distinct manifolds. The equivalence reverses the sign of the $U(1)$ charge of one $N = 2$ superconformal algebra, and therefore changes the sign of the Euler number of the manifold.

Möbius group the globally defined $SL(2, \mathbf{C})$ conformal symmetry of the sphere; or, the globally defined $SL(2, \mathbf{R})$ conformal symmetry of the disk.

mode operators the spatial Fourier components of a quantum field.

model see *vacuum*.

modular group the group of large coordinate transformations (often applied to a Riemann surface but also applicable to spacetime).

modular invariance the invariance of the string path integral under large coordinate transformations.

moduli 1. the parameters labeling the geometry of a manifold. Notable examples are the parameters for the complex structure of the string world-sheet, and the parameters for the geometry of compactification; 2. the parameters labeling a space of degenerate (and, usually, physically inequivalent) vacua in quantum field theory. This is closely related to the compactification example: in expanding around the classical limit, each compact solution of the field equations gives a vacuum of the quantum theory, to leading order; 3. the massless fields corresponding to position dependence of these parameters. Contrast *Goldstone boson*.

moduli space the space of geometries or vacua, whose coordinates are the moduli.

monodromy for a quantity which is locally single-valued, the multi-valuedness around nontrivial closed paths.

Montonen–Olive duality the weak–strong duality of $d = 4$, $N = 4$ Yang–Mills theory.

Nambu–Goto action a string action, which is proportional to the invariant area of the world-sheet in spacetime.

Narain compactification the abstract description of toroidal compactification in terms of the lattice of left- and right-moving momenta.

naturalness problem the problem of explaining why a constant of nature takes a value much smaller than estimated nonzero contributions. Examples are the Higgs scalar mass, the cosmological constant, and the QCD θ-angle.

Neumann boundary condition the condition that the normal derivative of a field vanish at a boundary; the value of the field is free to fluctuate.

Neveu–Schwarz algebra the world-sheet algebra of the Fourier modes of the supercurrent and energy-momentum tensor, in a sector where the supercurrent is antiperiodic and its moding therefore half-integer-valued.

Neveu–Schwarz (NS) boundary condition the condition that a fermionic field on the world-sheet be antiperiodic, in the closed string or in the double of the open string (see *doubling trick*). Its Fourier moding is then half-integer-valued.

Neveu–Schwarz 5-brane in the type I and type II superstring theories, the 5-brane that carries the magnetic charge of the NS–NS 2-form potential.

Neveu–Schwarz–Neveu–Schwarz (NS–NS) states in type I and type II superstring theories, the bosonic closed string states whose left- and right-moving parts are bosonic. These include the graviton and dilaton, and in the type II case a 2-form potential.

no-ghost theorem 1. the theorem that the OCQ or BRST Hilbert space has a positive inner product; 2. the further theorem that the string amplitudes are well defined and unitary in this space.

no-scale model a field theory that has, to some approximation, a line of degenerate vacua with broken supersymmetry.

Noether's theorem the theorem that an invariance of the Lagrangian implies a conserved quantity.

noncommutative geometry a generalization of ordinary geometry, focusing on the algebra of functions on a space. The noncommutative collective coordinates of D-branes suggest the need for such a generalization.

noncritical string theory 1. a Weyl-noninvariant string theory — one with a measure of world-sheet distance that is independent of the embedding in spacetime. These include strings with an independent world-sheet metric field, and strings with a short-distance cutoff; 2. more recently, the term has been applied to any string theory that does not have a weakly coupled limit with a Weyl-invariant world-sheet theory. In this form it includes various theories with stringlike excitations in which the coupling is fixed to be of order 1, such as the *(2,0) theory* and the *little string theories*. Such a theory does not have a well-defined world-sheet, because processes that change the world-sheet topology cannot be turned off.

nonlinear sigma model a scalar field theory in which the kinetic term has a field-dependent coefficient. This has a natural interpretation in terms of a curved field space, and corresponds to string theory in curved spacetime, or more generally one with position-dependent background fields.

nonrenormalization theorem a theorem restricting the form of quantum corrections to a given amplitude, or to the effective action. It may require that these corrections vanish, that they arise only at specific orders of perturbation theory, or that they arise only nonperturbatively.

normal ordering a prescription for defining products of free fields by specific subtractions of divergent terms. *Conformal normal ordering,* denoted : :, produces operators with simple conformal properties. *Creation–annihilation normal ordering,* denoted ⦂ ⦂, where lowering operators are put to the right of raising operators, produces operators with simple matrix elements. *Boundary normal ordering,* denoted ⁚ ⁚, is conformal normal ordering with an additional image charge subtraction to produce operators that are finite as they approach a boundary.

null state a physical state that is orthogonal to all physical states including itself. Or, a descendant in a conformal family which is orthogonal to all states in the family.

old covariant quantization (OCQ) a method of quantizing string theory, similar to the Gupta–Bleuler quantization of electrodynamics. It is equivalent to the light-cone and BRST quantizations.

one-loop the leading quantum correction, coming from surfaces of Euler number zero in string perturbation theory.

open string a string that is topologically a line segment.

operator equation in quantum theory, an equality between operators that holds in arbitrary matrix elements; equivalently, an equality that holds when inserted into a functional integral with arbitrary boundary conditions.

operator product expansion (OPE) the expansion of a product of operators as a sum of local operators. This provides an asymptotic expansion, as the separation of the operators vanishes, for an arbitrary expectation value containing the product. In CFT the expansion is convergent.

orbifold 1. (*noun*) a coset space M/H, where H is a group of discrete symmetries of a manifold M. The coset is singular at the fixed points of H; 2. (*noun*) the CFT or string theory produced by the gauging of a discrete world-sheet symmetry group H. If the elements of H are spacetime symmetries, the result is a theory of strings propagating on the coset space M/H. A *non-Abelian orbifold* is one whose point group is non-Abelian. An *asymmetric orbifold* is one where H does not have a spacetime interpretation and which in general acts differently on the right-movers and left-movers of the string; 3. (*verb*) to produce such a CFT or string theory by gauging H; this is synonymous with the second definition of *twist*.

oriented string theory a string theory in which the world-sheet has a definite orientation; world-sheet parity-reversal is not treated as a gauge symmetry.

orientifold a string theory produced by the gauging of a world-sheet symmetry group H, where H includes elements that combine the world-sheet parity-reversal Ω with other symmetries.

orientifold plane a plane (of any dimension p) consisting of fixed points of the orientifold group H (specifically, of an element of H that includes Ω).

p-brane a p-dimensional spatially extended object. Examples are *black p-branes, Dp-branes, M2-* and *M5-branes, NS5-branes, heterotic 5-branes,* and (in $d >$ 4) *Kaluza–Klein monopoles.*

p-form a fully antisymmetric p-index tensor, usually written in an index-free notation.

p-form gauge field a generalization of Abelian gauge theory, with a p-form potential A, a $(p-1)$-form gauge parameter λ, and a $(p+1)$-form field strength F. For $p = 0$ this is an ordinary massless scalar; for $p = 1$ it is an Abelian gauge field. A $(p+1)$-form potential couples naturally to a p-brane, through the integral of the form over the world-volume. A *self-dual p-form theory* is one where $*F = F$; this requires the spacetime dimension to be $d = 2p + 2$, and further d must be 2 mod 4 (in the Minkowskian case) in order that $** = 1$.

p-p' string an open string with one endpoint on a Dp-brane and the other on a Dp'-brane.

(p, q)-form on a complex manifold, a tensor that is completely antisymmetric in p holomorphic indices and q antiholomorphic indices.

(p, q) string a bound state of p F-strings and q D-strings in the IIB theory.

parafermion CFT a family of coset CFTs with $\mathbf{Z}_n$ symmetry, generalizing the $\mathbf{Z}_2$-invariant free fermion theory. These describe the generic critical behavior of a system with $\mathbf{Z}_n$ symmetry.

parity transformation an operation that reflects one spatial dimension, or any odd number. One distinguishes *spacetime parity, P* (or β) and *world-sheet parity* Ω.

partition function a sum over the spectrum of a quantum system, weighted by $e^{-H/T}$ where H is the Hamiltonian and T the temperature. Often additional charges are included in the exponent. This is the basic object in equilibrium statistical mechanics. In string theory it is given by a path integral on the torus or the cylinder, and so arises in one-loop amplitudes.

path integral a representation of the transition amplitudes of a quantum system as a coherent sum over all possible histories. In quantum mechanics the history is a particle path; in quantum field theory it is a path in field space; in first-quantized string theory it is the embedding of the string world-sheet in spacetime.

Pauli–Villars regulator a means of regulating quantum field theories by introducing a very massive wrong-statistics field.

Peccei–Quinn (PQ) symmetry an approximate symmetry, violated only by anomalies.

period matrix a $g \times g$ matrix characterizing the complex structure of a genus-g Riemann surface.

perturbation theory the expansion of the amplitudes of a quantum system in powers of the coupling.

physical state in a quantum system with constraints, a state annihilated by the constraints. In OCQ this is a state annihilated by the Virasoro lowering generators and having a specified L_0 eigenvalue.. In BRST quantization it is a state annihilated by the BRST operator. In both these cases, the true physical spectrum is the space of physical states with an additional equivalence relation, physical states differing by a null state being identified.

picture in the RNS superstring, one of several isomorphic representations of the vertex operators. The q-picture consists of vertex operators of $\beta\gamma$ ghost charge q. The natural pictures are $q = -1$ and $-\frac{1}{2}$, with higher pictures including partial integrations over supermoduli space. The *picture changing operator* increases q by one.

Planck length the natural length scale of quantum gravity, $L_P = M_P^{-1} = 1.6 \times 10^{-33}$ cm, constructed from $\hbar$, c, and G_N.

Planck mass the natural mass scale of quantum gravity, $M_P = 1.22 \times 10^{19}$ GeV, constructed from $\hbar$, c, and G_N.

plumbing fixture a procedure for constructing higher genus Riemann surfaces from lower ones by sewing in a handle. The construction includes a parameter q, such that when q goes to 0 the handle degenerates or *pinches*. This gives a canonical representation of the boundary of the moduli space of Riemann surfaces.

Poincaré dual a map from p-forms to $(d - p)$-forms, given by contraction with the completely antisymmetric tensor.

Poincaré invariance the invariance group of the flat metric, consisting of translations and Lorentz transformations.

point group the orbifold group H, with translations ignored (applicable only for orbifolds having a spacetime interpretation).

Polyakov path integral a representation of first-quantized string theory as a path integral with an independent world-sheet metric. A local Weyl symmetry guarantees that the classical degrees of freedom are the same as those of the Nambu–Goto theory.

primary field in CFT, a local operator annihilated by all of the lowering generators of a given algebra, such as the Virasoro algebra. The corresponding state is a *highest weight state*.

projective space a compact n-dimensional space constructed from a linear $(n + 1)$-dimensional space by identifying points under the overall rescaling $(x_1, \ldots, x_{n+1}) \cong (\lambda x_1, \ldots, \lambda x_{n+1})$. For x_i and λ real this produces RP^n, and for x_i and λ complex it produces CP^n (which has n complex $= 2n$ real dimensions).

pseudospin in a current algebra, an $SU(2)$ subalgebra not contained in the Lie algebra of the center-of-mass modes.

puncture a marked point on a Riemann surface, the position of a vertex operator.

QCD string a reformulation of non-Abelian gauge theory as a string theory, conjectured to exist at least in the limit of a large number of colors.

quaternionic manifold a $4k$-dimensional manifold with the holonomy group $Sp(k) \times SU(2) \subset SO(4k)$, with specific $SU(2)$ curvature. This is the geometry of the moduli space of hypermultiplets in $d = 6$, $N = 1$ or $d = 4$, $N = 2$ supergravity.

R symmetry a symmetry that acts nontrivially on the supercurrent(s).

raising operator operators that reduce the energy of a given state. In CFT, operators that increase the Virasoro generator L_0 (or $\tilde{L}_0$) by n units carry a grading (subscript) $-n$.

Ramond algebra the world-sheet algebra of the Fourier modes of the supercurrent and energy-momentum tensor, in a sector where the supercurrent is periodic and its moding therefore integer-valued.

Ramond (R) boundary condition the condition that a fermionic field on the world-sheet be periodic, in the closed string or in the double of the open string (see *doubling trick*). Its Fourier moding is then integer-valued.

Ramond–Neveu–Schwarz (RNS) superstring the formulation of type I and II superstrings that has superconformal invariance but not manifest spacetime supersymmetry. The latter emerges after imposing the GSO projection on the string Hilbert space.

Ramond–Ramond (R–R) states in type I and type II superstring theories, the bosonic closed string states whose left- and right-moving parts are fermionic. These include p-form potentials C_p, with p taking all odd values in the IIA string and all even values in the IIB string.

rank 1. the maximal number of commuting generators of a Lie algebra; 2. the number of indices on a tensor.

rational CFT a CFT with a finite number of primary fields under an extended chiral algebra, a generalization of the *minimal models*. Such CFTs are highly constrained.

refermionization after bosonization, the construction of new spin-$\frac{1}{2}$ fields from linear combinations of the bosonic fields. These fermions are nonlocal, nonlinear functions of the original fermions.

Regge behavior in scattering at large center-of-mass energy-squared s and fixed momentum transfer-squared $-t$, the scaling of the amplitude as $s^{\alpha(t)}$. The values of t where $\alpha(t) = j$ is a nonnegative integer correspond to exchange of a particle of spin j and mass-squared $-t$.

Regge slope denoted α', the square of the characteristic length scale of perturbative string theory. The tension of the fundamental string is $1/2\pi\alpha'$.

relevant interaction an interaction whose dimensionless strength increases with distance. In perturbation theory, this is equivalent to a *superrenormalizable* interaction.

renormalization group equation the differential equation governing the change of physics with length scale.

renormalization theory the calculus of path integrals.

Riemann–Roch theorem the theorem that the number of metric moduli minus the number of conformal Killing vectors on a Riemann surface is -3χ, with χ being the Euler number of the surface; and, generalizations of this result.

Riemann surface a two-(real)-dimensional complex manifold, equivalent to a Weyl equivalence class of Riemannian manifolds.

root a vector of the eigenvalues of the maximal set of commuting generators of a Lie algebra, associated with a state in the adjoint representation.

S-**duality** a duality under which the coupling constant of a quantum theory changes nontrivially, including the case of weak–strong duality. Important examples are the $SL(2,\mathbf{Z})$ self-dualities of IIB string theory and of $d = 4$, $N = 4$ supersymmetric Yang–Mills theory. More loosely, it is used for weak–strong dualities between different theories, such as IIA–M-theory (on a circle) duality, $SO(32)$ heterotic–type I duality, and E_8 heterotic–M-theory (on an interval) duality. In compactified theories, the term *S*-duality is limited to those dualities that leave the radii invariant, up to an overall coupling-dependent rescaling; contrast *T-duality* and *U-duality*.

S-matrix the overlap amplitude between states in the infinite past and states in the infinite future; the scattering amplitude. In coordinate-invariant quantum theories this is generally the simplest invariant. The term usually implies a basis of free particle states; this is problematic in theories with massless particles due to IR divergences, and meaningless in theories at nontrivial IR fixed points.

scale transformation a rigid rescaling of spacetime, or of the world-sheet.

Scherk–Schwarz mechanism the breaking of supersymmetry by dimensional reduction that includes a spacetime rotation.

Schottky group a discrete subgroup Γ of the $SL(2,\mathbf{C})$ Möbius transformations of the sphere S_2, with additional conditions such that S_2/Γ is a manifold and in particular a Riemann surface.

Schwarzian the combination of derivatives appearing in the finite conformal transformation of the energy-momentum tensor.

Schwinger–Dyson equation the operator equations of a quantum theory, expressed as equations for the expectation values.

short multiplet see *BPS state.*

sigma model metric the metric appearing in the string world-sheet action. Also known as the *string metric,* this differs from the *Einstein metric* by a dilaton-dependent Weyl transformation.

simple current an operator J such that, for any primary field $\mathcal{O}$, the operator product $J\mathcal{O}$ contains only a single conformal family.

soliton a state whose classical limit is a smooth, localized, and (usually) topologically nontrivial classical field configuration; this includes particle states, which are localized in all directions, as well as extended objects. By contrast, a state of ordinary *quanta* is represented near the classical limit by small fluctuations around a constant configuration. In a theory with multiple classical limits (*dualities*), solitons and quanta may exchange roles.

special Kähler geometry the geometry of the moduli space of vector multiplets in $d = 4$, $N = 2$ supergravity. It is most simply defined (section B.7) in terms of *special coordinates,* which are fixed up to a symplectic transformation. *Rigid special geometry* is obtained in the limit where gravity decouples and the supersymmetry becomes global.

spectral flow the adiabatic change in the spectrum produced by a continuous change in the boundary conditions.

spin the behavior of a field or a state under rotations. In a CFT this is given in terms of the conformal weights by $h - \tilde{h}$.

spin field the vertex operator for a Ramond ground state, which produces a branch cut in the spinor fields.

spin structure one of a set of inequivalent ways of defining a spinor field globally on a manifold. Roughly speaking, it corresponds to a choice of signs in the square roots of the transition functions.

spurious state in OCQ, a state produced by Virasoro raising operators.

state–operator isomorphism in CFT, a one-to-one isomorphism between states of the theory quantized on a circle and local operators. Also, a one-to-one isomorphism between states of the theory quantized on an interval and local operators on a boundary. In d dimensions the circle becomes a $(d-1)$-sphere and the interval a $(d-1)$-hemisphere.

string coupling the dimensionless parameter g governing the weights of different Riemann surfaces in string perturbation theory, the contribution from surfaces of Euler number χ being weighted by $g^{-\chi}$. The string coupling is related to the dilaton by $g = e^{\Phi}$. This definition corresponds to the amplitude to emit a closed string; the amplitude to emit an open string is proportional to $g^{1/2}$.

string field theory the representation of string theory as theory of fields, the fields being maps from a circle (or interval) into spacetime. This corresponds to an infinite number of ordinary quantum fields. This formalism can reproduce string perturbation theory, but it is unclear whether it can be defined beyond perturbation theory.

string metric see *sigma model metric*.

string scale the mass scale $\alpha'^{-1/2}$ characterizing the tower of string excitations.

string–string duality a term sometimes used to denote a weak–strong duality between different string theories, in particular between the heterotic string compactified on T^4 and the IIA string compactified on K3.

string tension the mass per unit length of a string at rest, related to the Regge slope by $1/2\pi\alpha'$.

Sugawara construction in current algebra, the construction of the energy-momentum tensor as a product of two currents. Originally proposed as a phenomenological model in four dimensions, this was later found to be an exact result in two dimensions.

superconformal algebra an extension of the conformal (Virasoro) algebra to include anticommuting spinor generators. The $(N, \tilde{N})$ superconformal algebra has N left-moving and $\tilde{N}$ right-moving supercurrents.

superconformal current algebra an extension of the conformal transformations to include both spin-$\frac{3}{2}$ and spin-1 currents.

superconformal field theory (SCFT) a quantum field theory that is invariant under superconformal transformations.

supercurrent a conserved spinor current. This includes the world-sheet current T_F associated with superconformal transformations, and the spacetime current associated with spacetime supersymmetry.

superfield a field on superspace, with specific transformation properties under a change of coordinates.

supergravity the union of general relativity and supersymmetry, implying also the promotion of supersymmetry to a local symmetry.

supermanifold (or *superspace*) a formal extension of the concept of manifold to include both commuting and anticommuting (Grassmann) coordinates.

supermoduli the anticommuting parameters characterizing a super-Riemann surface.

superpartner scale the mass scale of the superpartners of the Standard Model particles. This is expected to be between 10^2 and 10^3 GeV if supersymmetry solves the naturalness problem of the Higgs scalar mass.

superpotential in $d = 4$, $N = 1$ supersymmetry, the holomorphic function of the superfields that determines the nongauge interactions.

super-Riemann surface a supermanifold defined in terms of superconformal transition functions between patches.

supersymmetry a symmetry whose charge transforms as a spinor, which relates the masses and couplings of fermions and bosons.

supersymmetry breaking scale the mass scale of the expectation value that breaks supersymmetry. The superpartner scale is the supersymmetry breaking scale times the strength of the coupling of the Standard Model fields to the supersymmetry breaking fields.

T-**duality** a duality in string theory, usually in a toroidally compactified theory, that leaves the coupling constant invariant up to a radius-dependent rescaling and therefore holds at each order of string perturbation theory. Most notable is $R \to \alpha'/R$ duality, which relates string theories compactified on large and small tori by interchanging winding and Kaluza–Klein states. More generally it includes shifts of antisymmetric tensor backgrounds and large coordinate transformations in spacetime. Contrast *S-duality* and *U-duality*.

't Hooft–Polyakov monopole a classical solution with magnetic charge, which exists whenever a simple group is spontaneously broken to a group with a $U(1)$ factor.

tachyon a particle (almost always a scalar) with a negative mass-squared, signifying an instability of the vacuum.

tadpole an amplitude for creation of a single particle from the vacuum, induced by quantum effects.

target space the space in which a function takes its values. This is usually applied to the nonlinear sigma model on the string world-sheet, where the target space is itself spacetime.

Teichmüller parameters the moduli for the complex structure of a Riemann surface (strictly speaking, points in *Teichmüller space* are not identified under the modular group).

tensionless string theory an interacting theory with tensionless strings. These can arise as p-branes with $p-1$ directions wrapped on a collapsing cycle, as various 2-branes with one direction stretched between higher-dimensional branes when the latter become coincident, and on zero-size $E_8 \times E_8$ instantons. In general the coupling is fixed to be of order one, so there is no perturbation expansion. *Tensionless* implies that the string tension in units of the gravitational scale goes to zero; it is not applied to the fundamental string, which becomes noninteracting in that limit.

tensor multiplet 1. the multiplet of $d = 6$, $(1,0)$ supersymmetry whose bosonic content is one self-dual tensor and one scalar. This reduces to a vector multiplet of $d = 4$, $N = 2$ supersymmetry; 2. the multiplet of $d = 6$, $(2,0)$ supersymmetry whose bosonic content is one self-dual tensor and five scalars. This reduces to a vector multiplet of $d = 4$, $N = 4$ supersymmetry.

tensor operator in CFT, a local operator whose conformal or superconformal transformation involves only the first derivative of the transformation; synonymous with a *primary field* of the conformal or superconformal symmetry. Such an operator is mapped to a *highest weight state* by the state–operator isomorphism.

tetrad a basis of d orthonormal vector fields in d dimensions. The term tetrad originates in $d = 4$ (as does the equivalent term *vierbein*) but there is no other convenient term for general d.

the theory formerly known as strings see *M-theory*, second definition.

theory see *vacuum*.

theta functions the family of holomorphic functions having simple periodicity properties on a torus.

Thirring model the solvable quantum field theory of a single Dirac fermion with a quartic interaction in $1 + 1$ dimensions. This is equivalent under bosonization to a free scalar field at a general real radius.

threshold correction a correction to the low energy effective action, and in particular to the gauge coupling, due to virtual massive particles.

topological string theory a modification of string theory without local dynamics; all observables are topological.

toric geometry a generalization of the idea of projective space. Roughly speaking, this corresponds to the most general linear sigma model.

toroidal compactification the periodic identification of one or more flat dimensions.

torsion a term applied to various 3-form field strengths, so called because they appear in covariant derivatives in combination with the Christoffel connection.

tree-level the Feynman graphs which become disconnected if one propagator is cut, or the analogous string amplitudes, the sphere and disk. These correspond to classical terms in the effective action.

twist 1. (*verb*) to define a field in a periodic space to be aperiodic by a symmetry transformation h; 2. (*noun*) the aperiodicity h; 3. (*verb*) given a CFT or string theory, to construct a new theory using a symmetry group H. One adds closed strings twisted by any of the elements $h \in H$, and requires all states to be invariant under the transformations in H. This is equivalent to treating H as a world-sheet gauge symmetry. The term *orbifold* is also used as a synonym; 4. (*noun, archaic*) world-sheet parity.

twisted state a closed string with twisted periodicity.

type ... supergravity the low energy supergravity theory of the corresponding string theory. The Roman numeral signifies the number of $d = 10$ supersymmetries, and IIA and IIB distinguish whether the two supersymmetries have opposite or identical chiralities respectively.

type I superstring the theory of open and closed unoriented superstrings, which is consistent only for the gauge group $SO(32)$. The right-movers and left-movers, being related by the open string boundary condition, transform under the same spacetime supersymmetry.

type IIA superstring a theory of closed oriented superstrings. The right-movers and left-movers transform under separate spacetime supersymmetries, which have opposite chiralities.

type IIB superstring a theory of closed oriented superstrings. The right-movers and left-movers transform under separate spacetime supersymmetries, which have the same chirality.

type 0 string a pair of nonsupersymmetric string theories, which have tachyons and no spacetime fermions, with the same world-sheet action as the type II theories but with different projections on the Hilbert space.

U-**duality** any of the dualities of a string theory, usually of a toroidally compactified type II theory. This includes the *S-dualities* and *T-dualities*, but in contrast to these includes also transformations that mix the radii and couplings.

ultrashort multiplet see *BPS state*.

unit gauge a choice of coordinate and Weyl gauges in two dimensions, such that the metric is the identity.

unitary as applied to a quantum system, the property of having a conserved inner product in a positive-norm Hilbert space.

unoriented string theory a string theory in which world-sheet parity-reversal Ω is a discrete gauge symmetry. The perturbation theory includes unoriented world-sheets, and the spectrum is restricted to states with $\Omega = +1$.

UV divergence a divergence arising from short distances in spacetime, usually signifying a limit to the validity of a theory.

vacuum a stable Poincaré-invariant state. The novel feature of systems with unbroken supersymmetries is the frequent appearance of degenerate but physically inequivalent vacua.
While it is now clear that the different string theories are actually different *vacua* in a single theory, it is still common to use the term *theory* for each. Also, different CFTs within a single string theory are sometimes referred to as different *theories* rather than *vacua*. The term *model* is used to refer to string vacua whose low energy physics resembles the Standard Model.

vector multiplet in $d = 4$, $N = 1$ or $d = 6$, $N = 1$ supersymmetry, the multiplet whose bosonic content is a massless vector field. The $d = 6$ multiplet reduces to a vector field plus two real scalars in $d = 4$, $N = 2$ supersymmetry.

Veneziano amplitude the bosonic string tree-level amplitude for four open string tachyons.

Verma module the set of states obtained by acting on a highest weight state with Virasoro raising generators in all inequivalent ways, with the requirement that all such states be linearly independent (the generators act freely). This gives a representation of the Virasoro algebra that depends on the central charge c and on the weight h of the highest weight state. This is very similar to a *conformal family,* except that in the case of a degenerate representation some states in the latter may vanish; the Verma module would in this case give a reducible representation.

vertex operator a local operator on the string world-sheet, corresponding to a string in the initial or final state.

Virasoro algebra in a CFT, the infinite-dimensional Lie algebra of the Fourier modes of the energy-momentum tensor.

Virasoro–Shapiro amplitude the bosonic string tree-level amplitude for four closed string tachyons.

W algebras a family of extended chiral algebras with currents of spin greater than 2.

W string a proposed generalization of string theory with constraints of spin greater than 2. No complete construction exists.

Ward identity a relation between the divergence of the expectation value of a conserved current and the same expectation value without the current.

weight 1. in CFT, the L_0 or $\tilde{L}_0$ eigenvalue, which determines the behavior of an operator under scale transformations and rotations (= *conformal weight*); 2. in Lie algebra, a vector of the eigenvalues of the maximal set of commuting generators of a Lie algebra, in a state of a given representation.

Wess–Zumino consistency condition the condition that the second derivative of a functional integral with respect to the background fields be symmetric. This strongly constrains the form of possible anomalies in a theory.

Wess–Zumino–Novikov–Witten (WZNW) model a conformally invariant non-linear sigma model on a group manifold with an antisymmetric tensor background.

Weyl anomaly an anomaly in the Weyl transformation, this determines the critical dimension in the Polyakov formalism. This is sometimes called the *trace anomaly* or the *conformal anomaly.*

Weyl condition in even dimensions, the condition that a fermion have definite chirality. This defines a spinor representation of the Lorentz group, which contains half of the components of the Dirac representation of the gamma matrix algebra.

Weyl transformation a position-dependent rescaling of the metric.

Wilson line 1. a gauge field with vanishing field strength but with nontrivial parallel transport (holonomy) around nontrivial paths in spacetime. 2. the gauge-invariant operator that measures such a field: the trace of the path-ordered product of the line integral of the vector potential.

Wilsonian action the action in a low energy effective field theory, which incorporates the effects of higher energy virtual states.

winding state a closed string whose configuration is a nontrivial path in a non-simply-connected spacetime.

world-sheet 1. the two-dimensional surface in spacetime swept out by the motion of a string. 2. the abstract two-dimensional parameter space used to describe the motion of a string.

world-sheet duality the equivalence between Hamiltonian descriptions obtained by cutting open a string world-sheet along inequivalent circles. Important examples are the associativity condition, from the four-point sphere amplitude, and the equivalence of the open string loop and closed string tree descriptions of the cylinder.

wrapped refers to a p-brane, q of whose dimensions are wound on a nontrivial compact submanifold of spacetime, leaving a $(p - q)$-dimensional extended object.

Yang–Mills field non-Abelian gauge field.

Zamolodchikov metric the expectation value of a pair of local operators on the sphere, giving a natural inner product for the corresponding string states. This is the metric that appears in the kinetic term in the spacetime action.

zero modes 1. of a differential operator: eigenfunctions with zero eigenvalue. In a Gaussian functional integral, these would give an infinite factor in the bosonic case and a zero factor in the fermionic case. In general these have a physical origin and the functional integral has an appropriate integrand to give a finite result; 2. center-of-mass modes.

zero-point energy the energy due to vacuum fluctuations of quantum fields.

zero-slope limit the limit $\alpha' \to 0$. Only massless string states, described by low energy field theory, remain.

Index

Page numbers in italics refer to exercises